INTERNATIONAL ENERGY AGENCY

WORLD ENERGY OUTLOOK 2004

INTERNATIONAL ENERGY AGENCY
9, rue de la Fédération,
75739 Paris Cedex 15, France

The International Energy Agency (IEA) is an autonomous body which was established in November 1974 within the framework of the Organisation for Economic Co-operation and Development (OECD) to implement an international energy programme.

It carries out a comprehensive programme of energy co-operation among twenty-six* of the OECD's thirty member countries. The basic aims of the IEA are:

- to maintain and improve systems for coping with oil supply disruptions;

- to promote rational energy policies in a global context through co-operative relations with non-member countries, industry and international organisations;

- to operate a permanent information system on the international oil market;

- to improve the world's energy supply and demand structure by developing alternative energy sources and increasing the efficiency of energy use;

- to assist in the integration of environmental and energy policies.

* IEA member countries: Australia, Austria, Belgium, Canada, the Czech Republic, Denmark, Finland, France, Germany, Greece, Hungary, Ireland, Italy, Japan, the Republic of Korea, Luxembourg, the Netherlands, New Zealand, Norway, Portugal, Spain, Sweden, Switzerland, Turkey, the United Kingdom, the United States. The European Commission also takes part in the work of the IEA.

ORGANISATION FOR ECONOMIC CO-OPERATION AND DEVELOPMENT

Pursuant to Article 1 of the Convention signed in Paris on 14th December 1960, and which came into force on 30th September 1961, the Organisation for Economic Co-operation and Development (OECD) shall promote policies designed:

- to achieve the highest sustainable economic growth and employment and a rising standard of living in member countries, while maintaining financial stability, and thus to contribute to the development of the world economy;

- to contribute to sound economic expansion in member as well as non-member countries in the process of economic development; and

- to contribute to the expansion of world trade on a multilateral, non-discriminatory basis in accordance with international obligations.

The original member countries of the OECD are Austria, Belgium, Canada, Denmark, France, Germany, Greece, Iceland, Ireland, Italy, Luxembourg, the Netherlands, Norway, Portugal, Spain, Sweden, Switzerland, Turkey, the United Kingdom and the United States. The following countries became members subsequently through accession at the dates indicated hereafter: Japan (28th April 1964), Finland (28th January 1969), Australia (7th June 1971), New Zealand (29th May 1973), Mexico (18th May 1994), the Czech Republic (21st December 1995), Hungary (7th May 1996), Poland (22nd November 1996), the Republic of Korea (12th December 1996) and Slovakia (28th September 2000). The Commission of the European Communities takes part in the work of the OECD (Article 13 of the OECD Convention).

© OECD/IEA, 2004

Applications for permission to reproduce or translate all or part of this publication should be made to: Head of Publications Service, OECD 2, rue André-Pascal, 75775 Paris Cedex 16, France.

FOREWORD

World Energy Outlook 2004 appears at an extremely volatile and uncertain moment in modern energy history. Soaring oil, gas and coal prices, exploding energy demand in China, war in Iraq and electricity blackouts across the world are among the signs and causes of the profound transformations through which the energy world is passing. To this unsettling environment, *WEO-2004* brings a mass of statistical information, informed projections and focused energy analysis. It does not pretend to solve the problems, but it provides the indispensable information from which solutions will eventually be crafted.

It is my very great pleasure once again to present the International Energy Agency's flagship publication and to pay tribute to Dr. Fatih Birol and the *WEO* team under his direction. Acknowledgment is also due to many other members of the IEA staff who contributed to this effort, as well as to the scores of diligent "peer reviewers" from industry, government and academia who gave generously of their time and expertise.

As we say in French, "*La réputation du WEO n'est plus à faire*". The publication is acknowledged worldwide as the single most important source of energy statistics, projections and analysis. The *WEO-2002*, as well as last year's special *World Energy Investment Outlook*, received several awards for analytic excellence.

Ensuring the continuing security of energy supplies is the International Energy Agency's core mission and *raison d'être*. This book documents a large and growing array of potential threats to that security, including the spectre of disruption along vulnerable pipelines and sea-lanes, and especially at a number of narrow chokepoints on the oil routes which have been dubbed the "dire straits". The threats are all the more preoccupying in light of the *WEO*'s projections of oil demand, supply and trade. All the large consuming countries – now including China and India – are growing increasingly dependent on imports from an ever-smaller group of distant producer countries, some of them politically unstable. In consequence, oil markets are likely to become less flexible and prices more volatile.

Security is not the only issue this book raises. It records the continuing unacceptable contribution of the energy sector to climate-destabilising carbon dioxide emissions. It reminds the reader of the shameful fact that a billion-and-a-half of the world's poorest citizens totally lack access to electricity, and almost as many will lack it in the year 2030. It takes an in-depth look at Russia's emergence as a major world energy supplier, but poses some probing questions about Russia's energy future. And once again *WEO* draws attention to the staggering investments needed to meet rising energy demand over the next quarter century – and warns that the financing may well not be forthcoming, especially in the poorest countries.

Things could, of course, improve. The worst case is not inevitable. To make this point, the *WEO-2004* provides an Alternative Policy Scenario, including, for the very first time, developing as well as developed countries. This scenario demonstrates that we can indeed reduce our dependence on energy imports, cut our growing carbon emissions, burn our fuels more cleanly and efficiently. We can do it only if we can summon the required political will.

This work is published under my authority as Executive Director of the IEA. It does not necessarily reflect the views or policies of IEA member countries.

Claude Mandil

ACKNOWLEDGEMENTS

This study was prepared by the Economic Analysis Division of the International Energy Agency in co-operation with other divisions of the IEA. The Director of Long-Term Office, Noé van Hulst, provided guidance and encouragement during the project. The study was designed and managed by Fatih Birol, Head of the Economic Analysis Division. Other members of EAD who were responsible for bringing the study to completion include: Maria Argiri, Marco Baroni, Amos Bromhead, François Cattier, Laura Cozzi, Lisa Guarrera, Hiroyuki Kato, Trevor Morgan, Nicola Pochettino and Maria T. Storeng. Claudia Jones provided essential support.

The IEA's Carmen Difiglio, Fridtjof Unander, Dolf Gielen, Paul Waide, David Fyfe, Isabel Murray and Mabrouka Bouziane provided substantial contributions to this Outlook. Scott Sullivan carried editorial responsibility.

The study also benefited from input provided by other IEA colleagues, namely: Martina Bosi, Rick Bradley, John Cameron, Viviane Consoli, Doug Cooke, Sylvie Cornot, Muriel Custodio, Ralf Dickel, Lawrence Eagles, Jason Elliot, Meredydd Evans, Lew Fulton, Rebecca Gaghen, Jean-Yves Garnier, Dagmar Graczyk, Klaus Jacoby, Pierre Lefèvre, Jeffrey Logan, Lawrence Metzroth, Cédric Philibert, Loretta Ravera, Bertrand Sadin, Rick Sellers, Ulrik Stridbaek and Mike Taylor.

The work could not have been achieved without the substantial support provided by many government bodies, international organisations and energy companies worldwide, notably the Ministry of Environment of Italy, the Ministry of Petroleum and Energy of Norway, the U.S. Environmental Protection Agency/Argonne National Laboratory, the United Nations Environment Programme, the United Nations Development Programme, the Organization of the Petroleum Exporting Countries, IHS Energy Group, the Institute of Energy Economics of Japan and the World Coal Institute.

Many international experts commented on the underlying analytical work and reviewed early drafts of each chapter. Their comments and suggestions were of great value. All errors and omissions are solely the responsibility of the IEA.

Prominent contributors include:

Chapter 3 (Oil Market Outlook)

Thomas S. Ahlbrandt	U.S. Geological Survey, United States
Kenneth Chew	IHS Energy Group, Switzerland
Joel Couse	Total, France
Jean-Christophe Füeg	Swiss Federal Office of Energy, Switzerland
Dermot Gately	New York University, United States

Nadir Gürer	OPEC Secretariat, Austria
Alex Kemp	University of Aberdeen, United Kingdom
Timothy Klett	U.S. Geological Survey, United States
David Knapp	Energy Intelligence Group, United States
Jean Laherrere	Consultant, France
Alessandro Lanza	Eni S.p.A., Italy
Yves Mathieu	Institut Français du Pétrole, France
Peter Nicol	ABN AMRO, United Kingdom
Matthew R. Simmons	Simmons & Company, United States
Michael D. Smith	BP, United Kingdom

Chapters 4 to 8 (Other Fuels, Electricity, and Regional Outlooks)

William A. Bruno	Consol Energy Inc., United States
Marie-Françoise Chabrelie	Cedigaz, France
Christine Copley	World Coal Institute, United Kingdom
Eduardo Luiz Correia	Petrobras, Brazil
Reinhard Haas	Technical University of Vienna, Austria
Kokichi Ito	The Institute of Energy Economics, Japan
Tooraj Jamasb	University of Cambridge, United Kingdom
Jim Jensen	Jensen Associates, United States
Malcolm Keay	World Coal Institute, United Kingdom
Richard Lavergne	DGEMP, France
Maria Elvira Pinero Maceira	CEPEL, Brazil
John Paffenbarger	Constellation Energy, United States
Mark Radka	United Nations Environment Programme, France
Gustav Resch	Technical University of Vienna, Austria
Jonathan Stern	Oxford Institute for Energy Studies, United Kingdom
Yukari Yamashita	The Institute of Energy Economics, Japan

Chapter 9 (Russia – an In-Depth Study)

Rudiger Ahrend	OECD, France
Garegin S. Aslanian	Centre for Energy Policy, Russia
Igor Bashmakov	CENEF, Russia
Leonard L. Coburn	Department of Energy, United States
Bob Ebel	Center for Strategic and International Studies, United States
Jean-Christophe Füeg	Swiss Federation Office of Energy, Switzerland
Andrei Konoplyanik	Energy Charter Secretariat, Belgium
Vladimir Konovalov	Petroleum Advisory Forum, Russia

Catherine Locatelli	Laboratoire d'Économie de la Production et de l'Intégration Internationale – Département Économie et Politique de l'Énergie, France
Vladimir Milov	Institute of Energy Policy, Russia
Valery Nesterov	Troika Dialog, Russia
Stephen O'Sullivan	UFG, Russia
Jonathan Stern	Oxford Institute for Energy Studies, United Kingdom

Chapter 10 (Energy and Development)

Arnold Baker	Sandia National Laboratories, United States
Dermot Byrne	Electricity Supply Board, Ireland
Ananda Covindassamy	World Bank, United States
Laurent Dittrick	European Commission, Belgium
Christoph Frei	World Economic Forum, Switzerland
Mark Howells	University of Cape Town, South Africa
Céline Kauffmann	OECD, France
Philip Mann	University of Oxford, United Kingdom
Eva Paaske	Ministry of Petroleum and Energy, Norway
Prabodh Pourouchottamin	EDF, France
Kamal Rijal	United Nations Development Programme, United States
Jamal Saghir	World Bank, United States
Judy Siegel	Energy & Security Group, United States
Barrie Stevens	OECD, France
Minoru Takada	United Nations Development Programme, United States
Gordon Weynand	USAID, United States

Chapter 11 (World Alternative Policy Scenario)

Christo Artusio	Department of State, United States
Randall Bowie	European Commission, Belgium
John Christensen	United Nations Environment Programme, Denmark
Mario Contaldi	Italian Agency for the Protection of the Environment and Territory, Italy
Ananda Covindassamy	World Bank, United States
Zhou Dadi	Energy Research Institute, People's Republic of China
James Edmonds	Pacific Northwest National Laboratory, United States

Donald Hanson	Argonne National Laboratory, United States
Tom Howes	European Commission, Belgium
Kokichi Ito	The Institute of Energy Economics, Japan
John A. Laitner	U.S. Environmental Protection Agency/ Argonne National Laboratory, United States
Leonidas Mantzos	National Technical University of Athens, Greece
Eric Martinot	Worldwatch Institute, United States
Jonathan Pershing	World Resources Institute, United States
Kamal Rijal	United Nations Development Programme, United States
James Rockall	World LPG Association, France
Jamal Saghir	World Bank, United States
Roberto Schaeffer	Federal University of Rio de Janeiro, Brazil
Leo Schrattenholzer	International Institute for Applied Systems Analysis, Austria
Jim Steel	U.S. Department of State, United States
Yukari Yamashita	The Institute of Energy Economics, Japan
Yufeng Yang	Energy Research Institute, People's Republic of China

**Comments and questions are welcome
and should be addressed as follows:**

Dr Fatih Birol
Chief Economist
Head, Economic Analysis Division
International Energy Agency
9, rue de la Fédération
75739 Paris Cedex 15
France

Telephone: 33 (0) 1 4057 6670
Fax: 33 (0) 1 4057 6659
Email: Fatih.Birol@iea.org

TABLE OF CONTENTS

THE CONTEXT 1

GLOBAL ENERGY TRENDS 2

OIL MARKET OUTLOOK 3

NATURAL GAS MARKET OUTLOOK 4

COAL MARKET OUTLOOK 5

ELECTRICITY MARKET OUTLOOK 6

RENEWABLE ENERGY OUTLOOK 7

REGIONAL OUTLOOKS 8

RUSSIA – AN IN-DEPTH STUDY 9

ENERGY AND DEVELOPMENT 10

WORLD ALTERNATIVE POLICY SCENARIO 11

ANNEXES

Foreword	3
Acknowledgements	5
List of Figures	17
List of Tables	23
List of Boxes	26
Executive Summary	29

1 The Context — 39
- The Methodological Approach — 40
- The Reference Scenario — 41
  - *Government Policies and Measures* — *41*
  - *Macroeconomic Factors* — *41*
  - *Population* — *43*
  - *Energy Prices* — *46*
  - *Technological Developments* — *51*
- The World Alternative Policy Scenario — 52
- Main Uncertainties — 53

2 Global Energy Trends — 57
- Energy Demand — 58
  - *Primary Fuel Mix* — *58*
  - *Regional Trends* — *64*
  - *Sectoral Trends* — *66*
- Energy Production and Trade — 69
  - *Resource Availability and Production Prospects* — *69*
  - *Outlook for International Trade* — *71*
- Investment Outlook — 72
- Energy-Related CO_2 Emissions — 74
  - *Overview* — *74*
  - *Regional Emission Trends* — *76*
  - *Sectoral Emission Trends* — *77*
- World Alternative Policy Scenario — 78

3 Oil Market Outlook — 81
- Oil Demand — 82
- Oil Reserves and Resources — 87
  - *Classifying and Measuring Resources* — *87*
  - *Estimates of Proven Reserves* — *90*
  - *Turning Resources into Reserves* — *93*
- Oil Production — 105
  - *Summary of Projections* — *105*
  - *Conventional Oil Production Prospects by Region* — *106*
  - *Non-Conventional Oil Production Prospects* — *114*

Inter-Regional Oil Trade	115
Investment Outlook	119
Implications of High Oil Prices	122
Background and Assumptions	*122*
Results	*123*

4 Natural Gas Market Outlook — 129

Gas Demand	130
Gas Supply	135
Proven Reserves and Potential Resources	*135*
Production Prospects	*136*
Gas Trade	*140*
Investment Outlook	144
Price Developments	145
North America	*146*
Europe	*147*
Asia-Pacific	*147*
Regional Trends	148
North America	*148*
European Union	*153*
OECD Asia	*156*
OECD Oceania	*159*
Transition Economies	*159*
Developing Asia	*161*
Middle East	*163*
Africa	*165*
Latin America	*166*

5 Coal Market Outlook — 169

Coal Demand	170
Sectoral Demand	*171*
Impact of Environmental Policy and Technology	*172*
Coal Reserves and Production	174
Proven Reserves	*174*
Production Prospects	*175*
Hard Coal Trade	176
Price Developments	179
Investment Outlook	181
Regional Trends	182
OECD North America	*182*
OECD Europe	*182*
OECD Pacific	*184*
China	*185*

India	*187*
Africa	*188*
Indonesia	*188*

6 Electricity Market Outlook — 191

Electricity Demand — 192
  Drivers of Electricity Demand — *192*
  Sectoral Growth — *193*
Power Generation — 193
  Choice of New Plant — *194*
  The Electricity-Generation Mix — *196*
  Technology Outlook — *204*
  Impact on Fuel Markets — *206*
  Capacity Requirements and Investment Outlook — *207*
  CO_2 Emissions — *211*
Electricity Markets and the Status of Reforms — 212
Regional Trends — 216
  United States and Canada — *216*
  European Union — *218*
  OECD Pacific — *220*
  China — *220*
  India — *222*
  Brazil — *223*

7 Renewable Energy Outlook — 225

Renewable Energy Demand — 226
Renewables in Power Generation — 228
  Cost Developments — *232*
  Outlook by Source — *234*
  Capacity and Investment Outlook — *238*
Renewables in Industry and Buildings — 239
Biofuels — 240

8 Regional Outlooks — 243

OECD Regions and the EU — 244
  Overview — *244*
  OECD North America — *245*
  European Union — *249*
  OECD Asia — *254*
  OECD Oceania — *257*
Developing Countries — 259
  Developing Asia — *261*
  China — *263*

India	*269*
Indonesia	*271*
Latin America	*273*
Brazil	*274*
Middle East	*278*
Africa	*279*
Transition Economies	281

9 Russia – an In-Depth Study — 283

Energy Market Overview	284
Macroeconomic Context	286
Energy Policy Developments	291
Energy Demand Outlook	294
Overview	*294*
Sectoral Trends	*297*
Oil Supply Outlook	300
Resources	*301*
Crude Oil Production	*303*
Refining Capacity and Production	*305*
Export Prospects	*305*
Industry Structure and the Role of the State	*307*
Gas Supply Outlook	308
Resources and Production Trends	*308*
Export Prospects	*313*
Market Reforms	*314*
Coal Supply Outlook	315
Power and Heat Sector Outlook	317
Capacity Needs and Fuel Mix	*317*
Impact of Electricity Market Reforms	*320*
Implications of the Projections	323
Russia's Role in the Global Energy Market	*323*
Energy Investment Needs and Financing	*325*
Environmental Impact	*326*

10 Energy and Development — 329

The Role of Energy in Development	330
Energy and Economic Growth	*331*
Energy and Human Development	*334*
The IEA Energy Development Index	342
Prospects for Energy Development	346
EDI Projections to 2030	*346*
Energy Development and the Millennium Goals	*350*
Policy Implications	353
Appendix to Chapter 10: Electrification Tables	357

11   World Alternative Policy Scenario   367

Background and Approach	368
Why an Alternative Scenario?	*368*
Methodology	*369*
Key results	372
Energy Demand	*372*
Implications for Energy Supply	*374*
Carbon-Dioxide Emissions	*377*
Investment Outlook	*380*
Results by Region	384
OECD Regions and the EU	*385*
Non-OECD Countries	*389*
Results by Sector	396
Power Generation	*396*
Transport	*401*
Industry	*405*
Residential and Services Sectors	*408*
Beyond the Alternative Policy Scenario	411
Appendix to Chapter 11:	
Tables for World Alternative Policy Scenario Projections	415
Annex A Tables for Reference Scenario Projections	429
Annex B Energy Projections: Assessment and Comparison	519
Annex C World Energy Model 2004	531
Annex D The Precarious State of Energy Statistics	549
Annex E Definitions, Abbreviations and Acronyms	553
Annex F References	563

List of Figures

Chapter 1: The Context
1.1	World Primary Energy Demand and GDP, 1971-2002	42
1.2	Per Capita Income by Region	46
1.3	Average IEA Crude Oil Import Price	48
1.4	Regional Natural Gas Price Assumptions	50
1.5	China's Share of Incremental World Production and Energy Demand, 1998-2003	55

Chapter 2: Global Energy Trends
2.1	Increase in World Primary Energy Demand by Fuel	59
2.2	World Primary Energy Demand by Fuel	60
2.3	World Primary Natural Gas Demand	63
2.4	Primary Energy Intensity	63
2.5	Regional Shares in World Primary Energy Demand	65
2.6	Share of Developing Countries in World Primary Energy Demand by Fuel	65
2.7	Primary Energy Consumption per Capita by Region, 2030	67
2.8	Sectoral Shares in World Primary Energy Demand	68
2.9	Per Capita Energy Consumption in the Residential Sector, 2030	69
2.10	Increase in World Primary Energy Production by Region	70
2.11	Share of Inter-Regional Trade in World Primary Demand by Fossil Fuel	71
2.12	Cumulative Energy Investment, 2003-2030	73
2.13	World Energy-Related CO_2 Emissions by Fuel	75
2.14	Average Annual Growth in World Primary Energy Demand and Energy-Related CO_2 Emissions	76
2.15	World Energy-Related CO_2 Emissions by Region	76
2.16	Per Capita Energy-Related CO_2 Emissions by Region	77
2.17	Change in Primary Energy Demand and Energy-Related CO_2 Emissions in the Alternative Policy Scenario, 2010 and 2030	80

Chapter 3: Oil Market Outlook
3.1	Oil Demand and GDP Growth	83
3.2	Oil Intensity by Region	84
3.3	Incremental Oil Demand by Sector, 2002-2030	85
3.4	Share of Transport in Global Oil Demand and Share of Oil in Transport Energy Demand	85
3.5	Vehicle Stock by Region	86
3.6	Per Capita Transport Sector Oil Demand and GDP, 1971-2030	87
3.7	Hydrocarbon-Resource Classification	88

3.8	Crude Oil and NGL Reserves at End 2003	90
3.9	Regional Share of Proven Oil Reserves	91
3.10	Proven Oil Reserves by Region	92
3.11	Ultimately Recoverable Resources of Oil and NGL by Region	94
3.12	Ultimately Recoverable Oil Resources	96
3.13	Non-Conventional Oil Resources Initially in Place	96
3.14	Additions to World Proven Oil Reserves from the Discovery of New Fields and Production	97
3.15	Undiscovered Oil and Gas Resources, 1995 and New Wildcat Wells Drilled, 1993-2002	98
3.16	Cumulative Oil and Gas Discoveries and New Wildcat Wells Drilled, 1963-2002	99
3.17	World Oil Exploration Drilling Success Rate	100
3.18	Oil Reserves in the Troll Field	100
3.19	Oil Production from the Kingfisher Field in the UK North Sea	101
3.20	World Oil Production by Source	103
3.21	Change in Non-OPEC Production and Average IEA Crude Oil Import Price	105
3.22	OPEC Share of World Oil Supply	110
3.23	Canadian Tar Sands Oil Production Cost	114
3.24	Major Net Inter-Regional Oil Trade Flows	116
3.25	Oil Flows and Major Chokepoints, 2003	120
3.26	Cumulative Global Oil Investment, 2003-2030	121
3.27	World Oil Demand in the Reference Scenario and High Oil Price Case	123
3.28	Cumulative Reduction in Oil Demand in the High Oil Price Case by Region and Sector, 2002-2030	124
3.29	Share of OPEC in World Oil Production in the Reference Scenario and the High Oil Price Case	126
3.30	Cumulative Oil Investment in the Reference Scenario and the High Oil Price Case, 2003-2030	127

Chapter 4: Natural Gas Market Outlook

4.1	Incremental Demand for Natural Gas by Region	131
4.2	World Natural Gas Demand by Sector	132
4.3	GTL Production by Region	135
4.4	World Proven Reserves of Natural Gas	137
4.5	Proven Natural Gas Reserves	138
4.6	Natural Gas Production by Region	138
4.7	World Gas-Production Capacity Additions	139
4.8	Share of Middle East Gas in Total Gas Supply by Importing Region	141

4.9	Major Net Inter-Regional Natural Gas Trade Flows, 2002 and 2030	143
4.10	Sources of LNG	144
4.11	Cumulative Investment in Natural Gas, 2003-2030	145
4.12	Ratio of Natural Gas Prices to Oil Price	146
4.13	Natural Gas Demand in OECD North America	148
4.14	US Marketed Gas Production and Drilling	150
4.15	Sources of North American Gas Supply	151
4.16	North American Gas Balance	153
4.17	Gas Demand in the European Union by Sector	154
4.18	Gas Supply Balance in the European Union	155
4.19	Increase in Gas Consumption and Imports in the European Union, 2002-2030	156
4.20	Primary Gas Demand in OECD Asia	158
4.21	Primary Gas Demand in Transition Economies	160
4.22	Primary Gas Demand in Developing Asia	161
4.23	Gas Demand in the Middle East by Sector	163
4.24	African Gas Exports by Region	165
4.25	Latin American Gas Balance	166

Chapter 5: Coal Market Outlook

5.1	World Coal Demand by Sector	171
5.2	Reductions in CO_2 Emissions through Technological Innovation	174
5.3	Coal Production by Region, 2002-2030	176
5.4	Major Inter-Regional Coal Trade Flows, 2002-2030	177
5.5	International Hard Coal Trade, 1985-2030	178
5.6	Spot Steam Coal Prices and Freight Rates	179
5.7	OECD Europe Coal Balance, 1990-2030	183
5.8	Australian Coal Exports and Share of World Trade, 1990-2030	185
5.9	Coal Production and Exports in China	186
5.10	Coal Production, Demand and Exports in Indonesia	189

Chapter 6: Electricity Market Outlook

6.1	GDP and Electricity Demand Growth	192
6.2	Sectoral Growth in Electricity Demand	194
6.3	Indicative Mid-Term Generating Costs of New Power Plants	195
6.4	CO_2 Emissions by Type of Plant	196
6.5	World Electricity Generation, 2002 and 2030	198
6.6	Share of Natural Gas in Electricity Generation	199
6.7	Share of Natural Gas in Electricity Generation by Region	199
6.8	Share of Nuclear Power in Electricity Generation in OECD Countries, 2002	202
6.9	Nuclear Plant Capacity Additions by Region, 2003-2030	202

6.10	Renewables in World Electricity Generation	204
6.11	Commercial Availability and Efficiency Improvements of Key Technologies, 2002-2030	205
6.12	Fuel Requirements in Power Plants	207
6.13	Impact of Plant Age on OECD Capacity Requirements	209
6.14	Capacity Requirements by Region	209
6.15	World Power Generating Capacity Additions and Investment, 2003-2030	210
6.16	Investment Requirements in Electricity Generation, Transmission and Distribution by Region	210
6.17	Power-Sector CO_2 Emissions by Region	211
6.18	Power-Sector CO_2 Emissions of Coal, Oil and Gas-Fired Power Plants	212
6.19	US Capacity Additions since 1997	217
6.20	US Natural Gas Prices and Share of Gas in Electricity Generation	218
6.21	Growth in Installed Generating Capacity, Peak Load and Available Capacity in the European Union	219
6.22	Annual Growth Rates of Electricity Generation and Installed Capacity in China, 1980-2002	221

Chapter 7: Renewable Energy Outlook

7.1	World Renewable Energy Consumption by Region	227
7.2	Share of Renewables in Electricity Generation by Region, 2002	228
7.3	Shares of Non-Hydro Renewables in Power Generation in 2002 and 2030	229
7.4	Share of Non-Hydro Renewables in Electricity Generation in OECD Countries, 1997 and 2002	230
7.5	Electricity Generation from Non-Hydro Renewables in OECD Countries, 2002	231
7.6	World Electricity Generation from Non-Hydro Renewable Energy Sources	231
7.7	World Long-Term Renewable-Energy Potential for Electricity Generation	232
7.8	Capital Costs of Renewable Energy Technologies, 2002 and 2030	233
7.9	Electricity-Generating Costs of Renewable Energy Technologies, 2002 and 2030	233
7.10	Wind Power Costs	236
7.11	Share of Geothermal Power in Total Electricity Generation, 2002	238
7.12	Renewables Capacity Additions, 2003-2030	239

7.13	Projected Share of Solar in Energy Consumption for Hot Water in the OECD	240

Chapter 8: Regional Outlooks

8.1	Primary Energy Demand in OECD Countries	244
8.2	GDP and Primary Energy Demand Growth in North America	246
8.3	Net Imports of Oil and Gas as Share of Primary Demand in North America	249
8.4	Fuel Shares in Primary Energy Demand in the European Union	252
8.5	Fossil Fuel Net Imports in the European Union	253
8.6	Energy-Related CO_2 Emissions in the European Union	253
8.7	Changes in Electricity Generation by Fuel in OECD Asia	256
8.8	Increase in Energy-Related CO_2 Emissions by Sector in OECD Asia	257
8.9	Primary Energy Demand in Developing Countries	260
8.10	Household Energy Consumption in Developing Countries	261
8.11	Share of Developing Asia in World Incremental Energy Demand	261
8.12	China's Share in the Global Economy and Energy Markets	265
8.13	Vehicle Stock and Oil Demand for Road Transport in China	266
8.14	Oil Balance in China	266
8.15	Primary Energy Demand in India	270
8.16	Total Final Energy Consumption per Capita in India	271
8.17	Oil Balance in Indonesia	272
8.18	Oil Balance in Brazil	276

Chapter 9: Russia – an In-Depth Study

9.1	Share of Russia in World Energy, 2002	285
9.2	Primary Energy Demand and GDP in Russia	285
9.3	Primary Energy Demand by Fuel in Russia	286
9.4	Growth in Industrial Production in Russia	288
9.5	Contribution of Oil and Gas Sectors to GDP in Selected Countries, 2002	289
9.6	Real Average Electricity and Natural Gas Prices in Russia	293
9.7	Fuel Shares in Primary Energy Demand in Russia	295
9.8	Total Final Consumption by Fuel in Russia	296
9.9	Energy Intensity Indicators in Russia	297
9.10	Energy Services and GDP in Russia	298
9.11	Energy Intensity of Industrial Production in Selected Sectors and Countries, 2000	299
9.12	Energy Intensity of Iron and Steel Production in Selected Countries	299
9.13	Russian Oil Balance	301
9.14	Russian Oil Basins and Pipelines	302

9.15	Russian Gas Balance	308
9.16	Major Gas Reserves and Supply Infrastructure in Russia	310
9.17	Russian Coal Balance	316
9.18	Net Additions to Power-Generation Capacity in Russia	318
9.19	Proposed Structure of the Russian Electricity Industry after Planned Reforms	322
9.20	Russian Fossil-Fuel Exports as Share of World Inter-Regional Trade	324
9.21	Cumulative Energy Investment Needs in Russia, 2003-2030	326
9.22	Energy-Related CO_2 Emissions in Russia	327

Chapter 10: Energy and Development

10.1	HDI and Primary Energy Demand per Capita, 2002	334
10.2	Average Primary Energy Demand per Capita and Population Living on Less than $2 a Day, 2002	335
10.3	Final Energy Consumption per Capita by Fuel and Proportion of People in Poverty in Developing Countries, 2002	338
10.4	HDI and Electricity Consumption per Capita, 2002	339
10.5	Electricity and Improved Water Access, 2002	341
10.6	Selected Developing Countries Ranked on the Energy Development Index, 2002	343
10.7	EDI and HDI in Developing Countries, 2002	346
10.8	Outlook for Energy Development Index by Region	347
10.9	Electricity Deprivation	349
10.10	World Population without Electricity in Rural and Urban Settings	350
10.11	The Energy Implications of Halving Poverty in Developing Countries by 2015	352

Chapter 11: World Alternative Policy Scenario

11.1	Energy Demand in the Reference and Alternative Scenarios	373
11.2	Change in Energy Intensity in the Reference and Alternative Scenarios, 2002-2030	374
11.3	Reduction in Oil Demand by Sector in the Alternative Scenario, 2030	375
11.4	Net Natural Gas Imports in Selected Regions in the Reference and Alternative Scenarios, 2030	376
11.5	Global Energy-Related CO_2 Emissions in the Reference and Alternative Scenarios	377
11.6	Cumulative Reduction in Energy-Related CO_2 Emissions by Region in the Alternative Scenario, 2002-2030	378
11.7	Reduction in Energy-Related CO_2 Emissions in the Alternative Scenario by Contributory Factor, 2002-2030	379

11.8	Difference in Cumulative Energy Investment between the Reference and Alternative Scenarios by Region, 2003-2030	381
11.9	Energy-Supply Investment in the Reference and Alternative Scenarios, 2003-2030	383
11.10	Capital Intensity of Energy Supply in the Reference and Alternative Scenarios, 2003-2030	383
11.11	Reduction in Demand for Fossil Fuels in the Alternative Scenario by Region, 2030	384
11.12	OECD Energy-Related CO_2 Emissions in the Reference and Alternative Scenarios	385
11.13	Change in Energy Demand in the Alternative Scenario in the Largest Non-OECD Countries, 2030	391
11.14	Fuel Shares in Electricity Generation in the Reference and Alternative Scenarios, 2030	398
11.15	Share of Non-Hydro Renewables in Electricity Generation in the Reference and Alternative Scenarios by Region, 2030	400
11.16	CO_2 Emissions per kWh of Electricity Generated in the Reference and Alternative Scenarios	400
11.17	Average Vehicle Fuel Efficiency for New Light Duty Vehicles in Selected Regions, 2002	402
11.18	Oil Demand for Transport in the Reference and Alternative Scenarios by Region	404
11.19	Reduction in Industrial Energy Demand by Sector in the Alternative Scenario, 2030	407
11.20	Reduction in Electricity Demand in the Residential and Services Sectors in the Alternative Scenario, 2030	411
11.21	Global Energy-Related CO_2 Emissions in the Reference and Alternative Scenarios and the CCS Case	413

List of Tables

Chapter 1: The Context

1.1	Economic Growth Assumptions	44
1.2	Population Growth Assumptions	45
1.3	Fossil-Fuel Price Assumptions	47

Chapter 2: Global Energy Trends

2.1	World Primary Energy Demand	59
2.2	World Total Final Consumption	68
2.3	Energy-Related CO_2 Emissions	75

Chapter 3: Oil Market Outlook

3.1	World Oil Demand	82

3.2	Top 10 Countries with Proven Oil Reserves	91
3.3	USGS Estimates of Ultimately Recoverable Oil and NGL Resources	94
3.4	Impact of Different Oil-Resource Assumptions on Production Outlook	102
3.5	World Oil Supply	106
3.6	Planned Additions to Non-Conventional Oil Production Capacity in Canada and Venezuela	115
3.7	Oil-Import Dependence in Net Importing Regions	117
3.8	Oil and LNG Tanker Traffic through Strategic Maritime Channels	119
3.9	Oil Production in Reference Scenario and High Oil Price Case	125

Chapter 4: Natural Gas Market Outlook

4.1	World Natural Gas Primary Demand	130
4.2	Gas-Import Dependence	140
4.3	Existing and Planned LNG Capacity in North America, September 2004	153
4.4	Status of Gas Market Liberalisation in the European Union	157
4.5	LNG Projects in India	162

Chapter 5: Coal Market Outlook

5.1	World Coal Demand	170
5.2	Proven Coal Reserves at End-2002	175

Chapter 6: Electricity Market Outlook

6.1	Electricity's Share of Energy Demand by Sector	193
6.2	Top-Ten Electricity Markets in 2002	194
6.3	Market Shares in Electricity Generation	197
6.4	Electricity Generation from Renewable Energy Sources	204
6.5	New Electricity Generating Capacity and Investment by Region, 2003-2030	208
6.6	Nuclear Power Reactors in China, 2004	222

Chapter 7: Renewable Energy Outlook

7.1	World Renewable Energy Consumption	226

Chapter 8: Regional Outlooks

8.1	Primary Energy Demand in the OECD	245
8.2	Primary Energy Demand in North America	247
8.3	Primary Energy Demand in the European Union	251
8.4	Primary Energy Demand in OECD Asia	255
8.5	Primary Energy Demand in OECD Oceania	258
8.6	Primary Energy Demand in China	264

8.7	Electricity-Generation Mix in China	268
8.8	Primary Energy Demand in Brazil	276
8.9	Electricity-Generation Mix in Brazil	277
8.10	Primary Energy Demand in the Middle East	278
8.11	Population Structure in the Middle East	279

Chapter 9: Russia – an In-Depth Study

9.1	Key Economic and Energy Indicators for Russia and the World	287
9.2	Macroeconomic and Demographic Assumptions for Russia	290
9.3	Russian *Energy Strategy* and *WEO-2004* Projections to 2020	292
9.4	Total Primary Energy Demand in Russia	295
9.5	Russian Gas Production of Non-Gazprom Companies	312
9.6	Electricity Generation Fuel Mix in Russia	318

Chapter 10: Energy and Development

10.1	Contribution of Factors of Production and Productivity to GDP Growth in Selected Countries, 1980-2001	333
10.2	Commercial Energy Use and Human Development Indicators, 2002	336
10.3	Dominant Fuels in Developing Countries by End-Uses	337
10.4	Number of People without Electricity, 2002	340
10.5	Energy Development Index for Developing Countries, 2002	344
10.6	Population Relying on Traditional Biomass for Cooking and Heating	348
10.7	Electrification Rates by Region	348
10.8	Impact of Meeting MDG Poverty-Reduction Target on the Number of People without Electricity and Investment in Developing Countries	351

Chapter 11: World Alternative Policy Scenario

11.1	Changes in Energy-Related CO_2 Emissions by Sector and by Region in the Alternative Scenario, 2030	379
11.2	Additional Demand-Side Investment in the Alternative Scenario, 2003-2030	382
11.3	Main Policies Considered in the Alternative Scenario in OECD North America	387
11.4	Main Policies Considered in the Alternative Scenario in the EU	388
11.5	Main Policies Considered in the Alternative Scenario in OECD Pacific	390
11.6	Main Policies Considered in the Alternative Scenario in Russia	392
11.7	Main Policies Considered in the Alternative Scenario in China	393
11.8	Main Policies Considered in the Alternative Scenario in India	395
11.9	Main Policies Considered in the Alternative Scenario in Brazil	396

11.10	Changes in Electricity Generation by Fuel in the Alternative Scenario	399
11.11	Changes in Transport Energy Consumption and CO_2 Emissions in the Alternative Scenario	403
11.12	Changes in Transport Energy Consumption and CO_2 Emissions in the Alternative Scenario by Region	403
11.13	Change in Industrial Energy Consumption in the Alternative Scenario, 2030	406
11.14	Change in Residential and Services Sector Energy Consumption in the Alternative Scenario, 2030	410

List of Boxes

Chapter 1: The Context

1.1	Explaining High Oil Prices: The Risk Premium and Fundamentals	49
1.2	Impact of High Gas Prices on US Manufacturing Industry	51
1.3	The Impact of High Oil Prices on the Global Economy	54

Chapter 2: Global Energy Trends

2.1	What is Included in Primary Energy Demand	61
2.2	Recent Trends in Global Energy Demand	64
2.3	Status of the Kyoto Protocol	79

Chapter 4: Natural Gas Market Outlook

4.1	Status of GTL Projects	134
4.2	Sensitivity of US Gas Demand to Higher Prices	149

Chapter 5: Coal Market Outlook

5.1	FutureGen: Zero-Emission Technologies	173
5.2	Non-Physical Trading of Steam Coal	180

Chapter 6: Electricity Market Outlook

6.1	Nuclear Policy in OECD Countries	201
6.2	Major Blackouts in OECD Countries in 2003	214

Chapter 7: Renewable Energy Outlook

7.1	Economics of Wind-Power Integration	235

Chapter 9: Russia – an In-Depth Study

9.1	Russian Energy Strategy	292
9.2	Profile of Gazprom	309
9.3	The Outlook for Nuclear Power in Russia	319

Chapter 10: Energy and Development
10.1	UNDP Human Development and Poverty Indices	331
10.2	Assessing the Contribution of Energy to Economic Growth	332
10.3	The IEA Energy Development Index	342
10.4	The UN Millennium Development Goals	351

Chapter 11: World Alternative Policy Scenario
11.1	Example of OECD Policies Included in the Alternative Policy Scenario: the EU Renewables Target	370
11.2	Example of Non-OECD Policies Included in the Alternative Policy Scenario: Vehicle Efficiency in China	371

World Energy Outlook Series

World Energy Outlook 1993
World Energy Outlook 1994
World Energy Outlook 1995
World Energy Outlook 1996
World Energy Outlook 1998
World Energy Outlook 1999 Insights:
    Looking at Energy Subsidies: Getting the Prices Right
World Energy Outlook 2000
World Energy Outlook 2001 Insights:
    Assessing Today's Supplies to Fuel Tomorrow's Growth
World Energy Outlook 2002
World Energy Investment Outlook 2003 Insights
World Energy Outlook 2004
World Energy Outlook 2005 Insights (forthcoming):
    Middle East Energy Outlook: Implications for Global Energy Markets

EXECUTIVE SUMMARY

Energy Security in a Dangerous World

***World Energy Outlook 2004* paints a sobering picture of how the global energy system is likely to evolve from now to 2030.** If governments stick with the policies in force as of mid-2004, the world's energy needs will be almost 60% higher in 2030 than they are now. Fossil fuels will continue to dominate the global energy mix, meeting most of the increase in overall energy use. The shares of nuclear power and renewable energy sources will remain limited.

The Earth's energy resources are more than adequate to meet demand until 2030 and well beyond. Less certain is how much it will cost to extract them and deliver them to consumers. Fossil-fuel resources are, of course, finite, but we are far from exhausting them. The world is not running out of oil just yet. Most estimates of proven oil reserves are high enough to meet the cumulative world demand we project over the next three decades. Our analysis suggests that global production of conventional oil will not peak before 2030 if the necessary investments are made. Proven reserves of gas and coal are even more plentiful than those of oil. There is considerable potential for discovering more of all these fuels in the future.

But serious concerns about energy security emerge from the market trends projected here. The world's vulnerability to supply disruptions will increase as international trade expands. Climate-destabilising carbon-dioxide emissions will continue to rise, calling into question the sustainability of the current energy system. Huge amounts of new energy infrastructure will need to be financed. And many of the world's poorest people will still be deprived of modern energy services. These challenges call for urgent and decisive action by governments around the world.

A central message of this *Outlook* is that short-term risks to energy security will grow. Recent geopolitical developments and surging energy prices have brought that message dramatically home. Major oil- and gas-importers – including most OECD countries, China and India – will become ever more dependent on imports from distant, often politically-unstable parts of the world. Flexibility of oil demand and supply will diminish. Oil use will become ever more concentrated in transport uses in the absence of readily-available substitutes. Rising oil demand will have to be met by a small group of countries with large reserves, primarily Middle East members of OPEC and Russia. Booming trade will strengthen the mutual dependence among exporting and importing countries. But it will also exacerbate the risks that wells or pipelines could be closed or tankers blocked by piracy, terrorist attacks or accidents. Rapid worldwide growth in natural gas consumption and trade will foster similar concerns.

If current government policies do not change, energy-related emissions of carbon dioxide will grow marginally faster than energy use. CO_2 emissions will be more than 60% higher in 2030 than now. The average carbon content of energy, which fell markedly during the past three decades, will hardly change. Well over two-thirds of the projected increase in emissions will come from developing countries, which will remain big users of coal – the most carbon-intensive of fuels. Power stations, cars and trucks will give off most of the increased energy-related emissions.

Converting the world's resources into available supplies will require massive investments. In some cases, financing for new infrastructure will be hard to come by. Meeting projected demand will entail cumulative investment of some $16 trillion from 2003 to 2030, or $568 billion per year. The electricity sector will absorb the majority of this investment. Developing countries, where production and demand are set to increase most, will require about half of global energy investment. Those countries will face the biggest challenge in raising finance, because their needs are larger relative to the size of their economies and because the investment risks are bigger. The global financial system has the capacity to fund the required investments, but it will not do so unless conditions are right.

Reducing energy poverty is an urgent necessity. There will be some encouraging advances in energy development in non-OECD countries over the projection period. But even for the most developed among them in energy terms, the use of modern energy and the per capita consumption of every kind of energy will remain far below that of OECD countries. Little progress will be made in reducing the total number of people who lack access to electricity. And the ranks of those using traditional fuels in unsustainable and inefficient ways for cooking and heating will actually *increase* over the projection period. Developing countries are unlikely to see their incomes and living standards increase without improved access to modern energy services.

These trends, from our Reference Scenario, are, however, not unalterable. More vigorous government action *could* steer the world onto a markedly different energy path. This *Outlook* presents an Alternative Scenario, which analyses, for the first time, the global impact of environmental and energy-security policies that countries around the world are already considering, as well as the effects of faster deployment of energy-efficient technologies. In this scenario, global energy demand and carbon-dioxide emissions are significantly lower than in our Reference Scenario. Dependence on imported energy in major consuming countries and the world's reliance on

Middle East oil and gas are also lower. However, even in this Alternative Scenario energy imports and emissions would still be higher in 2030 than today.

It is clear from our analysis that achieving a truly sustainable energy system will call for technological breakthroughs that radically alter how we produce and use energy. The government actions envisioned in our Alternative Scenario could slow markedly carbon-dioxide emissions, but they could not reduce them significantly using existing technology. Carbon capture and storage technologies, which are not taken into account in either the Reference or the Alternative Scenario, hold out the tantalising prospect of using fossil fuels in a carbon-free way. Advanced nuclear-reactor designs or breakthrough renewable technologies could one day help free us from our dependence on fossil fuels. This is unlikely to happen within the timeframe of our analysis. The pace of technology development and deployment in these and other areas is the key to making the global energy system more economically, socially and environmentally sustainable in the long term. But consumers will have to be willing to pay the full cost of energy – including environmental costs – before these technologies can become competitive. Governments must decide today to accelerate this process.

Main Findings and Projections

Fossil Fuels will Still Meet most of the World's Energy Needs

World primary energy demand in the Reference Scenario is projected to expand by almost 60% between 2002 and 2030. But the projected annual rate of demand growth, at 1.7%, is slower than the average of the past three decades, which was 2%. Energy intensity – the amount of energy needed to produce a dollar's worth of GDP – will continue to decline as energy efficiency improves and the global economy relies less on heavy industry.

Fossil fuels will continue to dominate global energy use, accounting for some 85% of the *increase* in world primary demand. Oil will remain the single largest fuel in the primary energy mix, even though its percentage share will fall marginally. Among the fossil fuels, demand for natural gas will grow most rapidly, mainly due to strong demand from power generators. The share of coal will fall slightly, but coal will remain the leading fuel for generating electricity. Nuclear power's share will decline during the *Outlook* period.

Two-thirds of the increase in global energy demand will come from developing countries. By 2030, they will account for almost half of total demand, in line with their more rapid economic and population growth. More households will live in towns and cities and so will be better placed to gain access to energy services. The developing countries' share of global demand will increase for all of the primary energy sources except non-hydro renewables. Their share of nuclear-power production will increase fastest, because of strong growth in China and other parts of Asia. Their share of coal consumption will also increase sharply, mainly because of booming demand in China and India.

Oil-Supply Patterns will Shift as Demand and Trade Grow

Global primary oil demand is projected to grow by 1.6% per year, reaching 121 mb/d in 2030. Demand will continue to grow most quickly in developing countries. Most of the increase in world oil demand will come from the transport sector. Oil will face little competition from other fuels in road, sea and air transportation during the projection period. OPEC countries, mainly in the Middle East, will meet most of the increase in global demand. By 2030, OPEC will supply over half of the world's oil needs – an even larger share than in the 1970s. Net inter-regional oil trade will more than double, to over 65 mb/d in 2030 – a little more than half of total oil production. Huge investments will be needed in oilfields, tankers, pipelines and refineries, $3 trillion from 2003 to 2030. Most upstream investment will, in fact, offset production declines from already-producing fields. Financing will be a major challenge.

The International Energy Agency calls on all parties to work together to devise and implement a universally-recognised, transparent, consistent and comprehensive data-reporting system for oil and gas reserves. The reliability of reserves data reported by oil companies has been called into severe question. Doubts about the accuracy of reserve estimates – an issue highlighted in this *Outlook* – could undermine investor confidence and slow investment. Governments should be concerned about reserves-data problems, since the long-term security of energy supplies depends on the timely development of oil and gas reserves. The future availability and affordability of hydrocarbons affect decisions about what new policies and measures governments ought to adopt now to develop alternative sources of energy and to save energy.

As international trade expands, risks will grow of a supply disruption at the critical chokepoints through which oil must flow. A total of 26 million barrels currently pass through the Straits of Hormuz in the Persian Gulf and the Straits of Malacca in Asia every day. Traffic through these and other vital channels will more than double over the projection period. A disruption in supply at any of these points could have a severe impact on oil markets. Maintaining the security of international sea-lanes and pipelines will take on added urgency.

Future trends in oil prices are a major source of uncertainty. Prices of crude oil and refined products have risen sharply since 1999, hitting all-time highs in nominal terms in mid-2004. In a special analysis of sustained high oil prices, we have assumed that the price of crude oil imported into IEA countries would average $35 per barrel (in year-2000 dollars) from now to 2030 – about $10 more than in our Reference Scenario. In this high price case, global oil demand falls by 15%, or 19 mb/d in 2030, an amount almost equal to total US oil consumption today. Conventional and non-conventional oil production outside OPEC countries increases markedly at the $35 price, causing OPEC's market share to fall considerably. Cumulative OPEC revenues in 2003-2030 are about $750 billion, or 7% lower, than in the Reference Scenario. Plainly, OPEC would not benefit from higher prices in the long term.

Demand for Natural Gas will Overtake that for Coal

Worldwide consumption of natural gas will almost double by 2030, overtaking that of coal within the next decade. Gas demand is projected to grow most rapidly in Africa, Latin America and developing Asia. Yet the total volume increase in demand will be bigger in the mature markets of OECD North America, OECD Europe and the transition economies, where per capita gas use is much higher. Most of the increase in gas demand will come from power stations. Gas is often preferred to coal in new thermal plants for its environmental advantages, its lower capital costs and operational flexibility. Gas-to-liquids plants will emerge as a new market for natural gas, making use of reserves located far from traditional markets and meeting rising demand for cleaner oil products.

Gas reserves are easily large enough to meet the projected increase in global demand. Additions to proven reserves have outpaced production by a wide margin since the 1970s. Production will increase most in Russia and in the Middle East, which between them hold most of the world's proven gas reserves. Most of the incremental output in these regions will be exported to North America, Europe and Asia, swelling the surge in international energy trade. All regions that are currently net importers of gas will see their imports rise, and a growing number of countries and regions will become net importers for the first time. Liquefied natural gas, the bulk of which will be used for power generation, will account for most of the increase in traded gas. By 2030, just over half of all inter-regional gas trade will be in the form of LNG, up from 30% at present. OPEC countries will continue to dominate the supply of LNG. Cumulative investment needs for gas-supply infrastructure to 2030 will amount to $2.7 trillion, or about $100 billion per year from now to 2030. More than half will be for exploration and development of gas fields.

Even though coal's share of the global energy market will drop slightly over the *Outlook* period, coal will continue to play a key role in the world energy mix. In 2030, coal will meet 22% of all energy needs, essentially the same proportion as today. Virtually all the increase in coal consumption will be for power generation, and coal will remain that sector's main fuel – despite a loss of market share to natural gas. Coal demand will increase most in developing Asian countries. China and India alone will be responsible for 68% of the increase in demand over the period 2002 to 2030. Demand growth in the OECD will be minimal.

Carbon-free Energy Sources will Meet only a Small Part of Surging Electricity Needs

World electricity demand is expected to double between now and 2030, with most of the growth occurring in developing countries. By 2030, power generation will account for nearly half of world consumption of natural gas. It will also have absorbed over 60% of total investment in energy-supply infrastructure between now and then. The global power sector will need about 4 800 GW of new capacity to meet the projected increase in electricity demand and to replace ageing infrastructure. In total, electricity investment will amount to about $10 trillion, more than $5 trillion of that amount for developing countries alone. For many of them, investment will need to increase substantially. The electricity-supply industry is set for further restructuring and more far-reaching regulatory reforms. Reforms in the OECD have yielded positive results, but many challenges remain to be met. Blackouts in 2003 and 2004 highlighted the importance of adequate reserve margins, the need to improve the resilience of networks and the importance of providing adequate regulatory incentives for investment.

Worldwide nuclear capacity is projected to increase slightly, but the share of nuclear power in total electricity generation will decline. A substantial amount of capacity will be added, but this will be mostly offset by reactor retirements. Three-quarters of existing nuclear capacity in OECD Europe is expected to be retired by 2030, because reactors will have reached the end of their life or because governments plan to phase out nuclear power. Nuclear power generation will increase in a number of Asian countries, notably in China, South Korea, Japan and India.

Renewable energy sources as a whole will increase their share of electricity generation. The share of hydroelectricity will fall, but the shares of other renewables in electricity generation will triple, from 2% in 2002 to 6% in 2030. Most of the increase will be in wind and biomass. Wind power will be the second-largest renewable source of electricity in 2030, after hydroelectricity. Finding good sites for land-based wind turbines is becoming more difficult in some areas. The largest increases in renewables will occur in OECD Europe, where they enjoy strong government backing.

Russian Oil and Gas Exports are Poised for Further Growth in the Near Term

Russia will play a central role in global energy supply and trade over the *Outlook* period, with major implications for the world's energy security. The Russian energy sector has undergone a dramatic transformation in recent years. It has been the principal force behind the country's economic recovery since the late 1990s. The Russian economy's dependence on the oil and gas sectors has grown in recent years. Russia's long-term economic prospects hinge on improving the competitiveness and diversity of its other manufacturing sectors and internationally traded services.

The prospects for Russian oil production are very uncertain. Oil production has surged in recent years, mainly thanks to the rehabilitation of existing wells to enhance the recovery of reserves. Production is projected to continue its increase, though more slowly than in recent years. In the short to medium term, most of the extra production will be exported. But the share of Russian exports in world trade will fall after 2010, as Russian production stabilises, domestic demand expands and output picks up in the Middle East.

Russia's huge gas resources will underpin a continued increase in production. Higher output will not only meet rising domestic demand, but also provide increased exports to Europe and to new markets in Asia. Russia will still be the world's biggest gas exporter in 2030. But output from the country's old super-giant fields is declining, and huge investments in greenfield projects will be needed to replace them. The prospects of independent producers contributing more gas – and thereby allowing Russia to increase exports – will depend on whether Gazprom's network is effectively opened to them.

Developing Russia's huge energy resources, modernising existing infrastructure and improving efficiency will call for enormous investments. A stable and predictable business regime and market reforms are urgently required if these investments are to be financed. If gas-sector reform is delayed, worries about the security of future supply will increase. Large amounts of foreign capital are unlikely to be made available for energy projects that are not aimed at export markets.

Expanding Modern Energy Services in Poor Countries Will Remain Vital to Their Prospects for Development

Energy is a prerequisite to economic development. The prosperity that economic development brings, in turn, stimulates demand for more and better energy services. Energy services also help to meet such basic human needs as food and shelter. They contribute to social development by improving education and public health. Electricity plays a particularly important role in human development. Most developed countries have established a virtuous circle of improvements in energy infrastructure and economic growth. But in the world's poorest countries, the process has barely got off the ground.

Electrification rates will rise over the projection period, but the total *number* of people still without electricity will fall only slightly, from 1.6 billion in 2002 to just under 1.4 billion in 2030. Most of the net decrease in the number of people without electricity will occur only after 2015. The ranks of the electricity-deprived will fall in Asia, but will continue to swell in Africa. Access to electricity will remain easier in urban areas, but the absolute number of people without electricity will increase slightly in towns and cities, while it will fall in the countryside. The number of people using only traditional biomass for cooking and heating in unsustainable ways will continue to grow, from just under 2.4 billion in 2002 to over 2.6 billion in 2030.

Developing countries can look forward to further advances in energy and human development. According to the Energy Development Index, which the IEA presents for the first time in this *Outlook*, all developing regions can expect to experience increases in per capita energy use and improved access to modern energy services – including electricity. Yet only a few Middle East and Latin American countries will have reached the stage of energy development in 2030 that OECD countries had attained in 1971. Africa and South Asia will remain far behind.

Our analysis suggests that halving the proportion of very poor people will require much faster energy development than is projected in our Reference Scenario. The UN's Millennium Development Goals aim to reduce by 50% the proportion of people living on less than a dollar a day between 2000 and 2015. We estimate that this target will not be met unless access to electricity can be provided to more than half-a-billion people who, according to our Reference Scenario, will still lack it in 2015. To do that would require about $200 billion of additional investment in electricity supply. Meeting the target also implies a need to expand the use of modern cooking and heating fuels to 700 million more people by 2015 than projected in our Reference Scenario.

Governments must act decisively to accelerate the transition to modern fuels and to break the vicious circle of energy poverty and human under-development in the world's poorest countries. This will require increasing the availability and affordability of commercial energy, particularly in rural areas. Good governance in the energy sector and more generally will be critical to improving both the quantity and quality of energy services. The rich industrialised countries have clear economic and security interests in helping developing countries along the energy-development path.

New Policies Could Achieve a More Sustainable Energy System

This study presents a World Alternative Policy Scenario, which depicts a more efficient and more environment-friendly energy future than does the Reference Scenario. It analyses how global energy trends could evolve were countries around the world to implement a set of policies and measures that they are currently considering or might reasonably be expected to adopt. These policies would foster the faster deployment of more efficient and cleaner technologies. In this scenario, global primary energy demand would be about 10% lower in 2030 than in the Reference Scenario. The reduction in demand for fossil fuels would be even bigger, thanks largely to policies that promote renewable energy sources.

Demand for oil would be markedly lower than in the Reference Scenario. Global oil demand would be 12.8 mb/d, or 11%, lower in 2030 – an amount equal to the current combined production of Saudi Arabia, the United Arab Emirates and Nigeria. Stronger measures to improve fuel economy in OECD countries and the faster deployment of more efficient vehicles in non-OECD countries would contribute almost two-thirds of these savings in 2030. Oil-import dependence in the OECD countries and China would drop as a result. Coal demand would fall even more in percentage terms – by 24% in 2030. The amount saved would be around the current coal consumption of China and India combined. World natural gas demand would be 10% lower than in the Reference Scenario. Gas-import needs would be 40% lower in OECD North America and 13% lower in Europe. China's gas imports would be higher, after a shift from coal to gas.

By 2030, energy-related emissions of carbon dioxide would be 16% lower than in the Reference Scenario. This is roughly equal to the combined current emissions of the United States and Canada. Almost 60% of the cumulative reduction of CO_2 emissions would occur in non-OECD countries. In fact, OECD emissions would level off by the 2020s, and then begin to *decline*. More efficient use of energy in vehicles, electric appliances, lighting and industry account for more than half of the reduction in emissions. A shift in the power generation fuel mix in favour of renewables and nuclear power accounts for most of the rest.

The pattern of investment in energy supply and end-use equipment in the Alternative Scenario is substantially different from that in the Reference Scenario. The total amount of capital required over the projection period for the entire energy chain – from energy production to end use – does not differ much between the two scenarios. Larger capital needs on the demand side would be entirely offset by lower investment needs on the supply side – despite a 14% increase in the capital intensity of electricity supply in the

Alternative Scenario. Electricity prices would rise – for example, by 12% in the European Union. It is uncertain, however, whether all the investment invoked in the Alternative Scenario could actually be financed, especially in developing countries. This is mainly because end-users, who would have to invest more, are likely to find it harder to secure financing than would suppliers, who would need to invest less.

CHAPTER 1

THE CONTEXT

HIGHLIGHTS

- Global economic growth – the primary driver of energy demand – is assumed to average 3.2% per year over the period 2002-2030, slightly less than in the previous three decades. The rate will drop from 3.7% in 2002-2010 to 2.7% in the last decade of the projection period, as developing countries' economies mature and population growth slows. The economies of China, India and other Asian countries are expected to continue to grow most rapidly.

- The world's population is assumed to expand from 6.2 billion in 2002 to over 8 billion in 2030 – an increase of 1% per year on average. Population growth will slow progressively over the projection period, mainly due to falling fertility rates in developing countries. The share of the world population living in developing regions will nonetheless increase from 76% today to 80% in 2030.

- In the Reference Scenario, the average IEA crude oil import price is assumed to fall back from current highs to $22 (in year-2000 dollars) in 2006. The price is assumed to remain flat until 2010, and then to begin to climb steadily to $29 in 2030. Gas prices are assumed to move broadly in line with oil prices. Steam coal prices are assumed to average around $40/tonne through to 2010, and to rise very slowly thereafter, to $44 in 2030. In a High Oil Price Case, the crude oil price is assumed to average $35 over the entire projection period.

- The *WEO* projections of energy demand and supply are subject to a wide range of uncertainties, including macroeconomic conditions, resource availability, technological developments and investment flows, as well as government energy and environmental policies. The near-term energy outlook depends heavily on the prospects for economic growth – especially in China – and on oil-price trends.

- An Alternative Policy Scenario analyses how the global energy market would evolve were countries around the world to adopt a set of new policies and measures to tackle environmental problems and enhance energy security.

The Methodological Approach

As in the last two editions of the *Outlook*, a scenario approach has been adopted to analyse the possible evolution of energy markets to 2030. The primary objective is to identify and quantify the key factors that are likely to affect energy supply and demand. The central projections are derived from a Reference Scenario. They are based on a set of assumptions about government policies, macroeconomic conditions, population growth, energy prices and technology. The Reference Scenario takes into account only those government polices and measures that were already enacted – though not necessarily implemented – as of mid-2004. These projections should not be interpreted as a forecast of how energy markets are likely to develop. The Reference Scenario projections should rather be considered a baseline vision of how the global energy system will evolve if governments take no further action to affect its evolution beyond that which they have already committed themselves to. The World Alternative Policy Scenario, presented later in the book, takes into account a range of new policies to address environmental problems and enhance energy security that are currently under consideration by countries around the world. The first year of both sets of projections is 2003, as 2002 is the last year for which historical data are available for all countries.

The IEA's World Energy Model – a large-scale mathematical model that has been developed over several years – is the principal tool used to generate detailed sector-by-sector and region-by-region projections for both scenarios.[1] The model has been updated and revised substantially since the last *Outlook*. The main improvements include:

- Increased regional disaggregation, with the development of separate models for the enlarged European Union of 25 members, as well as for OECD Asia and OECD Oceania.
- Infrastructure-investment implications of the projected demand and supply trends.
- The inclusion of non-commercial biomass in the world energy balance.
- More detailed modelling of end-use sectors in developing countries.
- More sophisticated treatment of the final consumption of renewables and of energy use in combined heat and power production in all regions.
- A breakdown of inter-regional natural gas trade between pipelines and liquefied natural gas (LNG).
- Detailed modelling of coal production and inter-regional coal trade.
- A new resource-based model for oil production, which underpins the medium- to long-term projections.

---

1. See Annex C for a detailed description of the World Energy Model.

Another innovation for this *Outlook* is the Energy Development Index (EDI) – a simple composite indicator of the stage that developing countries have reached in their transition to modern fuels and of the degree of maturity in their energy end-use. It is intended to improve our understanding of the role that modern energy plays in economic and human development.

The Reference Scenario
Government Policies and Measures

The Reference Scenario takes account of those government policies and measures that were enacted or adopted by mid-2004, though many of them have not yet been fully implemented. Many of the most recent policies affecting the energy sector in OECD countries are designed to combat climate change and other environmental problems. Their impact on energy demand and supply does not show up in historical market data, which are available only up to 2002 for all countries. These initiatives cover a wide array of sectors and a variety of policy instruments. The Reference Scenario does not include possible, potential or even likely future policy initiatives. Major new energy-policy initiatives will inevitably be implemented during the projection period, but it is difficult to predict which measures will eventually be adopted and how they will be implemented, especially towards the end of the projection period.

Although the Reference Scenario assumes that there will be no change in energy and environmental policies through the projection period, the pace of implementation of those policies and the way they are implemented in practice are nonetheless assumed to vary by fuel and by region. For example, electricity and gas market reforms are assumed to move ahead, but at varying speeds among countries and regions. Progress will also be made in liberalising cross-border energy trade and investment, and in reforming energy subsidies, but these policies are expected to be pursued most rigorously in OECD countries. In all cases, the share of taxes in energy prices is assumed to remain unchanged, so that retail prices are assumed to change directly in proportion to international prices. Similarly, it is assumed that there will be no changes in national policies on nuclear power. As a result, nuclear energy will remain an option for power generation only in those countries that have not officially banned it or decided to phase it out.

Macroeconomic Factors

Economic growth is by far the most important driver of energy demand. The link between total energy demand and economic output remains close. In past decades, energy demand has risen in a broadly linear fashion along with gross domestic product: since 1971, each 1% increase in global GDP has been accompanied by a 0.6% increase in primary energy consumption.[2] However, the so-called income elasticity of energy demand – the increase in demand relative to GDP – has fallen over that period, from 0.7 in the 1970s to 0.4 from 1991 to 2002 (Figure 1.1). This is partly the result of warmer winter weather in the northern hemisphere, which has depressed demand for heating fuels, and improved energy efficiency. Demand for electricity and transport fuels remains very closely aligned with GDP. As a result, the energy projections in the *Outlook* are highly sensitive to underlying assumptions about GDP growth.

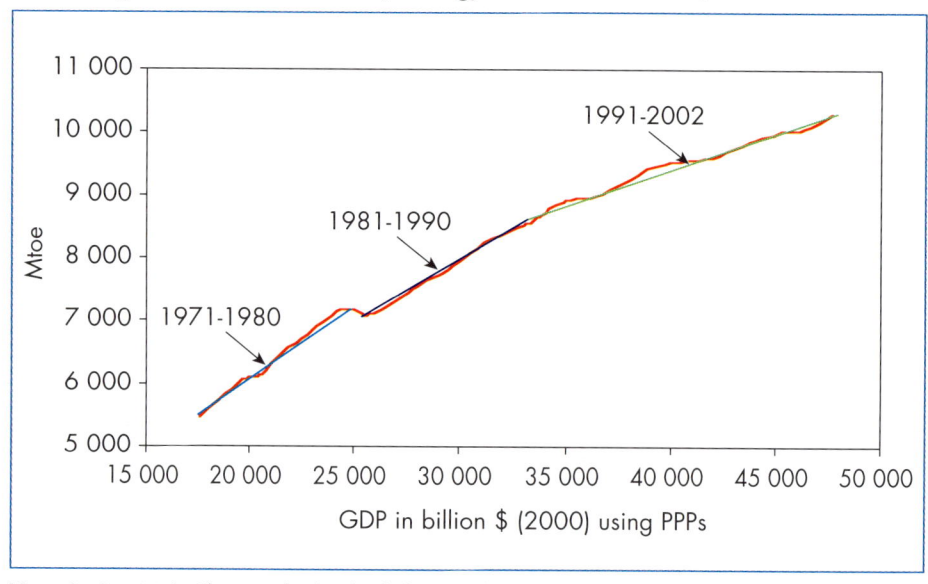

Figure 1.1: **World Primary Energy Demand and GDP, 1971-2002**

Note: See Box 2.1 in Chapter 2 for details of what is included in primary energy demand.

The global economy has rebounded strongly since 2002, following a major cyclical downturn in 2001. Industrial production has accelerated, global trade has increased sharply and business and consumer confidence has strengthened. Investment has also started to accelerate in almost all regions. In the second half of 2003, global GDP growth averaged close to 6% from the same period of 2002 – the fastest growth rate since 1999. This was partly due to one-off factors, notably a surge in consumption in the United States thanks to the

---

2. All GDP data cited in this report are expressed in year-2000 dollars using purchasing power parities (PPPs) rather than market exchange rates. PPPs compare costs in different currencies of a fixed basket of traded and non-traded goods and services and yield a widely-based measure of standard of living. This is important in analysing the main drivers of energy demand or comparing energy intensities among countries.

short-term impact of tax cuts and mortgage refinancing. In Asia, there was a robust rebound from the temporary economic slowdown related to the outbreak of Severe Acute Respiratory Syndrome.[3] Global GDP growth remained solid in the first half of 2004, according to preliminary data, but the pace of the current economic upturn varies markedly among countries and regions. It is strongest in developing Asia, especially China, and in the United States. While the Japanese economy is finally picking up, the recovery is much more tentative and gradual in the euro zone, where domestic consumption and investment remain weak.

This *Outlook* assumes that economic growth will accelerate further in most regions in 2004.[4] GDP for that year is expected to grow by almost 4% in the United States and Canada, by over 2% in OECD Europe and by 3.5% in both OECD Asia and Oceania. Growth in China, however, is assumed to fall back from an estimated 9% in 2003 to around 7% in 2004 as government measures to check the overheating of the economy take effect. The near-term economic outlook for China is a factor of particular significance for energy markets (see Main Uncertainties below). The short-term acceleration of global GDP will depend partly on oil prices falling back from the record levels reached in early to mid-2004.

GDP growth in all regions is expected to slow gradually over the next two decades. World GDP is assumed to grow by an average of 3.2% per year over the period 2002-2030 compared to 3.3% from 1971 to 2002. The rate will drop from 3.7% in 2002-2010 to 2.7% in the 2020s. China, India and other Asian countries are expected to continue to grow faster than other regions, followed by Africa and the transition economies. China is assumed to grow at 5%, the highest rate in the world, yet this figure is much lower than the 8.4% of the past three decades. The Chinese economy will undoubtedly slow as it becomes more mature, but will nonetheless become the largest in the world in PPP terms early in the 2020s. GDP in both OECD Europe and OECD Pacific is assumed to grow by 2% per year over the projection period. OECD North America will see a slightly higher growth, of 2.4%. All major regions are expected to experience a continuing shift in their economies away from energy-intensive heavy manufacturing towards lighter industries and services. Detailed GDP assumptions by region are set out in Table 1.1.

Population

Population growth affects the size and composition of energy demand, directly and through its impact on economic growth and development. Our

---

3. IMF (2004).
4. The economic growth assumptions in this *Outlook* are based on forecasts and studies by a number of organisations, including OECD (2004) and International Monetary Fund (2004).

Table 1.1: **Economic Growth Assumptions** (average annual growth rates, in %)

	1971-2002	2002-2010	2010-2020	2020-2030	2002-2030
OECD	**2.9**	**2.7**	**2.2**	**1.8**	**2.2**
North America	3.1	3.2	2.4	1.9	2.4
United States and Canada	*3.1*	*3.1*	*2.3*	*1.8*	*2.3*
Mexico	*3.7*	*3.7*	*3.7*	*3.0*	*3.5*
Europe	2.4	2.4	2.2	1.7	2.1
Pacific	3.5	2.5	1.9	1.7	2.0
Asia	*3.6*	*2.4*	*1.9*	*1.6*	*1.9*
Oceania	*3.0*	*3.1*	*2.3*	*1.7*	*2.3*
Transition economies	**0.7**	**4.6**	**3.7**	**2.9**	**3.7**
Russia	−1.1*	4.4	3.4	2.8	3.5
Other transition economies	−0.5*	4.8	3.9	3.0	3.8
Developing countries	**4.7**	**5.1**	**4.3**	**3.6**	**4.3**
China	8.4	6.4	4.9	4.0	5.0
East Asia	5.3	4.5	3.9	3.1	3.8
Indonesia	*5.9*	*4.5*	*4.1*	*3.3*	*3.9*
Other East Asia	*5.1*	*4.5*	*3.8*	*3.0*	*3.7*
South Asia	4.8	5.5	4.8	4.0	4.7
India	*4.9*	*5.6*	*4.8*	*4.0*	*4.7*
Other South Asia	*4.5*	*5.2*	*4.7*	*4.1*	*4.6*
Middle East	2.9	3.5	3.0	2.6	3.0
Africa	2.7	4.1	3.8	3.4	3.8
Latin America	2.9	3.4	3.2	2.9	3.2
Brazil	*3.8*	*3.0*	*3.1*	*2.8*	*3.0*
Other Latin America	*2.4*	*3.7*	*3.3*	*3.0*	*3.3*
World	**3.3**	**3.7**	**3.2**	**2.7**	**3.2**
European Union	*2.4*	*2.3*	*2.1*	*1.7*	*2.0*

* 1992-2002.
Note: See Annex E for regional definitions.

population growth-rate assumptions are drawn from the most recent United Nations population projections contained in *World Population Prospects: the 2002 Revision*. That report projects global population to grow by 1% per year on average, from an estimated 6.2 billion in 2002 to almost 8.1 billion in 2030. Population growth will slow over the projection period, in line with trends of the last three decades: from 1.2% in 2002-2010 to 0.8% in 2020-2030 (Table 1.2). Population expanded by 1.6% from 1971 to 2002. The

United Nations' growth expectations were slightly slower in 2002 than in the *2000 Revision* (the basis for our population assumptions in *WEO-2002*), partly because of the impact of the HIV/AIDS epidemic.

Table 1.2: **Population Growth Assumptions** (average annual growth rates, in %)

	1971-2002	2002-2010	2010-2020	2020-2030	2002-2030
OECD	**0.8**	**0.6**	**0.4**	**0.3**	**0.4**
North America	1.3	1.0	0.9	0.7	0.9
United States and Canada	*1.1*	*1.0*	*0.9*	*0.7*	*0.8*
Mexico	*2.3*	*1.3*	*1.0*	*0.7*	*1.0*
Europe	0.5	0.3	0.1	0.0	0.1
Pacific	0.8	0.2	0.0	−0.2	0.0
Asia	*0.8*	*0.2*	*−0.1*	*−0.3*	*−0.1*
Oceania	*1.3*	*0.8*	*0.7*	*0.5*	*0.7*
Transition economies	**0.5**	**−0.2**	**−0.2**	**−0.4**	**−0.3**
Russia	−0.3*	−0.6	−0.6	−0.7	−0.7
Other transition economies	0.0*	0.0	0.1	−0.1	0.0
Developing countries	**2.0**	**1.4**	**1.2**	**0.9**	**1.2**
China	1.4	0.7	0.5	0.1	0.4
East Asia	2.0	1.4	1.1	0.8	1.1
Indonesia	*1.8*	*1.2*	*0.9*	*0.6*	*0.9*
Other East Asia	*2.0*	*1.5*	*1.2*	*1.0*	*1.2*
South Asia	2.1	1.6	1.3	1.0	1.3
India	*2.0*	*1.4*	*1.1*	*0.8*	*1.1*
Other South Asia	*2.4*	*2.1*	*1.9*	*1.5*	*1.8*
Middle East	3.1	2.2	2.0	1.6	1.9
Africa	2.7	2.1	1.9	1.6	1.9
Latin America	1.9	1.3	1.1	0.8	1.0
Brazil	*1.9*	*1.1*	*0.8*	*0.6*	*0.8*
Other Latin America	*2.0*	*1.4*	*1.2*	*0.9*	*1.2*
World	**1.6**	**1.2**	**1.0**	**0.8**	**1.0**
European Union	*0.3*	*0.1*	*0.0*	*−0.1*	*0.0*

* 1992-2002.

The population of the developing regions will continue to grow most rapidly, by 1.2% per year from 2002 to 2030, though this is significantly lower than the average rate of 2% in the last three decades. Population in the transition economies is expected to decline: Russia's population will drop from 144 million in 2002 to 120 million in 2030, a cumulative fall of around 17%. The OECD's population is expected to grow by an average of 0.4% per annum over the *Outlook* period, with North America contributing much of the increase. Several European and Pacific countries, including Germany and Japan, will experience significant population declines and ageing. The share of the world population living in developing regions, as they are classified today, will increase from 76% now to 80%. All of the increase in world population will occur in urban areas (rural population will decline). As populations grow in developing countries, providing them with access to commercial energy will be an increasingly pressing challenge.

Combining our population and GDP growth assumptions yields an average increase in per capita income of 2.2% per annum, from $7 700 in 2002 to $14 200 in 2030. Per capita incomes will grow quickest in the transition economies and in developing countries, notably China. By 2030, incomes in OECD countries, which will increase by 65% to $39 800, will still be more than four times the average for the rest of the world (Figure 1.2).

Figure 1.2: **Per Capita Income by Region**

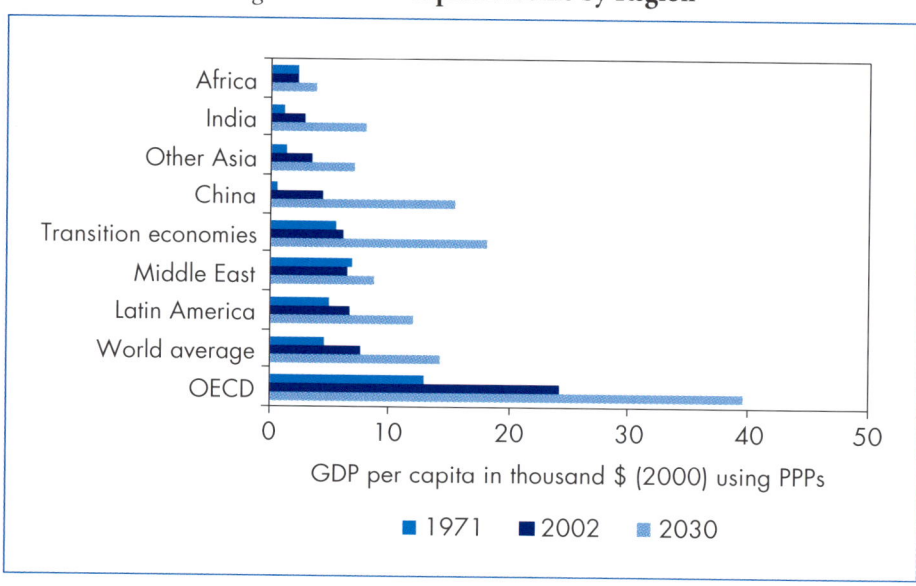

Energy Prices[5]

As in previous editions of the *WEO*, average end-user prices for oil, gas and coal are derived from assumed price trends on wholesale or bulk markets. Tax rates are assumed to remain unchanged over the projection period. Final electricity

---

5. All real prices in this *Outlook* are in expressed in year-2000 dollars unless otherwise specified.

prices are based on marginal power-generation costs. The assumed price paths, summarised in Table 1.3, should not be interpreted as forecasts. Rather, they reflect our judgement of the prices that will be needed to encourage sufficient investment in supply to meet projected demand over the *Outlook* period. Although the price paths follow smooth trends, this should not be taken as a prediction of stable energy markets. In fact, we are likely to see even more volatile prices in the future. However, we do not expect large divergences from the assumed price paths, such as the recent surge in oil prices, to be sustained for long periods in the Reference Scenario.

Table 1.3: **Fossil-Fuel Price Assumptions** (in year-2000 dollars)

	2003	2010	2020	2030
IEA crude oil imports ($/barrel)	27	22	26	29
Natural gas ($/MBtu):				
US imports	5.3	3.8	4.2	4.7
European imports	3.4	3.3	3.8	4.3
Japan LNG imports	4.6	3.9	4.4	4.8
OECD steam coal imports ($/tonne)	38	40	42	44

Note: Prices in the first column represent historical data. Gas prices are expressed on a gross-calorific-value basis.

International Oil Prices

The average IEA crude oil import price, a proxy for international prices, is assumed in the Reference Scenario to fall back from current highs to $22 (in real year-2000 dollars) in 2006. It is assumed to remain flat until 2010, and then to begin to climb in a more-or-less linear way to $29 in 2030 (Figure 1.4).

Uncertainty surrounding the near-term outlook for oil prices is unusually pronounced at present, complicating the analysis of overall energy-market trends. We assume in our Reference Scenario that the prices reached in mid-2004 are unsustainable and that market fundamentals will drive them down in the next two years. In June and September 2004, OPEC agreed to increase its production, moves which should help to bolster supply and replenish stocks, driving prices lower. But a continuing surge in demand and under-investment in production capacity combined with a large and sustained supply disruption could still result in a new price hike.[6]

The assumed rising trend in real prices after 2010 reflects an expected increase in marginal production costs outside OPEC and an increase in the market share of a small number of producers. A rising share of production will have

---

6. The impact on global energy markets of higher oil prices than assumed in the Reference Scenario is analysed in Chapter 3.

Figure 1.3: **Average IEA Crude Oil Import Price**

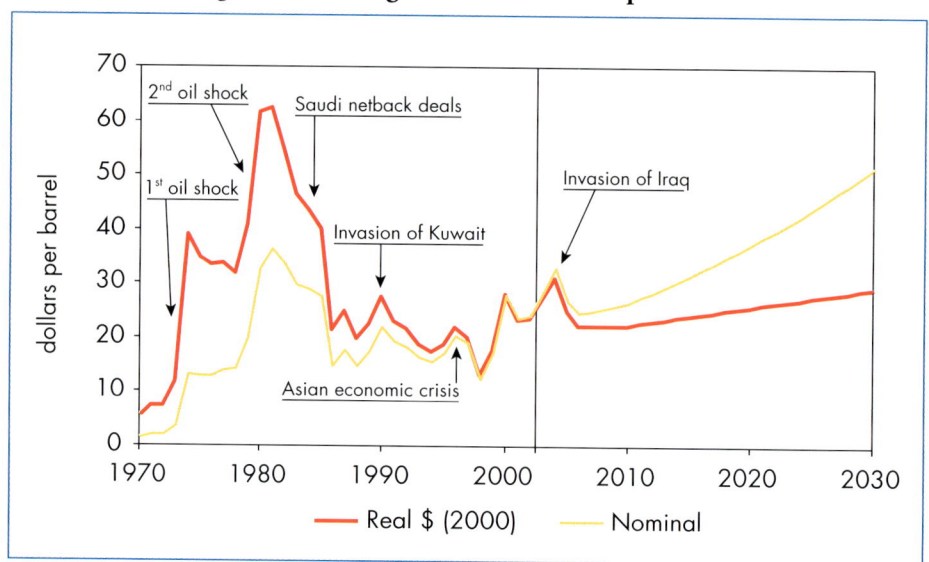

to come from smaller oilfields, where unit costs are higher. The share of existing giant fields, which still account for almost half of the world's crude oil supply, will continue to fall. The marginal cost of production in mature basins in non-OPEC countries – including North America and the North Sea – is rising, and this will deter investment and capacity additions in the long term. Non-conventional oil, such as synthetic and extra-heavy crude from oil sands in Canada and Venezuela is expected to play a growing role in total oil supply. So is gas-to-liquids conversion in countries with large low-cost gas reserves that are currently stranded. But most of the additional production capacity that will be needed over the projection period is expected to come from OPEC countries, mainly in the Middle East.

The increasing dependence of oil-importing regions on a small number of OPEC producers and Russia will increase those countries' market dominance and their ability to impose higher prices. Rapidly-growing populations in OPEC countries and the need in many exporting countries for higher public spending on welfare programmes and infrastructure will limit the amount of capital that they will be willing and able to spend on upstream development (IEA, 2003). Pressure on the governments of those countries to seek higher prices as a way of generating additional revenues is likely to grow. Yet it is in their interests to avoid prices rising so much that they depress global demand and encourage production of higher-cost oil in other countries. On balance, we remain of the view that the combination of these factors points to a moderate increase in prices in the longer term.

Box 1.1: **Explaining High Oil Prices: The Risk Premium and Fundamentals**

Oil prices rose strongly during 2003 and the first eight months of 2004: the price of West Texas Intermediate, a benchmark US crude oil, hit $49 per barrel in August 2004 – an all-time record in nominal terms and the highest price in real terms since 1986. Heightened fear of a supply disruption in the wake of the war in Iraq and the growing threat of terrorist attacks on oil facilities have certainly contributed to the price hike. Some of the increase can be seen, therefore, as a "geopolitical risk premium" – the additional price that the market is willing to pay to secure supply. Some analysts claim that the premium reached as much as $10 per barrel in mid-2004, though there is no generally accepted approach to measuring it.

But market fundamentals have also played a major role in driving up oil prices. At the time of writing, global oil-demand growth in 2004 was expected to hit 2.5 mb/d, or 3.2% – a 14-year high. China will account for more than half the increase. Crude oil production has not kept pace with rising demand and stocks have fallen to historically low levels. Spare production capacity has nonetheless diminished, because of a lack of investment in new capacity. By mid-2004, Saudi Arabia and the United Arab Emirates were the only exporters producing below capacity, and their remaining spare capacity was only around 1 mb/d – less than 2% of world demand. The shortage of spare capacity has exacerbated worries about the impact of a possible supply disruption. Bottlenecks in the supply of refined products – notably of gasoline in the United States – have added to the pressure on product and crude oil prices.

Natural Gas Prices

Unlike for oil, natural gas markets are highly regionalised, because it is expensive to transport gas over long distances. Prices often diverge substantially across and within regions. Nevertheless, regional prices usually move broadly in parallel with each other because of their link to the international price of oil, which reflects the keen competition between gas and oil products. Historically, Asian gas prices have been the highest and North American prices the lowest, with European prices generally lying mid-way between the two. Average prices in North America outstripped European import prices for the first time ever in 1999. They topped Asian prices in 2003. The dramatic surge in North American prices in recent years has been caused by tight constraints on production and import capacities. High gas prices are seriously affecting the profitability and output of manufacturing industry in the United States (Box 1.2).

In our Reference Scenario, gas prices are assumed to fall back in all three regions in 2006, and then to rise steadily from 2010 in line with oil prices.

Rising supply costs also contribute to higher gas prices from the end of the current decade in North America and Europe. Increased short-term trading in liquefied natural gas (LNG), which permits arbitrage among regional markets, will cause regional prices to converge to some degree over the projection period (Figure 1.4). LNG supply is becoming more flexible as international trade expands and downstream markets open up to competition. Lower unit pipeline-transport costs may also cause regional prices to converge. Price trends within each region are discussed in detail in Chapter 4.

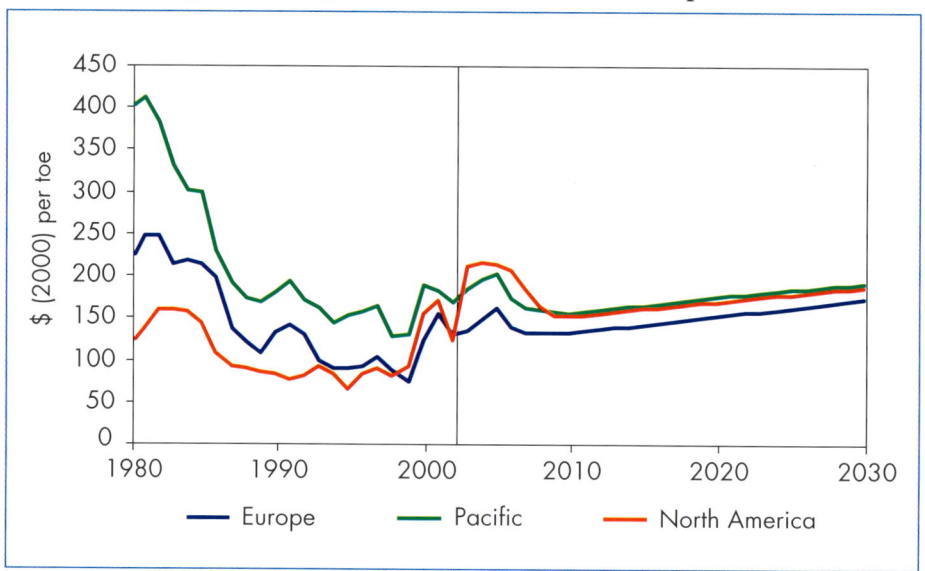

Figure 1.4: **Regional Natural Gas Price Assumptions**

Steam Coal Prices

International steam-coal prices have risen steadily in recent years, from $33.50 per tonne in 2000 to $38.50 in 2003 (in year-2000 dollars). Rising industrial production, especially in Asia, and higher gas prices have encouraged some power stations and industrial end-users to switch to coal. This has helped to boost overall coal demand and prices. Recently, serious production problems in several countries, including Australia, South Africa, Indonesia, Canada and the United States, have added to the upward pressure on prices. By mid-2004, spot steam-coal prices in nominal terms have surged to over $70 per tonne and coking coal to over $100.

Market fundamentals are expected to drive coal prices back down by 2006. OECD steam coal import prices are assumed to average $40 per tonne until 2010. Thereafter, prices are assumed to increase slowly and in a linear fashion, reaching $44/tonne in 2030. The increase is, nonetheless, slower than that of

Box 1.2: **Impact of High Gas Prices on US Manufacturing Industry**

Historically, North American industry has benefited from access to affordable and reliable natural gas supplies. The industry sector currently consumes more than 40% of the natural gas used in the region. Heavy users include the pulp and paper, metals, chemicals and petroleum refining. Since 2000, natural gas prices have risen sharply as demand, boosted by the needs of new power stations, has outpaced the expansion of domestic production capacity. At more than $6 per MBtu, prices are now the highest in the world. As a result, North American industry is facing increased competitive pressure from foreign factories.

The impact has been greatest on those involved in energy-intensive industries, especially those that use gas for fuel and as a raw material to make products. In the fertilizer industry, natural gas represents up to 80% of the cost of producing ammonia, the basis for all nitrogen-based fertilizers. Since 2000, around one-fifth of fertilizer capacity in the United States and Canada has been shut in (see Chapter 4). Other industrial activities, with the ability to switch to lower-cost fuels, have been less affected. A 2003 study by the American Chemistry Council, which represents America's largest industrial users of natural gas, estimates that sustained natural gas prices of $6 per MBtu would reduce US economic growth rate by 0.2% a year, or $200 billion. Although the region's energy intensity has fallen steadily in recent years, access to affordable and reliable energy supplies remains an important contributor to economic growth.

oil or gas. Rising oil prices will raise the cost of transporting coal and also make it more competitive for industrial users and power generators. This factor is assumed to offset an expected reduction in the cost of mining coal, as low-cost countries continue to rationalise their industries. Environmental regulations will restrict the use of coal, or increase the cost of using it, in many countries.

Technological Developments

Technological innovation and the rate of deployment of new technologies for supplying or using energy are important considerations. In general, it is assumed that available end-use technologies become steadily more energy-efficient. But the average efficiency of equipment actually in use and the overall intensity of energy consumption will depend heavily on the rate of retirement and replacement of the stock of capital. Since the energy-using capital stock in use today will be replaced only gradually, most of the impact of technological developments that improve energy efficiency will not be felt until near the end of the projection period.

The rate of capital-stock turnover varies considerably according to the type of equipment. Most cars and trucks, heating and cooling systems and industrial boilers will be replaced by 2030. On the other hand, most existing buildings, roads, railways and airports, as well as many power stations and refineries will still be in use then. The very long life of these types of energy-capital stock will limit the extent to which technological progress can alter the amount of energy needed to provide a particular energy service. Retiring these assets before the end of their normal lives is usually costly and would, in most cases, require major new government initiatives – beyond those assumed in the Reference Scenario. Refurbishment can, however, achieve worthwhile improvements in energy efficiency in some cases.

Technological developments will also affect the cost of energy supply and the availability of new ways of producing and delivering energy services. Power-generation efficiencies are assumed to improve over the projection period, but at different rates for different technologies. Towards the end of the projection period, fuel cells based on hydrogen are expected to become economically attractive in some power generation applications and, to a much smaller extent, in cars and trucks. Exploration and production techniques for oil and gas are also expected to improve, lowering the unit production costs and opening up new opportunities for developing resources. But the Reference Scenario assumes that no new breakthrough technologies beyond those known today will be used before 2030.

The World Alternative Policy Scenario

We have developed an Alternative Policy Scenario to analyse how the global energy market could evolve were countries around the world to adopt a set of policies and measures that they are either currently considering or might reasonably be expected to implement over the projection period. The purpose of this scenario is to provide insights into how effective those policies might be in addressing environmental and energy-security concerns.

In this scenario, it is assumed that OECD countries will adopt a range of policies and measures currently under active consideration. In non-OECD countries, the policies assessed in the Alternative Scenario include not just those under discussion but also a set of policies that could plausibly be adopted at some point in the future. These additional policies are assumed to bring about a faster decline in energy intensity than occurs in the Reference Scenario. Partly because of these policies, new, more efficient energy technologies are assumed to be deployed more rapidly in all sectors and in all regions. The assumptions on macroeconomic conditions and population are the same here as in the Reference Scenario. But energy prices are assumed to change in response to the new energy supply-demand balance.

Main Uncertainties

In common with all attempts to describe future market trends, the energy projections presented in the *Outlook* are subject to a wide range of uncertainties. Energy markets could evolve in ways that are much different from either the Reference Scenario or the Alternative Policy Scenario. The reliability of our projections depends both on how well the model represents reality and on the validity of the assumptions it works under.

Macroeconomic conditions are, as ever, a critical source of uncertainty. Slower GDP growth than assumed in both of our scenarios would cause demand to grow less rapidly. Growth rates at the regional and country levels could be very different from those assumed here, especially over short periods. Political upheavals in some countries could have major implications for economic growth. Sustained high oil prices – which are not assumed in either of our scenarios – would curb economic growth in oil-importing countries and globally in the near term (Box 1.3). The impact of structural economic changes, including the worldwide shift from manufacturing to service activities, is also uncertain, especially late in the projection period.

Uncertainty about the outlook for economic growth in China is particularly acute. With China's emergence as a major energy importer, any faltering of the country's economic development would have important implications for world energy markets. China has been responsible for a large share of the increase in world demand for raw materials – including energy – in the last few years (Figure 1.5). It is also becoming an important consumer of final goods, thereby contributing to economic growth in the rest of the world. There are increasing signs of overheating in the Chinese economy and the risk of a "hard landing" – an abrupt slowdown in economic activity – is growing as credit is tightened and investment drops. Such a development could have a major impact on global economic activity and, therefore, on energy consumption and import needs worldwide.

The effects of *resource availability and supply costs* on energy prices are very uncertain. Resources of every type of energy are sufficient to meet projected demand through to 2030, but the future cost of extracting and transporting those resources is uncertain – partly because of a lack of information about geophysical factors. Oil and gas producers, for example, do not usually appraise reserves in detail until they are close to actually exploiting them. How much of the world's resources can be produced economically will depend partly on production conditions and technological progress. Geopolitical factors will also affect the development of energy resources.

Changes in government *energy and environmental policies* and the adoption of new measures to address energy security and environmental concerns, especially climate change, could have profound consequences for energy

Box 1.3: **The Impact of High Oil Prices on the Global Economy**

Oil prices still affect the health of the world economy. High oil prices contributed to the global economic downturn in 2000-2001 and are dampening the current cyclical upturn. For as long as oil prices remain high and unstable, the economic prosperity of oil-importing countries – especially the poorest of them – will remain at risk.

The IEA has completed a study of the issue in collaboration with the OECD and with the assistance of the International Monetary Fund. That study indicates that a sustained $10 per barrel increase in oil prices from $25 to $35 (the average price, in fact, for the first eight months of 2004) would result in a loss to the OECD as a whole of about 0.4% of GDP in the first and second years of higher prices. Inflation would rise by half a percentage point, and unemployment would also increase. Euro-zone countries, which are highly dependent on oil imports, would suffer most in the short term, their GDP dropping by 0.5% and their inflation rising by 0.5% in 2004. The United States would suffer the least, with GDP falling by 0.3%, largely because indigenous production meets a large share of its oil needs. Japan's GDP would fall by 0.4%. Its relatively low oil intensity would compensate to some extent for its almost total dependence on imported oil. In all OECD regions, these losses would start to diminish in the following three years, as global trade in non-oil goods and services recovered.

The adverse economic impact of higher oil prices on oil-importing developing countries is even more severe than for OECD countries. This is because the developing countries are more dependent on imported oil, and because their economies are more energy-intensive. The loss of GDP caused by a $10 oil-price increase would average 0.8% in Asia and 1.6% in very poor highly indebted countries in the year immediately following the price increase. The loss of GDP in the sub-Saharan African countries would be more than 3%.

At least half of one per cent of world GDP would be lost – equivalent to $255 billion – in the year following a $10 oil-price increase. The economic stimulus provided by higher oil-export earnings in OPEC and other exporting countries would be more than outweighed by the depressive effect of higher prices on economic activity in the importing countries. A loss of business and consumer confidence, inappropriate policy responses and higher gas prices would amplify these economic effects in the medium term.

Source: IEA (2004).

Figure 1.5: **China's Share of Incremental World Production and Energy Demand, 1998-2003**

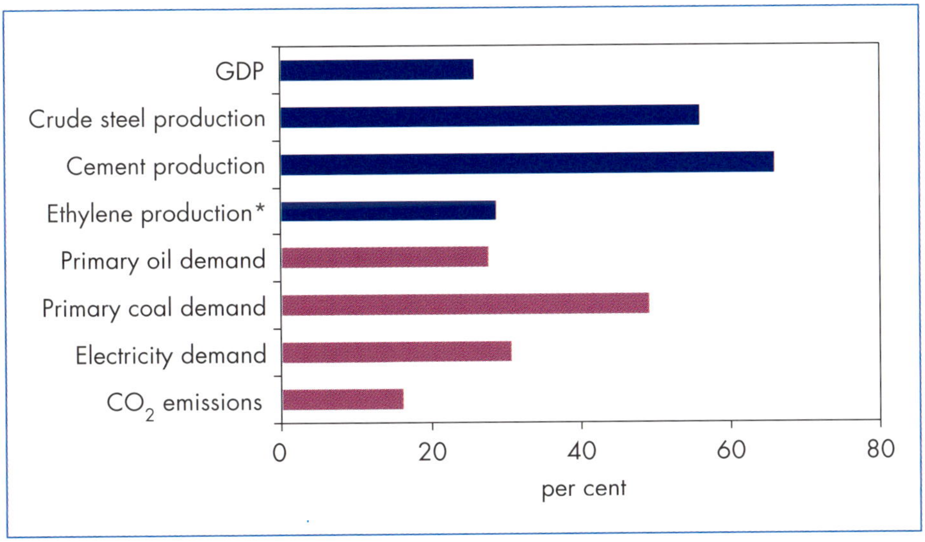

Category	Per cent
GDP	~25
Crude steel production	~55
Cement production	~65
Ethylene production*	~28
Primary oil demand	~27
Primary coal demand	~48
Electricity demand	~30
CO_2 emissions	~17

* 1998-2002.
Sources: IEA databases; Asian Development Bank (2004); BP (2004); Development Bank of Japan (2004); International Iron and Steel Institute (2004); US Geological Survey (2004).

markets. Among the leading uncertainties in this area are: the production and pricing policies of oil-producing countries, the future of energy-market reforms, taxation and subsidy policies, the possible introduction of carbon dioxide emission-trading and the role of nuclear power. The impact on energy markets of new environmental policies and other government actions aimed at enhancing energy security is analysed in the World Alternative Policy Scenario.

Improvements in the efficiency of current *energy technologies* and the adoption of new ones along the energy-supply chain are a key source of uncertainty for the global energy outlook. It is possible that hydrogen-based energy systems and carbon-sequestration technologies, which are now under development, could dramatically reduce carbon emissions associated with energy use. If they did so, they would radically alter the energy-supply picture in the long term. But these technologies are still a long way from ready to be commercialised on a large scale, and it is always difficult to predict when a technological breakthrough might occur.

It is uncertain whether all the *investment in energy-supply infrastructure* that will be needed over the projection period will be forthcoming.[7] Ample financial

---

7. See *World Energy Investment Outlook* (IEA, 2003) for a detailed assessment of global energy investment needs.

resources exist at a global level to finance projected energy investments, but those investments have to compete with other sectors. More important than the absolute amount of finance available worldwide, or even locally, is the question of whether conditions in the energy sector are right to attract the necessary capital. This factor is particularly uncertain in the transition economies and in developing nations, whose financial needs for energy developments are much greater relative to the size of their economies than they are in OECD countries. In general, the risks involved in investing in energy in non-OECD countries are also greater, particularly for domestic electricity and downstream gas projects. More of the capital needed for energy projects will have to come from private and foreign sources than in the past. Creating an attractive investment framework and climate will be critical to mobilising the necessary capital.

CHAPTER 2

GLOBAL ENERGY TRENDS

HIGHLIGHTS

- World primary energy demand in the Reference Scenario is projected to expand by almost 60% between 2002 and 2030, reaching 16.5 billion tonnes of oil equivalent. Two-thirds of the increase will come from developing countries. But the annual rate of projected world energy demand growth, at 1.7%, is slower than over the past three decades, when demand grew by 2% per year.

- Fossil fuels will continue to dominate global energy use. They will account for around 85% of the increase in world primary demand. Their share in total demand will increase slightly, from 80% in 2002 to 82% in 2030. Oil will remain the single largest fuel in the primary energy mix, even though its percentage share will fall marginally. The transport and power-generation sectors will absorb a growing share of global energy over the projection period, in line with past trends.

- The world's energy resources are adequate to meet the projected increase in energy demand to 2030. But the geographical sources of incremental energy supplies will shift markedly. International trade in energy will expand to accommodate the growing mismatch between the location of demand and that of production. Energy exports from non-OECD to OECD countries will increase by more than 80%, from some 1 500 millions tonnes of oil equivalent in 2002 to over 2 700 Mtoe in 2030.

- Oil will remain the most heavily traded fuel. Dependence on Middle East oil will continue to grow in the OECD regions and developing Asia. The world's vulnerability to a price shock induced by an oil-supply disruption will increase. Growing imports of natural gas in Europe, North America and other regions from the Middle East and the transition economies will heighten those concerns.

- The projected increase in global energy supply in this *Outlook* will entail cumulative infrastructure investment of some $16 trillion in year-2000 dollars over the period 2003-2030, or $568 billion per year. The electricity sector will absorb the majority of this investment. Financing the required investments in non-OECD countries is one of the biggest sources of uncertainty surrounding our energy-supply projections.

- Global carbon-dioxide emissions will increase by 1.7% per year over 2002-2030. Nearly 70% of the increase will come from developing countries. The average carbon content of primary energy – CO_2 emissions per tonne of oil equivalent – will hardly change. Power generation is expected to contribute about half the increase in global emissions.
- In the World Alternative Policy Scenario, global primary energy demand is 10% lower and energy-related CO_2 emissions are 16% lower than in the Reference Scenario in 2030. Fossil-fuel demand falls by 14% in 2030, while the use of nuclear power goes up by 14% and that of non-hydro renewable energy sources (excluding biomass) rises by 30%. The impact on energy demand of the policies considered in the Alternative Scenario grows throughout the projection period.

Energy Demand
Primary Fuel Mix

World primary energy demand is projected in the Reference Scenario to expand by almost 60% from 2002 to 2030, an average annual increase of 1.7% per year. Demand will reach 16.5 billion tonnes of oil equivalent compared to 10.3 billion toe in 2002 (Table 2.1). The projected *rate* of growth is, nevertheless, slower than over the past three decades, when demand grew by 2% per year.

Fossil fuels will continue to dominate global energy use. They will account for around 85% of the increase in world primary demand over 2002-2030 (Figure 2.1). And their share in total demand will increase slightly, from 80% in 2002 to 82% in 2030. The share of renewable energy sources will remain flat, at around 14%, while that of nuclear power will drop from 7% to 5%.

Oil will remain the single largest fuel in the global primary energy mix, even though its share will fall marginally, from 36% in 2002 to 35% in 2030. Demand for oil is projected to grow by 1.6% per year, from 77 mb/d in 2002 to 90 mb/d in 2010 and 121 mb/d in 2030. Oil use will become increasingly concentrated in the transport sector, which will account for two-thirds of the increase in total oil use. Transport will use 54% of the world's oil in 2030 compared to 47% now and 33% in 1971. Oil will face little competition from other fuels in road, sea and air transportation during the projection period. In OECD countries, the use of oil in the residential and services sector will decline sharply. In non-OECD countries, transport will also be the main driver of oil demand, though the industrial, residential and services sectors will also see steady oil-demand growth. In many developing countries, oil products will remain the leading source of modern commercial energy for cooking and heating, especially in rural areas.

Table 2.1: **World Primary Energy Demand** (Mtoe)

	1971	2002	2010	2020	2030	2002-2030*
Coal	1 407	2 389	2 763	3 193	3 601	1.5%
Oil	2 413	3 676	4 308	5 074	5 766	1.6%
Of which international marine bunkers	*106*	*146*	*148*	*152*	*162*	*0.4%*
Gas	892	2 190	2 703	3 451	4 130	2.3%
Nuclear	29	692	778	776	764	0.4%
Hydro	104	224	276	321	365	1.8%
Biomass and waste	687	1 119	1 264	1 428	1 605	1.3%
Of which traditional biomass	*490*	*763*	*828*	*888*	*920*	*0.7%*
Other renewables	4	55	101	162	256	5.7%
Total	**5 536**	**10 345**	**12 194**	**14 404**	**16 487**	**1.7%**

* Average annual growth rate.

Primary demand for **natural gas** will grow at a steady rate of 2.3% per year over the projection period. By 2030, gas consumption will be about 90% higher than now, and gas will have overtaken coal as the world's second-largest energy source (Figure 2.2). The share of gas in total primary energy use will increase from 21% in 2002 to 25% in 2030. The power sector will account for 60%

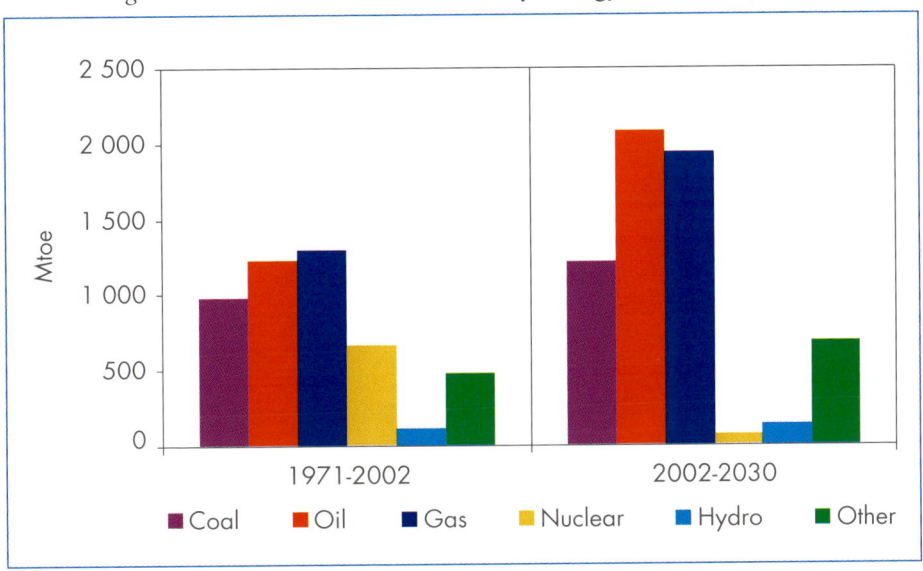

Figure 2.1: **Increase in World Primary Energy Demand by Fuel**

Chapter 2 - Global Energy Trends

of the increase in gas demand, with its share of the world gas market rising from 36% in 2002 to 47% in 2030. The power sector will be the main driver of demand in all regions. This trend will be particularly marked in developing countries, where electricity demand is expected to rise most rapidly.

Natural gas will remain the most competitive fuel in new power stations in most parts of the world, as it is the preferred fuel for high-efficiency combined-cycle gas turbines (CCGTs). A small but growing share of natural gas demand will come from gas-to-liquids plants and from the production of hydrogen for fuel cells.

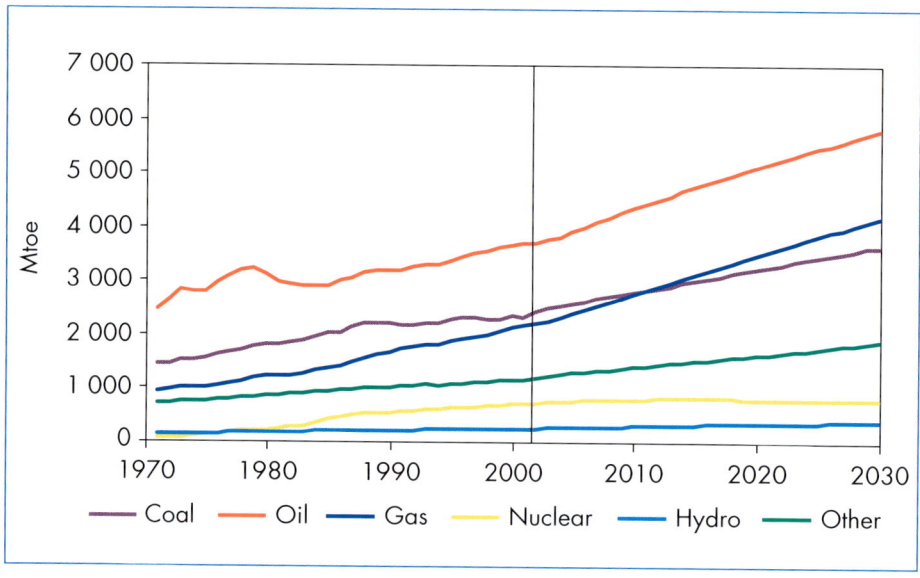

Figure 2.2: **World Primary Energy Demand by Fuel**

Coal use worldwide is projected to increase by 1.5% per year between 2002 and 2030. By the end of the *Outlook* period, coal demand, at just over 7 billion tonnes, will be just about 50% higher than at present. The share of coal in total primary energy demand will, nonetheless, fall slightly, from 23% to 22%. China and India, which both have ample coal supplies, will account for more than two-thirds of the increase in global coal use over the projection period. Power stations will absorb most of the increase in coal demand, though coal will continue to lose market share in power generation in all OECD regions and in some developing regions. Coal consumption will increase slowly in end-use sectors. Industry, households and services in non-OECD regions will use more coal, more than offsetting a continuing decline in OECD final consumption.

Box 2.1: **What is Included in Primary Energy Demand**

Total primary energy demand is equivalent to total primary energy supply (TPES). The two terms are used interchangeably throughout this *Outlook*. Primary energy refers to energy in its initial form, after production or importation. World primary demand includes international marine bunkers, which are excluded from the regional totals. Some energy is transformed, mainly in refineries, power stations and heat plants. Final consumption refers to consumption in end-use sectors, net of losses in transformation and distribution.

Total primary and final demand now includes traditional biomass and waste in all regions. Traditional biomass includes fuels that are not traded commercially: fuelwood, charcoal, dung and farm residues. This is a major change from past *Outlooks*, in which those fuels in non-OECD countries were not included in the regional and world totals. On average, traditional biomass represents 20% of total primary energy demand in developing countries and more than a quarter of final consumption, though the shares vary markedly among countries. As a result, the historical totals for primary and final energy demand for several non-OECD regions and for the world are significantly higher than before.

For convenience, the detailed tables showing the Reference Scenario projections for developing countries in Annex A show traditional biomass separately and total primary demand including and excluding it. In many cases, IEA statistics do not distinguish between traditional and other types of biomass. Where necessary, we have estimated the breakdown by assuming that all biomass and waste used in sectors other than power generation, transport and industry is of the traditional type. This approach probably overstates slightly the amount of energy produced from traditional biomass, as some charcoal and fuelwood used by households is in fact traded commercially.

The role of **nuclear power** will decline progressively over the *Outlook* period. The rate of construction of new reactors is expected to keep pace with the rate at which old reactors are retired. This is both because nuclear power will have trouble competing with other technologies and because many countries have restrictions on new construction or policies to phase out nuclear power. As a result, nuclear production will peak soon after 2010 and then decline gradually. Its share of world primary demand will drop from 7% at present to 6% in 2010 and to 5% by 2030. Nuclear output will increase in only a few countries, mostly in Asia. It is projected to fall in Europe. These projections are, however, subject to considerable uncertainty. Possible changes in government policies on nuclear power and public attitudes toward it could lead to nuclear power playing a much more important role than projected here.

Hydropower production will expand by 1.8% per year over the projection period, a slightly faster rate than that of global primary energy demand. The share of hydropower in global electricity generation will nonetheless drop, from 16% in 2002 to 13% in 2030. Most of the increase in output will occur in developing countries, where there are still considerable unexploited resources and where public opposition is less of an impediment to new projects.

The role of **biomass and waste**, the use of which is concentrated in developing countries, will gradually diminish over the projection period. Globally, its share of primary energy demand will drop from 11% in 2002 to 10% in 2030, as it is replaced with modern fuels. In absolute terms, the consumption of *traditional* biomass and waste in developing countries will continue to grow, but will slow over the projection period.

Other renewables – a group that includes geothermal, solar, wind, tidal and wave energy – will grow faster than any other primary energy source, at an average rate of 5.7% per year over the projection period. But other renewables' share of world demand will still be small in 2030, at 2% compared with 1% in 2002, because they start from a very low base. Most of the increase in the use of such renewables will be in the power sector. Their share in total electricity generation will grow from 1% in 2002 to 4% in 2030. Most of the increase will occur in OECD countries, many of which have policies to encourage the adoption of new renewable-energy technologies.

Global primary energy demand is projected to expand at almost the same average annual rate over the projection period as in the last edition of the *Outlook*. However, there are notable differences among fuels. In particular, demand for natural gas is now projected to grow less rapidly due to higher gas prices and to more use of renewables and nuclear power in power generation (Figure 2.3).

Global energy intensity, expressed as total primary energy use per unit of gross domestic product, will fall by 1.5% per year over 2002-2030. Intensity will fall in all regions, though at different rates (Figure 2.4). Significant differences in intensity among regions will persist, reflecting differences in the stage of economic development, the energy efficiency of end-use technologies, economic structure, energy prices, climate, geography, culture and lifestyles. Intensity will fall most steeply in the transition economies. This is due to more rapid improvements in energy efficiency in power generation and end-uses, as well as structural economic changes away from heavy industry towards lighter industry and services. Intensity is projected to fall by 2.2% per year in the transition economies. But it will still be about 90% higher than in OECD countries in 2030. Energy intensity will fall more quickly than in the past in

Figure 2.3: **World Primary Natural Gas Demand**

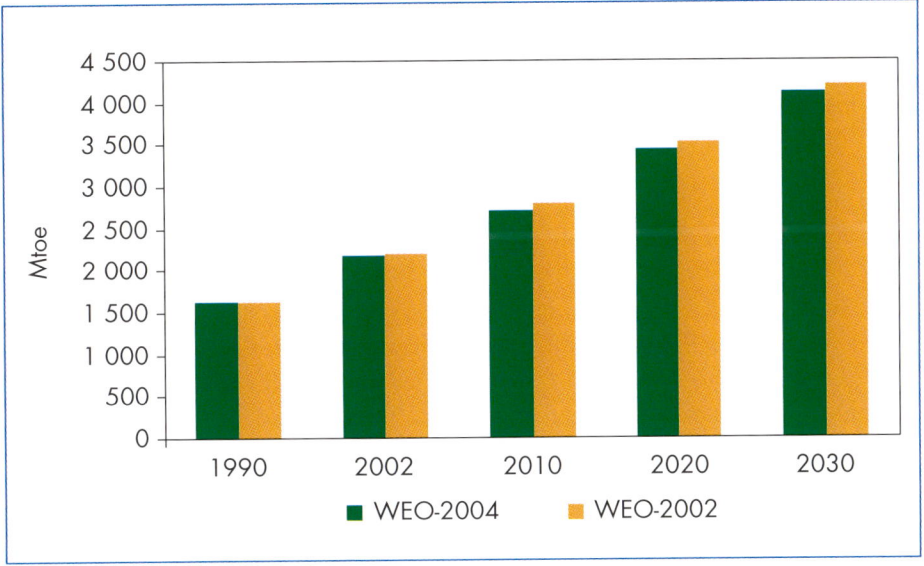

developing countries, by 1.6% per year from 2002 to 2030 compared with 0.9% from 1971 to 2002. In the OECD, intensity will fall by 1.2% per year, slightly slower than in the past. By 2030, energy intensity in the developing countries will be almost as low as in the OECD.

Figure 2.4: **Primary Energy Intensity**

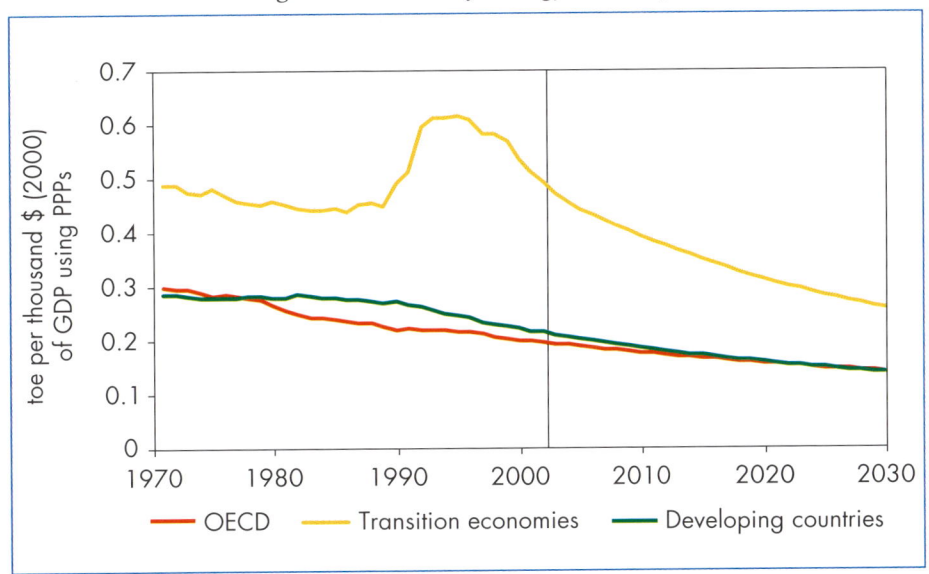

Chapter 2 - Global Energy Trends

Box 2.2: **Recent Trends in Global Energy Demand**

- Preliminary data for 2003 show that global primary energy demand grew by 2.9% in 2003 – the fastest increase since 1988 and twice as fast as in the previous five years. Developing Asian countries, whose economies are booming, were the main driving force. Demand in China alone surged by almost 14%. Energy use edged up in the mature OECD markets. The North American market expanded by only 0.2%, largely due to a sharp drop in US natural gas consumption – the result of high prices. Partial data for 2004 suggest that global energy demand may grow even faster than in 2003.
- Among the main primary fuels, coal use is rising most rapidly. It jumped by almost 7% in 2003. Of the biggest coal consumers, China registered the fastest growth, of over 15%, and accounted for 31% of world consumption. Demand surged by 7% in Russia and 5% in Japan, where unexpected closures of nuclear reactors led power generators to burn more coal. The rise was 2.6% in the United States, where high gas prices led power generators and industry to switch to coal.
- Globally, oil demand is estimated to have increased by 2.2% in 2003 – the highest rate since 1996. China contributed almost a third to the estimated 1.7 mb/d increase in global consumption. China's oil imports jumped by 0.6 mb/d to 2.1 mb/d. Demand was also strong in North America, jumping 0.53 mb/d or 2.2%.
- Natural gas consumption increased by 2% in 2003, with steady increases in Asia, Africa, the Middle East and some European countries offsetting a 3.5% drop in North America. Nuclear production fell by 2%, almost entirely due to the temporary reactor closures in Japan.

Regional Trends

Two-thirds of the increase in world primary energy demand between 2002 and 2030 will come from the developing countries. OECD countries will account for 26% and the transition economies for the remaining 8%. Consequently, the current 52% share of the OECD in world demand will decline to 43% in 2030, while that of the developing countries will increase, from 37% to 48% (Figure 2.5). The transition economies' share will fall from 10% to 9%.

The increase in the share of the developing regions in world energy demand results from their more rapid economic and population growth. Industrialisation and urbanisation will also boost demand. More people will live in towns and cities and will be better able to gain access to energy services. Increases in real prices to final consumers, a result of the gradual reduction in subsidies and rising international prices, are not expected to affect energy-demand growth in developing countries very much.

Figure 2.5: **Regional Shares in World Primary Energy Demand**

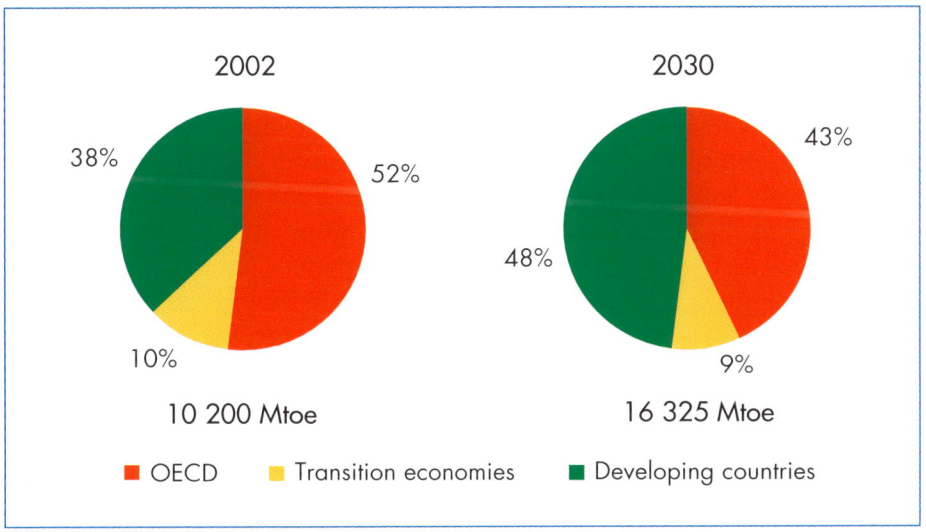

Note: Totals exclude international marine bunkers.

The developing countries' share of global demand will increase for all primary energy sources, except non-hydro renewables (Figure 2.6). The increase will be most pronounced for nuclear power, where production will fall in the OECD

Figure 2.6: **Share of Developing Countries in World Primary Energy Demand by Fuel**

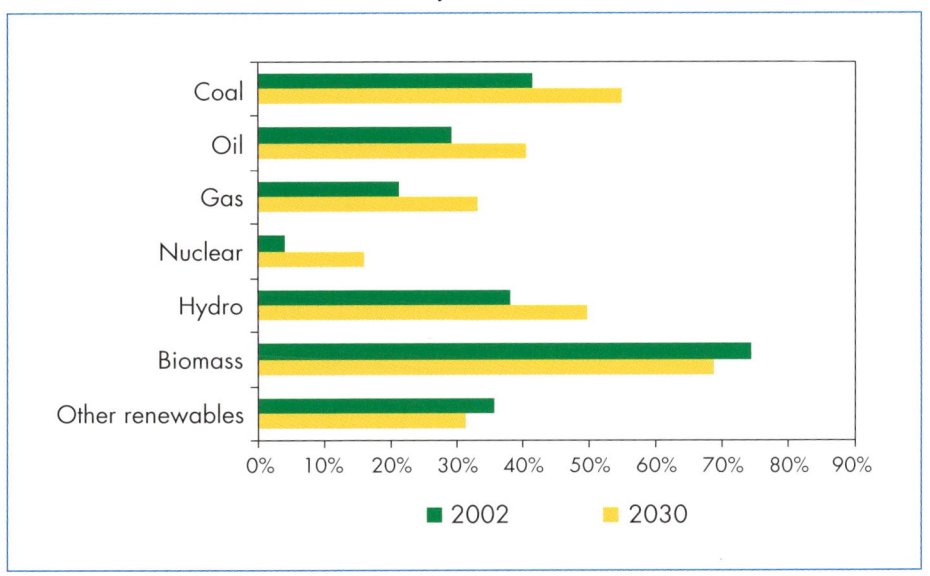

Chapter 2 - Global Energy Trends

while expanding in China and other parts of Asia. The developing countries will account for 18% of world nuclear output in 2030, compared with only 4% in 2002. Their share of world coal consumption also increases sharply, from 46% to 61%, mainly because of booming demand in China and India.

Developing countries will account for almost two-thirds of the 43-mb/d increase in global oil consumption between 2002 and 2030. Developing Asian countries will contribute 18 mb/d, and China alone will account for nearly half of that. Demand in the OECD countries will grow more slowly, yet North America will remain by far the largest single market for oil. The share of natural gas in the fuel mix is projected to increase in all regions. Growth rates are highest in developing Asian countries, Latin America and Africa. The biggest increase in volume terms will, however, occur in OECD North America. Coal demand will be driven primarily by the surging energy needs of developing Asia, particularly China and India. Coal use will increase modestly in North America and the OECD Pacific region and will fall marginally in OECD Europe.

Despite relatively strong growth in energy use in developing regions, per capita consumption there will remain much lower than in the rest of the world. By 2030, per capita primary energy consumption will average a mere 1.2 toe in developing regions, compared with 5.4 toe in the OECD and 4.7 toe in the transition economies. With a few exceptions, energy use will remain concentrated in the northern hemisphere (Figure 2.7).

Sectoral Trends

The transport and power-generation sectors will absorb a growing share of global energy over the projection period, in line with past trends (Figure 2.8). Together, their share will reach over 60% in 2030, compared with 54% now. Demand for mobility and electricity-related services will continue to grow broadly in line with GDP, but at a slower rate than in the past. Inputs to power stations worldwide will grow by 2% per year between 2002 and 2030.

Energy use in final sectors – transport, industry, households, services, agriculture and non-energy uses – will grow by 1.6% per year through to 2030. This is roughly the same rate as for primary energy demand. As a result, the share of final consumption in primary demand will hold steady at 68%. Transport demand will grow quickest, at 2.1% per year. Residential and services consumption will grow at an average annual rate of 1.5%, as will industrial demand.

The fuel mix in final energy uses will not change dramatically (Table 2.2). The share of electricity will rise, from 16% in 2002 to 20% in 2030. World electricity consumption will double over that period. The share of oil will also increase, from 43% to 45%, owing to rapid growth in transport demand,

Figure 2.7: **Primary Energy Consumption per Capita by Region, 2030**

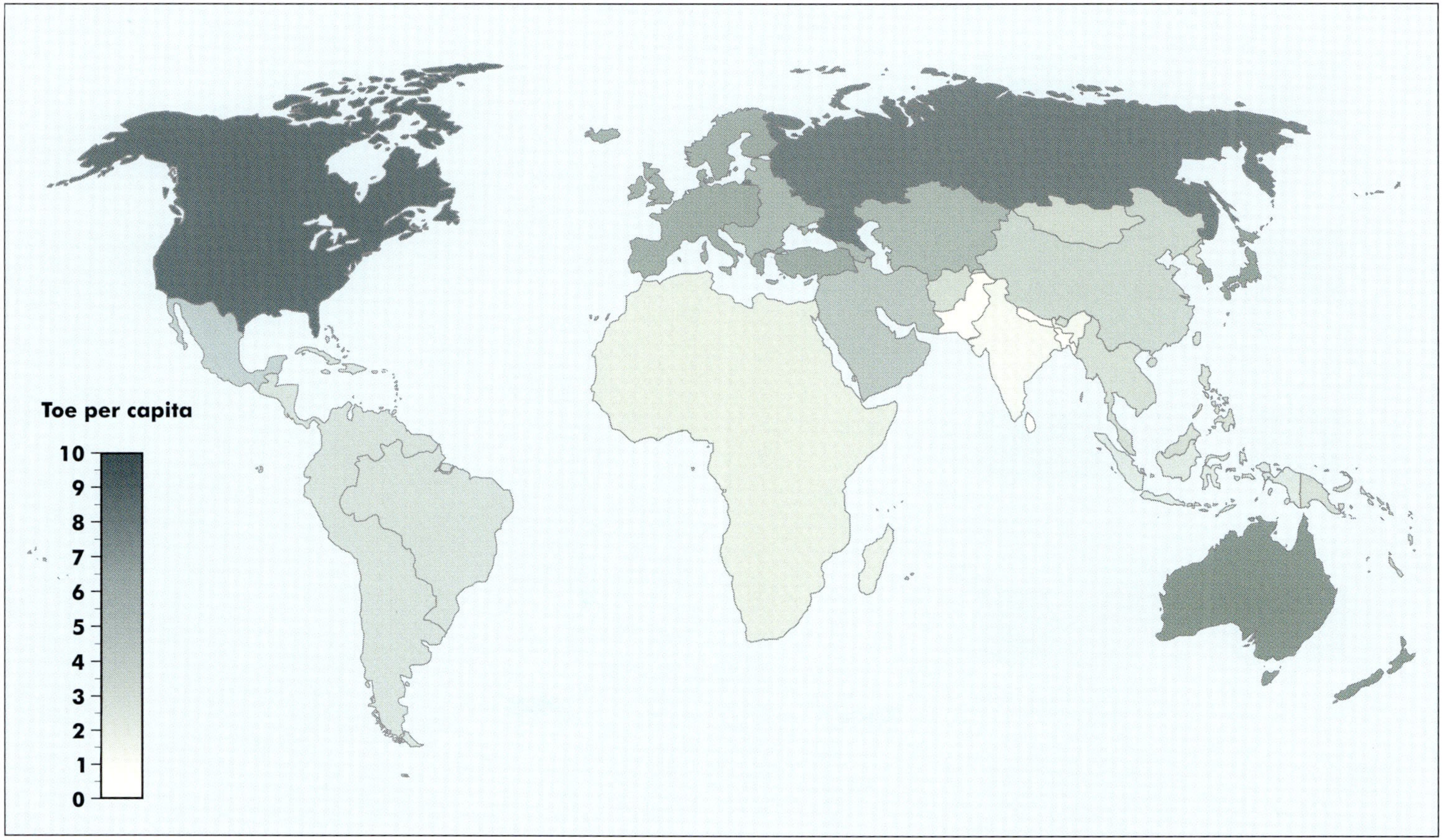

Chapter 2 - Global Energy Trends

Figure 2.8: **Sectoral Shares in World Primary Energy Demand**

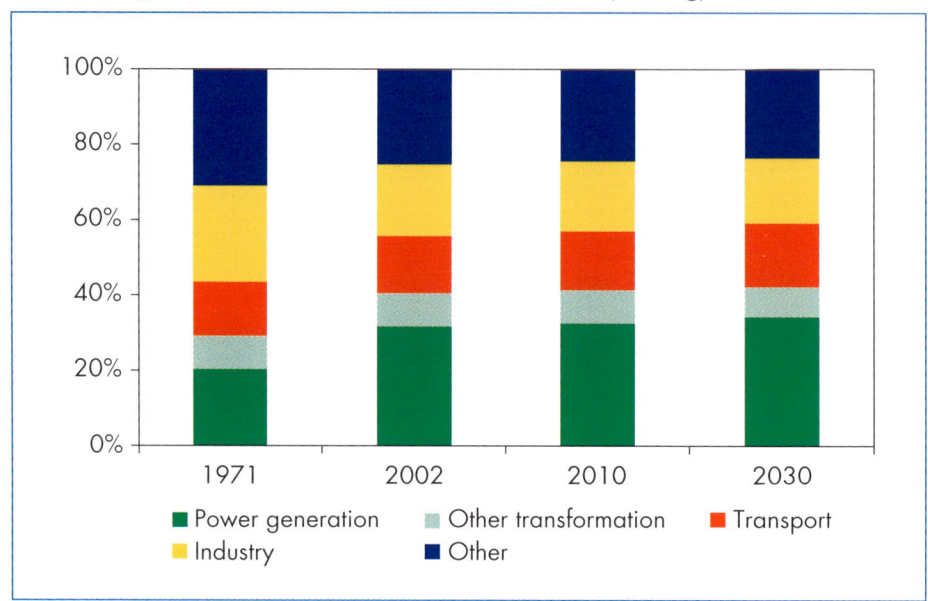

especially in developing countries. Biomass and waste will contribute less to meeting final energy needs, their share dropping from 14% to 12%. The share of coal, which has already fallen to 7% of total final consumption worldwide, will fall further to 5% in 2030. The share of natural gas will remain constant at 16%.

Table 2.2: **World Total Final Consumption** (Mtoe)

	1971	2002	2010	2030	2002-2030*
Coal	617	502	516	526	0.2%
Oil	1 893	3 041	3 610	5 005	1.8%
Gas	604	1 150	1 336	1 758	1.5%
Electricity	377	1 139	1 436	2 263	2.5%
Heat	68	237	254	294	0.8%
Biomass and waste	641	999	1 101	1 290	0.9%
Other renewables	0	8	13	41	6.2%
Total	**4 200**	**7 075**	**8 267**	**11 176**	**1.6%**

* Average annual rate of growth.

Per capita residential energy use in developing countries will remain much lower than in the OECD and in the transition economies (Figure 2.9). By 2030, households in developing countries will still use very little gas and electricity, and virtually no district heat. In contrast, biomass use will remain much higher than in the OECD and the transition economies. Lower incomes largely explain the much lower consumption of energy, especially electricity for household appliances and lighting. Warm climates in many developing countries contribute to lower per capita heating needs.

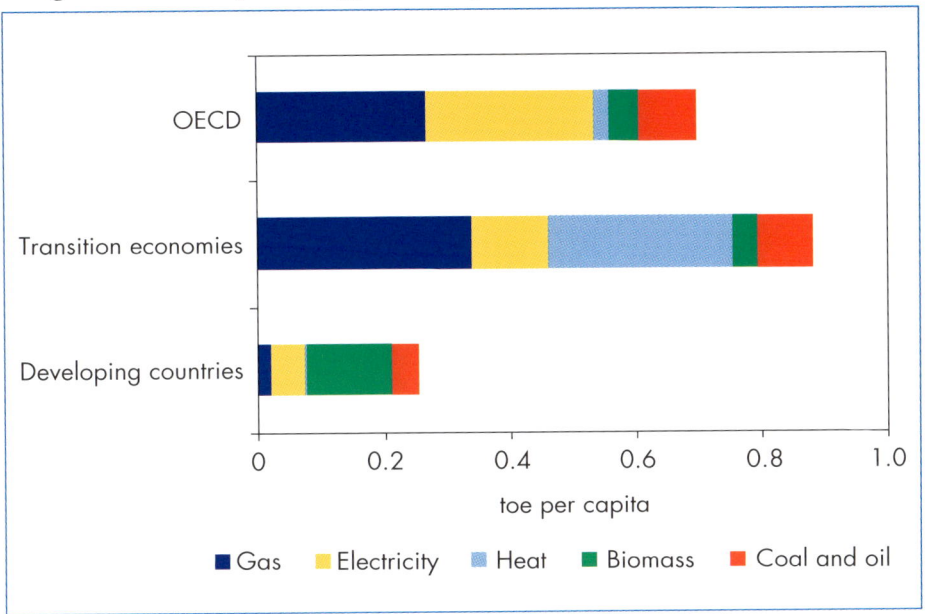

Figure 2.9: **Per Capita Energy Consumption in the Residential Sector, 2030**

Energy Production and Trade
Resource Availability and Production Prospects

The world's energy resources are adequate to meet the projected increase in energy demand until 2030. At issue is how much it will cost to develop and transport fossil-fuel and other resources to meet demand. We do not expect a supply crunch for at least the next three decades, though short-lived bottlenecks could constrain supply of a given fuel at a given time and drive up prices.

Proven reserves of gas and coal far exceed the cumulative amounts of both fuels that will be consumed over 2002-2030, and more reserves will certainly be added during that time. Proven conventional oil reserves today are sufficient to cover all the oil that will be needed until 2030. But additions to reserves from new discoveries and from "proving up" probable and possible reserves will be needed if production is not to peak before then. Non-conventional crude oil reserves are, in any case, very large and will replace any shortfall in conventional supplies as prices rise. There is also considerable potential for expanding oil production from gas-to-liquids plants, drawing on natural gas resources that otherwise would not find a market. Reserves of uranium for nuclear power production are also plentiful, while the technical potential for renewable energy sources is, in principle, almost limitless.

The geographical sources of incremental energy supplies will shift markedly over the projection period, mainly in response to cost factors and the location of resources. From 2002 to 2030, more than 95% of the increase in production will occur in non-OECD regions, against about 70% from 1971 to 2002 (Figure 2.10). Most low-cost fossil-fuel resources are located in non-OECD countries.

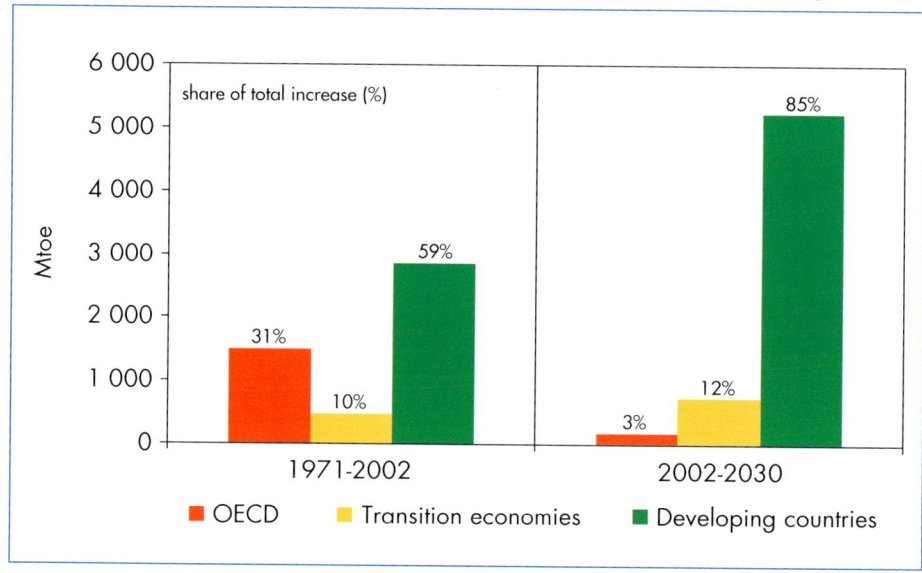

Figure 2.10: **Increase in World Primary Energy Production by Region**

OPEC countries, especially in the Middle East, will expand their production of **oil** most rapidly over the projection period. Their resources are large and their production costs are generally very low. Their share in total production will surge from 37% today to 53% in 2030. Over the rest of the current decade, however, *non-OPEC* countries are expected to contribute half of the increase in global production. World oil production will not peak before 2030, although output in some regions will already be in decline by then.

The biggest increases in production of **natural gas** are projected to occur in the transition economies and the Middle East. Africa and Latin America will experience the fastest rates of increase. The cost of non-associated gas production is believed to be lowest in the Middle East. Production costs are rising in the most mature producing regions, particularly in North America and Europe.

China will reinforce its position as the world's leading **coal** producer, accounting for around half the increase in global output over the projection period. The United States, India and Australia will remain the next biggest producers. Coal production in Europe will continue to decline as subsidies are reduced and uncompetitive mines are closed. Most of the increase in world coal production will be in the form of steam coal.

Outlook for International Trade

International trade in energy will expand to accommodate the growing mismatch between the location of demand and that of production (Figure 2.11). The OECD will account for 26% of the aggregate increase in demand but for only 3% of the growth in production. Energy exports from non-OECD to OECD countries will increase by more than 80%, from some 1 500 Mtoe in 2002 to over 2 700 Mtoe in 2030. Trade between countries within each grouping is also expected to grow.

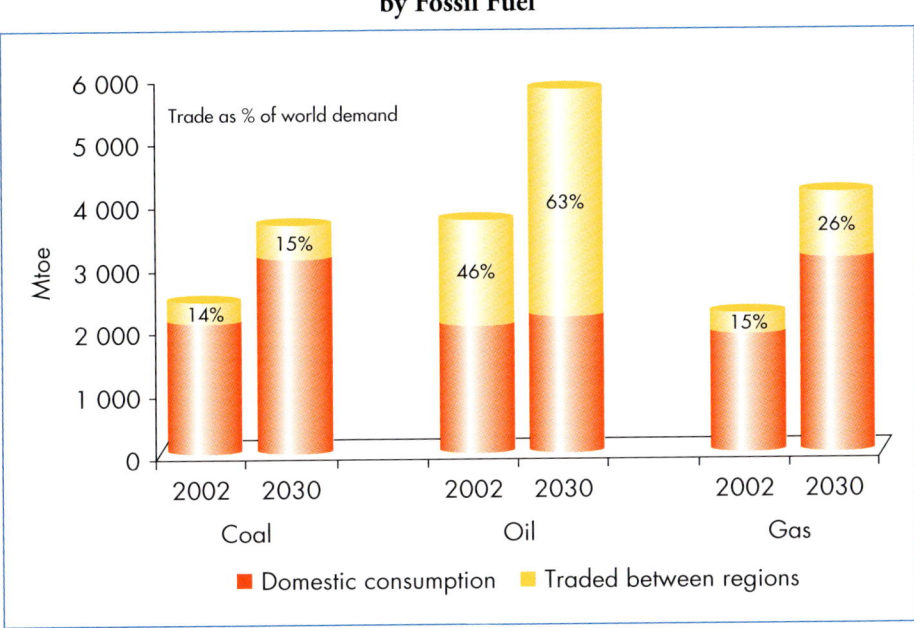

Figure 2.11: **Share of Inter-Regional Trade in World Primary Demand by Fossil Fuel**

Note: Percentages refer to the share of trade in total primary demand.

Oil will remain the most traded fuel. By 2030, 63% of all the oil consumed worldwide will be traded between the main regions covered in this *Outlook*, up from 46% in 2002. The volume of oil traded will more than double. Dependence on Middle East oil will continue to grow in the OECD and in

Chapter 2 - Global Energy Trends

developing Asia. These trends will have major geopolitical implications. Mutual economic dependence will increase between importing and exporting countries, but so will concerns about the world's vulnerability to a price shock induced by an actual or threatened disruption in supply. More attention will be paid to maintaining the security of international sea-lanes and pipelines, as more and more oil flows through sensitive chokepoints.

Growing flows of natural gas from and through politically unstable regions will heighten those concerns. Nonetheless, 60% of the growth in gas trade will be in the form of liquefied natural gas. Since LNG is shipped by sea, it avoids some of the risks of long fixed pipelines. The opening of gas markets to competition will encourage short-term trading. Together with falling unit transport costs, this will also enhance the ability of buyers to deal with a supply shortfall by seeking out alternative sources. The development of common operating standards covering safety, gas quality and other technical factors will help to foster international LNG trade.

The volume of hard coal traded internationally will expand steadily through to 2030, though the share of traded coal in total supply will increase only slightly and will remain much lower than for oil and gas. Most of the increase in trade will go to developing Asian countries and Europe.

Investment Outlook

The increase in global energy supply projected in this *Outlook* will call for cumulative infrastructure investment of $16 trillion (in year-2000 dollars) over the period 2003-2030, or $568 billion per year. This investment will be needed to expand supply capacity and to replace existing and future supply facilities that will be closed during the projection period. Projected annual investment needs are marginally higher than were shown in our 2003 *World Energy Investment Outlook*, where the figure was $550 billion. Although our aggregate demand projection is now slightly lower, there is a bigger contribution from renewables and nuclear power; both of which are more capital-intensive than fossil fuels.

The electricity sector will absorb most future energy investment. Power generation, transmission and distribution will absorb almost $10 trillion, or 62%, of total energy investment. If investment in the fuel chain to meet the fuel needs of power stations is included, electricity's share rises to more than 70%. More than half of the investment in the electricity industry will go to transmission and distribution networks. Total investments in the oil and gas sectors will each amount to almost $3 trillion, or around 18% of global energy investment. Exploration and development will take more than 70% of total investment in oil. The share is lower for gas, at 56%, because transportation infrastructure needs for gas are bigger. Coal investment will amount to only $400 billion, or 2.5%. Coal is about a sixth as capital-intensive as gas in producing and transporting a given amount of energy.

Developing countries, where production and demand will increase most rapidly, will require about half of global investment in the energy sector as a whole (Figure 2.12). China alone will need to invest $2.4 trillion – 15% of the world total and more than in all the other developing Asian countries put together. Investment needs amount to $1.1 trillion in Africa and $1 trillion in the Middle East. Upstream oil and gas development will absorb about three-quarters of total investment in both these regions. Russia and other transition economies will account for 10% of global investment and OECD countries for the remaining 40%. Investment needs will be largest in OECD North America – $3.4 trillion. Over 40% of total non-OECD investment in the oil-, gas- and coal-supply chains will go to provide fuels for export to OECD countries. These investments will be easier to finance than investment in projects to supply domestic markets, where payments are in local currencies.

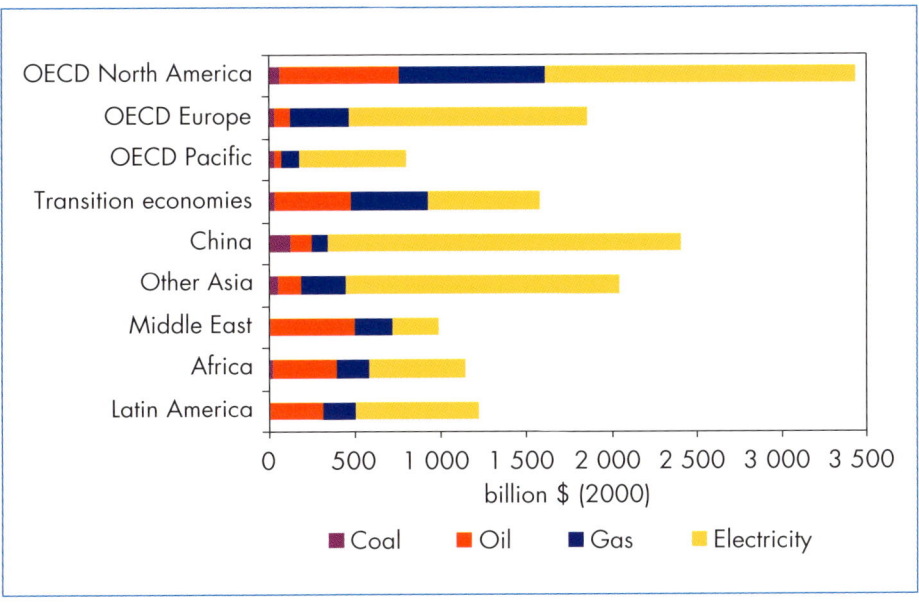

Figure 2.12: **Cumulative Energy Investment, 2003-2030**

More than half of all this investment will go simply to maintain the present level of supply. Most of the world's current production capacity for oil, gas and coal will need to be replaced by 2030. Indeed, much of the new production capacity brought on stream in the early years of the projection period will itself need to be replaced before 2030. Some electricity-generation, transmission and distribution infrastructure will also need replacing or refurbishing.

Enough money exists to finance this projected energy investment. Worldwide, domestic savings are much larger than the capital required for energy projects. But in some regions, those capital needs represent a very large share of total savings. In Africa, for example, the share is about a half. Although we judge that sufficient capital will be available from domestic and international sources, it is far from certain that all the infrastructure needed in the future will be fully financed in all cases. Mobilising the investment required will depend on whether returns are high enough to compensate for the risks involved. And, of course, energy will need to compete against other sectors of the economy for capital. More of the capital needed for energy projects will have to come from private sources than in the past, as governments continue to withdraw from the provision of energy services. Foreign direct investment is expected to become an increasingly important source of private capital in non-OECD regions.

Financing the required investments in non-OECD countries is the biggest challenge to the industry and the main source of uncertainty surrounding our energy-supply projection. The financial needs in transition economies and developing regions are much bigger relative to the size of their economies than is the case in OECD countries. In general, investment risks are also greater in these regions, particularly for domestic electricity and downstream gas projects. Few governments could fully fund the necessary investment, even if they wanted to. Raising private finance will depend critically on governments' creating an attractive investment framework and climate.

Energy-Related CO_2 Emissions

Overview

The projected trends in energy use in the Reference Scenario imply that global energy-related CO_2 emissions will increase by 1.7 % per year over 2002-2030. They will reach 38 billion tonnes in 2030, an increase of 15 billion tonnes, or 62%, over the 2002 level (Table 2.3). More than two-thirds of the increase will come from developing countries. By 2010, energy-related CO_2 emissions will be 39% higher than in 1990. Projected emission trends are similar to those shown in *WEO-2002*.

Oil will account for 37% of the increase in energy-related CO_2 emissions over the projection period, coal for 33% and natural gas for 30% (Figure 2.13). Emissions from natural gas will increase most rapidly, doubling between 2002 and 2030. But they will still make up only 24% of total emissions in 2030, up from 21% now. Coal's share will fall by three percentage points, to 36%, and oil's share will drop by two points to 39%.

Table 2.3: **Energy-Related CO_2 Emissions** (million tonnes)

	OECD		Transition economies		Developing countries		World	
	2002	2030	2002	2030	2002	2030	2002	2030
Power sector	4 793	6 191	1 270	1 639	3 354	8 941	9 417	16 771
Industry	1 723	1 949	400	618	1 954	3 000	4 076	5 567
Transport	3 384	4 856	285	531	1 245	3 353	4 914	8 739
Residential and services	1 801	1 950	378	538	1 068	1 930	3 248	4 417
Other*	745	888	111	176	605	1 142	1 924	2 720
Total	**12 446**	**15 833**	**2 444**	**3 501**	**8 226**	**18 365**	**23 579**	**38 214**

* Includes international marine bunkers (for the world totals only), other transformation and non-energy use.

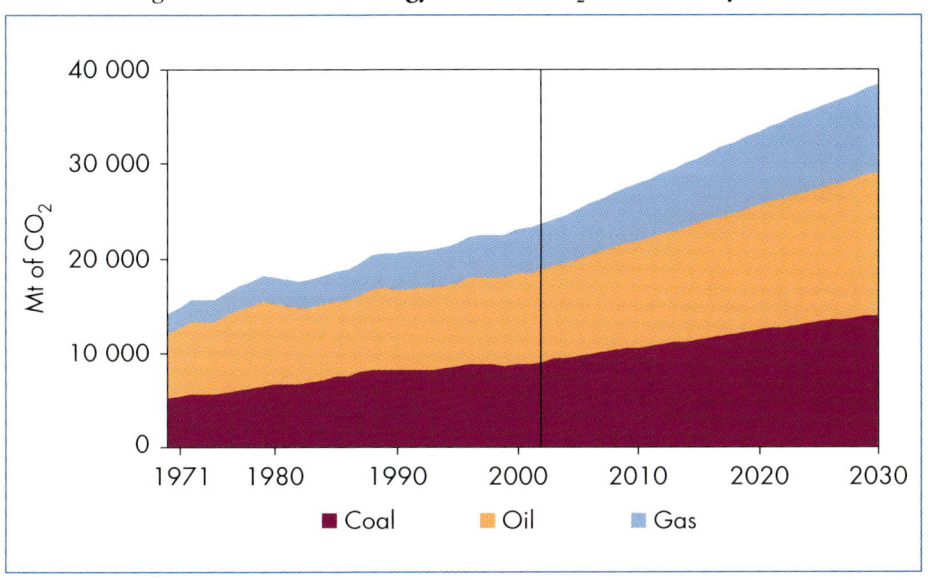

Figure 2.13: **World Energy-Related CO_2 Emissions by Fuel**

Over the past three decades, energy-related CO_2 emissions worldwide have grown less rapidly than has primary energy demand. Carbon emissions grew by 1.7% per year, while energy demand grew by 2%. Emissions and demand will grow at about the same rate, 1.7% per year, over the projection period (Figure 2.14). The average carbon content of primary energy consumption will remain more-or-less constant at about 2.3 tonnes of CO_2 throughout the projection period.

Chapter 2 - Global Energy Trends

Figure 2.14: **Average Annual Growth in World Primary Energy Demand and Energy-Related CO_2 Emissions**

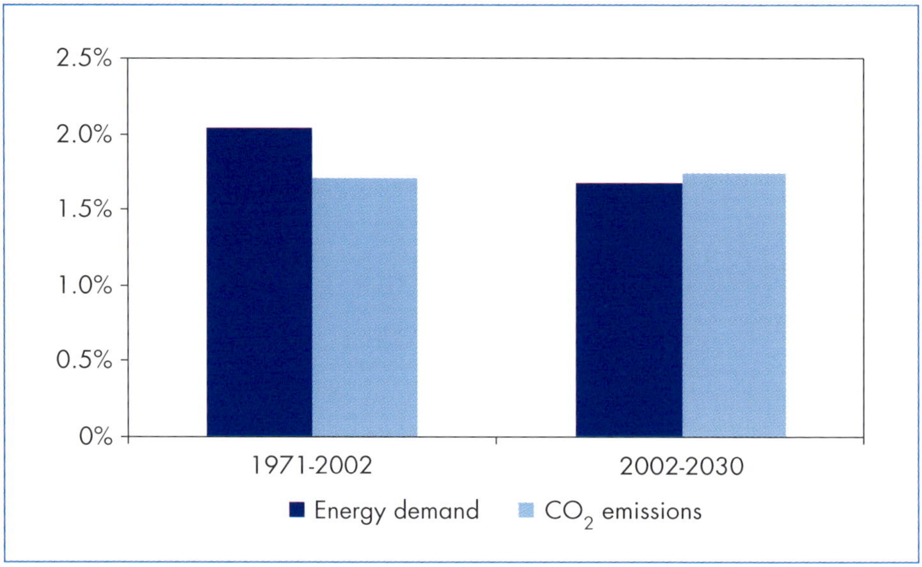

Regional Emission Trends

Developing countries will be responsible for 70% of the increase in global CO_2 emissions from 2002 to 2030 (Figure 2.15). They will overtake the OECD as the leading contributor to global emissions early in the 2020s. OECD

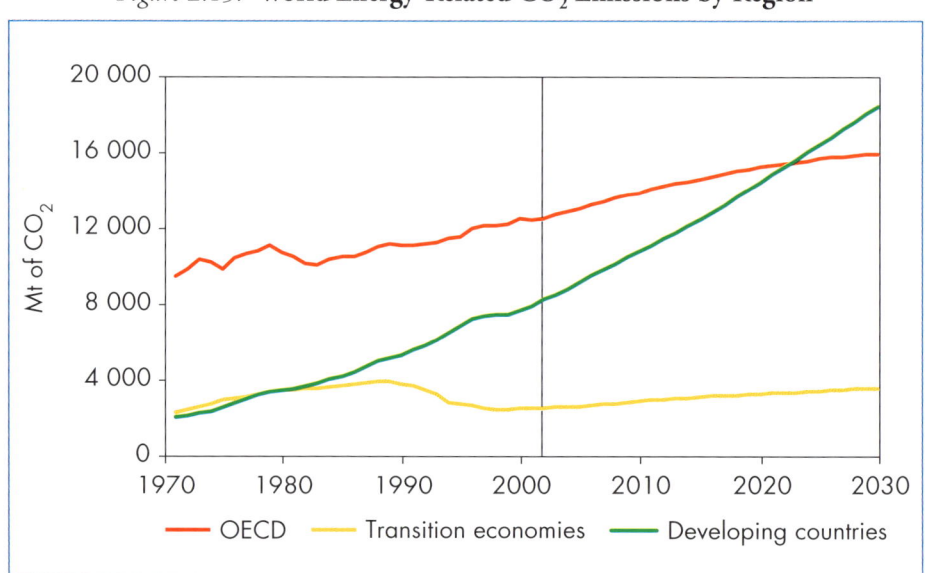

Figure 2.15: **World Energy-Related CO_2 Emissions by Region**

countries accounted for 54% of total emissions in 2002, developing countries for 36% and transition economies for 10%. By 2030, the developing countries will account for 49%, the OECD countries for 42% and the transition economies for 9%. Today, developing-country emissions are two-thirds of OECD emissions. By 2030, they will be 16% higher.

China's emissions alone will climb by 3 837 million tonnes, equal to more than a quarter of the increase in world emissions. Strong economic growth and heavy reliance on coal in industry and power generation drive this trend. Other Asian countries, notably India, also contribute heavily to the increase in global emissions. OECD emissions will increase by 3 386 million tonnes.

Despite the strong increase in emissions in developing countries, both the OECD and the transition economies will still have far higher *per capita* emissions in 2030 (Figure 2.16). Developing countries not only have lower per capita energy consumption; they also rely more heavily on biomass and waste, which are assumed to produce no emissions on a net basis.[1]

Figure 2.16: **Per Capita Energy-Related CO_2 Emissions by Region**

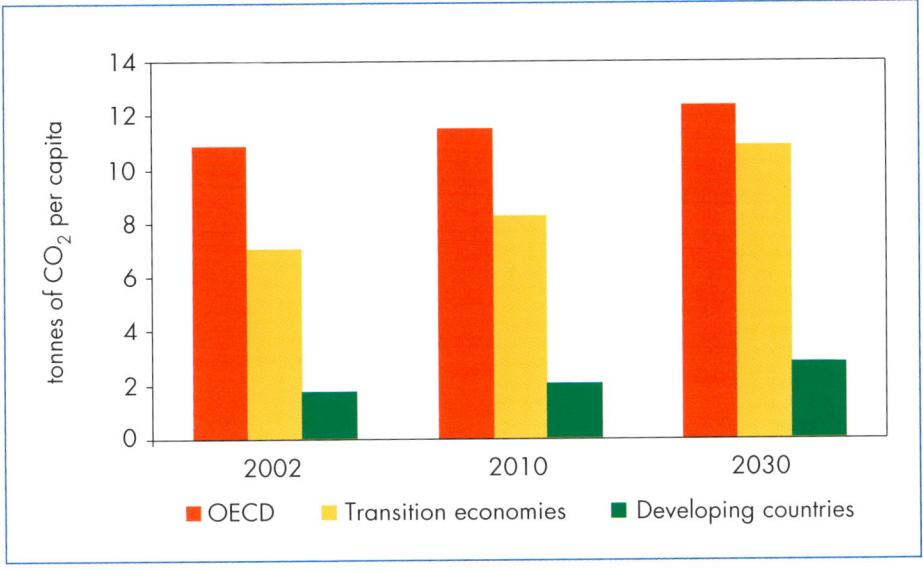

Sectoral Emission Trends

Power generation is expected to contribute around half the increase in global emissions from 2002 to 2030. Transport will contribute a quarter, with other uses accounting for the rest.

1. For the purposes of this analysis, all biomass is assumed to be replaced eventually. As a result, the carbon emitted when biomass fuels are burned is cancelled out by the carbon absorbed by the replacement biomass as it grows.

The share of the power sector in total emissions will rise, from 40% in 2002 to 44% in 2030, in line with the sector's increasing share in primary energy use. Power generation will also become more dependent on fossil fuels. Most of the increase in emissions from power stations will come from developing countries. Not only will their electricity production increase faster than in the OECD and the transition economies, but their reliance on coal will remain much higher too. Coal-fired plants will generate 47% of developing countries' electricity in 2030, up from 45% today. In the OECD, the share of coal will fall, from 38% to 33%, thanks to increased use of gas and renewables. Worldwide, average emissions per kWh of electricity produced will fall slightly, the result of continuing improvements in the thermal efficiency of power plants. The fall will be most pronounced in the transition economies.

Transport will consolidate its position as the second-largest sector for CO_2 emissions worldwide. Its share of total emissions will rise from 21% in 2002 to 23% in 2030. More than half the increase in the sector's emissions will occur in developing countries, where car ownership is expected to grow rapidly.

The energy-related CO_2 emissions presented in this *Outlook* give an indication of the efforts that countries with commitments to reduce their greenhouse-gas emissions under the Kyoto Protocol will need to make if the Protocol comes into effect. The Protocol covers six types of emissions and the contribution of sinks. Although our projections reflect only energy-related CO_2 emissions, these account for the bulk of greenhouse-gas emissions. The status of the Protocol is described in Box 2.3. If the Protocol comes into effect and if total greenhouse-gas emissions rise at the same rate as energy-related emissions, countries that have agreed to reduce emissions, known as Annex I countries, would not be able to meet the overall emission-reduction target – unless they adopt a new set of policies and measures. In 2010, the emissions of Annex I OECD countries are projected to be 30% above the target, while the emissions of Annex I transition economies will be 25% *below* target.

World Alternative Policy Scenario

The World Alternative Policy Scenario assesses how global energy markets could evolve were countries around the world to adopt a set of policies and measures that they are currently considering or might reasonably be expected to implement over the projection period. For OECD countries, the policies considered reflect current policy discussions. For non-OECD countries, where policy discussions are less advanced, energy efficiency and intensity are assumed to improve more rapidly than in the Reference Scenario as a result both of future policies and of faster transfer of technology from OECD countries.

In 2030, global primary energy demand in the Alternative Scenario would be 1 670 Mtoe, or 10%, less than in the Reference Scenario (Figure 2.17). Energy demand would grow by 1.3% per year, 0.4 percentage points less than in the

Box 2.3: **Status of the Kyoto Protocol**

The Kyoto Protocol calls for those industrialised countries and transition economies listed in its Annex I to reduce their overall greenhouse-gas emissions by at least 5% below their 1990 levels on average over 2008-2012. Annex I includes all the OECD countries except Korea and Mexico. Emission-reduction commitments vary among countries. To achieve their targets, countries can implement domestic emission reduction measures or use international "flexible mechanisms". The latter include emissions trading and joint implementation, a scheme in which countries invest in projects to reduce net emissions in other Annex I countries and thereby earn credits towards meeting their own national commitments. An additional approach is the clean development mechanism, a scheme similar to joint implementation, but involving projects in non-Annex I countries.

To take effect, the Protocol must be ratified by at least 55 nations, and the Annex I countries ratifying the Protocol must represent at least 55% of that group's total emissions in 1990. By July 2004, 123 countries, including all EU countries, had ratified the Protocol. Although the ratifying countries included all but eight Annex I countries, their aggregate emissions made up only 44% of total Annex I emissions. Two countries that had not ratified – the United States and Russia – made up almost all the remaining emissions. The Protocol cannot come into effect until one of those two countries ratifies it. The United States and Australia have announced that they do not intend to do so.[2]

Reference Scenario. The fuel mix is markedly different. Fossil-fuel demand as a whole would be 14% lower in 2030, while the use of nuclear power would go up by 14% and that of non-hydro renewables (excluding biomass) by 30%. The impact on energy demand would grow throughout the projection period, because new policies and technologies would take effect only gradually. By 2010, global energy savings would amount to just 2%.

Coal demand would be nearly a quarter lower in 2030 in the Alternative Scenario than in the Reference Scenario. The average annual rate of growth in coal demand would be 0.5%, down from 1.5% in the Reference Scenario. Almost 90% of the reduction in primary coal demand would come from power generation. Coal use in that sector would be driven down by lower electricity demand, increased thermal efficiency of coal power plants – especially in developing countries – and switching to other fuels. Primary oil demand is 11% lower in 2030. Two-thirds of the savings in oil are projected to come from the transport sector, where increased fuel efficiency and faster penetration of alternative-fuel vehicles push demand down.

2. On 30 September 2004, the Russian government approved the Protocol and submitted it to the state Duma for ratification. It remains to be seen whether the Duma will, in fact, ratify it.

Figure 2.17: **Change in Primary Energy Demand and Energy-Related CO_2 Emissions in the Alternative Policy Scenario*, 2010 and 2030**

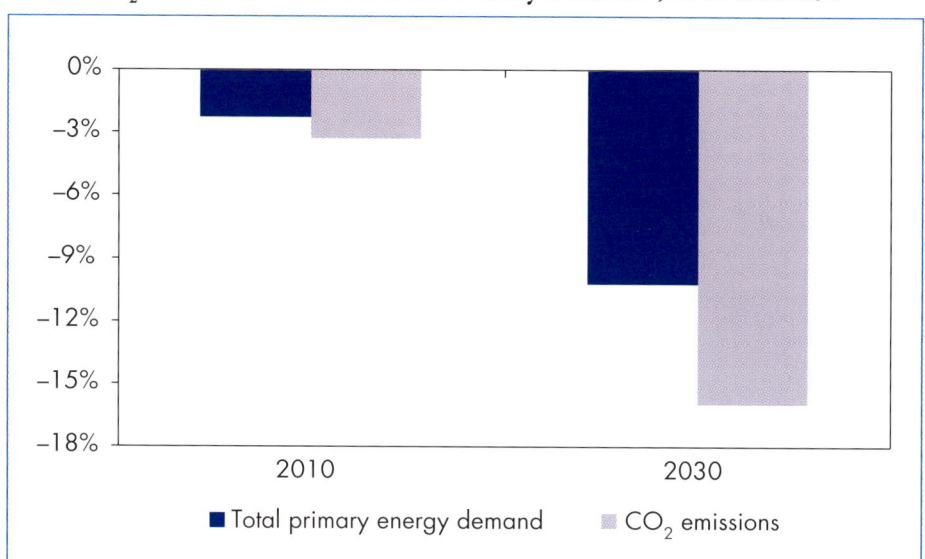

* Compared with the Reference Scenario.

Gas demand would be 10% lower in 2030. Again, the power sector would account for most of the savings, a full 72% in 2030. Renewables and nuclear power would displace part of fossil-fuel consumption.

As a result of the lower level of energy demand and of the different fuel mix, energy-related CO_2 emissions in the Alternative Policy Scenario are 6 000 Mt, or 16%, lower in 2030 than in the Reference Scenario. Emissions still increase by 8 600 Mt, or 37%, from 2002 to 2030. But the annual growth rate over the projection period falls from 1.7% to 1.1%. The gap is particularly wide in the third decade, when the annual growth rate is halved, from 1.4% to 0.7%. An increase in the share of carbon-free fuels in the fuel mix makes an important contribution to the reduction in emissions. By 2030, carbon-free fuels would account for 22% of global primary energy demand, four percentage points higher than in the Reference Scenario. Coal, the most carbon-intensive fuel, sees the biggest drop in market share. On average, emissions of CO_2 per unit of energy consumed in 2030 in the Alternative Scenario are 5% lower compared to 2002 and 6% lower compared to the Reference Scenario. Although CO_2 emissions are lower in the Alternative Scenario, they do not fall nearly enough to ensure that the concentration of carbon in the atmosphere is stabilised – the ultimate goal of the UN Framework Convention on Climate Change. Breakthrough technologies, such as carbon capture and storage and advanced nuclear, may be necessary to achieve the aim.

CHAPTER 3

OIL MARKET OUTLOOK

HIGHLIGHTS

- Global primary oil demand will grow by 1.6% per year, from 77 mb/d in 2002 to 121 mb/d in 2030. Demand will continue to grow most quickly in developing countries. Most of the increase in oil demand in all regions will come from the transport sector.
- Non-OPEC countries are expected to meet most of the increase in global demand over the rest of the current decade. In the longer term, however, production in OPEC countries, especially in the Middle East, will increase more rapidly. OPEC's worldwide market share will rise from 37% in 2002 to 53% in 2030 – slightly above its historical peak in 1973.
- Global oil production will not peak over the projection period so long as necessary investments in supply infrastructure are made. New capacity will be needed to offset production declines and to meet demand growth. About $3 trillion will need to be invested in the oil sector from 2003 to 2030. Financing that effort will be a major challenge.
- The reliability of reserves data reported by oil companies has been called into question by a series of large downward revisions. Uncertainty about reserves may undermine investor confidence and slow investment. There is an urgent need for all parties to work together to agree on and implement a universally-recognised, transparent, consistent and comprehensive data-reporting system for oil and gas reserves.
- Net inter-regional oil trade will more than double during the projection period, reaching over 65 mb/d in 2030 – more than half of global oil production. Exports from the Middle East, already the biggest exporting region, will rise most. The risk of a supply disruption at the critical chokepoints through which oil must flow will grow.
- This *Outlook* contains a High Oil Price Case, in which the average IEA crude oil import price is assumed to average $35 per barrel over the projection period. In this scenario, global oil demand is around 15% lower in 2030. Cumulative OPEC revenues in 2003-2030 are about $750 billion or 7% lower than in the Reference Scenario.

This chapter presents the Reference Scenario outlook for global oil markets to 2030. It first describes the projections of oil demand, by region and sector. It then assesses world oil reserves and the problems involved in measuring them. This is followed by a review of the outlook for oil production, inter-regional trade and investment needs. The chapter ends with an analysis of the impact on world oil demand and supply, OPEC revenues and oil investment of higher oil prices than assumed in the Reference Scenario.

Oil Demand

Global primary oil demand is projected to grow by 1.6% per year on average, from 77 million barrels per day in 2002 to 121 mb/d in 2030 (Table 3.1). Past trends in oil consumption have very closely followed the path of gross domestic

Table 3.1: **World Oil Demand** (million barrels per day)

	2002	2010	2020	2030	2002-2030*
OECD North America	22.6	25.5	28.7	31.0	1.1
United States and Canada	20.7	23.2	25.8	27.6	1.0
Mexico	2.0	2.3	2.9	3.4	2.0
OECD Europe	14.5	15.3	16.3	16.6	0.5
OECD Pacific	8.4	8.9	9.4	9.5	0.5
OECD Asia	7.5	7.9	8.3	8.3	0.4
OECD Oceania	0.9	1.0	1.1	1.2	1.2
OECD	**45.4**	**49.7**	**54.4**	**57.1**	**0.8**
Transition economies	4.7	5.5	6.5	7.6	1.8
Russia	2.7	3.1	3.6	4.2	1.6
Other transition economies	2.0	2.4	3.0	3.4	2.0
China	5.2	7.9	10.6	13.3	3.4
Indonesia	1.2	1.6	2.1	2.6	2.9
India	2.5	3.4	4.5	5.6	2.9
Other Asia	3.9	5.1	7.0	8.8	3.0
Latin America	4.5	5.4	6.8	8.4	2.3
Brazil	1.8	2.3	2.9	3.6	2.4
Other Latin America	2.7	3.2	3.9	4.8	2.1
Africa	2.4	3.1	4.4	6.1	3.4
Middle East	4.3	5.4	6.8	7.8	2.1
Non-OECD	**28.6**	**37.5**	**48.8**	**60.4**	**2.7**
Miscellaneous**	3.0	3.2	3.5	3.8	0.9
World	**77.0**	**90.4**	**106.7**	**121.3**	**1.6**
European Union	13.6	14.4	15.3	15.6	0.5

* Average annual growth rate.
** Includes bunkers and stock changes.

product (Figure 3.1). Demand has nonetheless grown less rapidly than GDP every year since 1976, resulting in a steady fall in global oil intensity. This trend will continue through the projection period: intensity will decline by a cumulative 34% in the period 2002-2030, compared to 46% in 1973-2002.

Figure 3.1: **Oil Demand and GDP Growth**

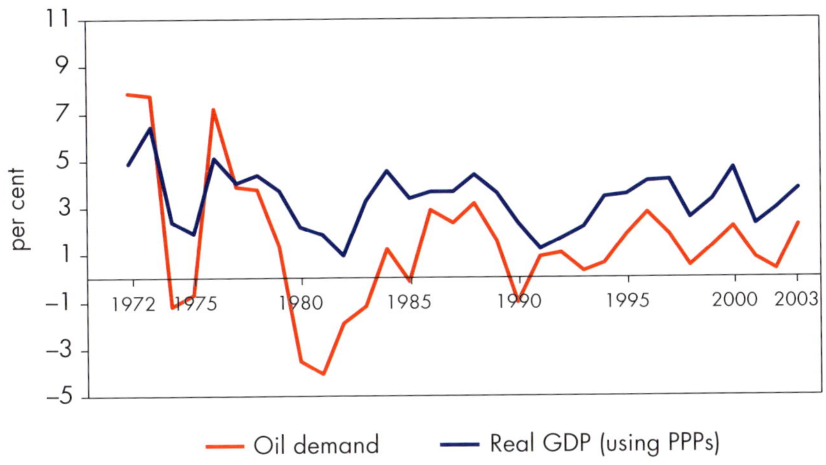

Oil demand will continue to grow most quickly in developing countries. Africa will see the fastest growth of 3.4% per year, though starting from a very low level. China's oil use will also rise by 3.4% per year, but still well below the extraordinary 11% increase in 2003. That surge, caused by a booming economy, contributed more than a third of the 1.74 mb/d increase in global oil consumption in 2003. That was the fastest year-on-year increase since the 1980s and a major cause of the recent hike in oil prices. Oil consumption in China increased by an additional 15% in the first half of 2004. Demand growth over the projection period will be almost as brisk in the rest of Asia as in China.

Demand in OECD countries, where economic and population growth will be relatively slack, will increase by only 0.8% per year. The North American market will rise faster than other OECD regions, boosted by robust economic growth in Mexico and population expansion throughout the region. Demand in OECD Europe and Pacific will rise by 0.5% per year. As a result of these trends, the non-OECD countries' share of global oil demand will grow from 39% in 2002 to just over half in 2030.

Developing countries as a group will remain more dependent on oil than OECD countries and the transition economies. Oil intensity, expressed as the amount of primary oil consumed per unit of GDP (using market exchange rates), will fall steadily over the projection period in all major regions. Oil intensity in developing countries and transition economies will decline faster than in OECD, reducing the gap between these regions. However, oil intensity in OECD countries will still be 28% lower in 2030 than in developing countries, and 31% than in transition economies (Figure 3.2).

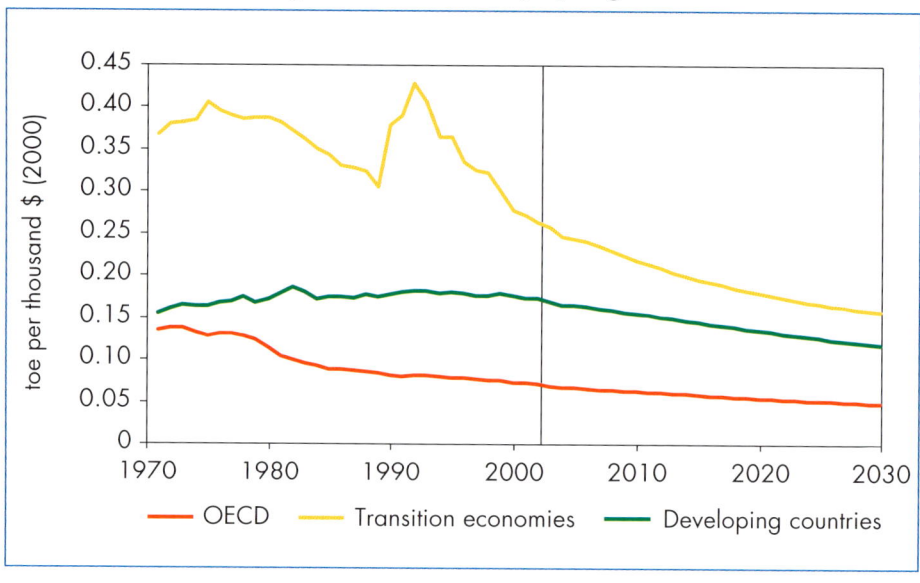

Figure 3.2: **Oil Intensity by Region**

Most of the worldwide increase in oil demand will come from the transport sector (Figure 3.3). In the OECD, oil use in sectors other than transport will hardly grow at all, and will even fall in the power sector. In non-OECD countries, the industrial, residential and services sectors will also contribute to the increase in oil demand.

The transport sector will account for 54% of global primary oil consumption in 2030 compared to 47% now and 33% in 1971 (Figure 3.4). Transport will absorb two-thirds of the increase in total oil use. Almost all the energy currently used for transport purposes is in the form of oil products. The share of oil in transport energy demand will remain almost constant over the projection period, at 95%, despite the policies and measures that many countries have adopted to promote the use of alternative fuels such as biofuels and compressed natural gas.

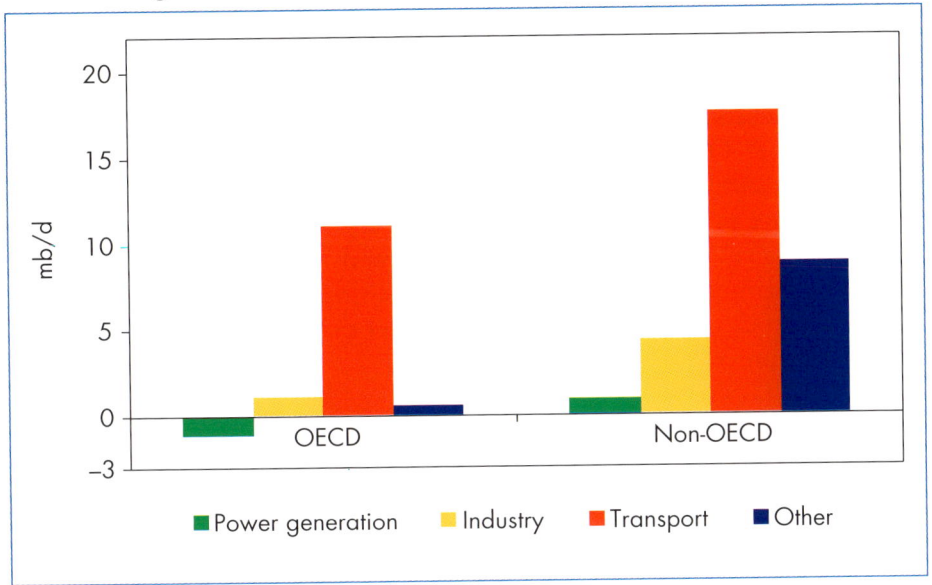

Figure 3.3: **Incremental Oil Demand by Sector, 2002-2030**

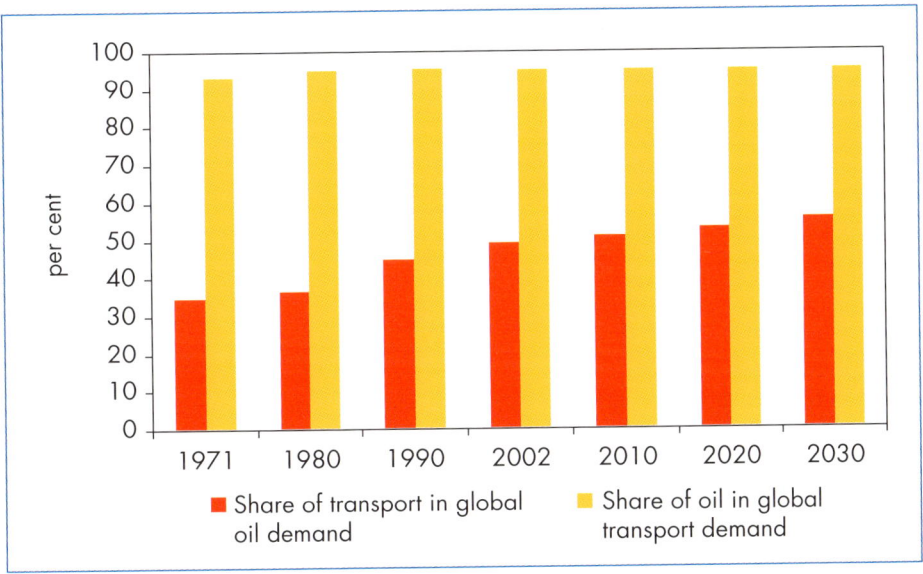

Figure 3.4: **Share of Transport in Global Oil Demand and Share of Oil in Transport Energy Demand**

Demand for road transport fuels is growing dramatically in many developing countries, in line with rising incomes and infrastructure development. The passenger-car fleet in China – the world's fastest growing new-car market –

Chapter 3 - Oil Market Outlook

grew by more than 9% per year in the five years to 2002, compared to just over 3% in the world as a whole (CPDP, 2003). Preliminary data show that more than two million new cars were sold in China in 2003. The scope for continued expansion of the country's fleet is enormous: there are only 10 cars for every thousand Chinese people compared with 770 in North America and 500 in Europe. Other Asian countries, including Indonesia and India, are also experiencing a rapid expansion of their car fleets. Freight will also contribute to the increase in oil use for transport in all regions. Most of the increased freight will travel by road, in line with past trends. The total vehicle stock in non-OECD countries is projected to triple over the projection period to about 550 million, but will still be about 25% smaller than that of OECD countries in 2030 (Figure 3.5).

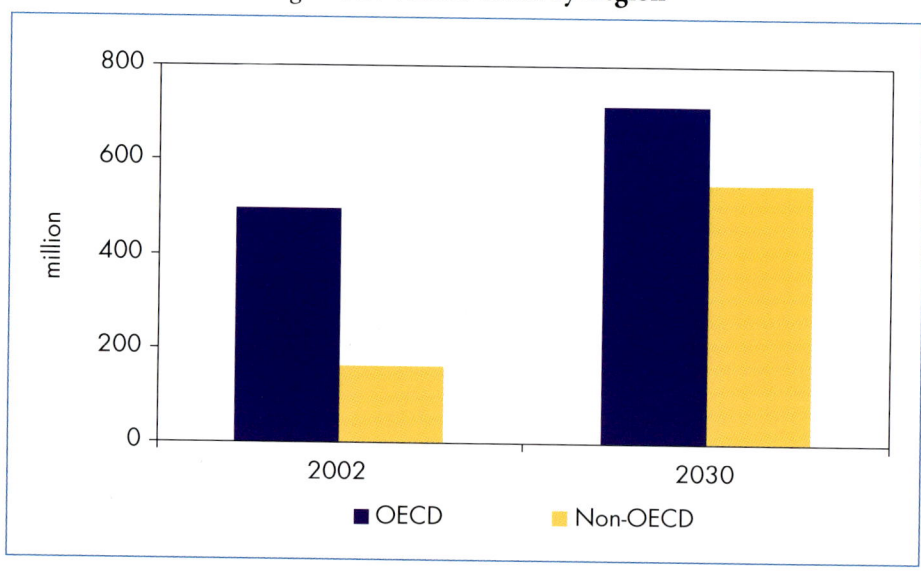

Figure 3.5: **Vehicle Stock by Region**

Despite the much faster rate of growth, per capita demand for transport oil will remain much lower in developing countries than in the OECD, especially North America. This is largely because incomes, even adjusted for purchasing power parity, will still be lower (Figure 3.6). By 2030, incomes in developing countries will approach those of OECD Europe in 1971, yet transport oil demand will be less than half what it was in the industrialised countries three decades ago. This is because vehicles will be much less fuel-intensive than they were in the past. Oil demand for transport will remain much higher in North America than in Europe relative to incomes, because of lower fuel taxes, long driving distances and cultural factors.

Figure 3.6: **Per Capita Transport Sector Oil Demand and GDP, 1971-2030**

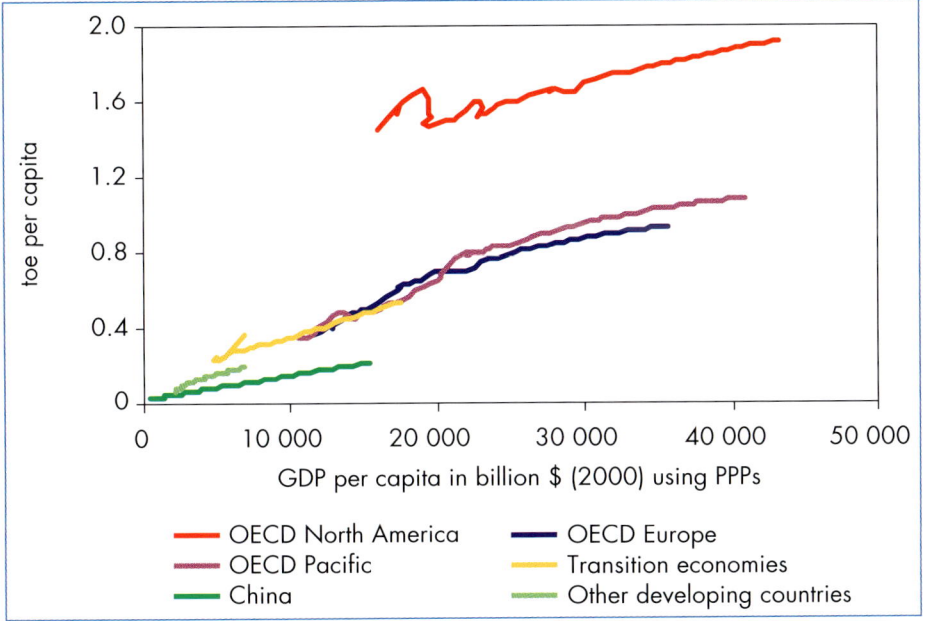

The continuing shift in oil demand towards lighter transport fuels will call for new investment in secondary processing facilities in refineries. The worldwide capacity of gas-to-liquids plants is projected to expand rapidly over the projection period, bringing a sharp increase in the supply of clean diesel – the main output of such plants (see Chapter 4).

Oil Reserves and Resources
Classifying and Measuring Resources

The earth contains a finite amount of fossil-hydrocarbon resources. Among the ways these resources can be categorised are the degree of certainty that they exist and the likelihood that they can be extracted profitably (Figure 3.7). Hydrocarbons that have been discovered and for which there is reasonable certainty that they can be extracted profitably (on the basis of assumptions about cost, geology, technology, marketability and future prices) are usually called "proven reserves". Estimates of proven reserves change over time as those assumptions are modified. "Probable reserves" include volumes that are thought to exist in accumulations that have been discovered and that are expected to be commercial, but with less certainty than proven reserves. "Possible reserves" include volumes in discovered fields that are less likely to be recoverable than probable reserves. There is, however, no internationally agreed benchmark or legal standard on how much proof is needed to demonstrate the existence of a discovery. Nor are there established rules about the assumptions to be used to determine whether discovered oil can be produced economically. This has created inconsistency and confusion over just how much oil can be produced commercially in the long term.

Figure 3.7: **Hydrocarbon-Resource Classification**

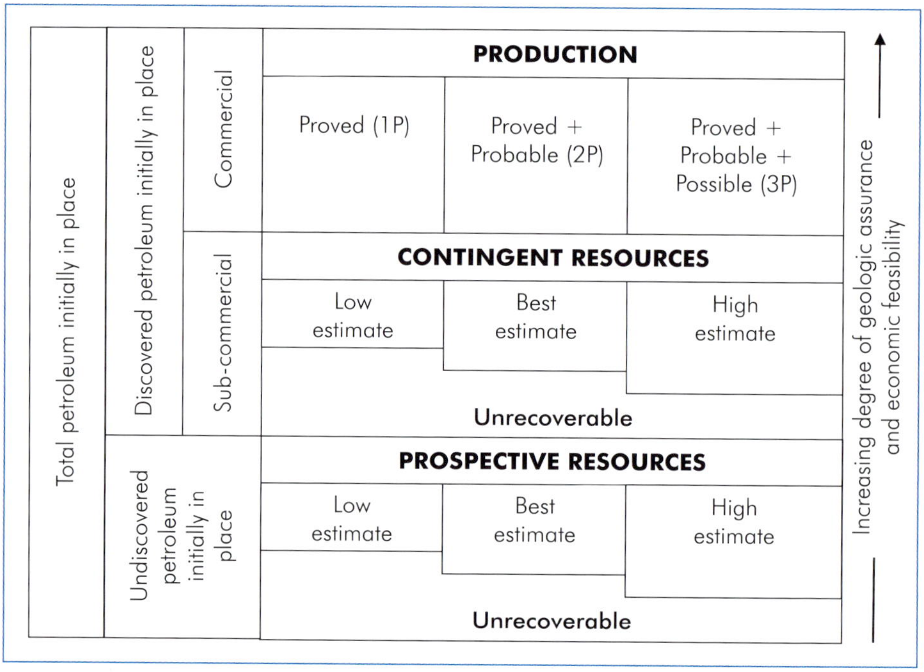

		PRODUCTION		
Commercial		Proved (1P)	Proved + Probable (2P)	Proved + Probable + Possible (3P)
Sub-commercial		**CONTINGENT RESOURCES**		
		Low estimate	Best estimate	High estimate
		Unrecoverable		
Undiscovered petroleum initially in place		**PROSPECTIVE RESOURCES**		
		Low estimate	Best estimate	High estimate
		Unrecoverable		

Source: SPE/WPC/AAPG (2000).

Attempts have been made in the past to harmonise definitions and methodologies. In 1997, the Society of Petroleum Engineers (SPE) and the World Petroleum Congress jointly adopted definitions for "proved", "probable" and "possible" reserves. Under these definitions "proved", or "1P", reserves are those with a probability of at least 90% that the estimated volumes can be produced profitably. "Proved plus probable", or "2P", reserves are required to have at least 50% probability, while "proved plus probable plus possible", or "3P", reserves are based on a probability of at least 10%. In 2000, the same two organisations, together with the American Association of Petroleum Geologists, approved definitions of petroleum resources. The US Securities and Exchange Commission (SEC), which lays down reserves-reporting standards for companies quoted on US stock exchanges, frequently updates its own guidelines (SEC, 2003). But there are no common information-disclosure requirements for reserves under the International Accounting Standards. The UN Economic Commission for Europe aims to develop a Framework Classification for Energy and Mineral Resources.

In practice, the methodologies used to measure different categories of reserves vary according to their purpose. Financial reporting standards are the strictest and lead to the lowest estimates. Typically, only those reserves that are proven with a confidence level above 90% may be reported in financial statements. SEC standards are the most restrictive and most detailed in the world. They nonetheless allow scope for some discretion. Reclassifications are tolerated. Regular adjustments are made – both up and down – as companies' knowledge of the technical characteristics of their fields improves and as technology and market conditions evolve.

Recent announcements by major oil and gas companies of steep cuts in their reserves estimates and differences in estimates by different companies with interests in the same field have raised questions about the reliability and comparability of industry data on reserves. They have also drawn attention to the lack of transparency in the reporting of reserves data. Shell stunned financial markets in January 2004 when it downgraded a fifth of its oil and gas reserves – equivalent to 3.9 billion barrels of oil – from "proven" to "probable" or to even less certain categories. Shell subsequently downgraded an additional 600 million barrels of oil equivalent of proven reserves. These moves occurred after Shell turned up errors in the classification of reserves at a number of fields discovered in the late 1990s. Several other companies have also announced major reserves downgrades since the beginning of 2004, including El Paso, which revised down its proven North American gas reserves by 41%, Canada's Nexen and Husky Energy, and the US independents, Forest Oil, Vintage Petroleum and Western Gas Resources. The El Paso and Forest revisions were apparently due to a reassessment of technical, geological and cost factors. The reliability of reserves estimates reported by national oil companies has long been a matter of great concern.

Estimating reserves remains as much an art as a science. Reservoir engineers typically apply a number of different formulas involving a range of variables and probabilities to test the sensitivity of a preferred approach or set of assumptions. Increasingly, companies rely on seismic data to map out hydrocarbon reservoirs, although SEC rules forbid the booking of reserves on undrilled acreage on the basis of seismic data alone. Auditing of reserves is far from universal. Many international oil companies and private Russian companies (under pressure from Western investors) use external auditors, but national oil companies generally do not. The price of oil or gas is a key parameter. As the price goes up, marginal fields become commercially viable and can be shifted into the probable or proven category. SEC rules, for example, require estimates to be based on the price of oil and gas on the last day of each year. Capital expenditure in exploration and development is also affected by prices and their volatility. The elasticity of exploration and production expenditures to the crude oil price has averaged 0.5 in the last

15 years (Appert, 2004). A 10% increase of the price has led to a 5% increase in exploration and production spending, boosting new discoveries.

Estimates of Proven Reserves

The most widely-quoted primary sources of global reserves data – the *Oil and Gas Journal* (O&GJ), *World Oil*, Cedigaz (for gas), and the US Geological Survey (USGS) – compile data from national and company sources. In addition, OPEC compiles and publishes data for its member countries. IHS Energy (formerly Petroconsultants) also compiles data but for "proven plus probable" reserves only. Other organisations, including BP, publish their own global estimates based mainly on data from the main primary sources. Despite the differences in approaches among these organisations, current estimates of remaining proven reserves worldwide do not vary greatly (Figure 3.8). BP puts global oil reserves at the end of 2003 at 1 148 billion barrels. Other estimates for the same year range from *World Oil's* 1 051 billion to IHS Energy and O&GJ's 1 266 billion barrels. Both BP and O&GJ include Canadian oil sands which increases their estimates substantially, although differences in their approaches produce very different results: BP includes 11 billion barrels of oil-sands reserves under active development. O&GJ includes *all* proven oil-sands reserves, which they estimate at 174 billion barrels. *World Oil* excludes natural gas liquids and Canadian oil sands. Differences among all the main sources are small in terms of the number of years of remaining reserves: 36 according to *World Oil* and 44 according to O&GJ.[1]

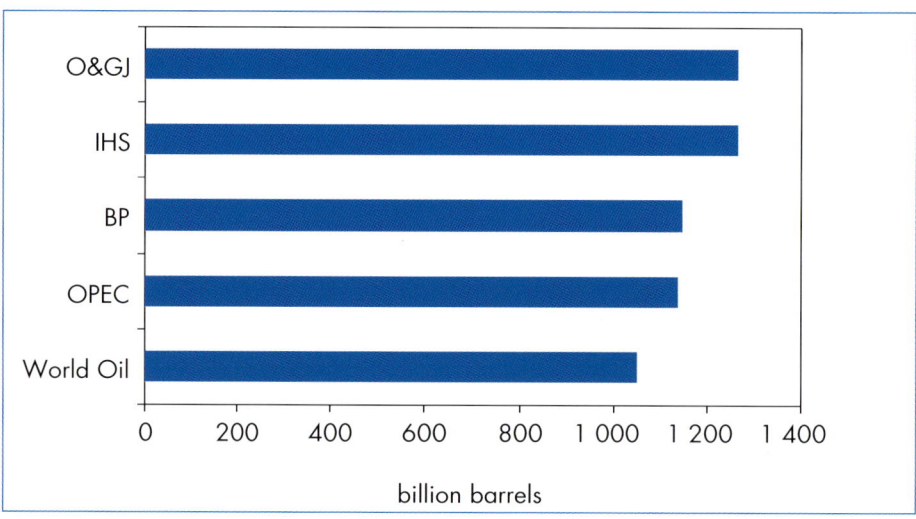

Figure 3.8: **Crude Oil and NGL Reserves at End 2003**

Note: The IHS Energy figures are for "proven and probable technically-recoverable resources". They include developed oil-sands reserves in Canada and certain developed Venezuelan extra-heavy oil resources in the Orinoco.

1. In 2003 based on production of 79.6 mb/d.

Global oil reserves are far from evenly distributed. All sources agree that reserves are concentrated in the Middle East, although the region's specific share of total reserves varies significantly among the different information sources (Figure 3.9). These differences are largely the result of varying estimates for some of the largest countries (Table 3.2).[2]

Figure 3.9: **Regional Share of Proven Oil Reserves**

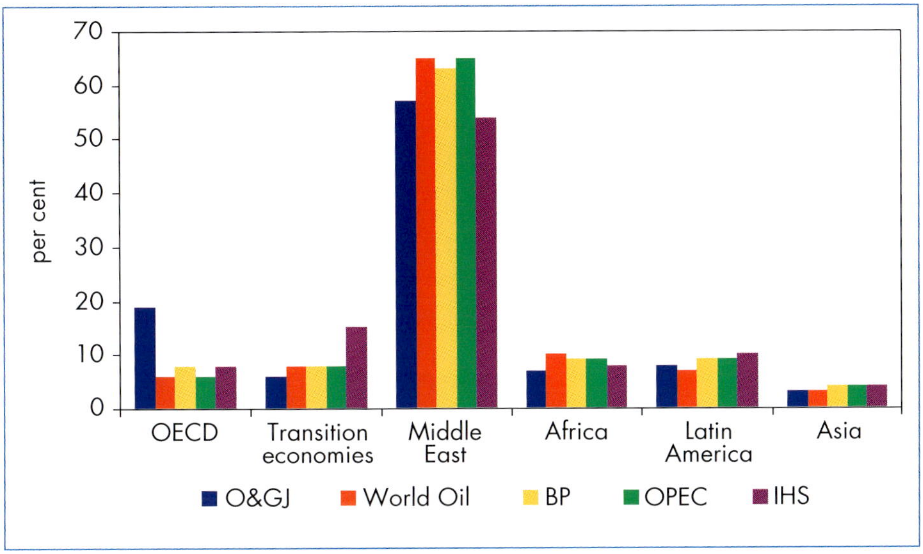

Table 3.2: **Top Ten Countries with Proven Oil Reserves*** (billion barrels, end of 2003)

	O&GJ	World Oil	BP	OPEC
Saudi Arabia	259	259	263	263
Iran	126	105	131	133
Iraq	115	115	115	115
United Arab Emirates	98	66	98	98
Kuwait	97	97	97	99
Venezuela	78	52	78	77
Russia	60	65	69	n.a.
Libya	36	31	36	39
Nigeria	25	33	34	35
United States	23	23	31	23

* According to O&GJ, excluding Canadian oil sands.

2. The IEA is undertaking a major study of energy supply and demand prospects in the Middle East, to be published in the *World Energy Outlook Insights* series in 2005. It will include a detailed analysis of oil and gas reserves and production costs.

Chapter 3 - Oil Market Outlook

According to BP, reserves increased dramatically in the 1980s and 1990s, from 670 billion barrels at the end of 1980 to 1 147 billion barrels at the end of 2003 (Figure 3.10).[3] But most of the increase occurred in OPEC countries, mainly in the Middle East, in the second half of the 1980s. Saudi Arabia and Kuwait revised their reserves upward by 50%, while Venezuelan reserves were boosted 57% by the inclusion of heavy oil in 1988. The United Arab Emirates and Iraq also recorded large upward revisions in that period. Total OPEC reserves jumped from 536 billion barrels in 1985 to 766 billion barrels in 1990. As a result, world oil reserves increased by more than 30%. This hike in OPEC countries' estimates of their reserves was driven by negotiations at that time over production quotas, and had little to do with the actual discovery of new reserves. In fact, very little exploration activity was carried out in those countries at that time. Total reserves have hardly changed since the end of the 1980s.

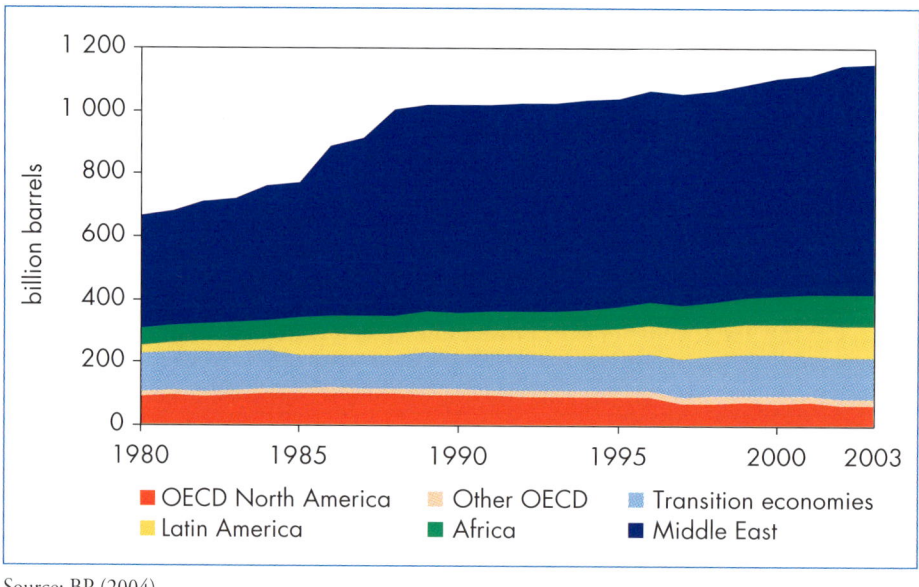

Figure 3.10: **Proven Oil Reserves by Region**

Source: BP (2004).

The primary sources of reserves data, such as O&GJ and *World Oil*, do not adjust reported data according to a standardised methodology. Nor do they adjust data for the large number of countries that rarely amend their official reserves estimates to account for actual production. Out of the 97 countries covered by O&GJ estimates at the end of 2003, the reserves of 38 countries were unchanged since 1998 and 13 more were unchanged since 1993, despite ongoing production. For example:

3. Part of this variation can be attributed to BP changing the sources of data they used to compile their estimates.

- Official reserves in Kuwait – the world's fourth-largest – stood unchanged at 94 billion barrels from 1991 to 2002, even though the country produced almost 8 billion barrels and did not make any important new discoveries during that period.
- Angola's proven oil reserves have been constant at 5.4 billion barrels since 1994, despite several deep-water discoveries.
- According to O&GJ, Algeria's proven reserves were unchanged at 9.2 billion barrels from 1990 to 2001, despite major exploration and development activity.

Reserves data from the main public sources are not "backdated". In other words, the historical time series of reserves for a given country is not revised in the light of new discoveries or increases in recovery rates.

Turning Resources into Reserves
Estimates of Ultimately Recoverable Resources

Reserves estimates give an indication of how much oil-production capacity could be developed in the near to medium term. They are a poor indicator, however, of how much oil remains to be produced and of the potential for raising productive capacity in the long term. For these purposes, it is more useful to consider remaining ultimately recoverable *resources*. This category includes proven, probable and possible reserves from discovered fields, as well as hydrocarbons that are yet to be discovered. The size of these resources will determine how much oil is likely to be produced in the long term. The larger the resource base, the more oil is likely to move into the proven category, the later the peak in production will be reached and the more oil will ultimately be produced.

Unfortunately, very few estimates of hydrocarbon resources are available. Those estimates that do exist are also subject to considerable uncertainty. The USGS is the most authoritative source of estimates of global ultimately recoverable resources of conventional oil. In its most recent study, released in 2000, the USGS estimates that ultimate conventional oil resources, including natural gas liquids (NGL), amounted to 3 345 billion barrels at the beginning of 1996 (Table 3.3).[4] This mean figure includes cumulative production to date, remaining reserves, undiscovered recoverable resources and estimates of "reserves growth" in existing fields. Reserves growth refers to the increase in reserves in oilfields that typically occurs through improved knowledge about the field's productive potential. Such growth accounts for around 28% of

---

4. USGS data cover only those parts of the world actually assessed. Their proven reserves and cumulative production data are for 1995 and were taken from Petroconsultants and NRG Associates. It considers a time horizon from 1995 to 2025. In this chapter, all projections and calculations take into account production from 1996 to 2003 and reserves updates since 1996.

remaining ultimately recoverable resources, whereas remaining reserves and undiscovered resources account for about 36% each in the USGS mean case. The USGS provides estimates for different degrees of probability. Undiscovered resources range from 495 billion barrels at 95% probability to 1 589 billion barrels at 5% probability. Reserves growth varies more widely, from 229 billion barrels to 1 230 billion barrels. Ultimate resources vary among regions, but, as is the case for proven reserves, the Middle East and the transition economies hold the majority of them (Figure 3.11).

Table 3.3: **USGS Estimates of Ultimately Recoverable Oil and NGL Resources**
(billion barrels)

Category/probability*	95%	50%	5%	Mean
Undiscovered	495	881	1 589	939
Reserves growth	229	730	1 230	730
Remaining reserves				959
Cumulative production				717
Total ultimately recoverable resources				3 345
Remaining ultimately recoverable resources				**2 628**

* Per cent chance of there being at least the amount indicated.
Source: USGS (2000). Data are as of 1 January 1996 and for those parts of the world actually assessed.

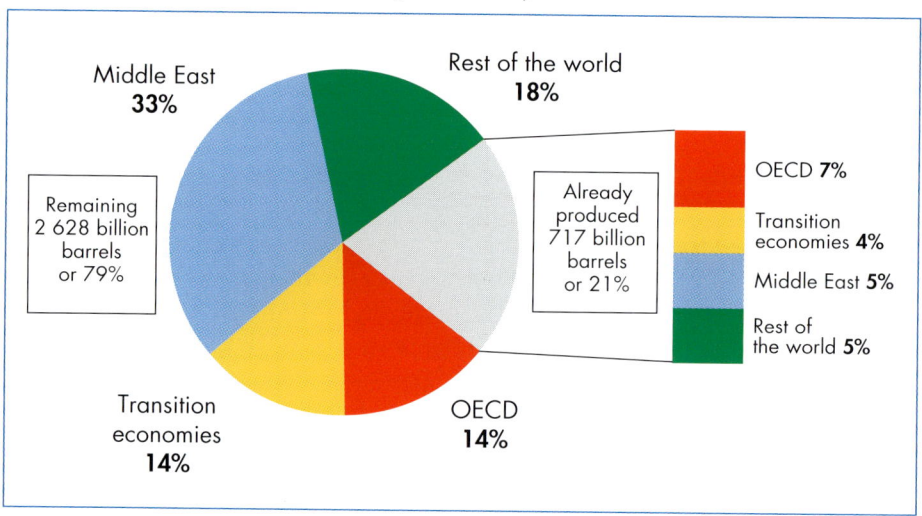

Figure 3.11: **Ultimately Recoverable Resources of Oil and NGL by Region**
(mean value)

Source: USGS (2000).

A widely used indicator of the adequacy of remaining oil reserves is the reserves-to-production (R/P) ratio, or reserves divided by annual production. According to BP, the R/P ratio is lower than 50 for all regions except the Middle East. In other regions, only five countries – Venezuela, Libya, Azerbaijan, Nigeria, and Trinidad and Tobago – have a ratio over 30. But reserves are constantly revised in line with new discoveries, changes in prices and technological advances. These revisions invariably add to the reserve base. On the other hand, production will also continue to expand, lowering the R/P ratio over time unless net additions to reserves outpace production. The ratio of remaining ultimately recoverable resources to the expected average production level over the projection period is a better guide to future production potential, and is much higher than the R/P ratio. Globally, remaining resources are sufficient to meet the projected average annual production, between now and 2030, 70 times over, based on USGS estimates and this *Outlook*'s projections. In comparison, the R/P ratio is 41 years according to BP and 36 years according to *World Oil*. The rate at which remaining ultimate resources can be converted to reserves, and the cost of doing so, is, however, very uncertain.

As is the case for reserves, estimates of ultimate resources change over time as more and better data become available and improved technologies for finding and producing oil are developed. Estimating resources involves making judgments about finding and development costs, oil prices and technological developments. The resulting assessments rely heavily on available geological, geophysical and production data. It is not surprising, therefore, that the assessments have changed markedly. The USGS has conducted five studies since 1984. The mean estimate of total world resources doubled between the 1984 study and the most recent study in 2000. Studies by other organisations have yielded very different results, most of them lower than the latest USGS assessment (Figure 3.12).

The estimates described above concern conventional resources. So-called "non-conventional oil" resources are thought to be almost as large and could play an increasing role in meeting future needs. Non-conventional oil initially in place could amount to as much as 7 trillion barrels (Figure 3.13). Extra-heavy oil in Venezuela, tar sands in Canada and shale oil in the United States, account for more than 80% of these resources. However, the amount of oil that could be recovered from these resources is very uncertain. IHS estimates that there were "only" 333 billion barrels of remaining recoverable bitumen reserves worldwide in 2003. This represents about 11 years of current total world oil production.

Sources of Reserve Additions

Additions to reserves come from two main sources: from discoveries of new oilfields and from revisions to estimates of the reserves in fields already in

Figure 3.12: **Ultimately Recoverable Oil Resources**

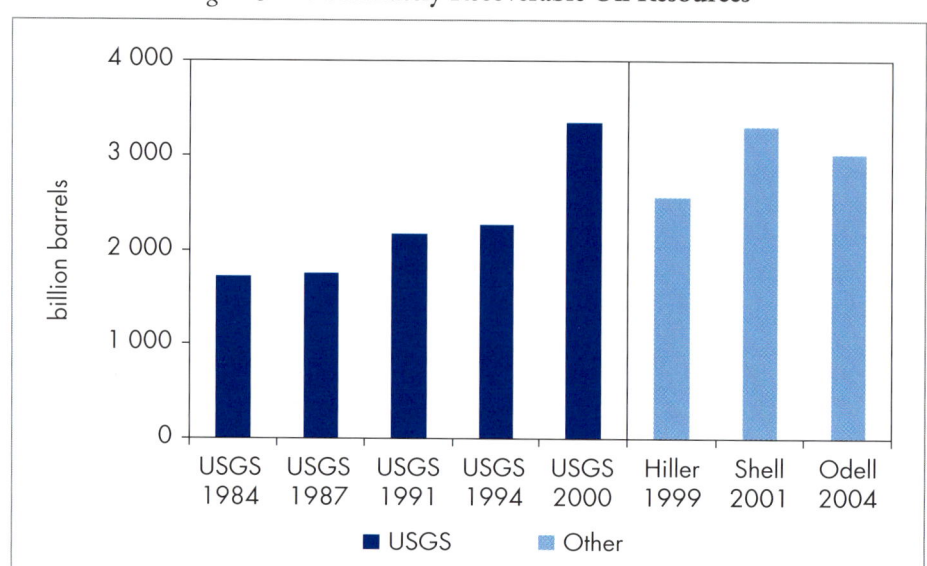

Figure 3.13: **Non-Conventional Oil Resources Initially in Place**

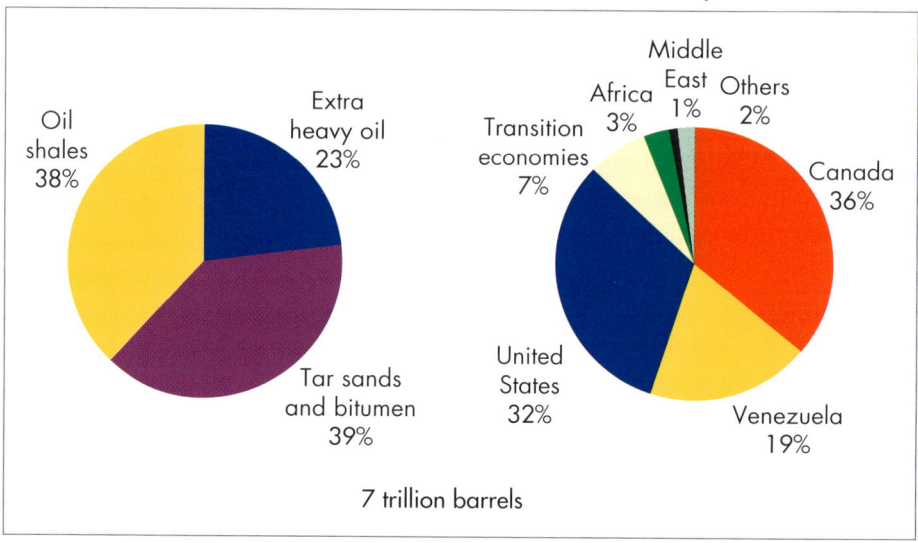

Sources: IEA analysis based on Cupcic (2003); Dyni (2003) and IHS Energy databases.

production or undergoing appraisal. In the last few decades, an increasing share of reserve additions has come from revisions based on improved operational data on known fields or revised cost and price assumptions. Moreover, the appraisal of known fields is increasingly carried out with seismic surveying and reservoir-simulation techniques in combination with drilling wells. The US Securities Exchange Commission does not, however, usually permit companies to book reserves solely on the basis of seismic data for undrilled acreage, as seismic techniques do not allow sufficiently accurate measurement of reservoir characteristics or the type of oil.

Reserve additions from discoveries of new oilfields have fallen sharply since the 1960s. In the last decade, discoveries have replaced only half the oil produced (Figure 3.14). By contrast, the amount of oil discovered in the 1970s was more than a third higher than that actually produced. The fall in oil discoveries has been most dramatic in the Middle East and the former Soviet Union. In the Middle East, discoveries plunged from 187 billion barrels in 1963-1972 to 16 billion barrels during the decade ending in 2002. The share of Africa, Latin America and Asia in new discoveries has increased sharply, though the absolute amount of oil discovered has fallen since the 1970s.

Figure 3.14: **Additions to World Proven Oil Reserves from the Discovery of New Fields and Production**

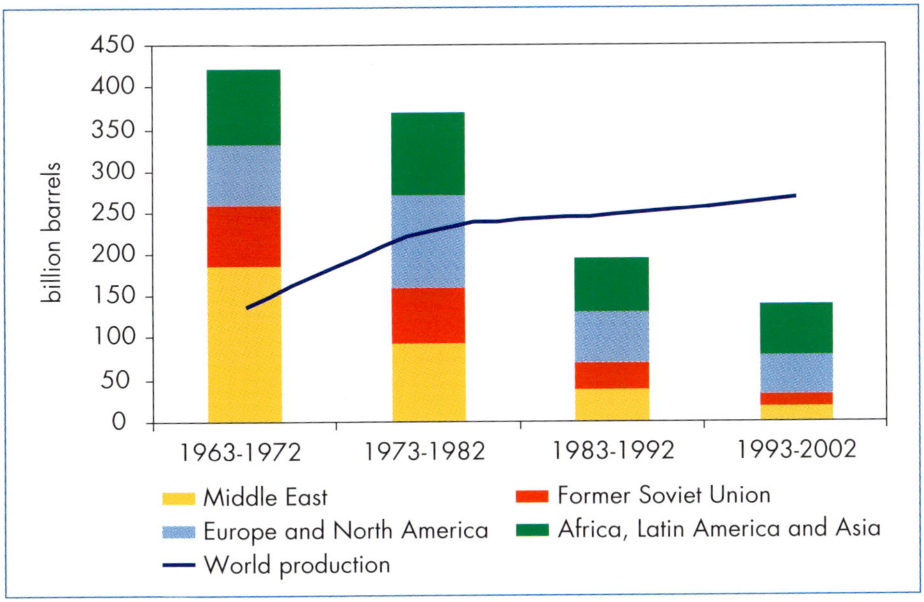

Source : IEA analysis based on IHS Energy database.

The drop in oil discoveries is largely the result of reduced exploration activity in those regions with the largest reserves, and of a fall in the average size of fields discovered. These factors have more than offset an increase in exploration success rates. There has been an overall decline in new-field wildcat drilling since the early 1990s. This reflects the difficulty that the industry has had in gaining access to prospective acreage, even outside those countries where drilling is controlled by state companies. Exploration drilling in the Middle East has been minimal for many years because existing proven reserves are already very large. Therefore, national oil companies in the region have had little incentive to appraise existing fields or explore for new ones. They prefer to focus investment on maintaining or increasing output at already producing fields. Drilling is now concentrated in North America and Europe – both mature producing regions with limited discovery potential. Only 3% of wildcat wells drilled in the ten years to 2002 were in the Middle East, even though the region is thought to hold over a quarter of the world's undiscovered resources of oil and gas (Figure 3.15). There is likely to be a rebound in exploration and appraisal drilling in the Middle East in the coming decades, as decline rates at existing fields rise and the number of undeveloped fields drops.

Figure 3.15: **Undiscovered Oil and Gas Resources, 1995 and New Wildcat Wells Drilled, 1993-2002**

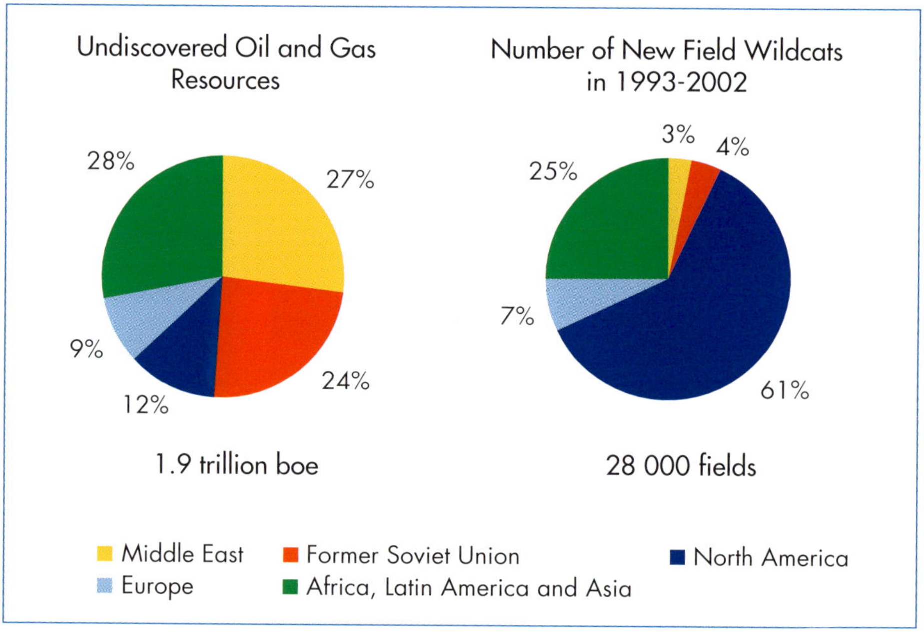

Undiscovered Oil and Gas Resources
- Middle East: 27%
- Former Soviet Union: 24%
- Europe: 9%
- North America: 12%
- Africa, Latin America and Asia: 28%

1.9 trillion boe

Number of New Field Wildcats in 1993-2002
- Middle East: 3%
- Former Soviet Union: 4%
- Europe: 7%
- North America: 61%
- Africa, Latin America and Asia: 25%

28 000 fields

Sources: IEA analysis based on USGS (2000) and IHS Energy databases.

There has been a marked reduction in the average size of fields discovered over the past four decades (Figure 3.16). This largely reflects the concentration of exploration activity in mature areas. International oil companies have had only limited access to acreage in many of the world's most promising regions. Exploration has, therefore, been concentrated in regions such as the North Sea, the Gulf of Mexico and deep waters offshore Brazil. This trend has been reinforced by higher oil prices since 1999. A high price makes exploration in high-cost regions economically feasible. A shift in drilling to the Middle East and other regions with greater long-term production potential is expected in the future.

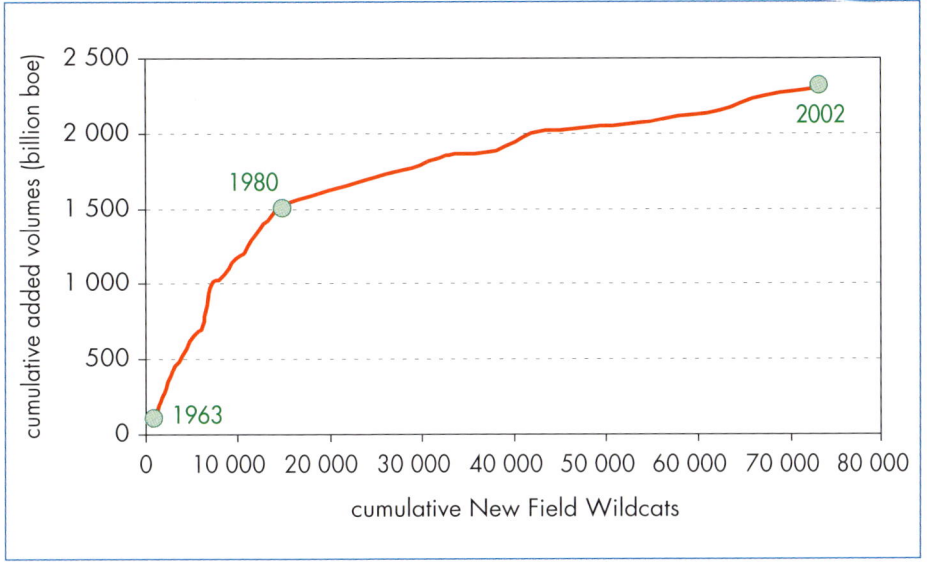

Figure 3.16: **Cumulative Oil and Gas Discoveries and New Wildcat Wells Drilled, 1963-2002**

Source: IEA analysis based on IHS Energy database.

The success rate for exploratory wells – the proportion of wells drilled that yield oil or gas – has increased sharply in recent years. New techniques and technology have boosted the average success rate worldwide from about 20% in the late 1940s to over 40% in recent years (Figure 3.17). Technological advances have also reduced the cost of drilling wells. The more widespread application of advanced oilfield technology is expected to lead to a substantial increase in proven reserves over the projection period.

The rate at which resources are "proved up" in the future – that is, moved into the category of proven reserves – will depend on a combination of factors. These include technological advances that facilitate the development of resources in geologically difficult or environmentally sensitive locations, lower development costs and improve recovery rates. A good example of how new technology has boosted proven reserves is the offshore Troll field in Norway. Troll, originally a gas field, contains oil in thin layers, making it hard to extract. At one time, it was not thought that any oil could be recovered profitably. But

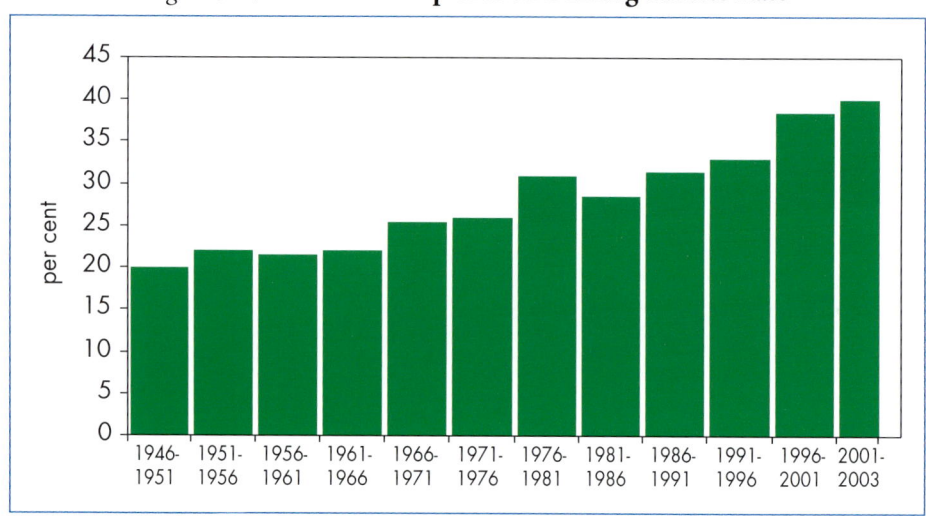

Figure 3.17: **World Oil Exploration Drilling Success Rate**

Source: IHS Energy database.

the incremental deployment of various techniques increased the field's oil reserves fivefold between 1990 and 2002. The recovery rate increased to 70% over this period (Figure 3.18). Some countries permit drilling only by their national oil companies. If these countries open up to drilling by foreign companies, exploration and development activity could increase significantly.

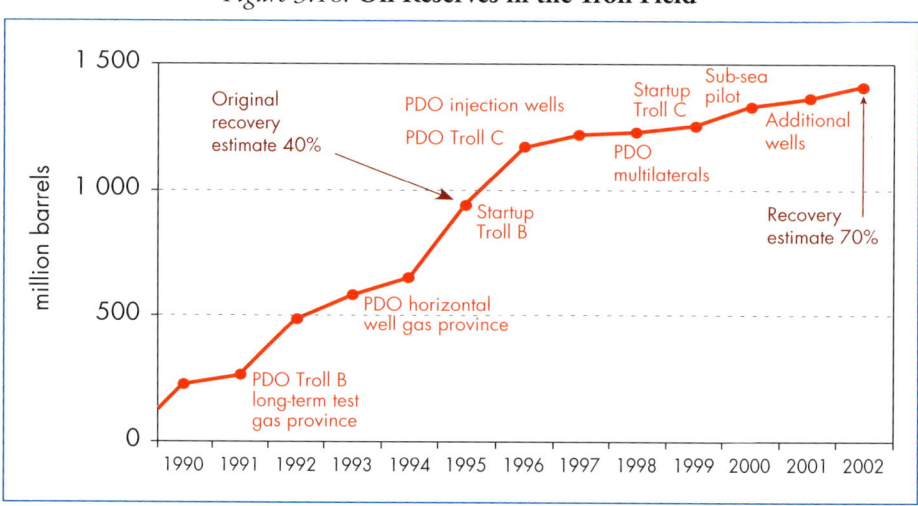

Figure 3.18: **Oil Reserves in the Troll Field**

Note: PDO is plan for development and operation.
Source: Ruud (2004).

Oil Resource Base and Future Oil Production

For a given estimate of ultimately recoverable conventional oil resources, many different production profiles can be envisaged. In all cases, production reaches a peak and then declines, but the timing of the peak and the rate at which it is reached differ sharply with different assumptions. The fall will be sharper than was the increase to peak if more than half of ultimate resources have already been produced when the peak is reached. The amount of ultimately recoverable resources is also uncertain, owing to incomplete information and to the possibility of technological improvements that could increase the share of oil that can be recovered commercially. In practice, the production profiles of individual oilfields vary considerably according to the characteristics of the field and how it is developed. Production from the Kingfisher field in the North Sea, for example, rebounded strongly in late 2000 in response to new investment (Figure 3.19).

Figure 3.19: **Oil Production from the Kingfisher Field in the UK North Sea**

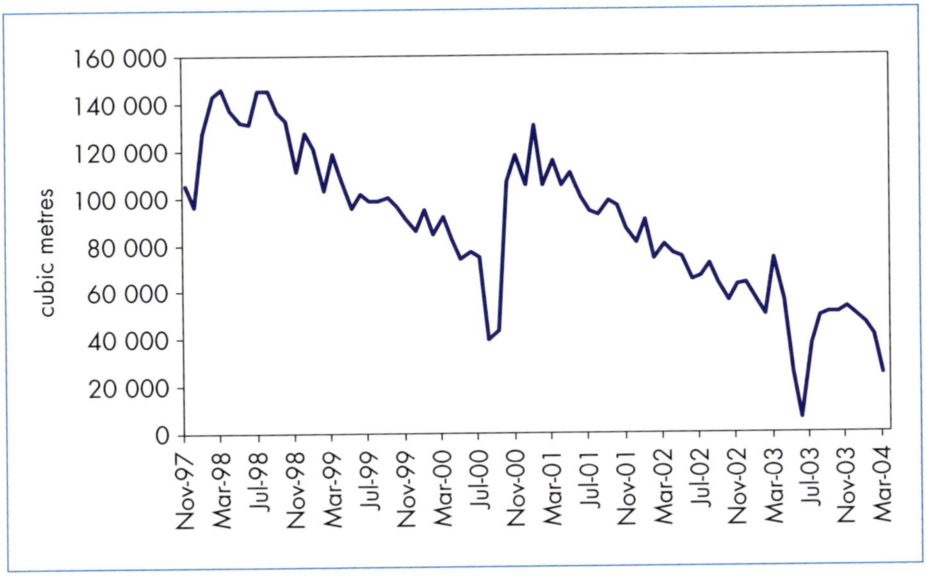

Source: UK Department of Trade and Industry (www.og.dti.gov.uk).

In the Reference Scenario, projected production for each region is derived from the USGS mean resource estimates, which total 3 345 billion barrels for the world. In order to test the sensitivity of production to this assumption, we have developed two cases based on a 10% probability of all the oil being recoverable (the "high resource case") and on a 90% probability (the "low resource case").[5] Oil demand is assumed to grow at slightly different rates in each case, on the assumption that prices change in response to different production levels. The impact on the aggregate production profile of the different assumptions is shown in Table 3.4. The peak of conventional oil production will depend on the demand and production profiles. In the low resource case, conventional production peaks around 2015. In response to higher prices, non-conventional oil production is more than three times higher in 2030 than in the Reference Scenario, making up for part of the sharp decline in conventional production. Non-conventional oil meets just under a third of the world's oil needs at that time. In the high resource case, conventional production does not peak until 2033, and non-conventional output is about 20% lower than in the Reference Scenario in 2030.

Table 3.4: **Impact of Different Oil-Resource Assumptions on Production Outlook**

	Reference Scenario	Low resource case	High resource case
Remaining ultimately recoverable resources base for conventional oil, as of 1/1/1996 (billion barrels)	2 626	1700	3200
Peak period of conventional oil production	2028-2032	2013-2017	2033-2037
Global demand at peak of conventional oil (mb/d)	121	96	142
Non-conventional oil production in 2030 (mb/d)	10	37	8

According to each of the primary data sources described above, proven reserves of conventional oil today are sufficient to meet cumulative world demand in the Reference Scenario, but a lot more oil will have to be "proved up" to prevent production from reaching its peak before 2030. The rate of additions to proven reserves will depend on how much oil is ultimately recoverable. If ultimately recoverable resources are at the low end of the current range of estimates and reserve additions are slower than expected, conventional oil production would peak within the next two decades. The

---

5. This has been done by the estimation of a probability function based on USGS different probability estimates.

sooner conventional oil production peaks, the less time there will be for new technology to raise recovery rates and thus temper the rate of decline in production. On the other hand, if resources turn out to be at the high end of current estimates, world conventional oil production would not peak for almost three decades.

A lack of reliable information on oilfield decline rates is a major cause of uncertainty about the rate at which new capacity will be needed. A high decline rate increases the need to develop reserves and to make new discoveries. Decline rates assumed in our analysis vary over time and range from 5% per year to 11% per year. Rates of decline are generally lowest in regions with the best production prospects and the highest R/P ratios, such as the Middle East, where they range from 4% to 6%. Decline rates are highest in mature OECD producing areas. By 2030, most oil production worldwide will come from capacity that is yet to be built (Figure 3.20).

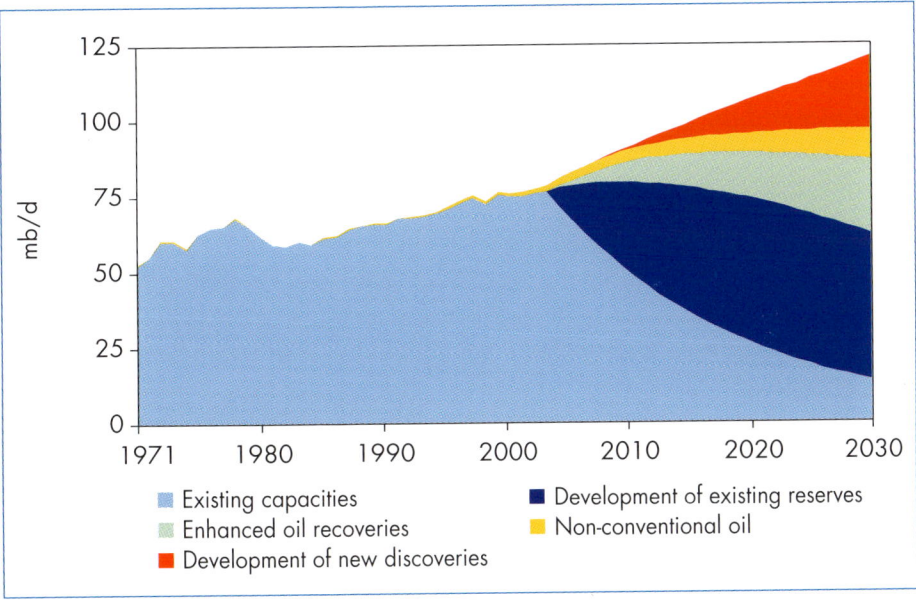

Figure 3.20: **World Oil Production by Source**

The Need for Reserves Data Reform

These uncertainties reinforce the case for stepping up efforts to improve the transparency, consistency and comprehensiveness of data on reserves and resources. The recent series of reserves reclassifications highlighted the urgent need to reform the whole process of reporting reserves. What is needed is a reliable and transparent system to monitor and report on a full range of indicators of future production potential, including reserves and the performance of fields already in production. The reliability and accuracy of reserve estimates is of growing concern for all who are involved in the oil industry.

There are two main issues that need to be addressed:

- Creating a common, universally-accepted set of definitions, methodological procedures and operational practices to determine what constitutes proven reserves. These standards would need to cover different categories of oil resources and ways of estimating how much oil can be recovered from a given field. If all companies were to estimate proven, probable and possible reserves using consistent standards, it would be possible to create a global database. This reform would provide a basis for building a more coherent picture of global reserves and for making comparisons of reserves among companies and countries.

- Collecting, compiling and publishing primary data on national reserves and production *by field*. Information on the actual performance of major fields already in production is critical to understanding production-decline profiles and the prospects for effective action to slow declines in the future. At present, the lack of verifiable data on specific oilfields makes it impossible to assess the quality of data on total reported proven reserves, whether by company or country.

Equity investors have been the most vociferous in demanding greater data transparency, as reserve estimates are a crucial measure of the value of company assets. Uncertainty about reserves may undermine investor confidence and slow investment. The chief executives and finance directors of upstream companies in the United States are demanding more reliable data from their own engineers, as they now have to sign off on official reserve estimates under the Sarbanes-Oxley Act, which was passed in the wake of the Enron collapse and other financial scandals.

Governments are also right to be concerned about reserve-data problems, since the long-term security of energy supplies depends on the development of oil and gas reserves. The greater the uncertainty over whether resources can be developed and at what cost, the less secure those supplies will be. This has profound implications for energy policy. The future availability of hydrocarbons at reasonable cost affects decisions that governments need to make now to encourage the development of alternative sources of energy and to save energy. Governments need to work closely with the oil industry to improve the transparency of reserves data.

Oil Production

Summary of Projections

World oil supply is projected to grow from 77 mb/d in 2002 to 121 mb/d in 2030. Non-OPEC countries will contribute most of the increase in global production over the rest of the current decade. High oil prices have stimulated increased development of reserves in those countries in recent years (Figure 3.21). Despite our assumption that oil prices will fall back after a peak in 2004, production is expected to continue to grow. Global oil production is not expected to peak before 2030, although output in most regions will already be in decline by then. The transition economies, West Africa and Latin America are expected to contribute most of the non-OPEC production increase. Russian output, which has soared in the last six years, would continue to rise, but at a slower rate. It will reach 10.4 mb/d in 2010, compared to about 9 mb/d in early 2004 (see Chapter 9).

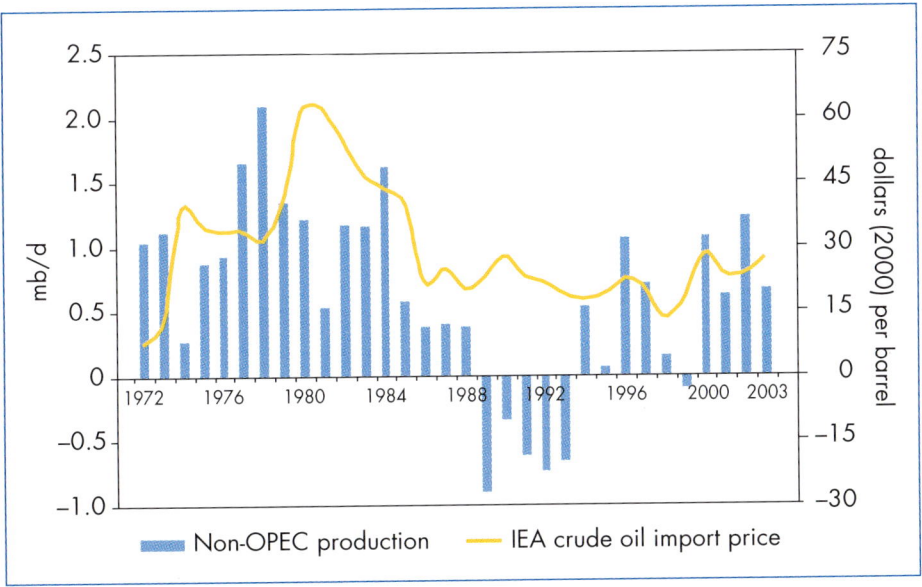

Figure 3.21: **Change in Non-OPEC Production and Average IEA Crude Oil Import Price**

In the longer term, production in OPEC countries, especially in the Middle East, will increase more rapidly because their resources are much larger and their production costs are generally lower than in other regions. OPEC's market share is projected to rise from 37% in 2002 to 53% in 2030, slightly above its historical peak in 1973. Higher oil prices would lower OPEC's market share by stimulating non-OPEC and non-conventional oil production (see High Price Case).

Table 3.5: **World Oil Supply** (million barrels per day)

	2002	2010	2020	2030	2002-2030*
Non-OPEC	45.3	51.3	47.9	43.4	-0.2
OECD Total	21.1	20.1	16.3	12.7	-1.8
OECD North America	13.7	14.8	12.6	10.0	-1.1
United States and Canada	10.1	10.6	8.7	7.2	-1.2
Mexico	3.6	4.2	4.0	2.8	-0.9
OECD Europe	6.6	4.8	3.1	2.2	-3.9
OECD Pacific	0.8	0.5	0.5	0.5	-2.0
Transition economies	9.5	14.6	15.4	15.9	1.8
Russia	7.7	10.4	10.6	10.8	1.2
Other transition economies	1.9	4.2	4.7	5.2	3.7
Developing countries	14.6	16.6	16.2	14.8	0.0
China	3.4	3.3	2.7	2.2	-1.5
India	0.8	0.7	0.6	0.5	-1.6
Other Asia	1.7	1.6	1.2	0.6	-3.4
Latin America	3.7	4.7	5.5	6.1	1.8
Brazil	1.5	2.5	3.3	4.0	3.6
Other Latin America	2.2	2.2	2.2	2.1	-0.2
Africa	3.0	4.6	4.9	4.4	1.4
Middle East	2.1	1.8	1.4	1.0	-2.7
OPEC	28.2	33.3	49.8	64.8	3.0
OPEC Middle East	19.0	22.5	37.4	51.8	3.6
Other OPEC	9.2	10.7	12.4	13.0	1.2
Non-conventional oil	1.6	3.8	6.5	10.1	6.7
of which GTL	0.0	0.4	1.5	2.4	16.0
Processing gains	1.8	2.0	2.5	3.0	1.9
World	77.0	90.4	106.7	121.3	1.6

* Average annual growth rate.

Conventional Oil Production Prospects by Region
Non-OPEC Countries

Oil production in *OECD North America* is expected to pick up slightly in the next few years, mainly thanks to a surge in output in the Gulf of Mexico. But, starting in the 2010s, the region is expected to resume its long-term downward trend, as output in Alaska, Western Canada and the lower 48 states tails off. Older fields in the Gulf of Mexico will also peak soon. New fields in ultradeep offshore waters will not be able to compensate for this decline. The lifting of a moratorium on drilling in the Arctic National Wildlife Refuge – not assumed in the Reference Scenario because Congress has not yet approved it – could temper the decline. Similarly, higher output from the Alaskan Naval Petroleum Reserve west of Prudhoe Bay may be possible if environmental difficulties can be overcome.

Mexico's proven oil reserves have fallen continuously over the past 19 years, because of under-investment in exploration. With the exception of some foreign oil services companies, no exploration permits have been awarded to companies other than Pemex, the national oil company. In 2003, Mexico produced around 3.8 mb/d of oil – 6% more than in the previous year. Most of the increase came from Mexico's main oilfield, Cantarel. In 2003, Mexico consumed slightly more than 2 mb/d of oil, and exported 1.8 mb/d, mostly to the United States. Mexican crude oil production is projected to peak at 4.2 mb/d around 2010, and then to remain almost flat during the 2010s. It is then projected to decline sharply, to 2.8 mb/d in 2030.

In *OECD Europe*, production from mature North Sea basins is already in long-term decline. Output is projected to drop to 4.8 mb/d by 2010, about three-quarters of current production, and dwindle to 2.2 mb/d in 2030. Despite higher oil prices, European exploration and development have declined sharply in the last few years, as prospects have deteriorated. Several of the leading international oil companies are shedding assets in the region. Independent operators are taking a bigger role in exploiting remaining reserves in mature fields and developing smaller fields which would otherwise not have been considered. However, production in the long term may be undermined, as the new entrants sometimes lack experience in dealing with the technical problems of ageing fields, and do not have the financial and technical clout of their predecessors in exploiting new prospects. Exploration activity in Norway has diminished recently, partly because of uncertainties over environmental restrictions on drilling in the Barents Sea.

At the end of 2003, *Australia's* oil reserves were 3.5 billion barrels.[6] Prospects for finding more oil, especially in Western and Southern Australia, are good. However, Australia's oil production has fallen from a peak of 0.8 mb/d in 2000 and is expected to continue to fall in the coming decades. The country will, therefore, depend more and more on imports.

In *Russia*, production has risen much more sharply than expected in the last two years. Output growth is expected to slow abruptly in the near term, but will continue to expand in the longer term. Almost all of the recent production increase came from established producing areas in western Siberia

---

6. All the reserve figures in this section are from the *Oil and Gas Journal* and refer to the end of 2003.

and the Volga-Urals region. These areas will still account for more than 80% of total supply in 2010, but new supplies will come increasingly from greenfield projects in Timan Pechora, Sakhalin, East Siberia and the Caspian region. Long-term production prospects for these regions will depend to a large extent on government licensing and fiscal policies, as well as on how quickly investment in pipeline and export infrastructure is forthcoming (see Chapter 9).

In the *other transition economies*, new export routes will allow production growth to accelerate. These new facilities include expansions to the Caspian Pipeline Consortium (CPC) pipeline that runs from Kazakhstan through Russia to Ukraine; its capacity will rise to over 1 mb/d by the end of the current decade. When completed, the 900 kb/d Baku-Tbilisi-Ceyhan pipeline will bring Azeri oil across Georgia to the deep-water port of Ceyhan on Turkey's Mediterranean coast. Northward shipments via the Russian Transneft system and increased Caspian/Iranian oil swaps may also be possible. In the longer term, a pipeline to China could be built. In Kazakhstan, there is potential for production growth at the three main existing fields, Tengiz, Kashagan and Karachaganak. Combined output there could reach 1 mb/d by 2010. In Turkmenistan, production doubled between 1997 and 2003 to 207 kb/d. The country's proven reserves are only half a billion barrels, compared to Kazakhstan's 9 billion barrels. But its ultimately recoverable resources are 7 billion barrels. So there is significant potential for boosting Turkmen production in the longer term; it could triple by 2030.

New oilfield developments in *China* will not be sufficient to offset declines in mature onshore areas, including the super-giant Daqing field, the world's fourth-largest, which has been in production since the 1960s. Surging imports and high oil prices have stimulated some production increases in recent years. Production at five major offshore fields – Peng Lai, Cheng Dao Xi, Zhou Dong, Panyu and Caofeidan – is scheduled to increase from 50 kb/d in 2003 to 300 kb/d by 2007, dipping to around 250 kb/d by 2010. Production from the Changqing area, Tarim Basin, and Xinjiang are expected to remain flat or grow modestly in the near term. Chinese production will fall to around 2.2 mb/d in 2030. In 2004, a huge oil and gas field was discovered in the Xifeng Gansu province. It is the largest oilfield found in China in the last ten years. Proven reserves are estimated at about 800 million barrels, though this figure may be revised upwards. Chinese authorities announced a major new find in April 2004 in the area of the existing Shengli field, China's second-largest, in the north-east. Reserves are still being assessed.

Output will be broadly flat in the rest of *Asia* through to the end of the current decade, and will decline thereafter. Some new developments are expected in Vietnam, Malaysia, India and Thailand, but output will decline in most other producing countries. Rehabilitation of the Bombay High field in India will boost output, while production from the Kikeh and Guntong projects in Malaysia is expected to rise by 175 kb/d. Production of natural gas liquids is also expected to grow. In Vietnam, production from the offshore Rang Dong, Dai Hung and Ruby fields will be supplemented by new supplies from the Su Tu Den and Su Tu Vang fields, which together are scheduled to yield an average of 150 kb/d by 2010. In Thailand, liquefied petroleum gas and ethane from gas-processing plants will account for much of the increase in oil supply. The oil-production plans of several Thai producers, including Unocal, are now markedly more optimistic than was previously the case.

Our overall projection for production in *Africa* in 2010, at around 4.6 mb/d, is close to that of *WEO-2002*. Prospects for Angola, Chad, Congo and South Africa have deteriorated, but Egypt, Ivory Coast, Equatorial Guinea and Sudan are now expected to produce more in 2010. In Angola, production will grow, driven mainly by new deep-water fields. These include Jasmim, Xikomba, Kizomba, Dalia, Plutonio, and Belize. Increments from as many as six other fields are expected from 2008. However, higher decline rates than expected at already producing fields could slow the expected overall increase in production. The production plans of Exxon-Mobil in Equatorial Guinea and China National Petroleum Corporation in Sudan are markedly more ambitious than two years ago. In both cases, however, further production growth will depend on the success of exploration and appraisal drilling in firming up reserves. With the completion of a pipeline through Cameroon, production from the Doba Basin in Chad has risen to over 200 kb/d in 2004. The basin holds around 900 million barrels of reserves. More extensive exploration is likely to boost Chad's reserves and production potential in the longer term.

Production in *Latin America* is expected to grow from 3.7 mb/d in 2003 to 4.7 mb/d in 2010 and then to 6.1 mb/d in 2030. Brazil, the region's largest non-OPEC producer, has proven oil reserves of 8.5 billion barrels and a huge potential for further discoveries. According to the USGS, undiscovered resources amount to 55 billions barrels. The country will account for much of the increase in oil production in Latin America over the projection period. Brazilian production, which reached 1.7 mb/d at the start of 2004, is projected to reach 4 mb/d in 2030. Brazil is currently an oil importer, but will soon become self-sufficient.

OPEC Countries

OPEC crude oil supply, which is assumed to meet the portion of global oil demand not met by non-OPEC producers, will need to increase from 28 mb/d in 2002 to 33 mb/d in 2010 and to 65 mb/d in 2030. These projections assume that there are no major disruptions in supply and that current high oil prices will not be sustained. A special analysis of the effect of a much higher oil price is presented at the end of this chapter.

The strong projected growth in OPEC production – particularly in the second half of the projection period – will boost OPEC's market share significantly. In the near term, the cartel's share, which currently stands at 36%, will remain roughly stable owing to rapid production increases in several non-OPEC regions, notably Russia and other transition economies (Figure 3.22). As prices return to a level closer to the average of the last two decades, incentives to raise output in non-OPEC regions will diminish, increasing the call on oil from OPEC producers. The second and third decades of the projection period will see more rapid growth in OPEC's market share. By 2030, OPEC production will reach 65 mb/d, or 53% of world oil supply.

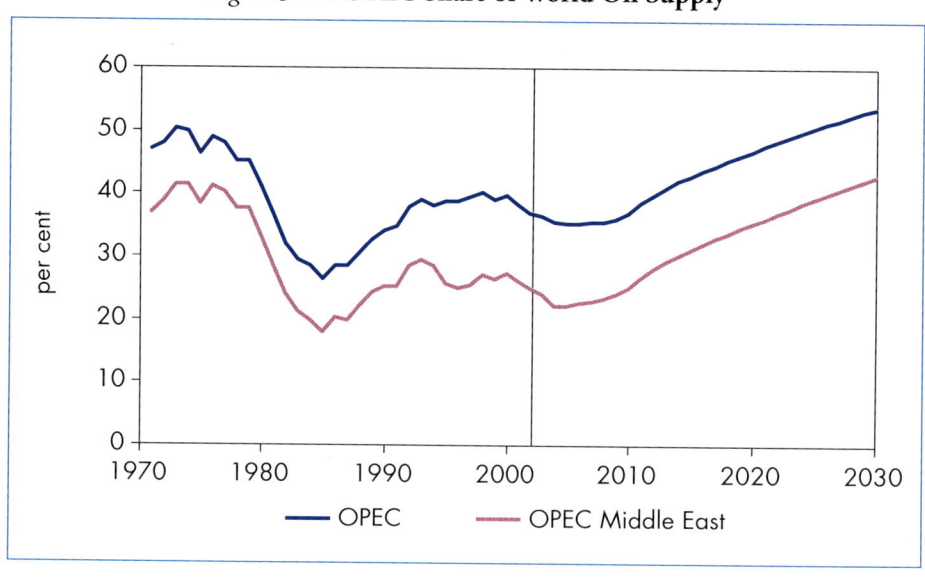

Figure 3.22: **OPEC Share of World Oil Supply**

Within OPEC, Middle East countries will account for most of the production increase. Of the projected 31-mb/d rise in world oil demand between 2010 and 2030, 29 mb/d will come from OPEC Middle East. This region holds about 60% of the world's proven oil reserves. Production costs there are among the lowest anywhere in the world, averaging less than $2 per barrel in the Arab Gulf countries. Investment costs are also very low, at less than $5 000 per barrel/day of capacity. Saudi Arabia, Iraq and Iran are likely to contribute most of the increase in Middle East production.

Saudi Arabia will continue to play a vital role in the global oil-market balance. Its willingness and ability to make timely investments in oil-production capacity will be a major determinant of future price trends. Saudi Arabia has produced 9.5 mb/d in August 2004.[7] There are some 15 fields in production, including a large number of super-giant fields. Seven of these have 7.25 mb/d of capacity: Ghawar, Abqaiq, Shaybah, Safaniyah, Zuluf, Berri, and Marjan. Approximately 80% of Saudi capacity is Arab Light and Extra Light oil. The Ghawar field – the world's largest – contributes 5 mb/d all by itself. Almost half Ghawar's 115 billion barrels of reserves have already been extracted. As the field matures, costs will undoubtedly increase and sustaining production will become more difficult. Despite the problems of ageing, however, recent work at Ghawar has reduced the water cut there from 36.5% to 33% and added oil and gas reserves (Baqi, 2004).

Maintaining and expanding Saudi capacity will require more investment in the future. Saudi Aramco, the national oil company, has estimated that the natural rate of decline at existing fields will be of the order of 6% over the next five years, so that some 600 kb/d of capacity will have to be replaced each year just to maintain overall capacity over that period. This will be achieved mostly through enhanced development of existing fields, which have been managed very conservatively over several decades in order to extend their plateau production for as long as possible. But as these mature fields age, their production will decline slowly, and new fields will have to contribute an increasing share of production. There has been little exploration effort in recent years and much of the country is unexplored, including the region close to the border with Iraq, the Red Sea and the Empty Quarter in the south-east.

At present, 70 fields, with 50% of the country's proven reserves, await development. Saudi Aramco estimates that it could raise production capacity to 12 mb/d and maintain it at that level until 2033 without finding any additional reserves. The company believes that future investments can be financed solely out of its own cash flow.

Saudi Arabia plays a central role in OPEC and in balancing the world market, not only because of the size of its production but also because of its spare capacity. In mid-2004, it was the only country in the world with an appreciable amount of sustainable capacity in reserve. Much of this capacity is in three offshore fields – Safaniyah (with a capacity of 1 mb/d), Zuluf (200 kb/d) and Marjan (300 kb/d) – all of which produce medium or heavy crude oil. Saudi Arabia is expected to remain the primary source of spare capacity. The country's official policy is to maintain from 1.5 to 2 mb/d of spare capacity for the foreseeable future. Saudi Arabia will undoubtedly account for a major share of the increase in Middle East production.

---

7. Including half of Neutral-Zone production.

The near-term prospects for oil production in *Iraq* remain very uncertain. Output has recovered since the war in 2003, but not as quickly as expected. Disruptions caused by acts of sabotage to vital facilities have become more frequent in recent months. In May 2004, production fell back to 2.1 mb/d, well below Iraq's estimated sustainable capacity of 2.5 mb/d, after an attack on the pipeline feeding the Basra Oil Terminal. Iraqi industry sources now target exports of 2.2 mb/d to 2.3 mb/d by the end of 2004.

Iraqi production has always been low relative to the size of the country's reserves. Iraq holds the world's third-largest remaining oil reserves, after Saudi Arabia and Iran, with 115 billion barrels, according to OPEC data. Some analysts estimate that exploration in the largely unexplored Western Desert could lift proven Iraqi reserves to 180 billion barrels. Iraqi production peaked in 1979 at 3.7 mb/d, of which 3.5 mb/d was exported. Production fell dramatically at the beginning of the 1980s as a result of the Iraq-Iran war and has barely recovered since then. Iraq has the potential to increase its production capacity by several millions of barrels per day. How quickly it will be able to do so will depend on three key factors:

- How effective the recently appointed Iraqi government proves to be and how successful it will be in restoring law and order.
- The opportunities for foreign oil companies to invest in Iraq, and the commercial and fiscal terms they might be offered.
- OPEC policy towards Iraq and Iraq's willingness to accept production quotas. Iraq is still a member of OPEC, but it has not been bound by any OPEC quotas since it resumed oil exports in December 1996.

Crude oil production potential in *Iran* is also very large. Output increased to 3.8 mb/d in 2003, making Iran the second oil producer in OPEC and the fourth-largest exporter in the world. But Iranian production remains far from its historical peak of 6 mb/d in 1974. Current sustainable capacity is estimated at around 4 mb/d. Maintaining this capacity will require large investments in existing fields, where decline rates are thought to be rising, and in new fields as well. Proven reserves are 126 billion barrels.

Developing new projects remains difficult in Iran, because the National Iranian Oil Company (NIOC) is strapped for cash and foreign participation is restricted. International oil companies have been able to invest in Iranian oil projects only under buy-back contracts. US companies are not allowed to invest at all under sanctions imposed by the US Congress in 1995. A new type of contract was launched by NIOC in 2002 offering combined exploration and development projects. These contracts are limited to areas in the north and west of the country and in southern offshore zones, but the area will be extended if these contracts prove successful in stimulating investment. An ambitious plan prepared by the Ministry of Petroleum in November 2003 sets a production target of 5 mb/d for 2010 and 8 mb/d for 2020. These goals could be met in several ways: by increasing oil recovery in mature fields through gas injection; by developing new fields in the Azadegan region; or by exploiting recently discovered fields around Bushehr. There is also good offshore potential in the Persian Gulf and Caspian Sea. Iranian production could double by 2030.

The *United Arab Emirates* has proven crude oil reserves of 98 billion barrels, around 10% of the world total. Crude oil production at the beginning of 2004 was 2.25 mb/d, some 250 kb/d below the Emirates' total production capacity. Only Saudi Arabia has more spare capacity than the UAE at present. The UAE plans to increase production capacity to 3 mb/d by the end of 2006. Several projects are under way or are planned to expand capacity at existing oilfields.

Outside the Middle East, *Venezuela* is the OPEC country with the largest reserves and the greatest potential for expanding production. But boosting crude oil capacity from its current level of 2.25 mb/d will require large amounts of investment. A significant share of that investment will be needed simply to maintain existing capacity, which has been neglected. PDVSA, the Venezuelan national oil company, has an ambitious production target of 5 mb/d in 2009. However, the political situation has remained tense since the oil-industry strikes in 2002, which severely disrupted operations. The Hydrocarbon Law, which came into effect at the start of 2002, opened up the country's oil industry to private investment, but it is unclear whether the terms and conditions will be acceptable to private investors in view of the large risks involved.

Nigeria's proven oil reserves are estimated at 25 billion barrels. The country produced 2.15 mb/d of crude oil in 2003, about 0.4 mb/d under capacity. The start-up of the Bonga, Ehra, Agbami Akbo and Amenam deep-water offshore fields will boost capacity to around 3 mb/d by the end of 2004. The federal government has set a goal of 40 billion barrels of proven reserves and 4.1 mb/d of production capacity by 2006. But securing the necessary finance to accomplish these goals will be difficult. Investment risk in Nigeria is high, because of political instability and corruption. Nigeria is seeking an increase in its share of the OPEC production ceiling.

Libya has proven oil reserves of 36 billion barrels, even though exploration in the country has been minimal. Libya is considered a highly attractive oil province because production costs are low, it is close to Europe and it has well-developed infrastructure. Libya produces high-quality, low-sulphur crude oil. Costs are as low as $1 per barrel at some fields. In September 2003, the United Nations lifted economic sanctions against Libya, a move which could pave the way for opening up the upstream industry to foreign investment. Libya plans to increase its production capacity from 1.6 mb/d now to 2 mb/d by 2010. Further increases are possible in the longer term.

Non-conventional Oil Production Prospects

Total non-conventional oil production is projected to grow from 1.6 mb/d in 2002 to 3.8 mb/d in 2010 and 10.1 mb/d in 2030. By the end of the projection period, it will make up 8% of global oil supply. Non-conventional production technologies are already economic in some locations at the prices assumed in the Reference Scenario and will remain so throughout the *Outlook* period. Production gains will come primarily from synthetic crude derived from oil sands in the Canadian province of Alberta, and from the Orinoco extra-heavy crude oil belt in Venezuela. Canadian oil-sands production has become much more competitive in recent years, especially from mining and upgrading projects.[8] In-situ recovery techniques involve the introduction of heat, normally via steam, into the oil sands to allow the bitumen to flow to well bores and then to the surface. The cost of production from such projects has been less than $10 per barrel for the last decade and a half (Figure 3.23).

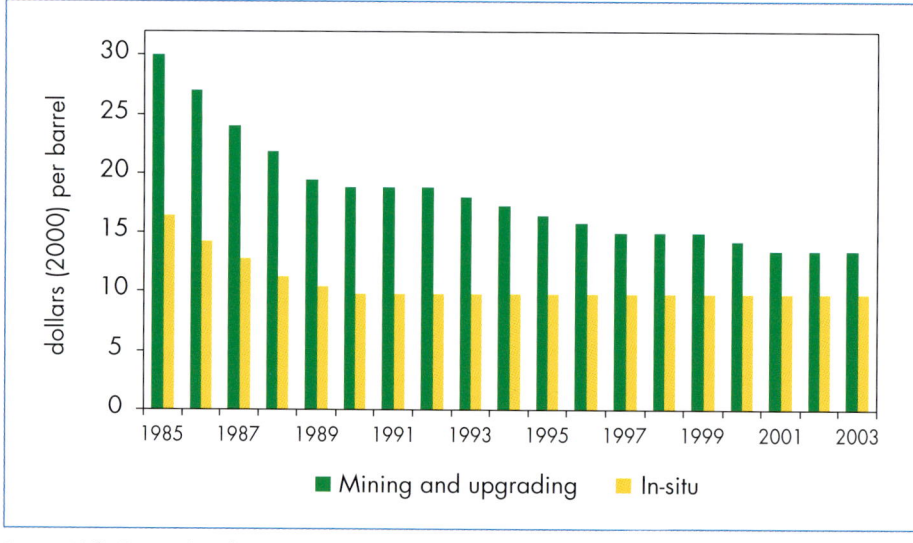

Figure 3.23: **Canadian Tar Sands Oil Production Cost**

Source: Utilis Energy (2004).

Planned additions to capacity in Venezuela and Canada will boost output by 2 mb/d by 2010 and by a further 1.8 mb/d in the 2010s (Table 3.6). In this *Outlook*, total capacity is projected to be just under 6 mb/d in 2030. The development of Canadian tar sands may be held back by the large amounts of natural gas that will be needed to produce steam.

---

8. These projects involve extracting raw bitumen from mined tar sands and upgrading it into a high-quality synthetic oil.

Table 3.6: **Planned Additions to Non-Conventional Oil Production Capacity in Canada and Venezuela** (mb/d)

	By 2010	After 2010
Canada		
Athabasca oil sands area mining projects	0.95	1.51
Athabasca in-situ projects	0.38	0.24
Cold Lake in-situ projects	0.14	0.00
Peace River in-situ projects	-	0.02
Venezuela		
Orinoco belt	0.18	-
Carribean coast	0.40	-
Total	**2.05**	**1.77**

Sources: National Energy Board (2004); Energy Intelligence Group (2004).

The rest of the increase in non-conventional production will come mainly from gas-to-liquids plants. GTL production is expected to grow very quickly, reaching 2.4 mb/d in 2030. Production from oil shales is also expected to grow. Biofuels derived from agricultural products and coal-to-liquids will also increase (see Chapter 7 and 5 respectively). Their share in world non-conventional oil supply will, however, remain small.

Inter-Regional Oil Trade

International oil trade will increase sharply during the projection period. Net *inter-regional* trade[9] will reach 65 mb/d in 2030 – over half of global oil production and more than twice as much as at present. This trend results from the steady growth in demand in all regions and the increasing concentration of production in a small number of countries. The Middle East, already the biggest exporting region, will see its net exports rise from 17 mb/d in 2002 to 46 mb/d in 2030 (Figure 3.24). Exports from Africa, Russia and other transition economies will also continue to expand steadily in the short to medium term, but all of them will have started to decline by 2020. Caspian countries will experience the fastest growth: their exports will rise from 1 mb/d in 2002 to 4 mb/d in 2030. Despite rapidly growing exports from Venezuela, net exports from Latin America as a whole will increase only modestly until 2010 and will stabilise thereafter.

9. Includes trade between main *WEO* regions only. Total international trade is considerably higher.

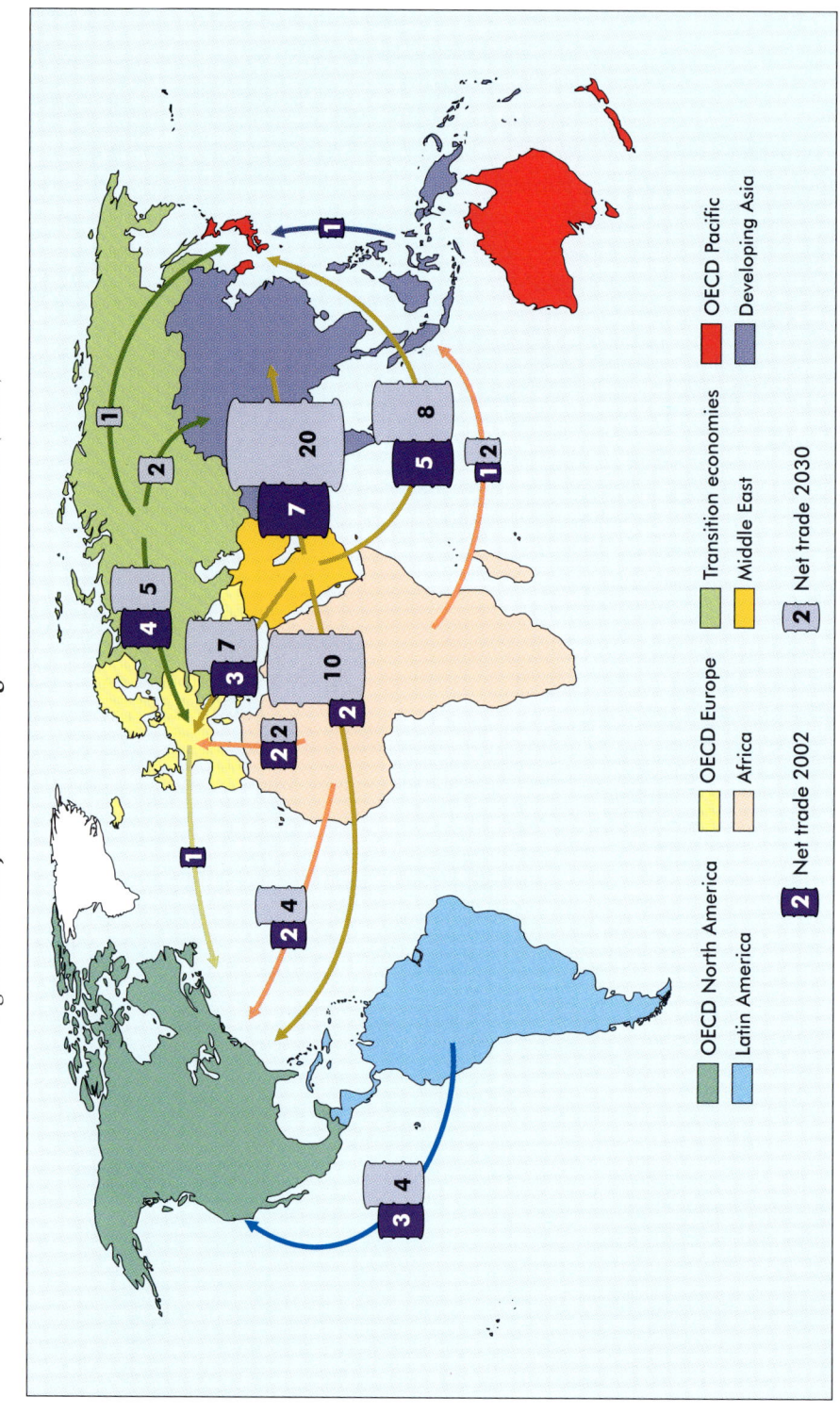

Figure 3.24: Major Net Inter-Regional Oil Trade Flows (mb/d)

The increase in trade will be particularly marked after 2010, reflecting the growing share of the Middle East in world oil supply. At 46 mb/d, exports from the Middle East will represent more than two-thirds of global trade in 2030. The region will increase its exports to all the major consuming regions, especially developing countries. Flows to developing Asian countries will increase the most.

The average distance over which oil will be transported will increase. Consequently, oil tankers will remain the main form of oil transport: pipelines will generally remain competitive only where they can be built onshore. The oil-tanker fleet is projected to expand by nearly 90%, to about 500 million dead-weight tonnes, by 2030. In addition to increasing capacity to meet demand growth, new tankers will be required to replace old ships that will have to be scrapped to meet the requirements of international pollution-prevention regulations adopted in recent years.

All the net oil-importing regions — notably the three OECD regions and developing Asia — will become even more dependent on imports over the projection period (Table 3.7). Imports as a share of total oil demand will jump from 63% in 2002 to 85% in 2030 in the OECD as a whole. The OECD Pacific region will remain the most dependent on imports, at 95% in 2030 compared with 90% currently. The increase in dependence will be most dramatic in China, which only became a net oil importer in 1993. By 2030 imports will meet 74% of China's oil demand. This is equal to 10 mb/d, the current volume of imports into the United States.

Table 3.7: **Oil-Import Dependence in Net Importing Regions** (%)

	2002	2010	2020	2030
OECD total	63	68	79	85
OECD North America	36	35	47	55
OECD Europe	54	68	80	86
OECD Pacific	90	94	94	95
Developing Asia	43	59	72	78
China	34	55	68	74
India	69	80	87	91
Other Asia	40	54	68	76
European Union	76	85	91	94

Note: Imports include non-conventional oil.

Increased trade, especially from the Middle East, will strengthen the mutual dependence among exporting and importing countries. But it will also intensify worries about the world's vulnerability to oil-supply disruptions, as much of the additional trade will involve transport along routes that are at risk of sudden closure. Some of the principal maritime routes that oil and LNG tankers are obliged to follow have narrow sections that are susceptible to piracy, terrorist attacks or accidents. The world will also rely more on oil shipped through pipeline systems, which are also vulnerable to accidental or deliberate disruptions. The main strategic oil transportation channels through which much of the world's oil flows are:

- The Straits of Hormuz, at the mouth of the Persian Gulf. This is the world's most important maritime oil-shipping route, through which more than 15 mb/d currently flows. Inbound and outbound lanes are around 3 km wide with a 3-km buffer between them. Only a small proportion of the oil shipped through the Straits could currently be transported along other routes.

- The Straits of Malacca, between Indonesia, Malaysia and Singapore. This is the principal oil route in Asia, handling 11 mb/d. At its narrowest point, it is only 2.5 km wide. Piracy and accidents have already disrupted shipping at times. A major blockage would force tankers to take much longer routes. Rising demand in China and other East Asian countries will lead to a substantial increase of traffic through the Straits, boosting its strategic importance.

- The Suez Canal, connecting the Red Sea with the Mediterranean. The canal can currently handle 1.3 mb/d. Its closure would force tankers to take the much longer route around the southern tip of Africa.

- The Sumed pipeline, linking the Red Sea with the Mediterranean. This two-line system has a capacity of 2.5 mb/d.

- Bab el-Mandab passage, connecting the Red Sea with the Gulf of Aden. Around 3.3 mb/d is currently shipped through the passage en route to the Suez Canal and the Sumed pipeline. In 2002, an attack on the French tanker, Limburg, off the coast of Yemen highlighted the importance of this chokepoint to world oil supplies.

- The Bosporus/Turkish Straits. This is a narrow 30-km-long waterway that connects the Black Sea with the Mediterranean. Oil traffic – mostly crude oil from ports on the Black Sea – is currently around 3 mb/d. The Straits are less than 1 km wide at the narrowest point. Although commercial shipping has the right of free passage under the 1936 Montreux Convention, the Turkish authorities have imposed restrictions on oil-tanker transit for safety and environmental reasons.

- Other vital oil-transport routes include the Panama Canal (0.4 mb/d), the Druzhba pipeline, through which Russian crude oil flows to Europe (1.2 mb/d); and the Baltic Pipeline System, which carries Russian crude to Baltic Sea ports (1 mb/d by late 2004).

Today, more than 35 million barrels pass through the above channels every day. A disruption in supply at any of these points could have a dramatic impact on oil prices, especially if oil supplies were already very tight. As global oil trade expands, these strategic channels will become even more heavily used. Traffic through the Straits of Hormuz and Malacca and the Suez Canal is projected to more than double over the projection period (Table 3.8). The share of oil imports through the Straits of Malacca in total world oil demand will grow from 14% in 2002 to 20% in 2030 (Figure 3.25). LNG trade through these channels will rise even more.

Table 3.8: **Oil and LNG Tanker Traffic through Strategic Maritime Channels**

		2002		2030	
		Volume oil (mb/d) gas (bcm)	Share of global inter-regional net trade (%)	Volume oil (mb/d) gas (bcm)	Share of global inter-regional net trade (%)
Straits of Hormuz	Oil tankers	15	44	43	66
	LNG carriers	28	18	230	34
Straits of Malacca	Oil tankers	11	32	24	37
	LNG carriers	40	27	94	14
Suez Canal	Oil tankers	1	4	3	4
	LNG carriers	4	3	60	9

Sources: DOE/EIA (2004); IEA analysis.

Investment Outlook

Cumulative global investment[10] in the oil industry will amount to around $3 trillion over the period 2003-2030, or around $105 billion per year, on the basis of the production projections presented in this *Outlook*. Capital spending will have to increase steadily through the period as existing infrastructure becomes obsolete and demand increases. Investment in OECD countries will be high relative to their production capacity, because their unit costs and decline rates are higher than in other regions. Exploration and development will dominate oil-sector investment, accounting for over 70% of

---

10. Investment needs have been calculated using the methodology outlined in the *World Energy Investment Outlook* (IEA, 2003) and on the basis of *WEO-2004* demand-supply projections and recent market developments.

Figure 3.25: Oil Flows and Major Chokepoints, 2003

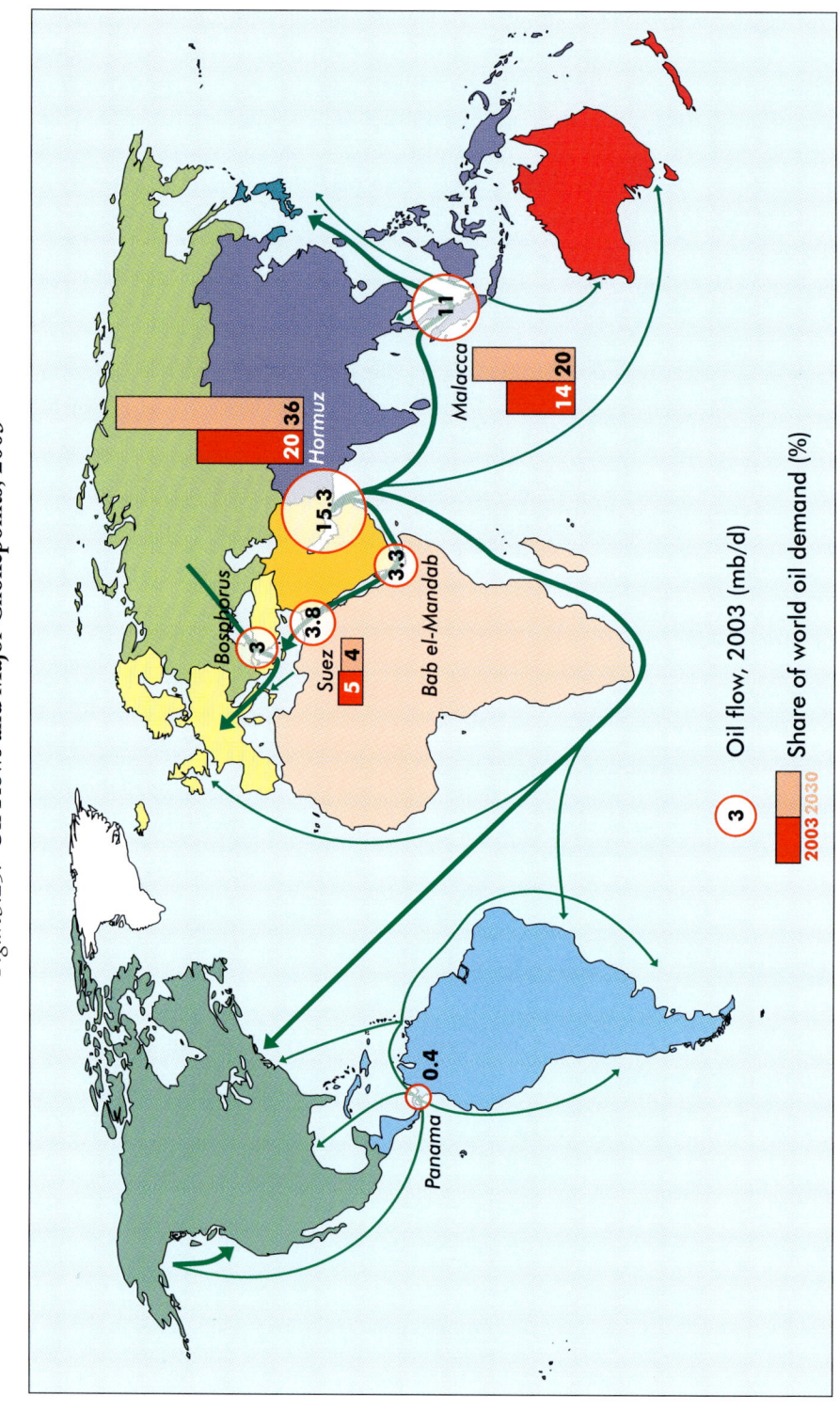

World Energy Outlook 2004

the total over the period 2003-2030 (Figure 3.26). Only a quarter of upstream investment will go to meet rising demand. The rest will be used to make up for the natural decline in production from wells already in production and those that will start producing in the future. At a global level, investment needs are, in fact, far more sensitive to changes in decline rates than to the rate of growth of oil demand. Investment in oil tankers and oil pipelines for international trade will amount to $234 billion from now to 2030. Supply chains will lengthen, so most of the investment in oil transport will be in tankers rather than pipelines.

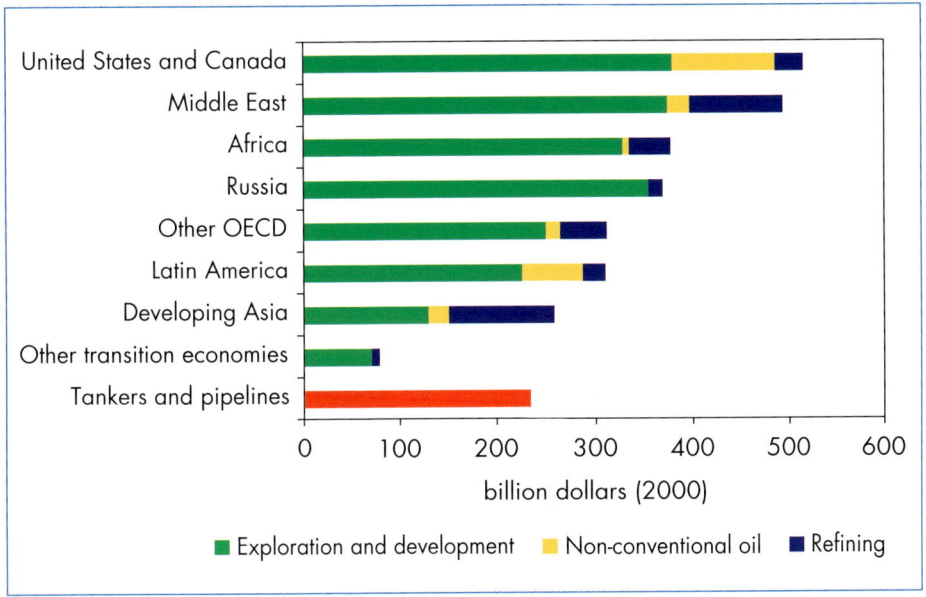

Figure 3.26: **Cumulative Global Oil Investment, 2003-2030**

Although the oil-investment needs we project through to 2030 will be large and will rise progressively, capital will be available to meet them. That does not mean, however, that all those investments will be made or that capital will flow to where it is needed. Several factors could discourage or dry up investment in particular regions or sectors. In other words, investment is more likely to be limited by a lack of profitable business opportunities rather than by any absolute shortage of capital.

Oil prices will play a key role in attracting investment to the sector. In recent years, upstream global oil and gas investment has tended to fluctuate with changes in oil prices. The openness of countries with large oil resources to foreign direct investment will be another important factor in determining how much upstream investment occurs and where. Today, three major oil-producing countries – Kuwait, Mexico and Saudi Arabia – remain totally closed to outside investment. Access to many others, such as Russia and Iran, is restricted.

Implications of High Oil Prices
Background and Assumptions

Crude oil and refined product prices have risen dramatically in recent years. By August 2004, crude prices in nominal terms were more than $30 per barrel higher than they were six years before. Forward prices several years out have also risen sharply, suggesting a profound shift in the industry's perception of the medium-term supply/demand balance and of OPEC pricing and production policies. In line with many other analysts, however, we assume in our Reference Scenario that the prices reached in mid-2004 are unsustainable and that market fundamentals will drive them down in the next two years. But a certain combination of factors could keep oil prices high in the years to come. These include:

- *Under-investment in supply infrastructure:* Restrictions on foreign access to resources or unattractive fiscal terms could limit investment in oil-producing countries and thus, the expansion of production capacity. National oil companies may identify more profitable uses of capital than investment in exploration and production. Producer governments might also deliberately limit investment in new production capacity. Such a policy could be intended to drive up international prices and boost revenues in the short term, or to preserve resources for future generations.[11]

- *Strong demand-side pressures:* Stronger-than-expected economic growth in Asian countries would drive up oil demand, especially for lighter products. The situation could be exacerbated by continuing tightness in North American gasoline markets.

- *Lack of resource availability:* If oil reserves prove to be smaller than current estimates or more difficult to recover than expected, prices could be higher.

- *Geopolitical factors:* Political instability, particularly in the Middle East, or disruptions in supplies at key chokepoints.

In view of the uncertainty surrounding oil prices, we have carried out a separate analysis to examine the effects of high oil prices on world oil supply and demand, OPEC revenues and global oil investment. In this analysis, the average IEA import price is assumed to average $35 per barrel in year-2000 dollars over the

---

11. See IEA (2003) for an analysis of the implications of such a policy.

projection period. To reflect inter-fuel competition in end-use markets, natural gas prices are also assumed to remain high. Coal prices are assumed to be unchanged.

Results

Impact on Oil Demand

In the High Oil Price Case, world oil demand in 2030 would be 19 mb/d, or 15%, lower than in the Reference Scenario. This amount is almost equal to US oil demand today (Figure 3.27). A sustained higher oil price would choke off energy demand generally and would prompt switching from oil to other fuels. Higher prices would induce behavioural changes – consumers would reduce energy waste and use fewer energy services – and promote the diffusion of more energy-efficient technologies. Demand would still rise over the projection period, at 1% per year, but more slowly than in the Reference Scenario (1.6%).

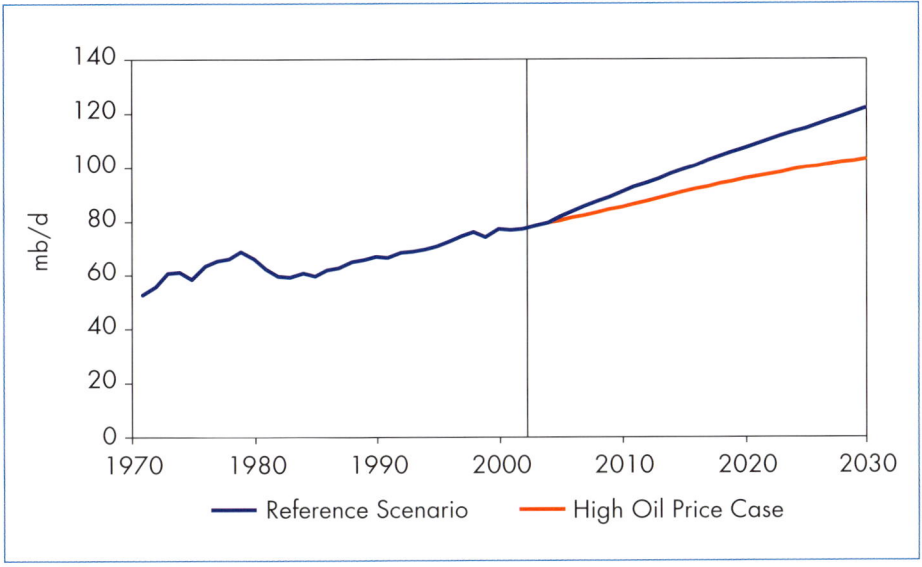

Figure 3.27: **World Oil Demand in the Reference Scenario and High Oil Price Case**

The magnitude of these effects would vary among regions. Demand would fall most sharply in developing countries, by 17% in 2030 compared to the Reference Scenario. Demand in OECD countries would drop by only 14%, because these countries use oil mostly for transportation. Moreover, the effect on demand of very high oil prices is blunted in OECD countries, and especially in Europe, by heavy taxation on gasoline and diesel. The assumed increase in the price of crude oil translates into a much smaller percentage increase in the price of fuel at the pump in most OECD countries. In the High Oil Price Case, demand in developing countries falls more steeply, partly because the share of transport in their oil use is slightly lower (48% in 2002, against 56% for OECD countries). In oil-producing countries, very high prices have little short-term effect on oil demand and on energy demand in general. High revenues from oil sales boost national income and spending, and offset the downward effect of shrinking volumes of sales. In any event, end-user prices in many of these countries are heavily subsidised.

Despite the lower price elasticity of oil demand in transport than in other sectors, the fall in oil demand comes mainly from the transport sector in all regions (Figure 3.28). Nearly two-thirds of the difference between the Reference Scenario and the High Oil Price Case is explained by weaker transport demand. This proportion rises to 70% in OECD regions. Oil use in power generation is hardly affected. In the OECD, few power stations still use oil, and those that do usually meet peak demand, which is relatively insensitive to price.

Figure 3.28: **Cumulative Reduction in Oil Demand in the High Oil Price Case* by Region and Sector, 2002-2030**

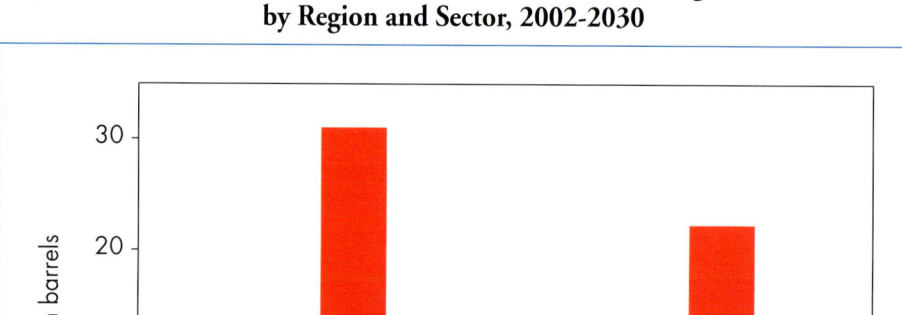

* Compared with the Reference Scenario.

Impact on Oil Supply

The drop in world oil demand that would result from persistent higher oil prices would lead to an equivalent fall in world production. The impact differs substantially among regions. Production in OPEC countries, which are assumed to be the residual source of supply, would be 38% lower in 2030 compared to the Reference Scenario. This effect would result from three main factors:

- Global oil demand would be 19 mb/d lower.

- OPEC production would face increasing competition from conventional oil production in non-OPEC regions. Non-OPEC production would be about 4 mb/d higher in 2030 than in the Reference Scenario. Higher oil prices would encourage the development of reserves outside OPEC, as marginal fields become commercial.

- Non-conventional oil production would increase even more rapidly than projected in the Reference Scenario. It is 15% higher in 2030. As a result, the share of non-conventional oil production in total world oil supply rises from 2% in 2002 to more than 11% in 2030.

The impact on non-OPEC regions is bigger in the short term, as reserves are developed more quickly. In the medium term, non-OPEC conventional oil production would fall more quickly than in the Reference Scenario, as it would reach its peak sooner and its reserves would be depleted more quickly. By the end of the projection period, OECD production is slightly lower than in the Reference Scenario (Table 3.9). As a result of lower oil demand and higher non-OPEC supply, OPEC's share in world oil supply increases at a slower rate in the High Oil Price Case, less than 40% in 2030 compared with over half in the Reference Scenario (Figure 3.29).

Table 3.9: **Oil Production in Reference Scenario and High Oil Price Case**
(mb/d)

	2002	2030		
		Reference Scenario	High Oil price Case	Difference (%)
World	77.0	121.3	102.5	-15
OPEC	28.2	64.8	40.4	-38
OECD	21.1	12.7	13.1	4
Other non-OPEC*	25.9	33.7	37.4	11
Non-conventional	1.8	10.1	11.6	15

* Including processing gains.

Figure 3.29: **Share of OPEC in World Oil Production in the Reference Scenario and the High Oil Price Case**

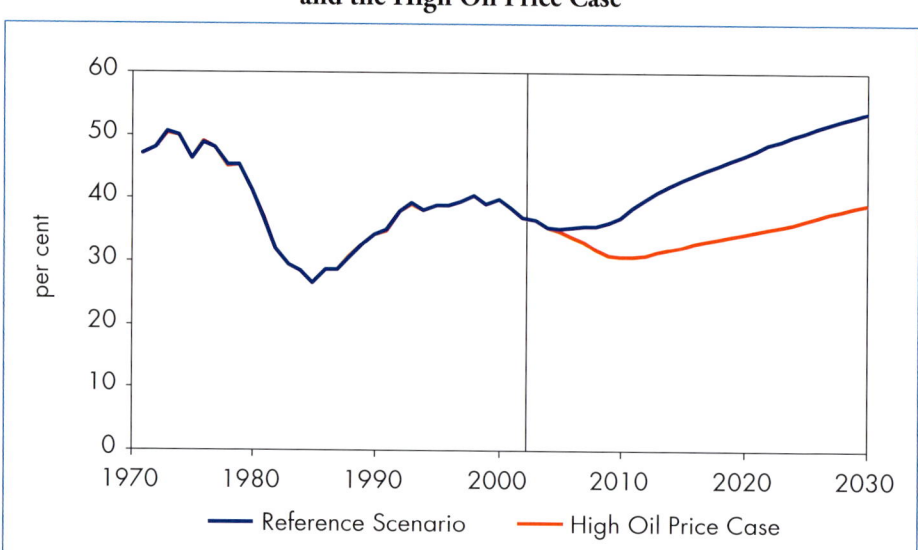

Impact on OPEC Oil Revenues and World Oil Investment

In the High Oil Price Case, OPEC's cumulative oil revenues over the projection period would be 7%, or $750 billion, lower. They are also lower when discounted at a rate of 10% per year. Higher oil prices are profitable for exporting countries in the short term, but lead to lower revenues in the longer term. The net loss of OPEC revenues would be even larger if the effects of higher oil prices on the global economy were taken into account. It is therefore in the interests of both consuming and producing countries to avoid high oil prices.

Despite the prospect of lower global production, upstream oil-investment requirements are similar to those shown in the Reference Scenario. Higher investments in all regions are compensated by the decrease in OPEC Middle East, where they are $116 billion lower (Figure 3.30). The bigger investment requirements in OECD countries and in Latin America are for non-conventional oil facilities, which become more commercially attractive. Global investments in non-conventional oil are 16% higher than in the Reference Scenario.

Figure 3.30: **Cumulative Oil Investment in the Reference Scenario and the High Oil Price Case, 2003-2030**

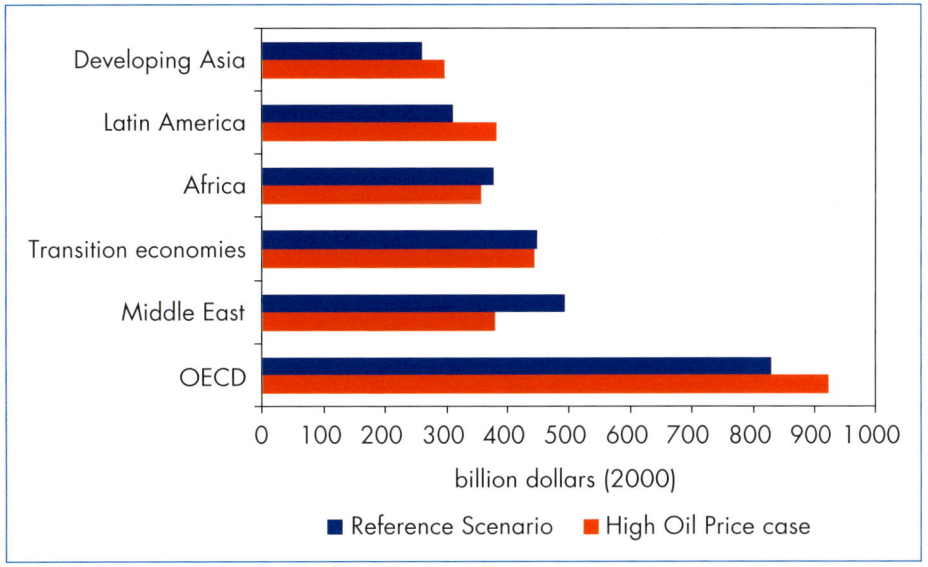

CHAPTER 4

NATURAL GAS MARKET OUTLOOK

HIGHLIGHTS

- Consumption of natural gas worldwide will almost double by 2030, driven mainly by power generation. The rate of increase – 2.3% per year – will be lower than in the past. Demand is projected to grow most rapidly in Africa, Latin America and developing Asia. Yet the total volume increase in demand will be bigger in the mature markets of OECD North America, OECD Europe and the transition economies.
- Gas-to-liquids plants will emerge as a major new market for natural gas, making use of reserves located far from traditional markets. Global capacity is projected to reach 0.4 mb/d in 2010 and 2.4 mb/d in 2030, though the rate of construction of GTL plants is hard to predict.
- Gas resources can easily meet the projected increase in global demand. Proven reserves have outpaced production by a wide margin since the 1970s and are now equal to about 66 years of production at current rates. Production will increase most in Russia and in the Middle East, which between them have most of the world's proven reserves. Most of the incremental output in these regions will be exported to North America, Europe and Asia, where indigenous output will fall behind demand.
- Inter-regional gas trade will triple over the projection period. All the regions that are currently net importers of gas will see their imports rise. Liquefied natural gas will account for most of the increase in global trade. LNG unit costs all along the supply chain are expected to continue to drop. By 2030, just over half of all inter-regional gas trade will be in the form of LNG, up from 30% at present. OPEC countries will continue to dominate the supply of LNG.
- Cumulative investment needs for gas-supply infrastructure to 2030 will amount to $2.7 trillion, or about $100 billion per year. Exploration and development of gas fields will absorb more than half of this investment.
- Gas prices have risen strongly in all regions in recent months, following the surge in oil prices. Tight supply has added to the upward pressure on gas prices in North America. Prices are assumed to drop back in the second half of the current decade. They will then recover steadily through to 2030. Gas-to-gas competition will put downward pressure on gas prices relative to oil prices, but this effect is expected to be largely offset by rising supply costs.

This chapter presents the Reference Scenario outlook for the natural gas market to 2030. It first details our gas-demand projections by region and by sector, and discusses the main factors and uncertainties. It then provides an overview of the world's natural gas reserves and the prospects for production. This is followed by an assessment of the implications of our demand and production projections for inter-regional gas trade and investment. The chapter ends with a detailed analysis of the prospects for natural gas demand and supply in nine separate world regions.

Gas Demand

Global consumption of natural gas is expected to increase more in absolute terms than that of any other primary energy source, almost doubling to 4 900 bcm (4 130 Mtoe) in 2030 (Table 4.1). Demand will grow at an average annual rate of 2.3%, a fraction lower than was projected in *WEO-2002*. Most of the increase will come from the power-generation sector. The share of gas in total world primary energy demand is projected to increase from 21% in 2002 to 25% in 2030.

Table 4.1: **World Natural Gas Primary Demand** (bcm)

	2002	2010	2020	2030	2002-2030*
OECD North America	759	866	1 002	1 100	1.3%
OECD Europe	491	585	705	807	1.8%
OECD Pacific	130	173	216	246	2.3%
OECD	**1 380**	**1 624**	**1 924**	**2 154**	**1.6%**
Russia	415	473	552	624	1.5%
Other transition economies	220	254	311	360	1.8%
Transition economies	**635**	**728**	**863**	**984**	**1.6%**
China	36	59	107	157	5.4%
Indonesia	36	53	75	93	3.5%
India	28	45	78	110	5.0%
Other Asia	109	166	242	313	3.8%
Brazil	13	20	38	64	5.8%
Other Latin America	89	130	191	272	4.1%
Africa	69	102	171	276	5.1%
Middle East	219	290	405	470	2.8%
Developing countries	**597**	**864**	**1 307**	**1 753**	**3.9%**
World**	**2 622**	**3 225**	**4 104**	**4 900**	**2.3%**
European Union	*471*	*567*	*684*	*786*	*1.8%*

* Average annual growth rate.
** World totals include stock changes and statistical differences.

The projected growth in gas demand is in line with historical trends. Global consumption rose by 2.5% per year from 1990 to 2002. Demand has faltered since the start of the current decade, increasing by only 1% in 2001 and, according to preliminary data, by 2.4% in 2003. The economic downturn and warmer winter weather across the northern hemisphere contributed to slower growth in 2001. A slump in gas use in the United States – the result of stagnating production and soaring prices – also played a role in 2001 and 2003. Since 2000, demand has grown most strongly in Latin America.

As in the previous edition of the *Outlook*, gas demand is projected to grow most rapidly in Africa, Latin America and developing Asia. The use of gas will grow by more than 5% a year in China and India, where gas will win market share from coal in the power sector and in industry. Demand will increase most in volume terms in developing Asia as a whole (Figure 4.1). The region's share of world demand will jump from 8% in 2002 to 14% in 2030. Per capita gas consumption will, nonetheless, remain highest in the mature markets of OECD North America and the transition economies. By 2030, OECD North America alone will still account for 23% of world gas consumption, OECD Europe for 16% and Russia for 12%.

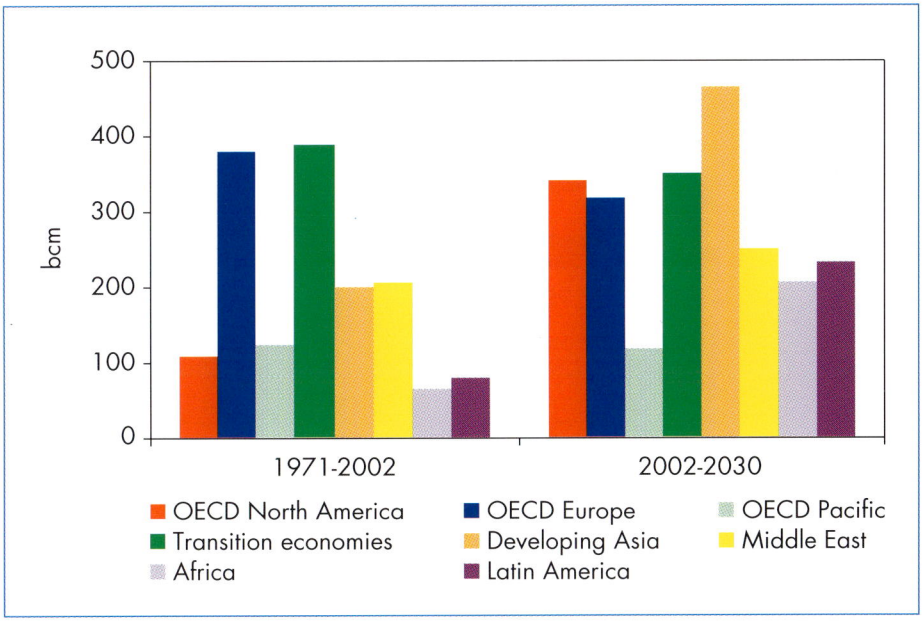

Figure 4.1: **Incremental Demand for Natural Gas by Region**

Power generation is expected to account for 59% of the increase in world gas demand over the projection period (Figure 4.2). As a result, the power sector's share of the world gas market will rise from 36% in 2002 to 47% in 2030. The power

Chapter 4 - Natural Gas Market Outlook

sector will be the main driver of demand in all regions, especially in developing countries where electricity demand is expected to rise most rapidly, while demand for gas in the residential and services sectors will remain relatively modest. Despite rising prices after 2010, natural gas will remain the most competitive fuel in new power stations in most parts of the world, as it is the preferred fuel for high-efficiency combined-cycle gas turbines (CCGTs). Natural gas has inherent environmental advantages over other fossil fuels, including lower carbon content and fewer emissions of noxious gases. Moreover, the capital costs and the construction lead-times of CCGTs are lower than for other thermal power plants. These factors, together with their smaller economies of scale, make gas-fired CCGTs particularly well-suited to competitive power markets. Electricity output from gas-fired stations will increase even more rapidly than gas inputs to generation because of continuing improvements in the thermal efficiency of CCGTs.

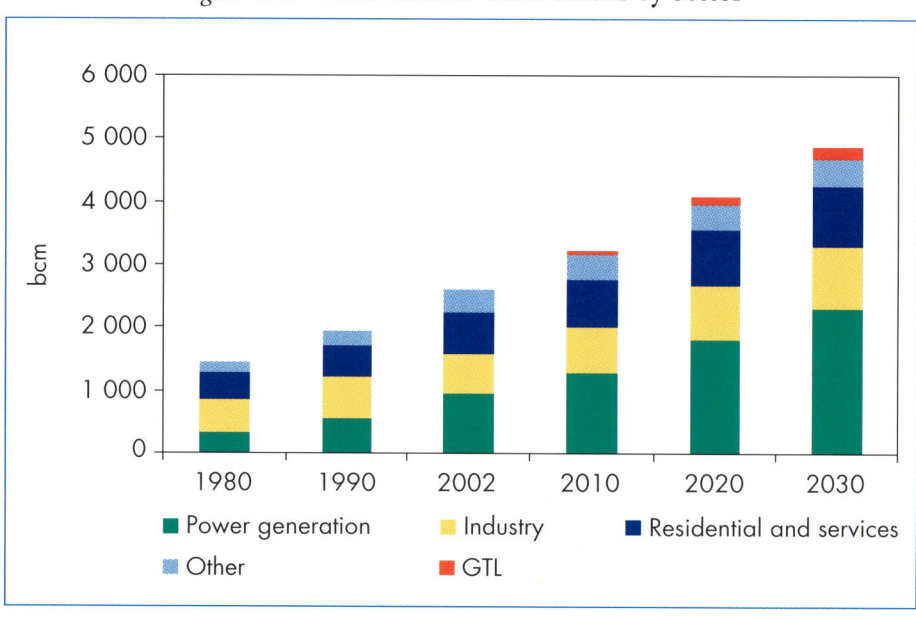

Figure 4.2: **World Natural Gas Demand by Sector**

The prospects for gas-fired generation are uncertain because of:

- Movements in relative fuel prices and their effect on plant dispatch and the economics of new plant construction.
- Shifts in the relative costs of building and operating new thermal plants.
- The possible difficulty of financing new power plants, especially in developing countries.

- Government policies on nuclear power, as well as the effect of new technology on the cost and acceptability of building new reactors, of extending the lives of existing reactors and of dealing with radioactive waste.
- Government policies and measures to promote the use of renewable energy sources.
- Environmental policies and measures to deal with emissions of noxious and greenhouse gases, such as the introduction of CO_2-emissions trading.

Gas-to-liquids (GTL) plants are expected to emerge as a major new market for natural gas, making use of cheap reserves located far from traditional markets. Interest in developing GTL projects has grown rapidly in recent years due to technological advances that have greatly reduced production costs and to higher oil prices. Holders of gas reserves that cannot be transported economically to market by pipeline may now be able to turn to GTL as an alternative or complement to LNG. In practice, where a choice will need to be made between GTL and LNG, it will be driven mainly by financial considerations. But GTL can help to diversify oil companies' activities and reduce their overall portfolio risk. This could be a decisive advantage for GTL in cases where the economics of the two technologies are very similar.

All GTL plants now in operation, under construction or planned use Fischer-Tropf technology, which converts natural gas into synthesis gas (syngas) and then, through catalytic reforming or synthesis, into very clean conventional oil products. The main fuel produced in most plants is diesel. Global demand for gas from GTL producers is projected to surge from just 4 bcm in 2002 to about 40 bcm in 2010 and 214 bcm in 2030. About 45% of the gas supplied to GTL plants is currently consumed in the conversion process, with the rest used as feedstock. As a result, GTL plants emit large amounts of carbon dioxide. The share of energy used in conversion is, however, assumed to drop by 2030 as a result of efficiency improvements.

Most GTL plants are expected to be built in the Middle East. A commercial-scale plant is already under construction in Qatar and several others are planned to be commissioned before 2010 (Box 4.1). Global GTL capacity is projected to reach 0.4 mb/d in 2010 and 2.4 mb/d in 2030 (Figure 4.3). The rate of construction of GTL plants is nonetheless hard to predict. Further technology improvements could reduce the energy intensity of GTL processes. On the other hand, further declines in LNG supply costs could undermine the attraction of GTL. Turbulence in the oil market and the possible impact of future policies to reduce carbon-dioxide emissions are also complicating factors.

At 1.5% per annum, final gas consumption will grow much more slowly than will primary use over the projection period. Industrial demand will grow faster than that of any other sector, and industry will remain the largest end-

Box 4.1: **Status of GTL Projects**

There are currently only two commercial-scale GTL plants in operation: the 22 500-b/d Mossgas facility in South Africa, which started up in 1991, and Shell's 12 500 b/d Bintulu plant in Malaysia, which was commissioned in 1993. Several other plants are currently under construction or planned, most of them in Qatar using gas from its huge North Field. If all these projects come to fruition, global GTL capacity would exceed 800 000 b/d by 2011 – nearly 1% of world output of refined products. Oryx GTL, a 50-50 joint venture between South Africa's Sasol and ChevronTexaco, is building a 34 000 b/d plant in Qatar, which is due on stream in 2005. An expansion of the plant's capacity to 100 000 b/d is planned for 2009. In addition, Qatar Petroleum and Sasol/ChevronTexaco have agreed to develop a $6-billion, 130 000-b/d integrated project, which could begin operating in 2010.

Shell is also developing a much larger integrated project in Qatar, involving the construction of a 140 000-b/d plant based on its proprietary Shell Middle Distillate Synthesis technology. The plant will be built in two stages, with the first unit due on stream in 2009 and the second in 2011. The project is expected to cost around $5 billion in total, which Shell will finance on its own. Exxon-Mobil is also planning a 100 000-b/d plant in Qatar to be commissioned in 2008 at the soonest. Marathon and ConocoPhillips are also planning large plants in Qatar.

Two projects are planned outside Qatar: a 34 000-b/d plant in Nigeria integrated with the Escravos Gas Project Phase 3 upstream development, which will cost a total of $2 billion and which could be on stream by 2007; and a 67 000-b/d plant being developed by Sasol/ChevronTexaco in Australia, for completion in the second half of this decade. The plant in Nigeria will process associated gas that would otherwise be flared.

consumer of gas. Industrial demand is expected to increase most rapidly in developing countries, by 2.9% annually. But this will happen only if the needed gas-supply infrastructure is built. In the transition economies, expected improvements in energy efficiency will hold the growth in gas demand down to less than 2% per annum. There is tremendous scope for efficiency gains in Russian manufacturing industry – especially in chemicals and in iron and steel, which use large amounts of gas. Industrial gas demand in OECD countries is projected to grow by less than 1% per year, roughly the same rate as over the past three decades.

Gas demand in other final sectors – mainly residential and services – will grow by 1.4% per year. Growth in the use of gas for space and water heating will be

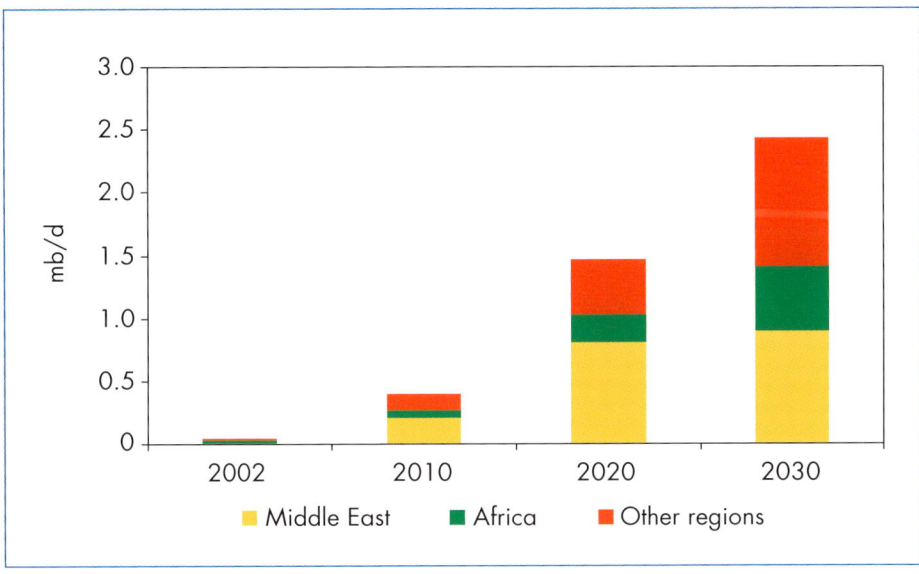

Figure 4.3: **GTL Production by Region**

limited by saturation effects in many OECD countries. There is little scope for establishing and extending local distribution networks in many parts of the developing world, because heating needs are small or because incomes are too low. The share of gas in overall final energy use in these sectors will nonetheless remain broadly constant at about one-fifth.

Gas Supply

Proven Reserves and Potential Resources[1]

Natural gas resources are easily large enough to meet the projected increase in global demand described above. Proven reserves of gas have increased steadily since the 1970s, as reserve additions have outpaced production by a wide margin. According to estimates by Cedigaz, an international centre for gas information, reserves stood at 180 trillion cubic metres at the beginning of 2004 – almost twice as high as twenty years ago. Reserves are equivalent to 66 years of production at current rates. Were production to grow at our projected annual rate of 2.3%, reserves would last 40 years. The increase in reserves has resulted both from sustained exploration and appraisal activity in many parts of the world and from advances in technology that have allowed existing reserves to be upgraded. The majority of the gas that has been discovered so far has been found in the course of oil exploration. As with

1. See Chapter 3 for a detailed discussion of methodological issues concerning oil and gas reserves estimates.

oilfields, new gas fields that have been discovered recently are generally smaller than in the past.

Three countries, Russia, Iran and Qatar, hold 55% of global gas reserves. Nonetheless, gas is more widely distributed geographically than is oil (Figure 4.4). The former Soviet Union holds almost a third of global reserves, but its share has decreased steadily over the past decade, as a result of low exploration activity in Russia (Figure 4.5). Reserves in the region still amount to more than 77 years of production at current rates. The Middle East holds 40% of all reserves and its share is growing as new pockets of gas are discovered and reserves in existing fields are upgraded, notably in Iran, Saudi Arabia and Qatar. The region's reserves-to-production ratio is around 200 years. Gas reserves in OECD countries, at 18 tcm, are equal to 10% of the world total, or 16 years of current production.

Potential gas resources are much greater than proven reserves. According to the last survey of global hydrocarbon resources carried out by the US Geological Survey in 2000, undiscovered gas resources are estimated at 147 tcm, of which 25% are associated with oil and 75% non-associated. Just over half of undiscovered gas resources are thought to be in the former Soviet Union and the Middle East – a lower share than for proven reserves. Remaining *discovered* resources, including proven, probable and possible reserves, are estimated at 136 tcm.[2] "Reserves growth" – increases in known gas reserves that occur as fields are developed and exploited – is estimated at 104 tcm, which is nearly as large as estimated undiscovered resources. According to USGS, ultimate gas resources amount to 436 tcm, of which slightly more than 10% have already been produced (the figure is almost 25% for estimated world oil resources).

Production Prospects

The regional outlook for gas production will depend largely on the proximity of reserves to markets, as well as on production costs. Despite substantial unit cost reductions in recent years, gas transportation remains very expensive, whether by pipeline or in the form of LNG, and usually represents most of the overall cost of gas delivered to consumers. Much of the world's gas resources are located far from the main centres of demand, so that only a small proportion can as yet be exploited profitably.

Production is projected to grow most strongly in volume terms in Russia and the other transition economies and in the Middle East (Figure 4.6). Latin America and Africa will experience the fastest *rates* of increase. Most of the incremental output in these regions will be exported to North America, Europe

---

2. The USGS estimates are for 1 January 1996. Cedigaz estimates remaining reserves at 180 tcm at 1 January 2004.

Figure 4.4: World Proven Reserves of Natural Gas

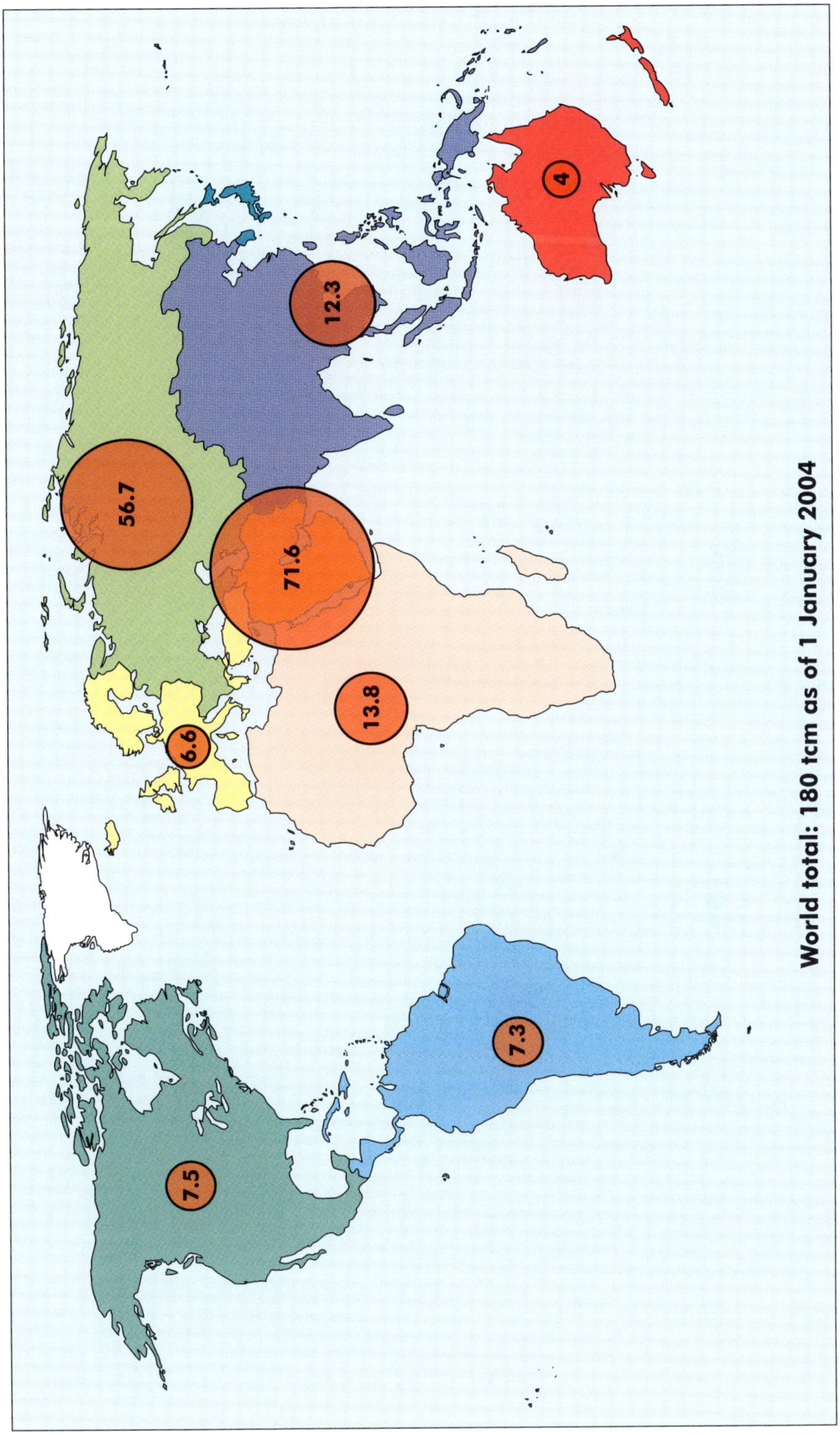

World total: 180 tcm as of 1 January 2004

Source: Cedigaz (2004).

Chapter 4 - Natural Gas Market Outlook

137

Figure 4.5: **Proven Natural Gas Reserves**

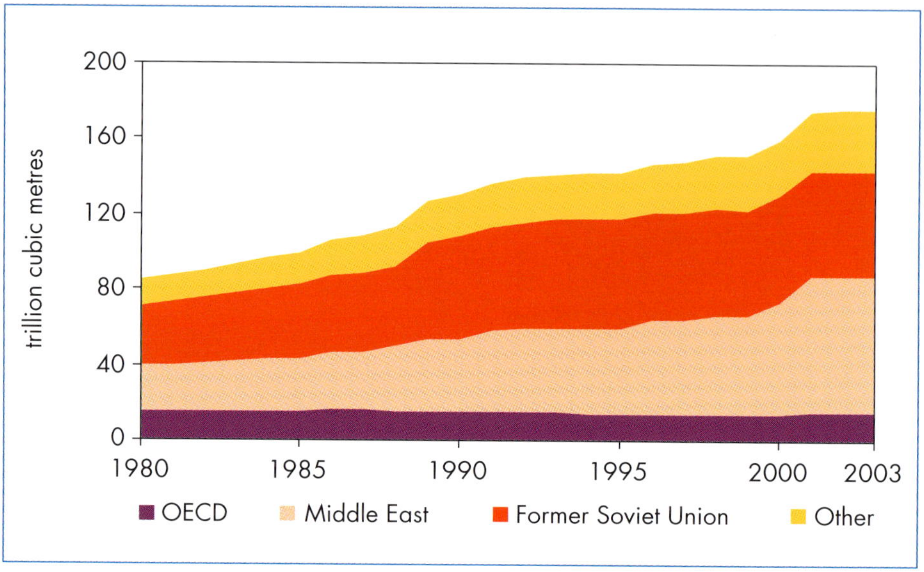

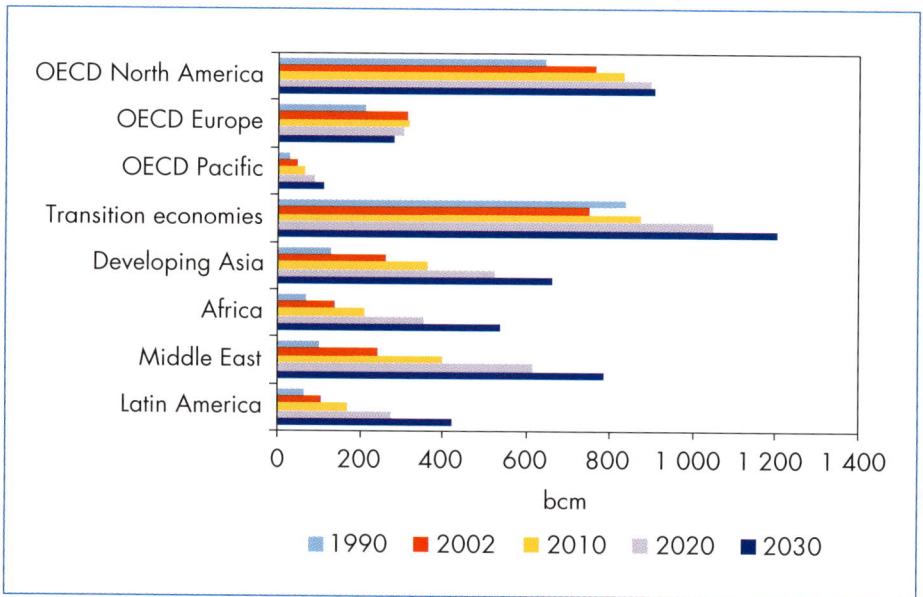

Source: Cedigaz (2004).

and Asia, where indigenous output will not keep pace with demand. The cost of producing gas that is not associated with oil is believed to be lowest in the Middle East. In some cases, such as the North Field in Qatar, selling the

Figure 4.6: **Natural Gas Production by Region**

condensates and other liquids contained in the gas that is extracted covers much of the cost of developing the field. Depletion rates and production costs are rising in the most mature producing regions, including North America and Europe. The cost of developing new fields in Russia is much higher than it was for the large existing fields that have been in production since the 1970s and 1980s. In the Middle East and Africa, associated gas that is currently flared will make a growing contribution to total marketed gas output. Iran, Abu Dhabi, Algeria and Nigeria are implementing programmes to reduce gas flaring.

Non-conventional gas – mostly extracted from coal beds (coal-bed methane), from low-permeability sandstone (tight sands) and from shale formations (gas shales) – could make an increasingly important contribution to gas supply, especially in North America. These sources of gas have become an important component of US gas supply since the late 1980s, accounting today for around a quarter of total gas production. In the rest of the world, unconventional gas production is still modest. Although resources are thought to be abundant in many parts of the world, they are generally costly to produce and so have not been appraised in detail.

Worldwide, 7.3 trillion cubic metres of new gas-production capacity will be needed over the next three decades, around 260 bcm a year. Less than a third of this new capacity will go to meet rising demand. The rest will compensate for declining production from wells that are already in operation and from others that will come on stream and decline during the projection period (Figure 4.7). The rate of new capacity additions will reach around 320 bcm

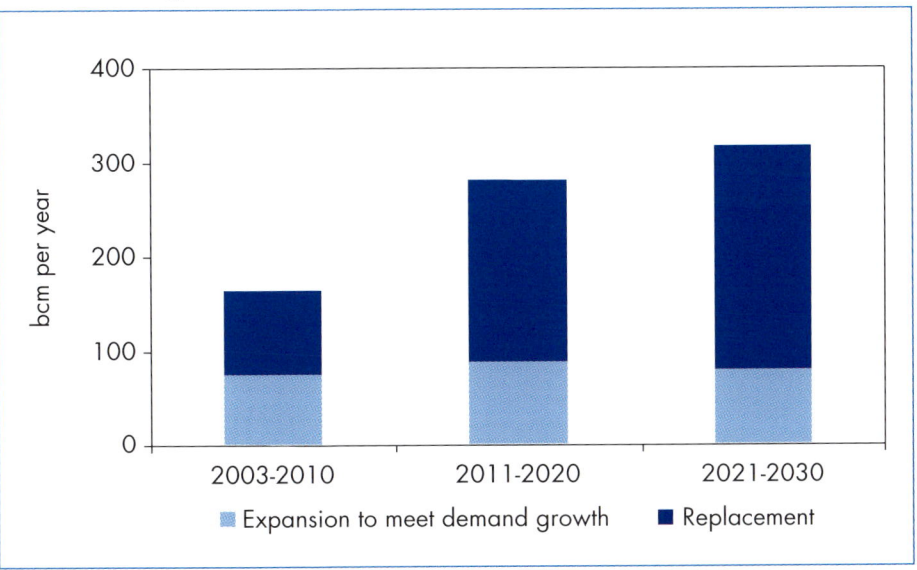

Figure 4.7: **World Gas-Production Capacity Additions**

Chapter 4 - Natural Gas Market Outlook

per year in the third decade. A quarter of these additions will be in North America, where decline rates are high because of the advanced age of existing fields, the falling size of new discoveries and extraction technologies that tend to maximise initial production. Additions to production capacity will also be large in Russia and in the Middle East.

In 2002, 71% of all the natural gas produced in the world came from onshore fields. This share is expected to drop to 64% in 2030, as exploration and development shifts to more lucrative offshore sites. North America, the transition economies and the Middle East will account for two-thirds of the onshore capacity brought on-stream over the projection period. The Northwest Europe Continental Shelf and the Gulf of Mexico together will account for almost a third of new offshore capacity. Asian countries will account for almost a quarter.

Gas Trade

Inter-regional trade[3] in natural gas will more than triple over the projection period, from 417 bcm in 2002 to 1 265 bcm in 2030, as a result of the geographical mismatch between resource location and demand. All the regions that are currently net importers of gas will see their imports rise, both in volume and as a share of their total gas consumption (Table 4.2). The biggest increase in import volumes will occur in the European Union. By 2030, the Union will rely on imports for 80% of its gas needs compared with 50% at present. Most of the increase will be met by Russia, Africa, the Middle East

Table 4.2: **Gas-Import Dependence**

	2002		2010		2030	
	Bcm*	%**	Bcm*	%**	Bcm*	%**
OECD North America	0	0	33	4	197	18
OECD Europe	162	36	267	46	525	65
OECD Asia	98	98	130	97	183	94
China	0	0	9	15	42	27
India	0	0	10	23	44	40
European Union	*233*	*49*	*342*	*60*	*639*	*81*

* Net imports.
** Per cent of primary gas supply.

3. Trade between major regions – OECD North America, OECD Europe, OECD Asia, OECD Oceania, China, the transition economies, East Asia, South Asia, Middle East, Africa and Latin America – only. Total world trade is much larger because it includes trade between countries within each region.

and the Caspian/Central Asian region. North America will be the second-largest importing region by the end of the projection period, ahead of OECD Asia.

The Middle East will be the world's largest exporting region in 2030. Net exports from the transition economies and Africa will also grow substantially, but at a slower rate. Net exports from the Middle East will increase most in absolute terms, by 274 bcm, from just 30 bcm in 2002 to 304 bcm in 2030. Most of these exports will be in the form of LNG. All the world's importing regions will become more dependent on Middle East gas (Figure 4.8).

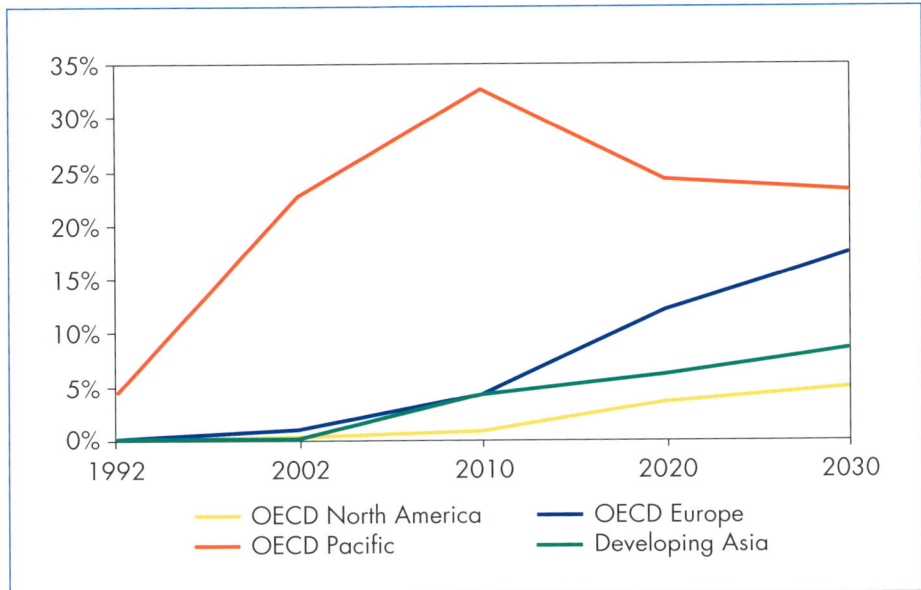

Figure 4.8: **Share of Middle East Gas in Total Gas Supply by Importing Region**

Inter-regional LNG trade, which totalled 150 bcm in 2002, will expand rapidly over the projection period, reaching 250 bcm in 2010 and 680 bcm in 2030. By 2030, more than 50% of all inter-regional gas trade will be in the form of LNG. Almost 70% of all cross-border gas trade is now shipped by pipeline. LNG flows have doubled in the past decade, reaching 150 bcm in 2002 – about 6% of total world consumption of natural gas. At the beginning of 2004, there were 15 LNG export terminals, 43 import terminals and 154 LNG tankers operating worldwide.

A continuing decline in unit costs all along the LNG chain will underpin this growth. At the beginning of 2004, eight liquefaction terminals were being expanded and five new ones were under construction. In addition, eight new import terminals were being built in the OECD region, 54 new LNG ships were on order and some 30 new LNG supply projects were planned (IEA, 2004). Most of the new terminals to be built in the next decade and a half will be in the United States and Europe. Thus, LNG trade, which has until now been largely focused on the Asia-Pacific region, will become much more widespread (Figure 4.9). OPEC countries will continue to dominate the supply of LNG (Figure 4.10).

While long-term contracts will continue to dominate the LNG business in the foreseeable future, spot sales – short-term or single-cargo sales – are expected to become more important. Spot trading represented almost 11% of global LNG trade in 2003[4], up from less than 2% in the late 1990s. Several ships now being built are not earmarked for particular projects, and so will be available for spot-trading opportunities. In addition, several older tankers will be freed from their current assignments when long-term contracts expire. Part of the capacity of some liquefaction plants built in recent years is not covered by long-term contracts, and so will be available to supply the spot market. Nonetheless, because of the high cost of producing and shipping LNG and the highly capital-intensive nature of the business, most new projects will still require long-term contracts covering most of their capacity (IEA, 2003).

More flexible pricing mechanisms and shorter-term contracts will become more common in liberalised markets. LNG suppliers are already adapting their pricing policies to the needs of individual buyers, including power generators who are starting to contract for their LNG purchases directly. Sales are generally pegged to spot gas prices in the United States – usually at Henry Hub – and to spot or futures prices in the United Kingdom. But in Asia, LNG prices are still indexed to crude oil prices, and in continental Europe to fuel-oil prices. Indexation to gas prices is likely to become more widespread as genuine gas-to-gas competition takes hold, hubs and market centres develop and liquidity grows. Buyers will also push for less onerous take-or-pay obligations because assessing their future needs will be harder in competitive markets. More contractual flexibility and more LNG trade will increase the scope for buyers in different countries to swap supplies for different time periods, to take advantage of differences in peak load and to minimise purchase costs. Power generators in Japan, where peak demand is in the summer, and in Korea, where the peak is in the winter, swapped a dozen cargoes in 2003 and early 2004 (IEA, 2004). More integration between LNG producers and buyers is likely, in response to increased price risk in competitive downstream markets.

---

4. International Group of LNG Importers, cited in *World Gas Intelligence*, 19 May 2004. Spot trading includes both one-off sales of cargoes as well as contract-balancing transactions between buyers that find themselves with a temporary shortage or surplus of LNG.

Figure 4.9: Major Net Inter-Regional Natural Gas Trade Flows, 2002 and 2030 (bcm)

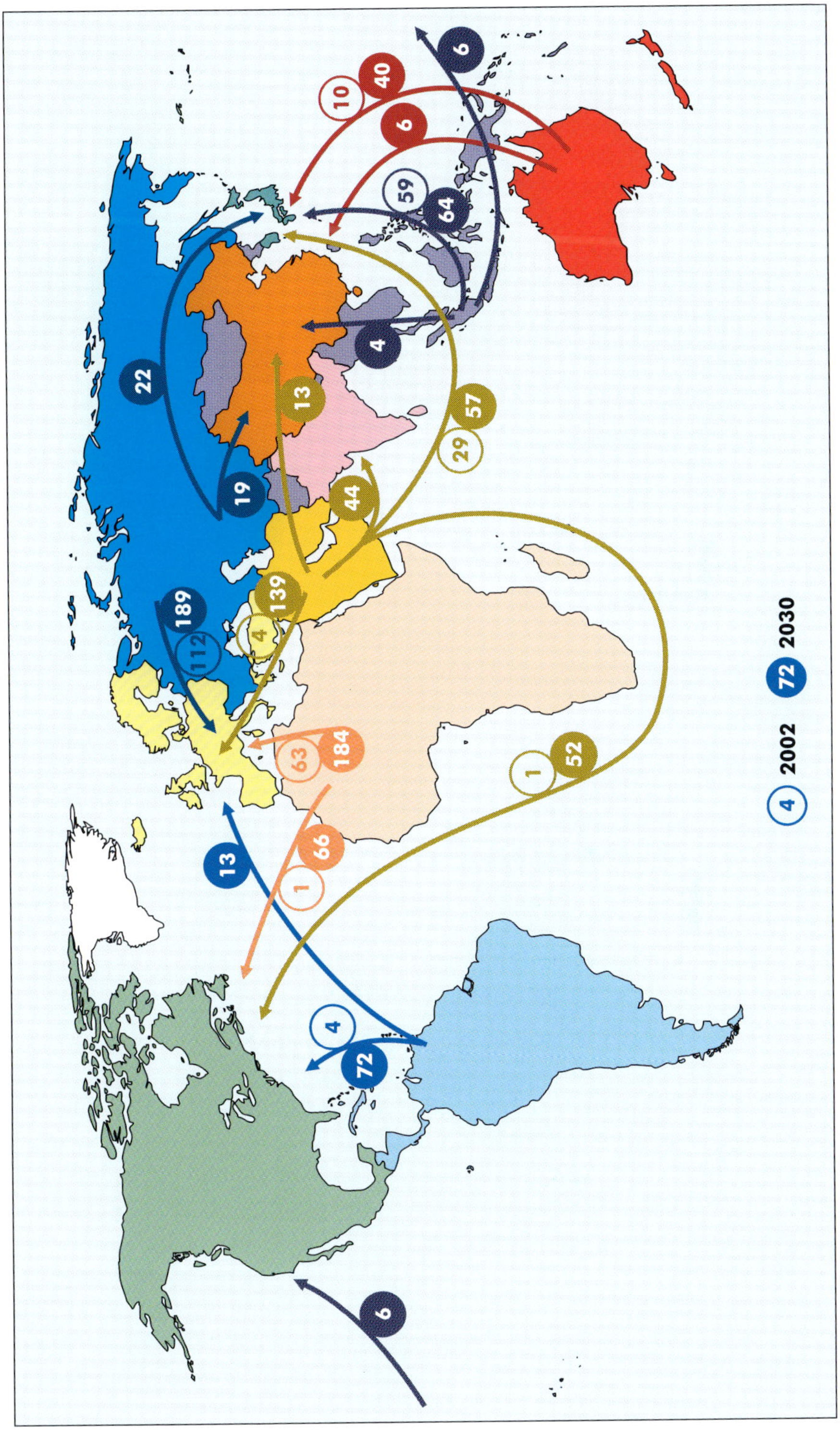

Chapter 4 - Natural Gas Market Outlook

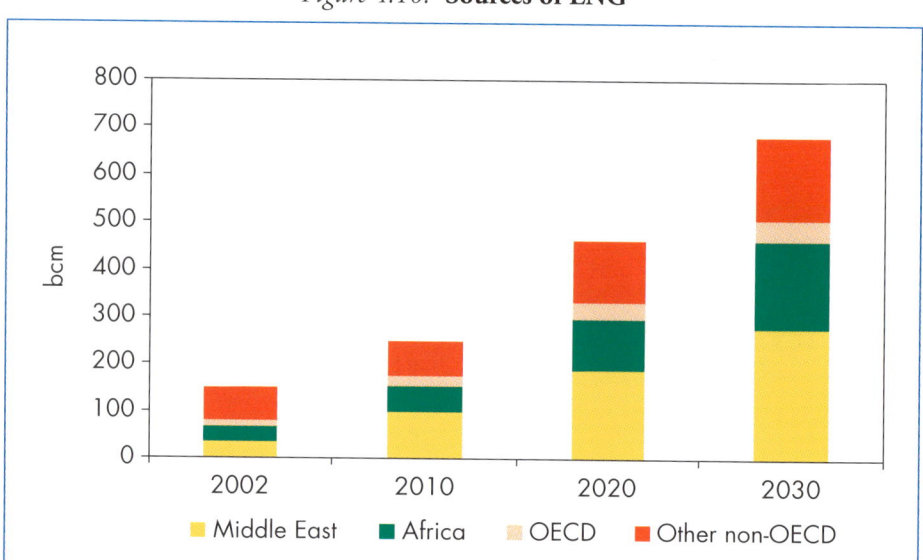

Figure 4.10: **Sources of LNG**

Investment Outlook

Projected gas-supply trends over the period 2003-2030 will entail cumulative global investment of $2.7 trillion (in year-2000 dollars), or about $100 billion per year. This investment will be needed to replace existing capacity that will be shut down during the projection period as well as to expand capacity to meet a near-doubling of demand. Exploration and development of gas fields will absorb more than half of total gas investment. Building downstream infrastructure – high-pressure transmission pipelines, local distribution networks, storage facilities, LNG liquefaction and regasification plants and LNG carriers – will account for the rest. An increasing share of investment will go to LNG supply.

The OECD as a whole will account for almost half of global gas investment (Figure 4.11). North America alone will claim more than a quarter of new investment. Unit capital costs and production-decline rates are much higher in the industrialised countries than in other parts of the world. The main exporting regions – Russia, the Caspian region, the Middle East and Africa – will attract most investment outside the OECD. Although a bigger share of drilling will occur in lower-cost regions, a doubling of global production and a shift in drilling to offshore fields will cause an overall increase in upstream investment. Gas-processing costs, included in exploration and development, may also rise, as the quality of reserves deteriorates. The Middle East will have the largest requirement for LNG investment, while the transition economies, including Russia, will account for the largest share of investment in transmission networks.

Figure 4.11: **Cumulative Investment in Natural Gas, 2003-2030**

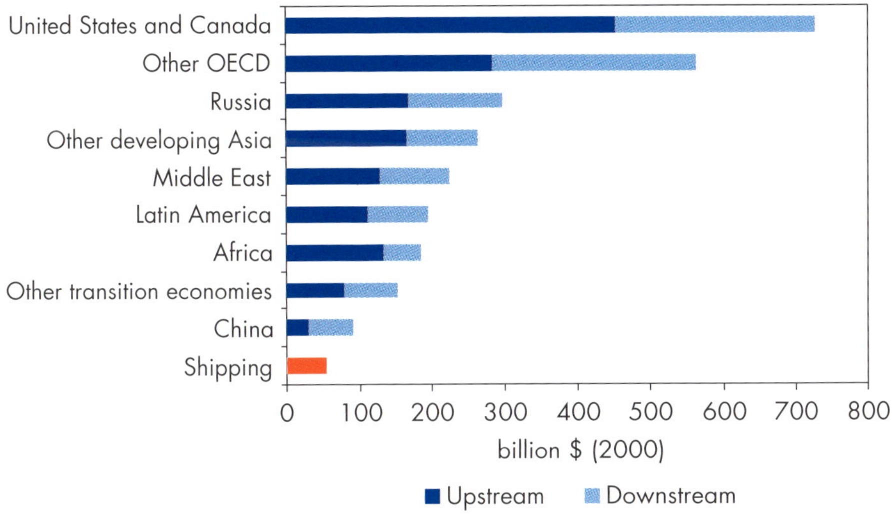

Price Developments

Average gas prices to end-users are derived from assumed trends in wholesale or bulk gas prices (see Chapter 1). These trends reflect our underlying assumptions about future oil prices, which will remain a major determinant of gas prices. They also reflect our judgment of the prices that will be needed to stimulate investment in replacing and expanding supply infrastructure, as well as the impact of increasing competition on the relationship between oil and gas prices. Tax rates are assumed to remain unchanged.

Most countries with well-established gas markets have adopted policies aimed at opening up their markets to competition in supply, usually through third-party access to transmission and distribution networks. Market reforms are most advanced in North America, Great Britain and Australia. All the other OECD countries that use gas, as well as many developing countries, are planning or are implementing similar moves. Competitive markets are expected to result in a more efficient allocation of resources, capacity and investment and, thus, to lower the cost of supply. This will help to drive down prices, especially where gas supply exceeds demand and competition is intense. Gas and oil prices will decouple to some degree, as spot or futures gas prices replace oil prices as the basis for indexing gas prices in long-term contracts. Electricity prices will also be used increasingly to index gas prices for sales to power generators. Yet oil prices will continue to influence gas prices on spot and futures markets, because of competition between gas and oil products in non-power sectors. But other factors, including the cost of developing new sources of supply and possible shortfalls in production or transportation capacity could offset all or part of the effect of competition. In any event, prices will become more volatile.

Gas prices rose strongly in all regions during the first half of 2004 due to the surge in oil prices. Tight supply added to the upward pressure on gas prices in North America. We assume that prices drop back in 2006, remaining broadly flat through to the end of the current decade. They then recover steadily through to 2030 in line with oil prices. Rising supply costs also contribute to higher gas prices in Europe from the end of the current decade. As a result, European prices rise only slightly relative to oil prices (Figure 4.12). Regional prices are expected to converge to some degree over the next three decades as increased spot trading of LNG allows arbitrage between markets. This will erode the linkage between gas and oil prices.

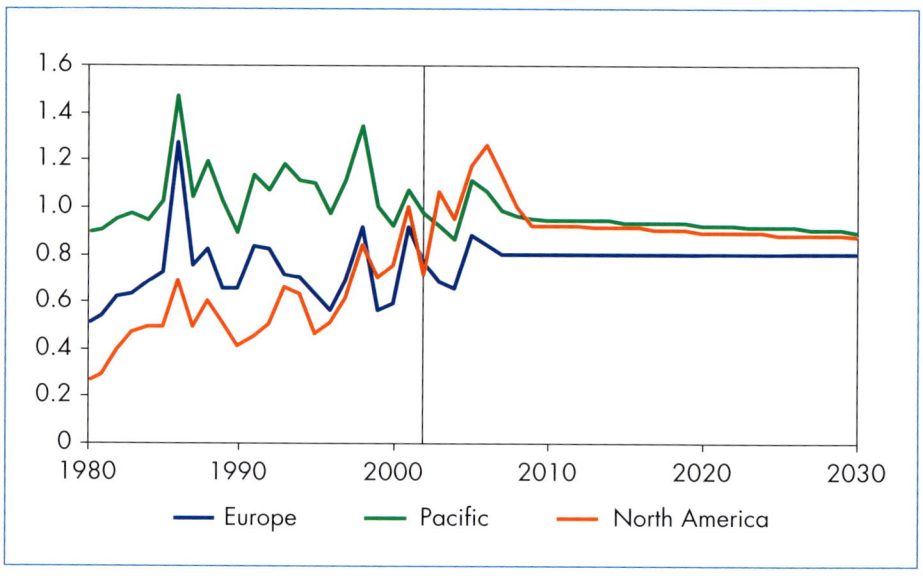

Figure 4.12: **Ratio of Natural Gas Prices to Oil Price**

North America

North American spot gas prices have soared since the end of the 1990s, initially due to higher oil prices but also, increasingly, because of tighter supplies. Despite record drilling activity, gas production in the United States has stagnated. Pipeline imports from Canada and LNG imports have been insufficient to relieve the upward pressure on prices. Higher prices have been inducing many gas customers to switch away from gas. Some manufacturers have been forced simply to stop producing. A large number of new LNG import terminals are planned, but they will take several years to bring on stream.

Bulk natural gas prices in North America are assumed to fall back from an average of $5.30/MBtu (in year-2000 prices) in 2003-2005 to $3.80 by the end of the current decade. Even so, gas prices continue to rise relative to oil prices. Lower oil prices and increased LNG import capacity after 2007 help to ease the pressure on gas markets. Prices start to rise after 2010 but continue to decline slightly in relation to oil prices. North America will become increasingly reliant on LNG imports and on new sources of gas supply, including fields in northern Canada and Alaska and non-conventional sources. Prices reach $4.70 by 2030.

Europe

European gas prices rose to more than $4/MBtu in the first half of 2004. Yet they have not kept up with oil prices over the last two years because of lags in the price-indexation clauses in long-term contracts. Most gas in continental Europe is still traded under long-term contracts, with prices indexed to oil prices over the previous six to twelve months. Weaker-than-expected demand growth and some large additions to pipeline import capacity have also helped to ease the pressure on gas prices.

The average gas import price in Europe is assumed to peak in 2005 in lagged response to high oil prices in 2004. Prices are assumed to fall back to around $3.30/MBtu (in year-2000 dollars) towards the end of the current decade and then rise gradually to $4.30 by 2030. Gas-to-gas competition is expected to exert some downward pressure on gas prices at borders as spot trade develops. But the cost of bringing new gas supplies to Europe is expected to rise as the distances over which the gas has to be transported lengthen and import costs increase. This is assumed to offset the impact of falling unit supply costs and of growing competition. On balance, gas prices are expected to rise slowly in relation to oil prices from 2008 on.

Asia-Pacific

Japanese LNG prices, our benchmark for gas prices in the Asia-Pacific region, are assumed to peak in 2005 and then fall back to around $3.90/MBtu (in year-2000 dollars) by 2010. Prices rebound slowly during the second and third decades of the projection period, reaching $4.80 in 2030. Gas prices fall slightly in relation to oil prices after 2006 owing to growing competitive pressures. Many of Japan's long-term LNG contracts will expire in the next few years, providing buyers with opportunities to press for lower LNG prices in new contracts and to seek out cheaper spot supplies. This will undermine the historical link between LNG and crude oil prices. Nonetheless, prices in Asia are expected to remain marginally higher than in North America in 2030, because of the region's continuing heavy reliance on distant sources of gas imports.

Regional Trends

North America

Gas demand in OECD North America[5] is projected to grow by 1.3% per year from 2002 to 2030. The power sector will absorb almost two-thirds of the 341 bcm increase in demand, as the majority of new power stations will be gas-fired CCGTs (Figure 4.13). Primary gas demand is expected to grow much more rapidly in Mexico, at 3.5% per year, than in the United States and Canada, at 1.2%. The Mexican gas market is expected to more than double in size over the projection period. Projected demand in North America as a whole is significantly lower than that presented in *WEO-2002*, partly because prices are assumed to be higher, choking off demand. Preliminary data show that demand dropped by 3.6% in 2003 (4.6% in the United States) after a sharp rise in gas prices (Box 4.2).

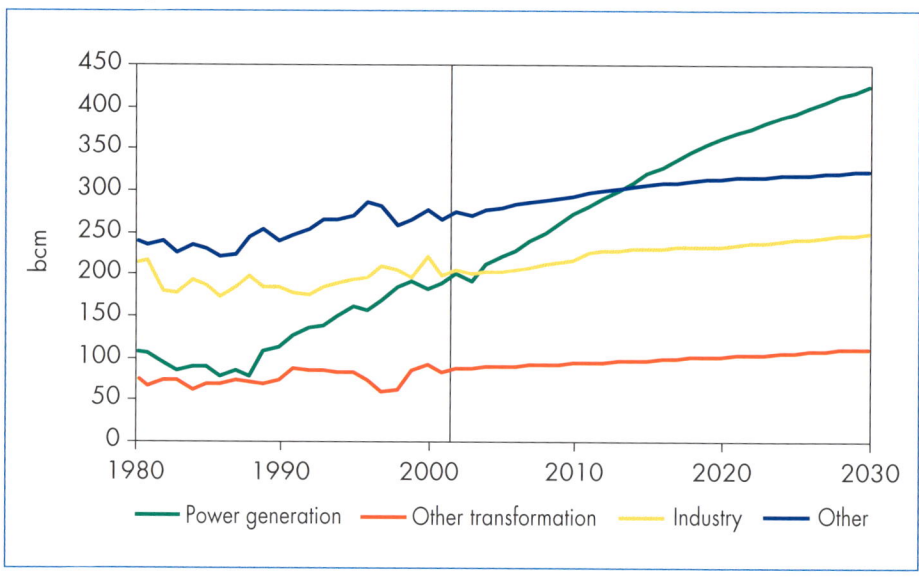

Figure 4.13: **Natural Gas Demand in OECD North America**

A growing share of North America's gas needs will have to be imported in the future. As predicted in past *Outlooks*, North American natural gas production is struggling to keep pace with demand, which has pushed up prices and increased the use of the four LNG terminals in the United States. Gas resources in the region are meagre compared to most other parts of the world. Proven reserves amounted to 7.5 trillion cubic metres at the beginning of 2004, of which 5.4 tcm were in the United States. These reserves are equal to only 4% of the world total and to less than ten years of production at current rates.

5. Canada, Mexico and the United States.

> *Box 4.2:* **Sensitivity of US Gas Demand to Higher Prices**
>
> US consumption of natural gas plunged 30 bcm, or 4.6% in 2003, according to preliminary data. The drop came after a surge in gas prices caused by poor drilling results. Imports of piped gas from Canada and of LNG failed to make up the difference. Average wellhead prices peaked at $6.69/MBtu in January 2003 and averaged almost $5 over the year as a whole. Prices remained above $5 in the first half of 2004.
>
> Reduced gas use by industry and power generators accounted for the entire drop in demand in 2003, more than offsetting a continuing rise in residential and commercial consumption. Power stations used 21 bcm, or 13%, less gas, as generating companies switched to cheaper coal and heavy fuel oil (in multi-fired or backup plants). Coal deliveries to power stations increased by almost a quarter, to over 1 million short tons in 2003.
>
> Industry used 16 bcm, or 7%, less gas. Higher gas prices drove industrial firms to use cheaper heavy fuel oil and distillate, to shut factories temporarily or to shift some production overseas. Some gas-intensive firms that could not switch fuels, such as ammonia and steel producers, stopped or reduced domestic production, though higher prices for a wide range of industrial products helped some to stay open. A weaker dollar also blunted commercial incentives to shut plants. The chemicals industry, which accounts for 40% of industrial gas use, shut down large amounts of capacity. Around a fifth of fertilizer capacity in the United States and Canada has been mothballed since 2000 because of high gas prices.[6] Other industries have switched to cheaper fuels. Short-term fuel-switching capacity in US industry remains large. About a quarter of all companies that usually use gas maintain multi-firing capability.

Gas drilling has been intense since the start of the decade compared to the previous decade, reflecting much higher prices. In the United States, almost 1 000 rotary rigs were in operation in the first quarter of 2004, close to the highs reached in 2001 and well above the average of 872 in 2003, of 691 in 2002 and of 441 in the 1990s. The number of exploration and development wells drilled jumped from 830 per month in the 1990s to 1 644 in 2003 and 1 994 in July 2004. But the results have been disappointing: US production has hardly risen since the end of the 1990s, fluctuating between 561 bcm and 583 bcm per year (Figure 4.14). Canadian production dropped, from 188 bcm in 2002 to 182 bcm in 2003, despite record drilling.

The disappointing results of recent drilling are largely due to geological factors. Mature gas fields in the main producing basins are approaching exhaustion,

---

6. World Gas Intelligence, *US Chemicals Wilt Under Hot Gas Prices* (21 April 2004).

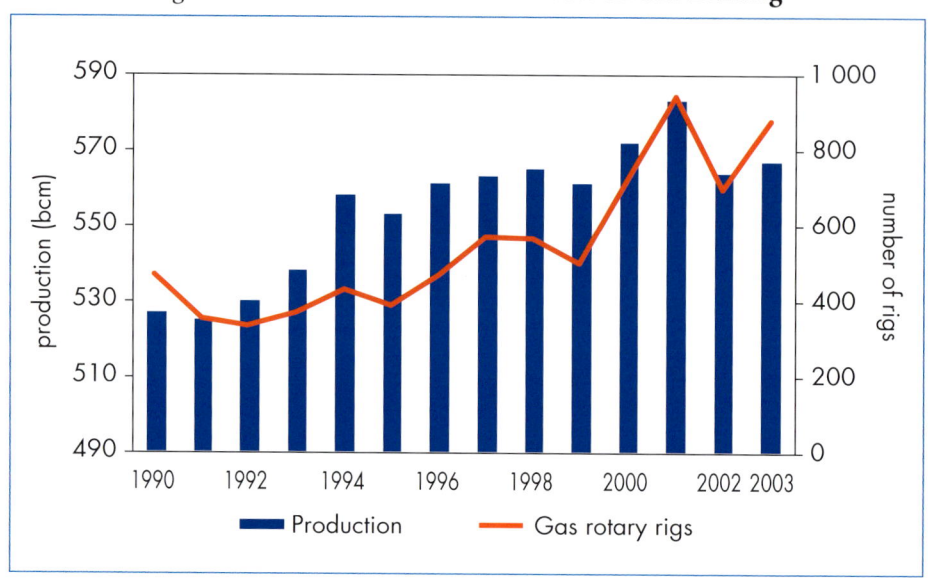

Figure 4.14: **US Marketed Gas Production and Drilling**

Source: US Department of Energy/Energy Information Administration (www.eia.doe.gov).

development costs are rising and production-decline rates per well are accelerating. The average observed rate of decline from producing wells is now about 20% per year. In other words, one-fifth of current production has to be replaced every year just to keep overall production flat. Decline rates at newly drilled wells in the United States are now in excess of 50% and more than 80% in the shallow waters of the Gulf of Mexico. As a result, many more wells need to be drilled now than in the past to compensate for natural production declines. Finding and development costs have risen sharply in recent years and are now thought to average in excess of $2.50/MBtu. Although average wellhead prices have risen even more, from about $2 in 1998 to $5 in 2003, the volatility of prices and the limited availability of drilling rigs are holding back what might otherwise be even higher rates of drilling.

In the next two decades or so, production from relatively undeveloped basins and new areas is expected to offset fully declines in the main established basins in the United States and Canada. These new sources include deepwater locations in the US Gulf of Mexico, Canadian offshore reserves in Labrador, Newfoundland and Nova Scotia, and the undeveloped Mackenzie River Delta/Beaufort Sea region in northern Canada. Non-conventional reserves – coal-bed methane, tight gas and shale gas – mostly found in the US Rocky Mountains will also provide a major new source of supply (Figure 4.15). The Alaskan North Slope is also expected to augment supply, although delivering Alaskan gas – and gas from northern Canada – to market will require the construction of a large-diameter, long-distance pipeline. How soon that line can be built will depend on regulatory

Figure 4.15: **Sources of North American Gas Supply**

Source: IEA based on NPC (2003).

Chapter 4 - Natural Gas Market Outlook

approvals and financing. The US Congress is considering loan and price guarantees for a proposed 35-bcm/year pipeline, which could cost as much as $20 billion. We assume that the line will be commissioned during the 2010s.

The natural gas industry is pushing the US Administration to ease restrictions on exploring for and developing reserves on federal lands, especially in the Rocky Mountains. The National Petroleum Council estimates that 7.7 trillion cubic metres of gas reserves in the lower-48 states are effectively stranded because of public access and environmental restrictions on drilling on federal lands.[7]

The prospects for increasing gas production in Mexico are very uncertain. Reserves in the Burgos basin, the country's large non-associated gas field, are large. But Pemex, the national oil and gas company, has so far been unable to finance the field's development. The government has tried to attract investment from foreign companies under multiple-service contracts, whereby Pemex retains ownership of the gas and the contractors take responsibility for financing, developing and operating projects for a set fee. But this formula has met with little interest. As a result, only modest increases in production are expected over the rest of the current decade. Mexico is expected to become more dependent on piped gas imports from the United States and, later, on LNG imports from Latin America and Asia.[8] In the longer term, we expect Mexican production to catch up demand.

We project that aggregate North American gas production will rise slowly from 766 bcm in 2003 to 833 bcm in 2010 and 904 bcm in 2030. Despite these increases, there will be a widening gap between indigenous production and demand, which will have to be filled with imports of LNG (Figure 4.16).

High gas prices continue to spur interest in developing LNG projects to supply North American markets. As of September 2004, six new regasification terminals had been approved by the US Federal Energy Regulatory Commission and the US Coastguard. Another 27 projects were awaiting approval in the United States, including expansions of existing terminals. Canada was considering seven applications for projects and Mexico, five (Table 4.3). Clearly, not all these projects will proceed. Local opposition will block some of them, especially new plants located on the east and west coasts. But other projects will undoubtedly emerge later. We foresee that three new terminals, each with a capacity of about 10 bcm (7 Mt) per year, will be operational by 2010, with the first commissioned by 2008. We project that imports will reach 197 bcm in 2030. The longer gas markets remain tight and spot prices high, the greater the number of LNG projects that will go ahead.

---

7. About 5.5 tcm of that amount are in the Rocky Mountains and mid-continent regions (NPC, 2003).
8. Most of the gas from planned projects in Baja on the Pacific Coast of Mexico would, nonetheless, go to US markets.

Figure 4.16: **North American Gas Balance**

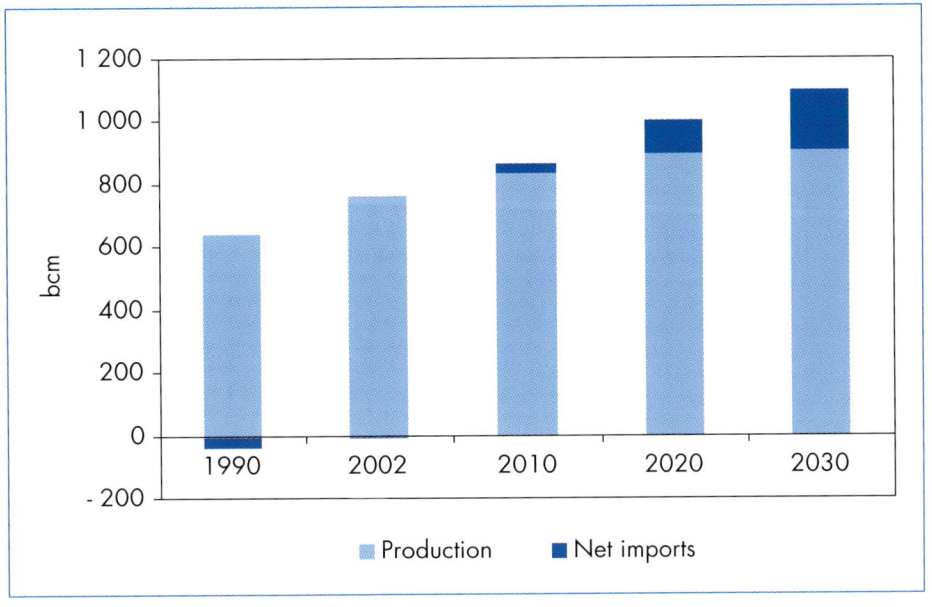

Table 4.3: **Existing and Planned* LNG Capacity in North America, September 2004**

Status	Country	Number of terminals	Capacity (million cubic metres/day)
Existing	United States	4	126
Approved	United States	6	192
Awaiting approval		39	994
of which	*United States*	*27*	*712*
	Canada	*7*	*131*
	Mexico	*5*	*151*
Total		**49**	**1 312**

* Projects at the filed or pre-filed stage of the authorisation process.
Source: Federal Energy Regulatory Commission (www.ferc.gov).

European Union

Natural gas demand in the European Union[9] is projected to grow by an average 1.8% per year over the projection period – the most rapid growth rate of any fuel other than non-hydro renewables. This will still be below the 4.7% rate of

9. In this *Outlook*, the European Union comprises 25 member states, including the ten new members that joined in May 2004.

growth in EU gas demand over the past three decades. The share of gas in total primary demand will continue to rise, from 23% at present to 32% in 2030. Gas-demand growth is projected to slow progressively throughout the projection period, from 2.3% in 2002-2010 to 1.4% in the 2020s. The power sector will be the main driver of gas demand, especially in the first half of the projection period (Figure 4.17). Gas is expected to account for the bulk of incremental power generation. The share of gas in power production is projected to surge from 15% in 2002 to over 35% in 2030, including hydrogen fuel cells based on gas. The EU power sector's use of gas will increase by 3.7% per year from 2002 to 2030. Demand in end-use sectors will also increase steadily: by around 0.9% per year in the residential and services sectors and by 1% per year in industry.

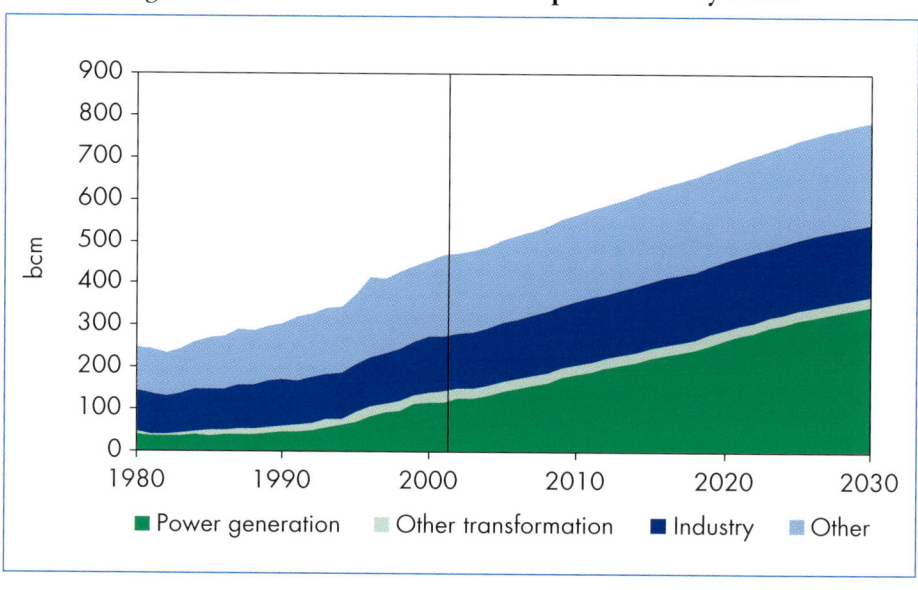

Figure 4.17: **Gas Demand in the European Union by Sector**

EU gas production amounted to around 240 bcm in 2003. Two producers – the United Kingdom (108 bcm) and the Netherlands (73 bcm) – accounted for 76% of the total, mostly from offshore fields in the North Sea – a mature producing region. Germany, Italy, Denmark and Poland are the only other significant producers. There is limited potential for increasing gas production in the region as resources are small. Proven reserves are less than 3.4 tcm, or 2% of the world's total. Production from the North Sea is expected to decline steadily over the projection period. The United Kingdom will become a major net importer of gas before the end of the current decade. Production in the Netherlands is also expected to continue to fall gradually, but the country will remain a net exporter of gas. Total EU gas production is projected to decline down to 225 bcm in 2010 and 147 bcm in 2030 (Figure 4.18).

Figure 4.18: **Gas Supply Balance in the European Union**

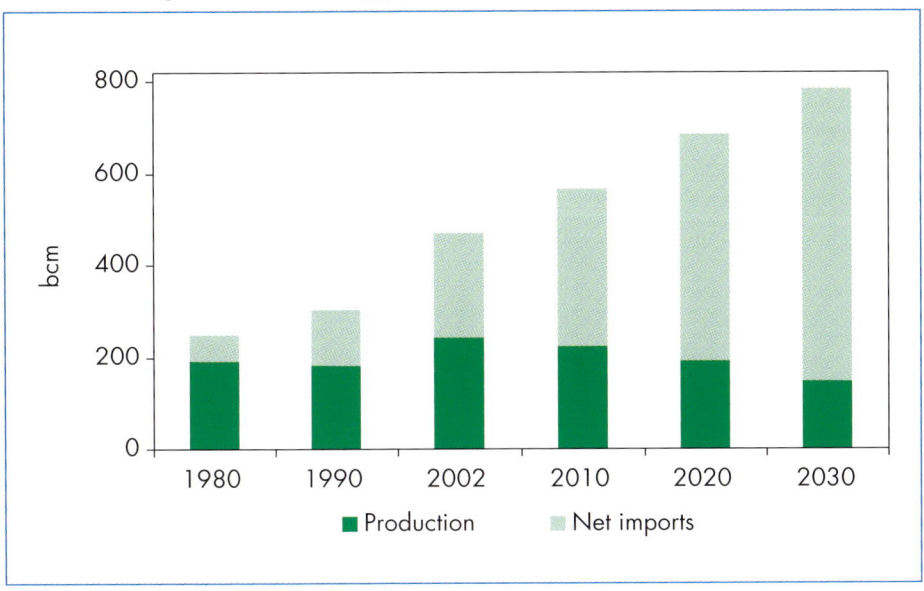

Rising demand and stagnating production will result in a surge in net imports, from 233 bcm in 2002 to 342 bcm in 2010 and 639 bcm in 2030. The bulk of this gas will go to meet new power-sector needs (Figure 4.19). The share of imports in the region's total gas demand will rise from 49% in 2002 to over 81% by the end of the projection period. Incremental imports are expected to come from the Union's three main current suppliers, Russia, Norway and Algeria. Production in Norway, most of which is exported to the European Union, is expected to continue to grow, from 77 bcm in 2003 to 94 bcm in 2010 and 135 bcm in 2030. Most of the increase will come from the Norwegian Sea and the Norwegian sector of the Barents Sea.

Europe will also import a mixture of piped gas and LNG from other African and former Soviet Union countries, the Middle East and Latin America. Russia will remain the largest single supplier in 2030, exporting around 155 bcm compared with 105 bcm in 2002. But the biggest increase in supplies will be from the Middle East, mostly in the form of LNG, although increasing quantities of gas are expected to be transported to Europe by pipeline from Iran and possibly Iraq towards the end of the projection period. Imports of LNG from Trinidad and Tobago and from Nigeria are set to rise. Other new sources of gas are expected to include the Caspian region (by pipeline), Libya (via under-sea pipeline), Egypt and Qatar (both as LNG). Venezuela could also emerge as an LNG supplier in the long term. Spot shipments from other LNG exporters in the Middle East, Latin America and Africa and possibly further afield could play an increasingly important role if a global short-term market in LNG develops. Turkey, which has over-contracted for gas supplies for the next several years, is expected to sell its surplus volumes to EU countries once a pipeline link has been built.

Figure 4.19: **Increase in Gas Consumption and Imports in the European Union, 2002-2030**

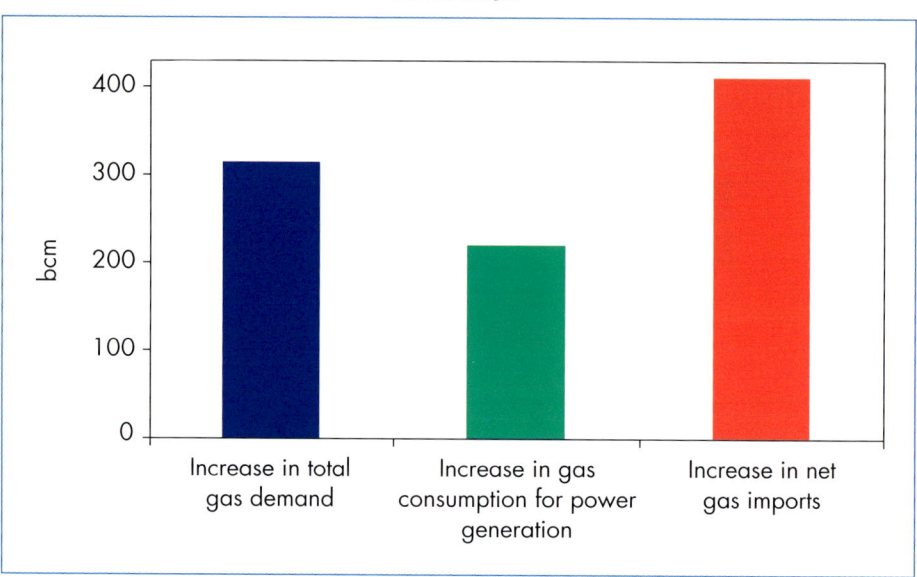

Progress in liberalising EU gas markets varies markedly among member states (Table 4.4). The implementation of a second EU gas directive, adopted in 2003, should give impetus to the development of competition in several countries. The directive allows all industrial and commercial consumers to choose their supplier starting in July 2004 and all other consumers to do so by July 2007. It also requires vertically integrated gas utilities to unbundle their transmission operations by July 2004 and their distribution operations by July 2007. According to a recent European Commission report, the main obstacles to the achievement of a truly competitive EU gas market are delays in opening up retail markets, ineffective regulation of network services and the concentration of market power in a small number of large companies (EC, 2004). The Commission has proposed a new regulation, similar to one already adopted for electricity, to promote cross-border trade.

OECD Asia

Gas demand in the OECD Asia region – Japan and Korea – will grow from 99 bcm in 2002 to 195 bcm in 2030, an increase of 2.5% per year. Trends differ between the two countries: demand will grow by 1.9% per year in Japan

Table 4.4: **Status of Gas Market Liberalisation in the European Union**

	Declared retail market opening (%)	Size of open retail market (bcm)	Switching from initial supplier, 2002 (%)		Concentration in wholesale market
			Large eligible industrial users	Small commercial and domestic users	
Austria	100	8	6	0	Yes
Belgium	83	9	Not known	Not applicable	Yes
Denmark	100	5	17	Not applicable	Yes
France	37	15	20	Not applicable	Moderate
Germany	100	90	5	<2	Moderate
Ireland	85	4	100	Not applicable	No
Italy	100	69	10	0	Yes
Luxembourg	72	<1	0	Not applicable	Yes
Netherlands	60	25	15	Not applicable	Moderate
Spain	100	20	38	1	Yes
Sweden	51	<1	0	Not applicable	Yes
United Kingdom	100	105	16	19	No
Estonia	80	<1	0	Not applicable	Yes
Latvia	0	0	0	Not applicable	Yes
Lithuania	80	2	0	Not applicable	Moderate
Poland	34	4	0	Not applicable	Yes
Czech Republic	0	0	0	Not applicable	Yes
Slovak Republic	33	2	<5	Not applicable	Yes
Hungary	0	0	Not known	Not applicable	Yes
Slovenia	50	<1	0	Not applicable	Yes

Source: EC (2004).

and by a whopping 3.9% in Korea (Figure 4.20). In both markets, power generation will be the main driver of demand, especially in Korea where gas use in power stations is expected to grow by 5.4% per year.

Japan will remain the largest gas consumer and importer of LNG, which currently meets 94% of Japanese gas needs. Primary gas demand is projected to grow from 76 bcm in 2002 to 128 bcm in 2030, with power generation accounting for 40 bcm, or just over three-quarters, of the increase. Power-sector demand could grow even faster if the government's plans to boost nuclear capacity are delayed or blocked by local opposition.

There is no prospect of any significant increase in gas production in Japan, as the country's reserves are minimal. LNG will continue to account for the bulk

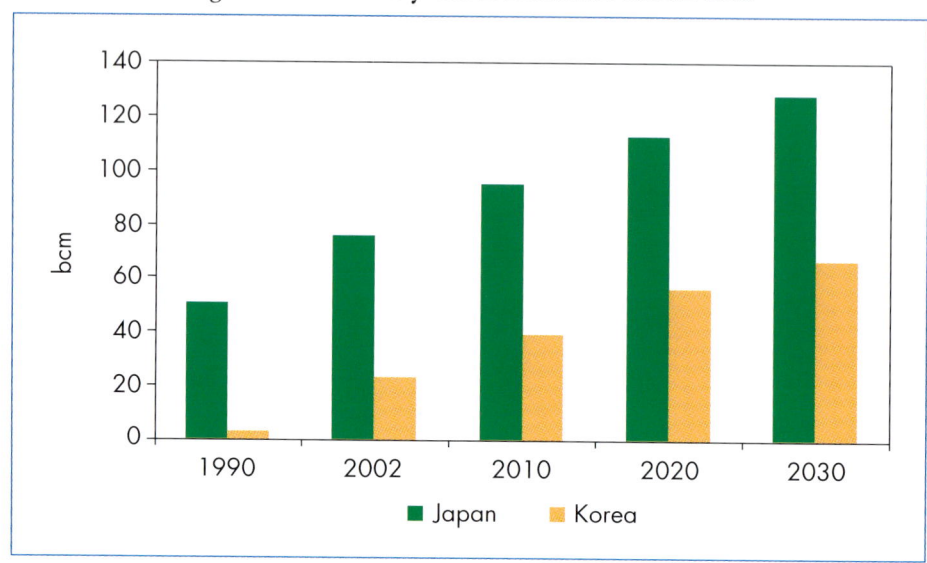

Figure 4.20: **Primary Gas Demand in OECD Asia**

of the country's gas imports. Many of the long-term purchase contracts signed in the 1980s have reached their renewal dates or soon will. Gas and electricity utilities are finding it harder than before to negotiate rigid long-term take-or-pay contracts, because of uncertainty surrounding future gas demand and because of increased competition between buyers in Asia. Some contracts have already been renewed, generally for shorter periods and with more flexible delivery terms than before, while others are still being negotiated. In addition, the first-ever LNG imports into Japan from Russia will begin in 2007, with gas coming from the Sakhalin II project. Japan also plans to import LNG from a planned plant at Darwin in North Australia. Regasification capacity at the country's 24 existing terminals is sufficient to meet the increase in imports. Imports of gas by pipeline, from Sakhalin in Russia, are expected to begin towards the end of the projection period. The timing of this project is very uncertain, in view of the high cost of building an undersea line to northern Japan, as well as onshore lines.

Korea's gas market will remain the most dynamic in the OECD. Demand is expected to surge from 23 bcm in 2002 to 67 bcm in 2030. Two-thirds of the increase will come from the power sector, with the residential sector accounting for most of the rest. The long-delayed restructuring and privatisation of the Korean gas sector is assumed to proceed. Like Japan, Korea relies on imported LNG for all its gas needs. It is currently the second-largest LNG importer in the world. Piped gas, initially from Russia's Kovykta field in eastern Siberia, is expected to make a growing contribution to Korea's imports starting in the second decade of the projection period. The timing of that and of other Russian pipeline projects will depend critically on China's gas needs, as Korean demand alone will be insufficient to support such hugely expensive projects. It will also depend on Gazprom's role in the development of eastern Siberian gas resources.

A growing proportion of any new LNG demand in Japan and Korea is likely to be met by spot purchases. Long-term contracts will, nonetheless, continue to be used for new greenfield projects. The supply and purchase contracts with Japanese buyers for LNG from Sakhalin are for a period of 21 years, while those for Darwin LNG will run for 17 years (IEA, 2004).

OECD Oceania

The only OECD country in the Asia-Pacific region with significant gas resources is *Australia*, which exports gas in the form of LNG to Japan, Korea and other regional markets. Australia's proven gas reserves were 3.9 tcm at the end of 2003, equal to more than 100 years of production at current rates. The commonwealth and state governments have introduced gas-to-gas competition through mandatory open access to pipelines. Most Australian states now allow large and medium-sized customers to choose their own gas supplier. The removal of regulatory barriers and the interconnection of state networks have led to a sharp increase in inter-state trade. Further reforms are planned. In *New Zealand*, growing output from small fields is expected to make good the imminent depletion of the Maui gas field, which still produces 80% of New Zealand's 6 bcm of gas supply, though the country may need to import LNG in the longer term.

Primary gas demand in Australia and New Zealand together is projected to rise from 31 bcm in 2002 to 38 bcm in 2010 and 52 bcm in 2030. Combined production in the two countries is projected to increase even more, from 41 bcm in 2002 to 58 bcm in 2010 and 98 bcm in 2030. As a result, Australian LNG exports are projected to increase from 10 bcm in 2002 to about 20 bcm in 2010 and close to 50 bcm in 2030. A fourth LNG train being built at the North West Shelf project will boost export capacity from 7.5 Mt/year at present to around 11.7 million tonnes when completed in 2004. A fifth train is planned. In addition, a second 3.5-Mt/year liquefaction plant at Darwin will be completed in 2006. Four other projects – Gorgon, Sunrise, Browse Gas and Scarborough – are planned.

Transition Economies

The transition economies as a whole will remain the world's second-largest gas market and one of the largest exporters. Primary gas demand will grow from 635 bcm in 2002 to 984 bcm in 2030 – an increase of 1.6% per year. Russia will remain the largest consumer and producer.[10] Eastern European countries and, to a lesser extent, the Central Asian countries, will account for a growing share of the transition economies' overall gas demand over the projection period (Figure 4.21).[11]

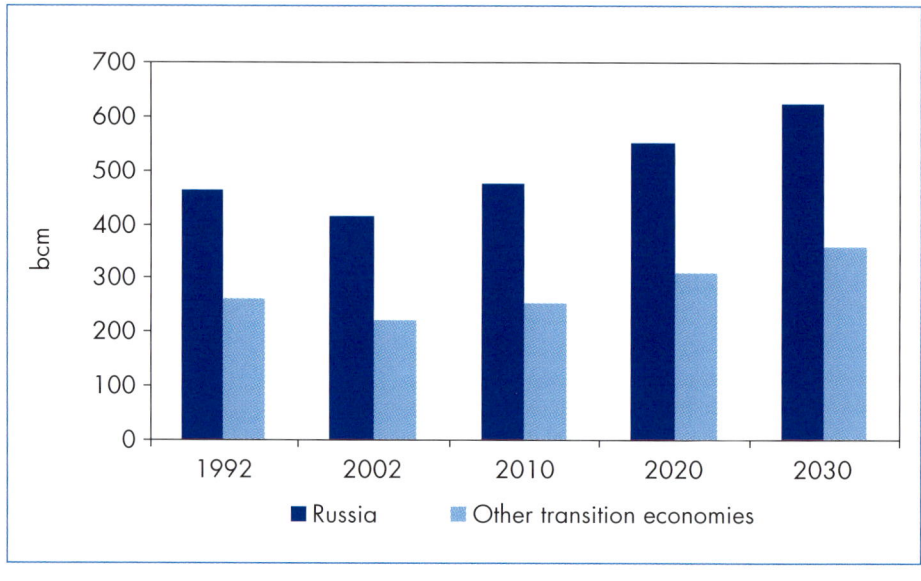

Figure 4.21: **Primary Gas Demand in Transition Economies**

Gas production in the Caspian region, comprising Turkmenistan, Uzbekistan, Kazakhstan and Azerbaijan, is expected to grow steadily over the projection period. Over 1 tcm of the region's gas reserves of 8 tcm have already been sold to Russia's Gazprom under long-term deals signed in 2002 and 2003. Those deals will postpone the need for Gazprom to invest in developing more expensive indigenous supply sources. They will also effectively eliminate those countries as direct competitors for export sales to Europe. Gazprom has also restricted their national gas companies' access to its transmission system, impeding their ability to sell directly to European buyers.

A new pipeline system is being built to carry gas from the Shah Deniz field in Azerbaijan to Turkey via Georgia, bypassing Russia. The first segment of the $2.5-billion line, which will ultimately have a capacity of 16 bcm/year, is expected to be completed in 2006. A long-distance line from Turkmenistan to Turkey is assumed to be built towards the end of the projection period. This will, however, depend on Turkmenistan's success in proving up more reserves,

10. Gas-market prospects in Russia are discussed in detail in Chapter 9.
11. See Annex E for regional definitions.

and on geopolitical developments in the region. Another proposed pipeline from Turkmenistan to China, which could also pick up gas from Kazakhstan and Uzbekistan, is assumed to proceed in the last decade of the projection period. The timing of this project is very uncertain because of geopolitical factors.

Developing Asia

Consumption of natural gas is expected to grow faster in developing Asia than in any other major region. Primary demand is projected to expand more than threefold, from 208 bcm in 2002 to 322 bcm in 2010 and 672 bcm in 2030. Power generation will account for over 50% of incremental demand, but industry will contribute 23% – a large share compared to most other regions. East Asia will remain the main market, but the importance of China and India will increase markedly (Figure 4.22).

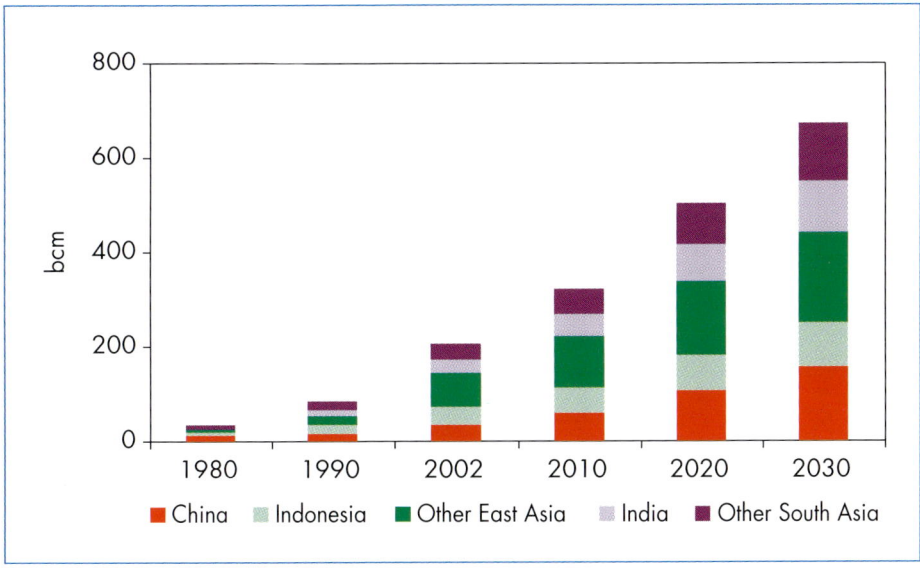

Figure 4.22: **Primary Gas Demand in Developing Asia**

China's natural gas market will take off in the second half of this decade as major new gas-infrastructure projects are completed. The first section of the ambitious West-East Pipeline project, which will bring gas from the central and western provinces to energy-hungry markets in and around Shanghai, is almost complete. First gas from the Chanqing field in the Ordos basin in central China is expected to flow at the end of 2004. PetroChina, which is leading the project, has signed more than 20 take-or-pay contracts covering the majority of the pipeline's initial

12-bcm/year capacity. China will also begin importing LNG from Australia in 2006 at the 3.7-Mt/year terminal which the China National Offshore Oil Corporation is building in the Guandong province. Capacity may be expanded to 5.7 Mt/year by 2008. CNOOC is planning a second 2.6-Mt/year terminal in Fujian to be commissioned in 2007 at the soonest. The government has refused to approve a third terminal that CNOOC wants to build at Zhejiang. It will have to wait until the first two plants are operational. The pace of gas-market expansion will depend on the effectiveness of regulatory reforms.

Table 4.5: **LNG Projects in India**

Project	Location (state)	Capacity (Mt/year)	Supplier	Status
Petronet LNG	Dahej (Gujarat)	5.0	Qatar	Commercial sales began April 2004.
Metropolis Gas	Dabhol (Maharashtra)	5.0	Oman, Abu Dhabi	Complete; commissioning delayed by contractual dispute.
Shell Hazira LNG	Hazira (Gujarat)	2.5	Shell portfolio	Under construction; first gas due end-2004.
Petronet LNG	Koch (Kerala)	2.5	Qatar	Planned.
Dakshin Bharat Energy	Ennore (Tamil Nadu)	2.5	Qatar	Planned.
Gujarat Pipavav LNG	Pipavav (Gujarat)	2.6	Yemen	Planned.
Kakainda Indian Oil LNG	Kakinda (Andhra Pradesh)	2.5	Malaysia	Planned.
Gopalpur LNG	Gopalpur (Orissa)	5.0	Australia	Planned.
Reliance LNG	Jamnagar (Gujarat)	5.0	Not known	Proposed prior to the company's offshore gas find.

Sources: IEA database; *Gas Matters* (February 2004).

Gas consumption is poised to grow strongly in *India*, fuelled largely by LNG imports. Bolstered by Reliance Industries' recent discovery of a large deepwater deposit, indigenous output is expected to grow from 27 bcm in 2003 to 66 bcm in 2030, but that will not be fast enough to meet demand. The industrial and power sectors are expected to push primary gas consumption up from 28 bcm in 2003 to 45 bcm in 2010 and 110 bcm in 2030. Indian LNG imports began in early 2004 with the arrival of the first cargo at Petronet's terminal at Dahej in

Gujarat. Volumes are expected to build quickly to 5 Mt, or 8 bcm, per year. Another 5-Mt/year plant at Dabhol has been completed but has yet to begin commercial operations because of a contractual dispute over supplies to an adjacent power plant. One other terminal at Hazira in Gujarat is under construction and several others are planned (Table 4.5). LNG imports are expected to reach at least 10 bcm by 2010 and 30 bcm by 2030.

A project for a pipeline to bring gas from Bangladesh to eastern India has stalled because of nationalist opposition to exporting gas that could otherwise be used to supply local markets. A recent study by BHP Billiton, a diversified resource conglomerate, has confirmed the technical and economic feasibility of a proposed pipeline from Iran to India, but our projections assume that the project will not proceed before 2030 because of political tensions in the region. A 290-kilometre line from Myanmar to India, with an initial capacity of about 5 bcm, is assumed to be built after 2010.

Middle East

Middle East gas demand will more than double between 2002 and 2030, from 219 bcm to 470 bcm, led by the power sector (Figure 4.23). Iran and Saudi Arabia, which together accounted for more than 60% of the region's gas consumption in 2002, will remain the main markets. By 2030, gas will have overtaken oil as the region's main energy source, meeting around 50% of primary needs. It is used mainly in industry as a petrochemical feedstock and

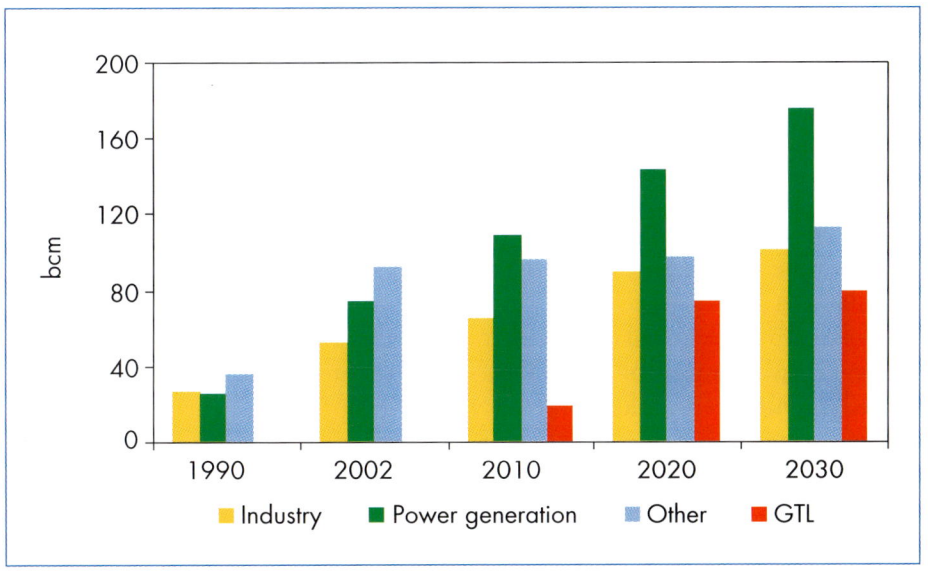

Figure 4.23: **Gas Demand in the Middle East by Sector**

to fuel water-desalination plants, but the power sector's share of primary demand – currently a little over one-third – is growing rapidly. By 2030, power generation will absorb almost 40% of all the gas used in the region. Gas-to-liquids plants will also take a growing share, reaching 17% by 2030.

Production in the Middle East is projected to increase by 4.2% between 2002 and 2030. The region holds 72 trillion cubic metres of proven gas reserves – 40% of the world total. Iran and Qatar alone have 52 tcm, or almost a third of global reserves. Most of the additional output will be exported, predominantly as LNG. Exports are projected to jump from 30 bcm in 2002 to about 110 bcm in 2010 and just over 300 bcm in 2030. The share of exports in total production will grow from 14% in 2002 to 42% in 2030.

Most of the increase in exports will come from Iran, Qatar, Oman, the United Arab Emirates and Yemen. Qatar – by far the largest exporter in the region already – is planning several new projects to add to the 14.9 Mt/year of capacity at the country's three existing liquefaction plants. The planned two-train, 15-Mt/year RasGas-3 and QatarGas-2 projects, together with the four-train, 18-Mt/year QatarGas-3 project could boost total export capacity to over 65 Mt/year by early in the second decade. All the gas will come from the giant offshore North Field. Iran has stepped up its efforts to get into the LNG business, using gas from the same field (known as South Pars in Iran), which straddles its maritime border with Qatar. Pars LNG, a joint venture of Total, Petronas and the National Iranian Oil Company could be completed as early as 2008. A project to export 4 bcm of gas by pipeline to Kuwait starting in 2007 is also under discussion. Iran is targeting total exports of about 60 bcm/year (45 Mt/year) by 2014, though this appears a very optimistic goal. Iraq could emerge as a major exporter towards the end of the projection period.

Europe and North America will displace Asia as the main markets for Middle East gas near the end of the projection period. Export pipelines to Europe from Iran and possibly from Iraq are expected to be built towards the end of the projection period, though their timing will depend on geopolitical developments. One export pipeline opened in 2001 runs from Iran to Turkey.

Whether the required investment in new production and export projects can be mobilised is highly uncertain. Project risks in the region vary according to geopolitical and technical factors. Cross-border pipeline projects are considered extremely risky in view of regional political tensions. So far, however, access to capital has not been a major problem for new gas development projects. Most projects have been funded out of a mixture of retained earnings, state budget allocations and, in the case of most LNG projects, project finance and/or international bond issues. But there are signs that financing may be more difficult in the future, especially if geopolitical risk remains high. In some cases, governments' ability to finance new projects could be limited by budget deficits and by competing demands on their financial resources.

Africa

Africa will experience very rapid growth in primary gas demand, averaging 5.1% a year from 2002 to 2030, but from a very low base. Despite a fourfold increase to 276 bcm in 2030, African demand will still be much lower in absolute and per capita terms than anywhere else in the world. Poverty, a lack of manufacturing industry and the limited potential for residential use because of climatic factors will continue to hold down gas consumption in most parts of the continent.

Africa is well endowed with gas resources, with 14 trillion cubic metres or 8% of total world proven reserves. African gas production and exports are set to rise strongly in the next few years, mainly as a result of greenfield projects and expansions of existing facilities in North and West Africa. Mozambique will also emerge as a major new producer. Europe will remain the main market for African gas exports, but North America will take a growing share (Figure 4.24).

GTL is expected to provide a new outlet for gas reserves in West Africa that would otherwise be stranded. The planned Escravos GTL plant in Nigeria, which would initially process around 4 bcm/year of gas, is expected to be operational before the end of the current decade (see Box 4.1).

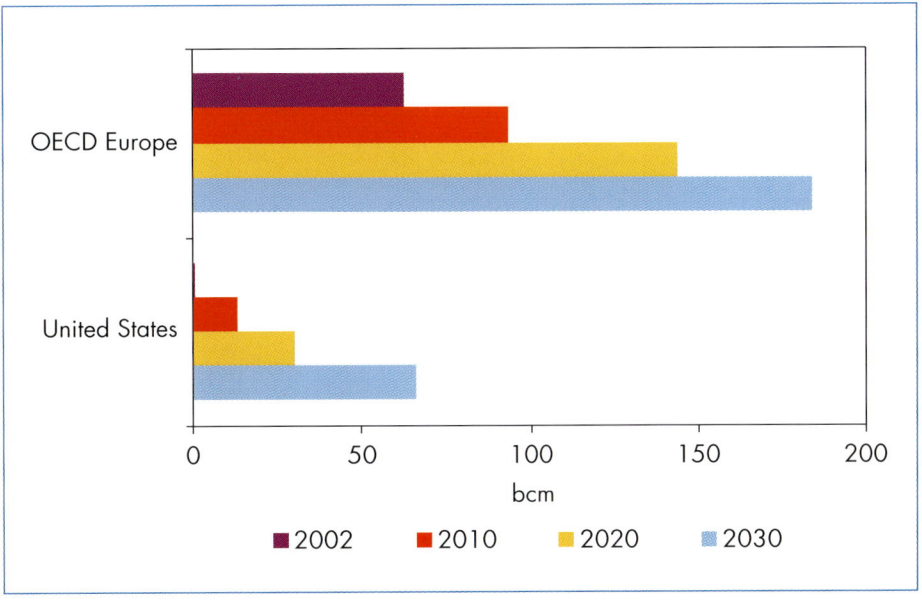

Figure 4.24: **African Gas Exports by Region**

North African export capacity will increase substantially in the second half of this decade, with the commissioning of new pipelines from Algeria to Spain and from Algeria and Libya to Italy. New LNG projects and expansions in Algeria, Libya and Egypt are also planned. Total capacity is expected to jump from around 63 bcm/year now to more than 120 bcm/year by 2010. The three trains at the Skikda LNG plant in Algeria, which were damaged in an explosion in January 2004, are assumed to be rebuilt or repaired. Nigerian LNG capacity will rise from 9 Mt/year at present to 17 Mt/year in 2005 when two new trains are commissioned at the Bonny Island plant in 2005. A sixth 4-Mt/year train is expected to be approved soon. In addition, a new two-train, 10-Mt/year plant, Brass River LNG, is expected to become operational by the end of the decade. There are also plans to build liquefaction plants in Equatorial Guinea and Angola.

Latin America

Primary gas demand in Latin America is projected to grow at a brisk 4.4% per year, from 102 bcm in 2002 to 335 bcm in 2030 – driven mainly by the power sector (Figure 4.25). Brazil, in particular, is poised for rapid growth in gas use, with demand projected to rise from only 13 bcm in 2002 to 64 bcm in 2030. Argentina and Venezuela currently account for well over half of the region's gas use.

Production will grow even more rapidly than demand. New LNG export projects are expected to be developed in several countries, including Trinidad and Tobago – the only country in Latin America that already produces LNG –

Figure 4.25: **Latin American Gas Balance** (bcm)

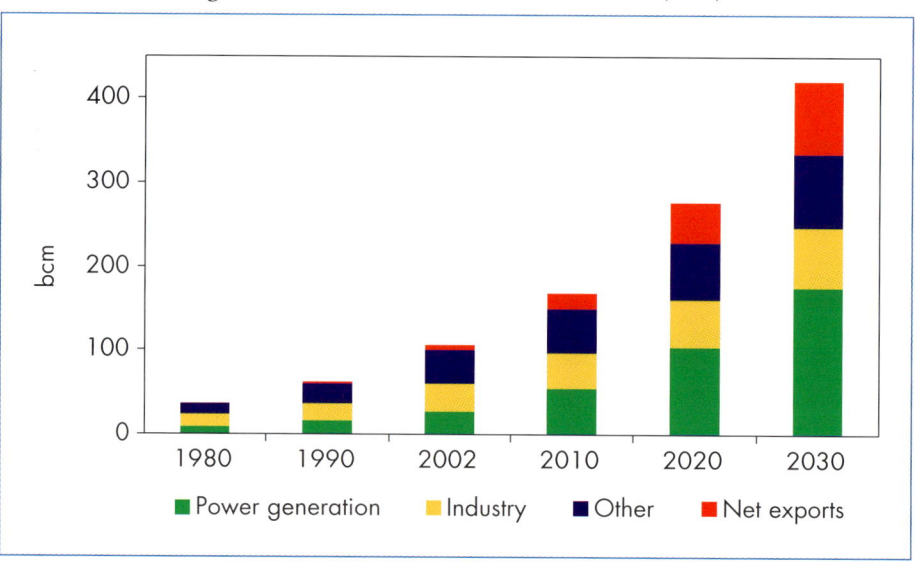

Venezuela, Peru and possibly Bolivia and Brazil. In July 2004, a referendum in Bolivia gave qualified approval to the government to proceed with a controversial plan to export gas to the United States or Mexico via an LNG liquefaction plant, probably in Peru. Overall, Latin American LNG exports are expected to reach 19 bcm by 2010 and close to 90 bcm by 2030. Trinidad and Tobago is adding a fourth train at its Point Fortin plant, which will raise capacity from 9.9 Mt/year to 15.1 Mt/year. Two further trains are planned. A project to build a 4.4 Mt/year LNG plant in Peru, using gas from the Camisea field, is expected to be completed by the end of the decade. Venezuela has enough gas reserves to become a major LNG exporter, but the country's plans to build liquefaction facilities have stalled because of political uncertainty and financing difficulties. A consortium involving foreign companies, led by the Venezuelan national oil company, PDVSA, is trying to develop a one-train 4.7-Mt/year plant in the Gulf of Paria.

CHAPTER 5

COAL MARKET OUTLOOK

HIGHLIGHTS

- Coal will continue to play a key role in the world energy mix. In 2030, coal will meet 22% of global energy needs, essentially the same as today. The electricity sector will be responsible for over 95% of the growth in demand as coal remains the leading fuel for power generation. High quality coking coal will remain the leading reductant in steel-making.
- Coal demand will increase most in developing Asian countries. China and India alone will be responsible for 68% of the increase in demand to 2030. Demand growth in the OECD will be minimal.
- Despite the rapid escalation of coal prices in recent years, many market fundamentals remain unchanged, so that prices are likely to moderate in the long term. There are many existing and potential suppliers, the market is still highly competitive and coal prices remain low relative to the prices of other primary energy commodities.
- World proven coal reserves are enormous. Compared with oil and natural gas, they are widely dispersed. Over 40% of the world's 907 billion tonnes of coal reserves – equal to almost 200 years of production at current rates – is located in OECD countries.
- China will consolidate its position as the world's largest coal producer. Steam coal production will remain widely dispersed geographically, but coking coal production will be increasingly concentrated in China, Australia, the United States and Canada.
- Seaborne inter-regional coal trade will continue to grow faster than global coal demand. Increased trade will meet demand from countries in Asia, which lack high-quality indigenous resources, and from OECD Europe, where production is declining. Australia will remain the world's leading exporter of both steam and coking coal.
- In the OECD, demand prospects for coal depend greatly on climate change polices and on the development and deployment of advanced clean coal technologies. Such factors will have less influence in non-OECD countries, which often place a higher value on economic growth and security of supply than on environmental objectives.
- Some $400 billion needs to be invested in the coal industry, globally, over the period 2003-2030. If investment in coal-fired power generation capacity is included, that amount rises to $1.7 trillion.

This chapter presents the Reference Scenario outlook for the coal market to 2030. It first details our coal-demand projections by region and by sector and discusses the main factors and uncertainties that will influence consumption patterns, including the important role of clean coal technologies. It then provides an overview of the world's coal reserves and our expectations for their development. This is followed by analysis of the implications of our projections on world coal trade and coal-sector investment. The chapter ends with analysis of the prospects for coal demand and supply in seven important regions.

Coal Demand

Coal will continue to play a key role in the world energy mix, with demand projected to grow at an average annual rate of 1.4% to 2030. At that time, coal will meet 22% of global energy needs, only 1% less than it does today. There is significant variation between regions in the demand prospects for coal (Table 5.1).

Table 5.1: **World Coal Demand*** (Mt)

	2002		2030		Average annual rate of growth in demand, 2002-2030 %
	Million tonnes	Coal's share of electricity generation %	Million tonnes	Coal's share of electricity generation %	
OECD North America	1 051	46	1 222	40	0.5
OECD Europe	822	29	816	24	0.0
OECD Pacific	364	36	423	29	0.5
OECD	**2 237**	**38**	**2 461**	**33**	**0.3**
Russia	220	19	244	15	0.4
Other transition economies	249	27	340	18	1.1
Transition economies	**469**	**22**	**584**	**16**	**0.8**
China	1 308	77	2 402	72	2.2
East Asia	160	28	456	49	3.8
South Asia	396	60	773	54	2.4
Latin America	30	4	66	5	2.8
Middle East	15	6	23	5	1.6
Africa	174	47	264	29	1.5
Developing countries	**2 085**	**45**	**3 984**	**47**	**2.3**
World	**4 791**	**39**	**7 029**	**38**	**1.4**
European Union	*767*	*31*	*716*	*25*	*-0.2*

* Including hard coal (steam coal and coking coal), brown coal and peat.

Coal demand will be driven primarily by the surging energy needs of developing Asia, particularly China and India, in both of which countries domestic coal supplies are abundant. In OECD North America and the OECD Pacific region, coal use will grow at a slower rate. Coal demand in OECD Europe will increase slowly in the first half of the *Outlook* period but then decline to slightly less than current use by 2030.

Consumption of steam coal, which is principally used for generating electricity and process heat, will grow by 1.5% per year over 2002-2030. Demand for coking coal, which is mainly used for making iron and steel, will increase by 0.9%. Lignite or brown coal, a fuel with low calorific value which is used in power generation, will grow by 1.0%. The use of brown coal is limited by its high moisture content, which makes long-distance transportation uneconomic, and also by its propensity to self-ignition.

Sectoral Demand

The power sector's share of global coal demand will rise from 69% in 2002 to 79% by 2030 (Figure 5.1). Despite this growth, coal's share of global electricity production will decline slightly, from 39% at present to 38% in 2030. The main contributor to demand growth will be the rapid expansion of coal-fired generating capacity in China and other parts of developing Asia. Renewed interest in coal-fired power plants is also becoming apparent in several mature markets, particularly the United States, where the price of natural gas has risen sharply. In the long term, coal use in the power sector will

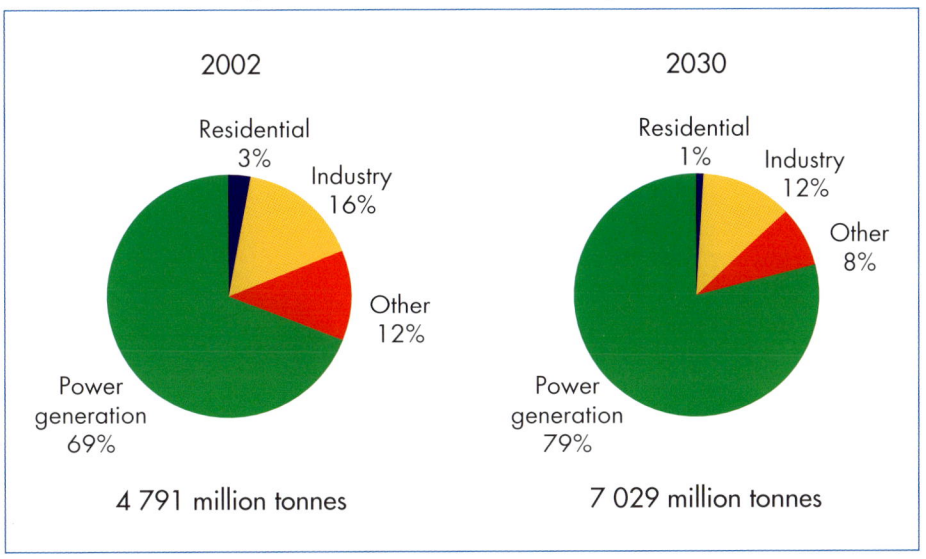

Figure 5.1: **World Coal Demand by Sector**

be boosted by an assumed reduction in its price relative to gas, as well as by the gradual development and deployment of advanced coal technologies. The main impediment to investment in coal-fired capacity will be the cost of meeting climate change targets and other environmental requirements.

Industrial coal use, principally the use of coking coal for the manufacture of iron and steel, will increase by 0.5% per year over the projection period. This modest growth reflects increased use of recycled steel and continuing improvements in the efficiency of iron production and blast-furnace technology. As with steam coal, growth in the coking coal market will be most robust in developing Asian countries, where construction, car production and demand for household appliances will increase as incomes rise. Demand for coking coal will continue to decline in the OECD, as it has since the 1980s, as coke and pig iron production shifts to developing regions. Coal for industrial and residential heating will continue to lose market share to gas.

Impact of Environmental Policy and Technology

The main uncertainty surrounding the outlook for coal demand in OECD regions is the impact of government policies and measures to address environmental concerns. Typical existing coal-fired power plants emit more pollutants and more carbon dioxide than do oil or natural gas-fired plants. Partly because of such uncertainty, few coal-fired power stations have been built in recent years in developed countries outside Asia. Potential investors have shied away from coal-fired power stations and boilers for fear that new environmental rules could limit their use and increase costs.

Implementation of clean coal technologies[1] (CCT), which would improve the thermal efficiency of coal production and use and reduce emissions, could minimise investment risks and give a major boost to the prospects for coal demand. While attention is usually focused on power generation technologies, continuous technological advances are being made along the entire coal chain. New techniques have been developed for coal mining and the preparation of coal for use in power stations, as well as for coal combustion, emissions control and the disposal of solid waste. Technologies on the horizon such as carbon capture and storage could achieve near-zero emissions of *all* pollutants from coal-fired power plants (Box 5.1 and Figure 5.2).

Cost is the major barrier to the adoption of clean coal technologies. Government actions, including increased research and development, could help reduce costs. If they do, coal could remain a low-cost source of electricity generation in a carbon-constrained environment.

---

1. Clean coal technologies are defined as technologies designed to enhance the efficiency and the environmental acceptability of coal extraction, preparation and use.

> *Box 5.1:* **FutureGen: Zero-Emission Technologies**
>
> Coal-fired power stations can emit several noxious pollutants including particulates, mercury and sulphur and nitrogen oxides. Available commercial technology can significantly reduce these emissions. It is now thought technically possible virtually to eliminate all these emissions, as well as those of carbon dioxide.
>
> Using available technology and technology under development, the United States government, in partnership with industry, plans to build a prototype zero-emission coal plant. The programme, known as the Integrated Sequestration and Hydrogen Research Initiative, or FutureGen, will cost $1 billion. The plant will have a capacity of 275 MW and will produce hydrogen as well as electricity using integrated gasification combined-cycle technology. It will serve as a large-scale engineering laboratory for testing new technologies, including clean power, carbon capture, and coal-to-hydrogen techniques. If built, it would be the cleanest power plant based on fossil fuels in the world.
>
> The plant will convert coal into synthesis gas made up primarily of hydrogen and carbon monoxide. This gas will be made to react with steam to produce additional hydrogen and a concentrated stream of carbon dioxide. The hydrogen will then be used as a fuel for electric power generation in turbines, fuel cells or hybrid combinations of the two. The hydrogen could also be used as a feedstock for refineries and, in the future, as a fuel for cars and trucks.
>
> The captured carbon dioxide would be separated from the hydrogen by membranes currently under development and then permanently stored underground in a geological formation, such as depleted oil and gas reservoirs, deep saline aquifers, or basalt formations. Sulphur dioxide and nitrogen oxides would be cleaned from the coal gases and converted to useable by-products such as fertilisers and soil enhancers. Mercury pollutants would also be removed. The project could begin as early as 2005 and is expected to be partly funded by industry.

Because of the long life of coal-fired power plants and boilers, and the higher cost of building advanced plants, new infrastructure will come into operation only very gradually. Large numbers of conventional plants are still being built, and many existing conventional plants are being retrofitted with equipment to limit emissions and extend the plants' economic lives.

Figure 5.2: **Reductions in Emissions of CO_2 through Technological Innovation**

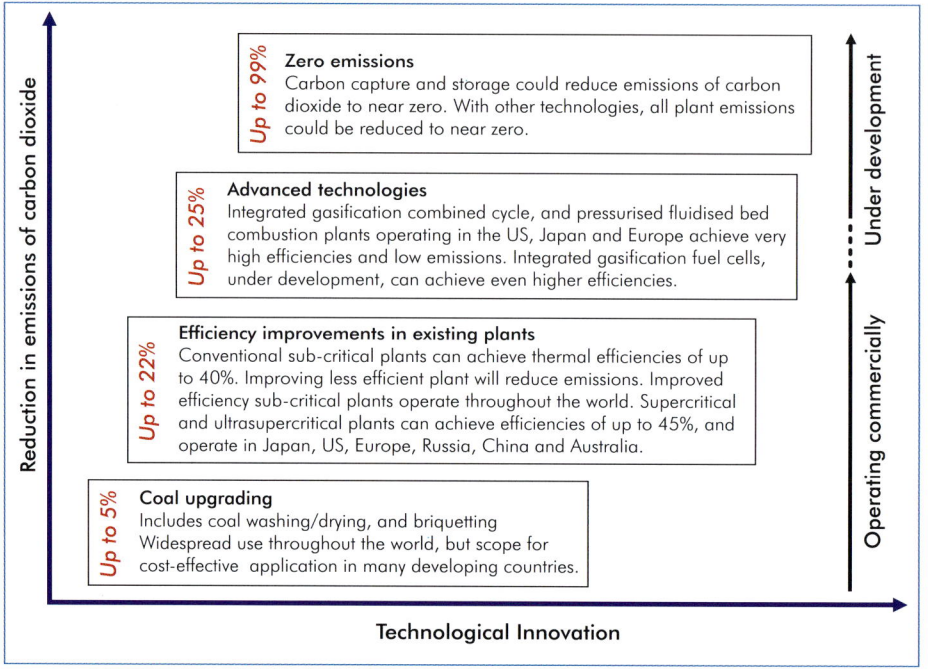

Source: Based on World Coal Institute (2003).

Coal Reserves and Production

Proven Reserves

Proven coal reserves worldwide total 907 billion tonnes or almost 200 years of production at current rates (Table 5.2). In energy equivalent terms, this exceeds the combined proven reserves of both oil and gas by a very wide margin. Hard coal – coking coal and steam coal – makes up 83% of proven reserves. The rest is brown coal.

Countries which rely heavily on coal for domestic needs or export revenue generally have large reserves of coal. The reason for this is that as coal production, consumption and transportation infrastructure expands, resources in proximity to already exploited reserves often become economically viable and enter the proven reserve classification. The largest reserves are found in the United States, Russia and China. Although steam coal reserves are widespread, mining costs and quality vary. Coking coal reserves are more limited, with the best-quality deposits found in the United States, Australia, Canada and China. Brown coal is typically used by power

Table 5.2: **Proven Coal Reserves at End-2002** (Mt)

	Hard Coal	Brown Coal	Total
OECD Europe	22 420	17 041	39 461
OECD North America	218 818	35 614	254 432
OECD Pacific	39 677	38 033	77 710
OECD	**280 915**	**90 688**	**371 603**
Transition economies	**208 762**	**38 872**	**247 634**
of which Russia	*146 560*	*10 450*	*157 010*
China	95 900	18 600	114 500
East Asia	3 053	4 330	7 383
South Asia	90 146	5 350	95 496
of which India	*90 085*	*2 360*	*92 445*
Latin America	19 769	124	19 893
of which Brazil	*10 113*	*–*	*10 113*
Africa	50 333	3	50 336
Middle East	419	–	419
World	**749 297**	**157 967**	**907 264**

Source: World Energy Council (2003).

plants situated close to where the coal is mined, because its energy content is low and it is costly and difficult to transport.

Production Prospects

Global coal production will increase by 1.4% per annum over the Outlook period, reaching 7 billion tonnes in 2030. China will reinforce its position as the world's leading producer, accounting for around half the increase in global output over the period (Figure 5.3). The other major producers in 2030 will be the United States, India and Australia. Coal production in Europe will continue to decline as subsidies are reduced and uncompetitive mines are closed.

Some 80% of incremental coal production during 2002-2030 will be steam coal. By 2030, steam coal production will reach 5 212 Mt, compared with 3 417 Mt in 2002. Steam coal output will continue to be widely dispersed geographically. Production of coking coal will grow more slowly, from 485 Mt in 2002 to 624 Mt in 2030. Coking coal production will become increasingly concentrated in China and Australia. These countries will account for about 60% of global supply in 2030. Brown coal production will increase at a rate of 1.1% per year, reaching 1 175 Mt in 2030.

Chapter 5 - Coal Market Outlook

Figure 5.3: **Coal Production by Region, 2002-2030**

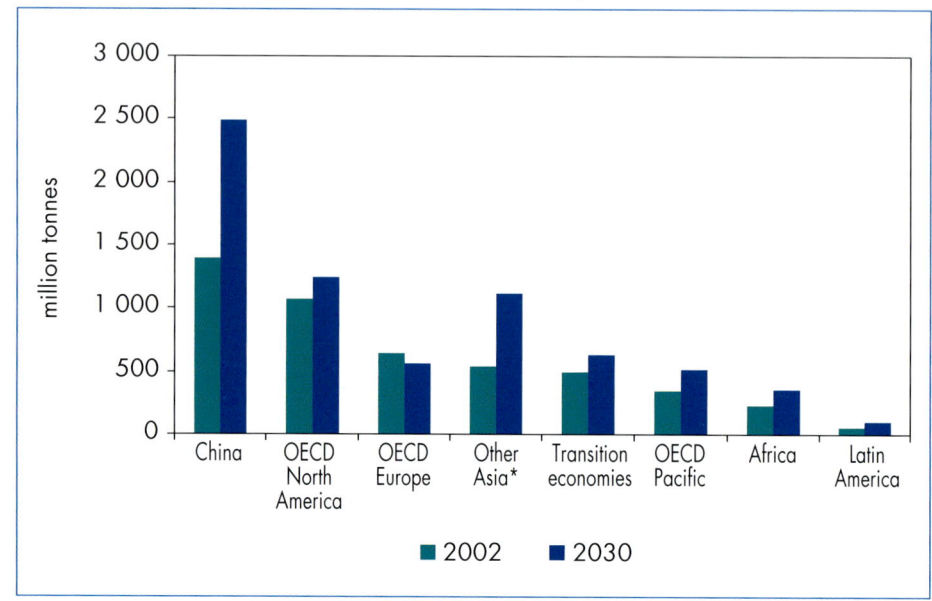

* Other Asia comprises East Asia and South Asia.

Hard Coal Trade

The volume of hard coal traded[2] internationally will increase steadily to 2030. Main drivers of the increase will be continuing industrialisation of developing Asian countries and the decline of coal mining in Europe, which will provide enlarged markets for exporters. Total trade is expected to increase from 688 Mt in 2002 to 1063 Mt in 2030. Regional trade patterns and the breakdown of trade by coal type will change markedly (Figure 5.4).

As a share of total hard coal demand, trade has increased rapidly, from 10% in the early 1990s to 17% in 2002. By 2030, this share is expected to have risen slightly more, to 18%. Although some regions will import more of their coal needs, much of the increase in production in major producing regions, including China, North America and India, will be destined for domestic markets.

The share of steam coal in world coal trade will continue to rise, stimulated by strong demand from electricity generators in Asia and by a growing need for imports in Europe (Figure 5.5). By 2030 trade in steam coal will account for 76% of total hard coal trade versus 69% now. Trade in coking coal will grow slowly, reflecting the growing use of steel-making technologies such as pulverised coal injection (which uses steam coal quality), the application of

2. International trade of brown coal and peat are negligible.

Figure 5.4: **Major Inter-Regional Coal Trade Flows, 2002-2030** (Mt)

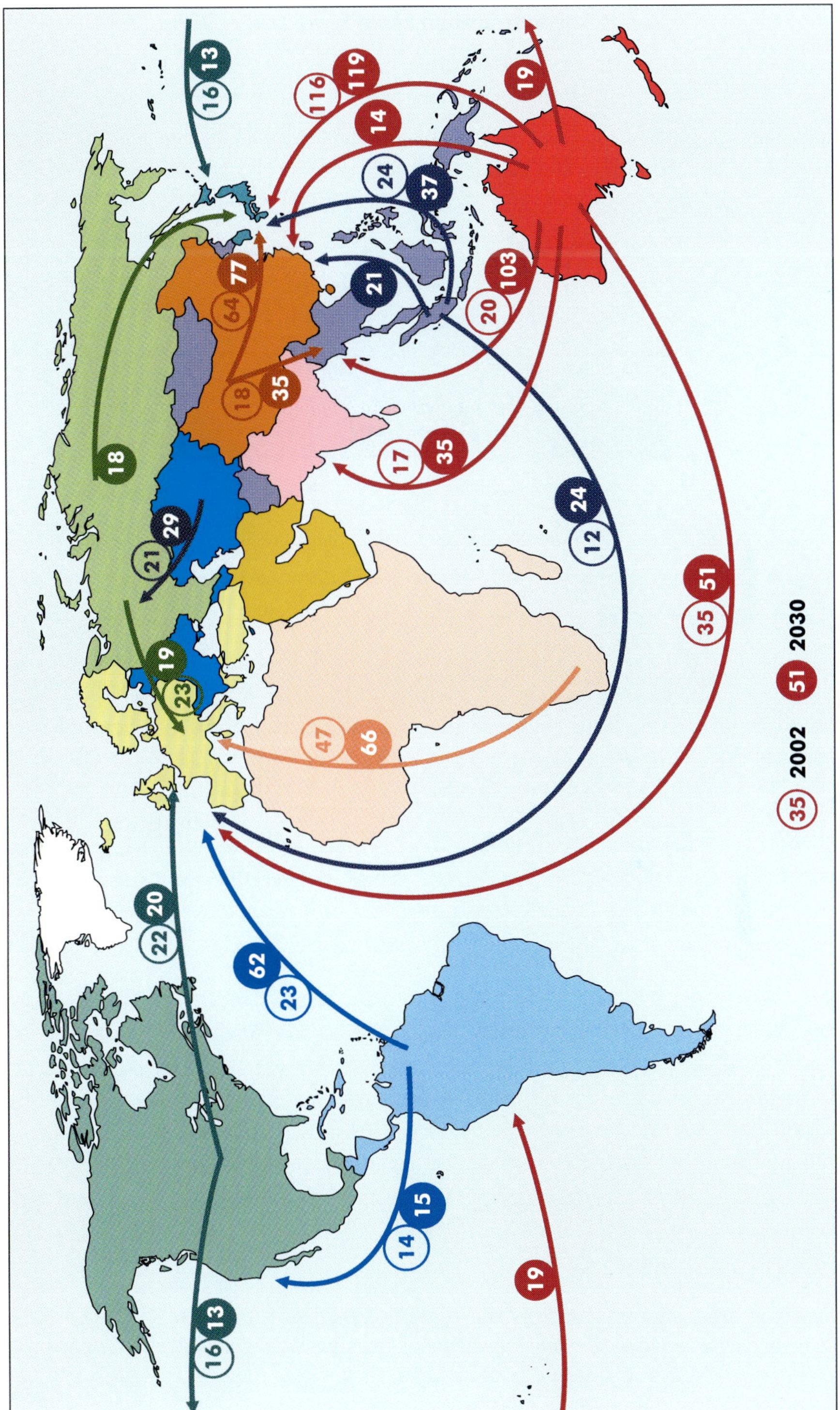

Chapter 5 - Coal Market Outlook

Figure 5.5: **International Hard Coal Trade, 1985-2030**

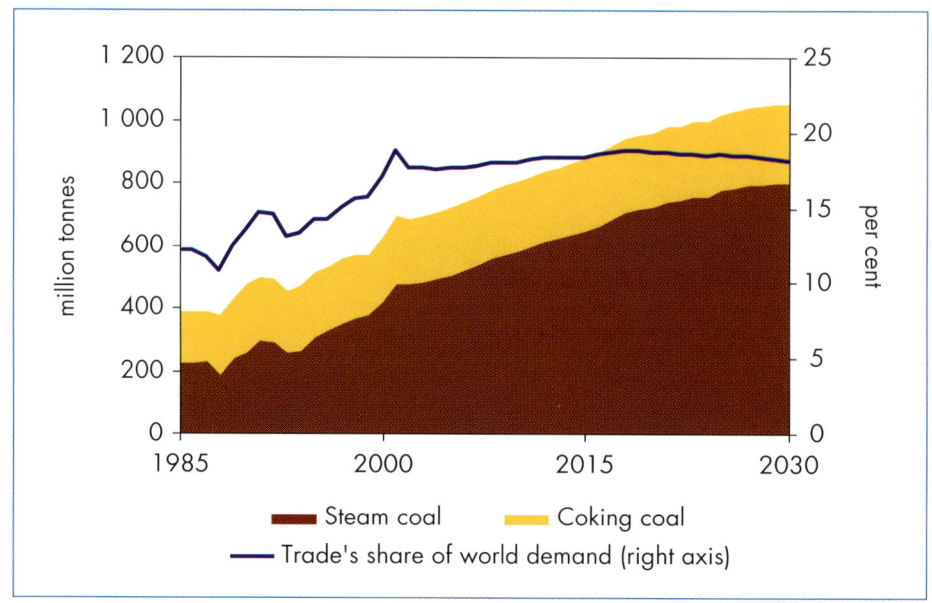

advanced steel-making technologies not requiring high-quality coking coal, such as direct smelting and ongoing steel recycling.

Because transport costs account for a large share of the total delivered price of coal, international trade in steam coal is effectively divided into two regional markets – the Atlantic and the Pacific. Markets overlap when prices are high and supplies are plentiful. South Africa is the natural point of convergence of the two markets and plays an important role in transmitting price signals between them.

The Atlantic market is made up of importing countries in Western Europe, notably the United Kingdom, Germany and Spain. The Pacific market consists of developing and OECD Asian importers, notably Japan, Korea and Chinese Taipei. The Pacific market currently accounts for about 60% of world steam coal trade. Through the *Outlook* period, it is expected that the Pacific market will be supplied mainly by Australia, Indonesia and China. South Africa, the United States, Colombia and Venezuela will be the primary suppliers to the Atlantic market. South Africa will be well placed to continue supplying Europe, Asia and the Americas.

Sources of internationally traded coking coal will remain limited. Australia is by far the largest current supplier, accounting for 51% of world exports in 2002. The United States and Canada are also significant exporters. Recently, China has also emerged as an important supplier of coking coal to world markets. Because coking coal is more expensive than steam coal, Australia can afford the high freight costs involved in exporting it around the globe. Growth in coking coal trade will be led by the needs of the Asian steel industry.

Price Developments

Since 2003, coal prices around the world have risen sharply after having moved in a fairly narrow band through the preceding decade. The spot price of steam coal delivered to northwest Europe jumped from around $36 per tonne in January 2003 to $79 in July 2004 (Figure 5.6). Contract prices, which are typically well below spot prices, have risen less. In the Reference Scenario, the IEA steam coal import price averaged $38 per tonne in 2003 ($41/t in nominal terms). Prices are assumed to increase into 2005 before falling back to and stabilise at around $40 until 2010. After that, we assume they will rise slowly and in a linear fashion, reaching $44 in 2030. The increase is, nonetheless, slower than those of oil or gas. Rising oil prices will raise the cost of transporting coal and also make it more competitive for industrial users and power generators. This factor is assumed to offset an expected reduction in the cost of mining coal, as low-cost countries continue to rationalise their industries.

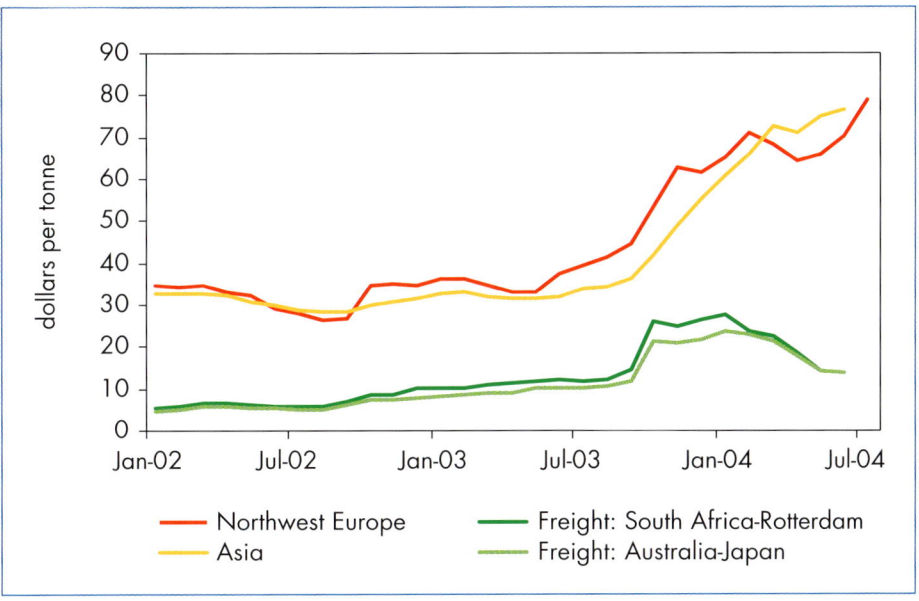

Figure 5.6: **Spot Steam Coal Prices and Freight Rates**

Note: Prices shown include cost, insurance and freight.
Source: SSY Consultancy & Research and McCloskey Group.

> *Box 5.2:* **Non-Physical Trading of Steam Coal**
>
> Steam coal is increasingly traded on electronic trading platforms. Compared with oil and gas, this method of trading has been slow to develop, largely because power generators require a detailed assessment of coal specifications – often necessitating pilot testing – prior to use. The development of coal indexes that offer standardised definitions of the origin, quality and other conditions for traded coal has alleviated this constraint. Improved coal indexes have also enabled trading in derivatives, which has increased market transparency, enabled coal producers and customers to hedge their floating price exposure and allowed speculation on future price movements. Derivatives are now largely used to set the spot price for steam coal in the Atlantic market.
>
> Electronic coal trading has developed primarily in the Atlantic market, partly because of the fierce competition in supplying power generators that has emerged since deregulation of electricity markets. GlobalCOAL, a coal-trading group, estimates that by mid-2004 the volume of coal traded electronically in the Atlantic market was between four and seven times the size of the underlying physical market. In the Pacific market, non-physical trade has yet to reach the size of physical trade.
>
> The future development of non-physical steam coal trading is unpredictable. Improved understanding of the performance of coal in different uses could permit further transition to this method of sale. Non-physical trade in coking coal has not developed because of the large number of quality parameters involved and the different ways they are assessed by customers.

Strong demand has been the primary cause of the recent jump in prices. World coal consumption increased by close to 7% in 2003 (BP, 2004). Demand increased most sharply in China, where industrial production and electricity demand are both booming. In Japan, the unscheduled closure of several nuclear reactors has increased the use of coal in power generation. Coal producers have been hard pressed to respond to increased demand. In North America, coal mines are currently operating at close to full capacity as in recent years there has been little expansion of new production capacity and a slow down in the rate of productivity improvements. Upward pressure on prices has been further compounded by Chinese government restrictions on coal exports to ensure that domestic needs are met.

Another factor in the recent price rise has been a dramatic increase in maritime freight rates. The cost of transporting coal from the port of Newcastle in

Australia to Japan rose from around $4.5 per tonne in January 2002 to over $22 during February 2004. Freight rates have since subsided but remain at historically high levels. An important factor in the run-up in freight rates was China's huge demand for imports of commodities such as iron ore, which compete directly with coal for space on dry-bulk cargoes. In 2002, coal accounted for about 42% of the world's seaborne dry-bulk trade. A further contributing factor to high freight rates has been bottlenecks at some major coal-loading ports. The appreciation of the currencies of several major coal-exporting countries, especially the Australian dollar and the South African rand, has also contributed to higher prices denominated in US dollars.

Investment Outlook

Cumulative investment[3] needs in the global coal industry, including financing for mining, shipping and ports, are expected to be just under $400 billion over the period 2003-2030. This much capital will be needed to replace production capacity that will close during the period, to meet rising demand and to accommodate growing trade. If investment in coal-fired power generating capacity is included, investment needs increase to $1.7 trillion.

Investment requirements will be increasingly concentrated in developing Asian countries. China alone will account for $129 billion, over 35% of the global total. This is just slightly more than the combined investment requirements of the coal industries in all OECD countries. Around 8% of the investment in non-OECD countries will go to supply infrastructure to export coal to the OECD.

Investment in mining, at around $350 billion, represents close to 90% of the projected total. Investment in the dry-bulk cargo fleet (for coal transportation) will amount to $34 billion, and coal-related investment in ports will be $13 billion.

Advanced technologies now available and under development could dramatically alter investment patterns by boosting coal demand, particularly in OECD regions. In the Pacific market, Japan is the only major coal user with a commitment to cut its carbon dioxide emissions under the Kyoto Protocol. Other countries in the region often place a relatively higher value on economic growth and security of supply than on environmental objectives. As a result, coal use in the region is likely to remain strong regardless of changes in technology.

---

3. Investment needs have been calculated using the methodology outlined in the *World Energy Investment Outlook* (IEA, 2003a) and on the basis of *WEO-2004* demand-supply projections and recent market developments.

Regional Trends[4]

OECD North America

OECD North America is expected to remain the world's second-largest coal market in 2030, after China. By that time, coal will account for 18% of the region's total energy needs, down from 21% in 2002. Demand will rise by a mere 0.5% per year, from 1 051 Mt to 1 222 Mt, over the projection period. In 2003, the region consumed a record level of coal and – for the first time ever – produced less coal than it consumed.

Coal use will be driven mainly by the needs of the power sector and by the prices of alternative fuels. Demand will be encouraged by the rising cost and reduced availability of natural gas, particularly in the United States. By 2030, coal is expected to fuel 40% of the region's power generation compared to 46% in 2002. The region's demand for coking coal will increase at a modest 0.8% per annum thanks to efficiency gains and technological changes in steel manufacturing.

Most North American coal demand will be met domestically. The combined proven coal reserves of the United States, Canada and Mexico amount to 254 billion tonnes – more than a quarter of the world total. In recent years in the United States, there has been a trend towards increasing coal production in the western states. This follows on from the strict sulphur dioxide emission limits for power plants introduced in 1995 by the Clean Air Act. Coals mined in the western states tend to have lower sulphur content than those found in the eastern states, reducing the need for investment in flue gas desulphurisation.

The United States was the largest coal exporter in the world from 1984 until the early 1990s, but its exports have since fallen precipitously in the face of competition from lower-cost producers in South America, South Africa and Australia. United States reserves of export-quality coal are declining and domestic markets are absorbing most of the remaining low-sulphur coal. Reserves of export-quality coal are extensive in some areas, but they are located far from ports for shipment to export markets. Both the United States and Canada are expected to continue to play an important, yet diminishing, role in supplying high-quality coking coal to the steel industry, particularly in Europe. Their share of the global coking coal market is expected to fall from 21% to 14% over the projection period.

OECD Europe

Coal demand in OECD Europe will increase slowly in the first half of the *Outlook* period and then decline in the later years. Demand is expected to be

---

4. See Chapter 9 for an analysis of Russian coal market prospects.

816 Mt in 2030, compared to 822 Mt in 2002. Coal will continue to lose market share to natural gas in the power sector if gas prices remain relatively low and stable, and if gas supply is secure. The relative cost of coal will also be influenced by environmental policies.

OECD Europe's proven coal reserves total 39 billion tonnes or about 4% of the world total. Around forty per cent of these reserves is brown coal. In most cases, Europe's hard coal reserves are expensive to mine. Estimates of European coal reserves have been revised downwards by more than 60% since the end of 1999 following an economic re-evaluation of what makes a viable mine. The largest downgrades have occurred in Germany and to a lesser extent in Poland. Coal production in OECD European countries has declined significantly since 1990, from around 1 036 Mt to 647 Mt in 2002. The largest decline occurred in Germany, where output fell by 230 Mt between 1990 and 2002. Production dropped by 62 Mt in the United Kingdom, 54 Mt in Poland and 48 Mt in the Czech Republic over the same period. The fall in production from European coal mines has exceeded a parallel slump in demand. Imports have risen and this trend is set to continue, with imports projected to grow from 223 Mt in 2002 to over 300 Mt around 2020 before dropping back to 268 Mt in 2030. The share of imports in total coal demand will increase from 27% in 2002 to 33% in 2030 (Figure 5.7).

Much of the region's brown coal production is commercially competitive and, with a few exceptions, unsubsidised. In contrast, most European hard coal production remains uneconomic and depends on subsidies or other forms

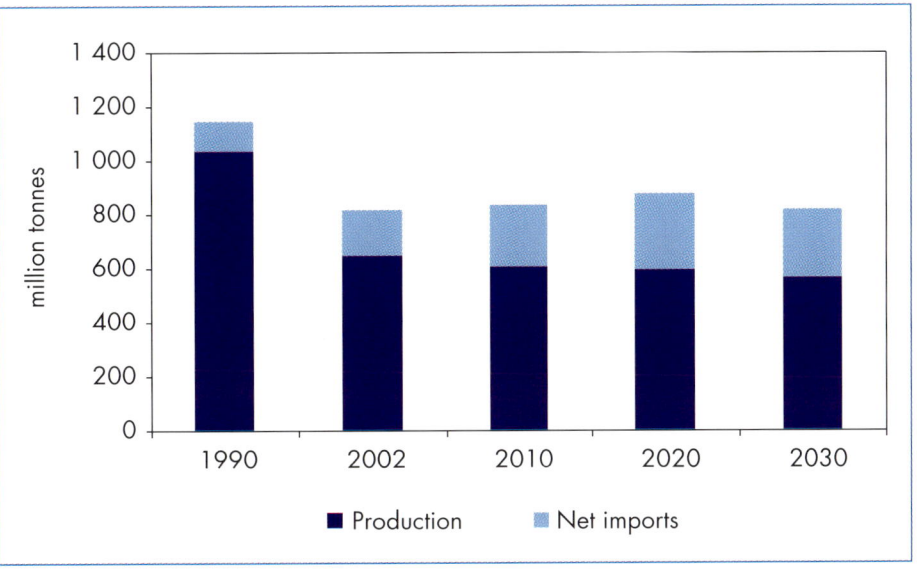

Figure 5.7: **OECD Europe Coal Balance, 1990-2030**

of protection. Subsidies are being phased out gradually in most European countries, although some of them may be retained on the grounds of security of energy supply. This will allow some high-cost mining capacity to be maintained, especially in Germany and Spain.

Two of the ten new members of the European Union, Poland and the Czech Republic, have large coal industries. Poland's hard coal production exceeds the total production of the 15 countries that made up the Union before the new members joined. The coal industries in both Poland and the Czech Republic have undergone considerable restructuring since the collapse of communism in 1989. Production has been cut and unprofitable pits closed. The process continues, and coal subsidies in the two countries are expected to be phased out slowly.

OECD Pacific

Coal demand in the OCED Pacific region will increase from 364 Mt in 2002 to 423 Mt in 2030, at an annual rate of growth of 0.5%. Prospects vary through the region. Japan will use slightly less coal in 2030 than it does today. But this will be more than offset by robust growth in Korea and slow growth in Australia and New Zealand. In all OECD Pacific countries, use of steam coal for power generation will be the main driver of increasing demand.

Australia is the only significant coal producer in the region. It is currently the world's largest exporter and sixth-largest producer. With 78 billion tonnes of proven coal reserves, and excellent rail and port infrastructure, Australia has the potential to greatly increase steam coal exports to the Pacific market. The extent to which Australian producers increase production to meet growth in Asian demand will depend largely on developments in China, which has emerged in recent years as a major exporter. Chinese exports are expected to continue to grow, but more slowly than in the recent past because of increasing domestic demand. Australia is projected to account for 26% of world steam coal trade in 2030, up from 21% in 2002.

Australia will face less competition in the coking coal market, where it is expected to extend its position as the leading exporter in both the Atlantic and Pacific markets (Figure 5.8). Australia's share of world coking coal exports is projected to reach 58% in 2030, up from around 50% in 2002.

Despite a projected decline in its consumption, Japan will remain the world's largest importer of coking coal, and, as with Korea, among the world's largest importers of steam coal. By 2030, Japanese coal imports will account for 13% of total world coal trade, down from 24% in 2002. Although Chinese producers have been gaining market share in Korea in recent years, much of the country's future demand is expected to be supplied from Australia.

Figure 5.8: **Australian Coal Exports and Share of World Trade, 1990-2030**

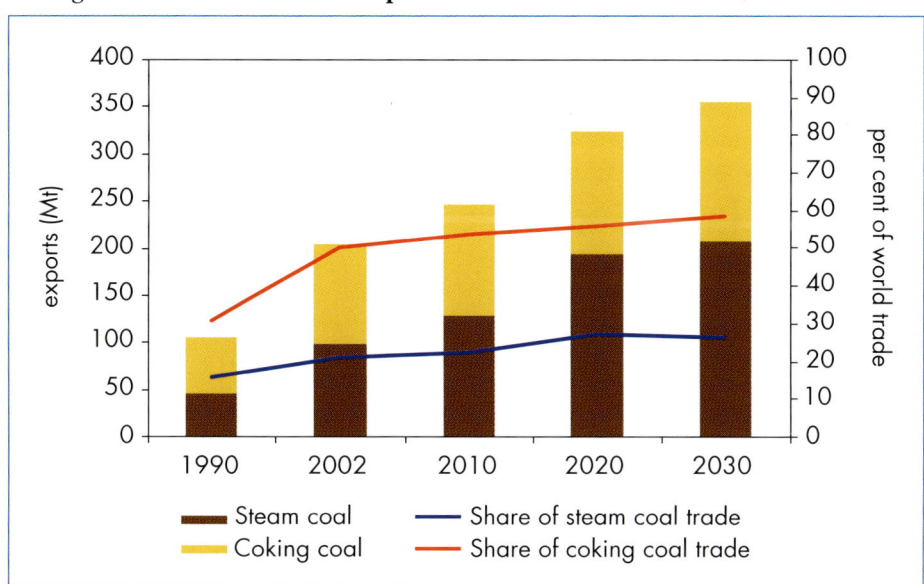

China

Chinese primary coal demand will grow from 1 308 Mt in 2002 to 2 402 Mt in 2030, an increase of 2.2% per year. China will thus extend its position as the world's largest coal consumer. Although coal will remain the dominant fuel in China's energy mix, its share of total primary energy consumption will drop from 57% in 2002 to 53% in 2030, owing to the growing use of natural gas in electricity generation and of oil in transportation.

The power sector will account for more than 73% of total Chinese coal consumption in 2030, compared with 52% in 2002, as coal remains the backbone of China's generating capacity. Industrial uses, mainly of coking coal in steel production, will rise at a more modest 0.7% per year. In response to rising incomes and urbanisation, coal use in the residential sector will decline as it loses out to more convenient and cleaner sources of energy.

China has an estimated 114 billion tonnes of proven coal reserves. The majority of these are found in northern China, particularly in the provinces of Hebei, Shaanxi and Inner Mongolia. Hard coal accounts for 84% of total proven reserves. The remainder consists of lower-quality coals, including lignite. China will remain the world's biggest producer of both steam and coking coal in 2030. In 2002, production totalled 1 398 Mt, around 29% of the world total. It is projected to increase to 2 490 Mt in 2030, or 35% of the world total (Figure 5.9).

Figure 5.9: **Coal Production and Exports in China**

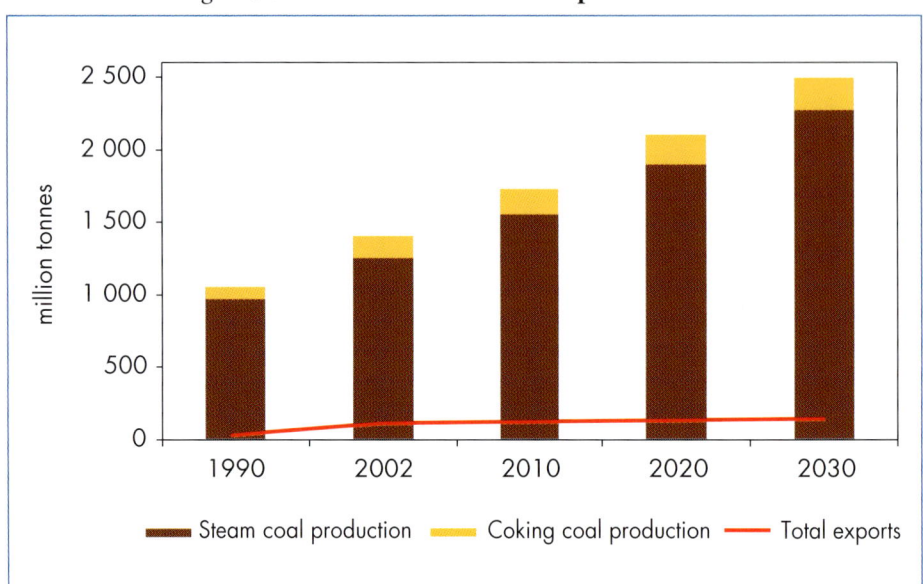

The Chinese coal industry has undergone a major rationalisation involving the closure of thousands of small mines, as well as the expansion of large mines operated by the central government. This programme has helped raise productivity levels, improved safety standards and given the government more control over production. Nonetheless, further reforms are needed. Illegal small-scale mining operations continue to pose problems; safety standards remain low and regulations are poorly applied. In addition, large state-owned mines are burdened with a wide range of social responsibilities, such as providing schools and hospitals, which distract management and undermine efficiency. The sector requires more investment, not only to expand capacity but also to modernise existing mines. Priorities include mechanising underground mines, building coal-preparation plants and improving rail transport, water supply and waste-disposal facilities.

China is an important, yet unpredictable, supplier of coal to world markets. China's exports totalled 97 Mt in 2002, making it the world's second-largest exporter after Australia. Exports are expected to increase steadily to 130 Mt in 2030. Currently, however, exports are being restricted by the government to alleviate local shortages of coal for electricity generation and steel production. Restrictions include a reduction in export price rebates and lowered export quotas. The export cap for 2004 is expected to be 80 Mt. These moves have disrupted world markets and raised prices, hitting European steel manufacturers particularly hard. Over the long term, however, China is expected to exploit its vast reserve base and its proximity to rapidly growing Asian markets to win market share away from more distant suppliers. Export prospects will depend to a large extent on whether the government continues to support coal producers. Other factors will be the development of rail and port infrastructure and the rate of growth in domestic demand.

Because of its vast coal reserves and increasing reliance on foreign crude oil imports, China's interest in developing projects that convert coal into synthetic liquid fuel is growing. Under a recent agreement, South Africa's Sasol, in partnership with a number of Chinese companies, plans to develop two large coal-to-liquids plants in coal-rich Ningxia and Shaanxi provinces. If these projects proceed as hoped, their combined capacity will amount to 60 Mt of oil per year. Sasol is the only company in the world currently operating commercial coal-to-liquids plants. Capital costs for such plants are much higher than costs for conventional oil projects and even for gas-to-liquids projects, but operating costs are moderate, particularly in the light of current world oil prices and if low-cost coal feedstocks are available. To reduce costs, China has established a centre in Shanghai to study new liquefaction techniques.

India

Coal will remain the dominant fuel in India's energy mix through 2030. Demand is projected to grow from 391 Mt in 2002 to 758 Mt in 2030, at an average rate of growth of 2.4% per year. Only China's demand for coal will outstrip India's in the *Outlook* period. As in other regions, the power sector will be the chief driver of Indian demand. Currently, 71% of India's electricity is generated from coal. This share will decline to 64% by 2030.

India's coal needs will be largely met domestically. Production totalled 364 Mt in 2002, and is projected to increase to 705 Mt in 2030. India has 92.4 billion tonnes of proven coal reserves, 10% of the world total. Coal is located mainly in the centre and east of the country, far from the main consuming areas. As a result, large quantities of coal have to be transported by rail over long distances. Smaller amounts are shipped by a combination of rail and sea, at very high cost. This has encouraged growth in coal imports into certain coastal areas in recent years. India imports much of its coking coal needs as indigenous supplies are of a low quality due to their high ash content and low calorific value.

Productivity in Indian coal mines is well below international standards, because of low levels of mechanisation and poor mine design. Investment is urgently needed along the whole coal chain from production to use. But the dire financial condition of the state electricity generators – the main users of coal in India – is holding back investment in coal-fired stations. Some electricity pricing reforms have been implemented, but much more needs to be done. Further liberalisation of the domestic coal market is also needed, including the removal of impediments to foreign investment. Competition from imported coal could be a stimulus to improving performance in the domestic industry and to raising its attractiveness for investors.

Africa

African coal demand is expected to increase at an average annual rate of 1.5%, from 174 Mt in 2002 to 264 Mt in 2030. Africa has about 50.3 billion tonnes of proven coal reserves, or around 6% of the world total. South Africa accounts for the majority of both reserves and production on the continent. It produced around 223 Mt in 2002, over 95% of the African total. South Africa is the world's largest producer of coal-based synthetic liquid fuels. This market accounted for about 21% of the country's total coal consumption in 2002.

South Africa was the world's fourth-largest coal exporter in 2002. Exports, mostly of steam coal, totalled 70 Mt, most of which went to Europe. Europe has become an even more vital market for South African steam coal in recent years as exports to the Pacific have suffered from Chinese competition. South African coal exports may taper off in the long term, because of the country's limited reserves of export-quality coal. About 90% of exports are handled by the Richards Bay Coal Terminal, which is now operating at close to full capacity, but the terminal is currently being enlarged. We expect coal exports from Africa to total 110 Mt in 2030, or around 10% of world coal trade, the majority of which will continue to go to the European Union.

Indonesia

Indonesia, the world's fourth most populous nation, will fulfil an important role in the coal market, particularly in the Asia-Pacific region. Indonesian coal demand is projected to grow by 4.6% per year, from 29 Mt in 2002 to 102 Mt in 2030 (Figure 5.10).

Exports remain the driving force behind Indonesian coal production. Producers have a strong incentive to export coal, because export prices are higher than domestic prices, which are held down by the government. Exports have increased rapidly, from 1 Mt per year in the mid-1980s to 74 Mt in 2002. Indonesia is now the third-largest hard coal exporter in the world, after Australia and China. Exports are expected to reach 146 Mt by 2030, most of which will continue to go to the Asia-Pacific market.

Proven coal reserves in Indonesia total 5 billion tonnes, equivalent to 48 years of production at current rates, and are more than ample to meet projected domestic and export demand for several decades to come. It is not certain, however, that the investment needed to develop these reserves can be raised. The investment climate for coal producers and for coal service providers is clouded by transportation problems and by political, social and economic

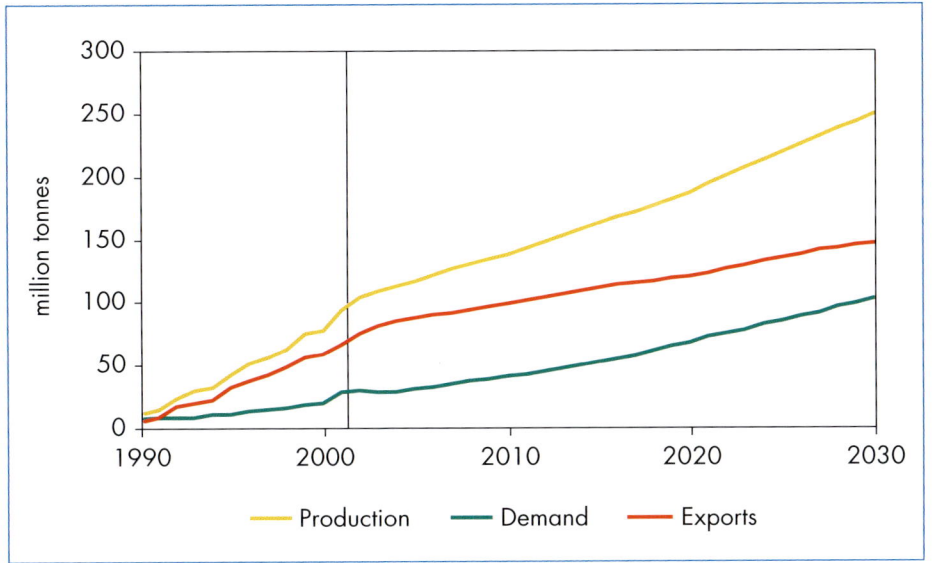

Figure 5.10: **Coal Production, Demand and Exports in Indonesia**

instability. As a result, investment has slowed over the past few years. Economic nationalism further discourages inward direct investment in Indonesia. Foreign firms operating in the country's coal-mining sector are obliged to sell a majority stake to Indonesian companies within ten years of the start of production. This obligation recently led to litigation between the east Kalimantan provincial government and Kaltim Prima Coal, a 50-50 joint venture between Rio Tinto and BP, which produced about 18 Mt per year of steam coal for export markets. The case involved the pace of divestiture and the value of the stake. Rio Tinto and BP finally ended the dispute in 2003 by deciding to sell their entire stake, rather than just the 51% required by law. This dispute and others of its kind have discouraged foreign investment in Indonesia's coal sector and hence jeopardise the prospects for export growth.

CHAPTER 6

ELECTRICITY MARKET OUTLOOK

HIGHLIGHTS

- World electricity demand will double between 2002 and 2030. Most of the growth will be in developing countries. By 2030, power generation will account for nearly half of world consumption of natural gas. It will also have absorbed over 60% of new investments in energy supply between now and then.
- The global power sector will need about 4 800 GW of new capacity between now and 2030 to meet the projected increase in electricity demand and to replace ageing infrastructure. Just over half this amount will be needed in developing countries. OECD countries will need nearly 2 000 GW, including replacements. Nearly a third of the current installed capacity in the OECD could be retired by 2030.
- Electricity markets in the OECD will need investments of over $2 trillion in power generation and $1.8 trillion in transmission and distribution networks. While electricity-market reform in the OECD has shown its first positive results, many challenges remain to be met. The blackouts in 2003 and 2004 highlighted the importance of adequate reserve margins and the need to improve networks.
- Developing countries will need some $5.2 trillion, more than half of world requirements. For many of them, investment will have to rise well above current levels if they are to meet their goals for economic growth and social development, but many obstacles remain. Overcoming these obstacles will require significant restructuring and reforming of developing-country electricity sectors.
- Natural gas and non-hydro renewables will increase their shares in the electricity-generation mix. Unless governments act much more vigorously to limit CO_2 emissions, coal will remain the world's largest single source of electricity generation. Nuclear power generation will increase in absolute terms, but its share will fall. Nearly 40% of existing nuclear plants will be retired.
- By 2030, the power sector could account for almost 45% of global energy-related CO_2 emissions. Carbon-dioxide emissions from power stations in developing countries will treble from 2002 to 2030. In 2030, coal plants in developing countries will produce more CO_2 than the entire power sector in the OECD in that year.

This chapter discusses the major trends in electricity markets over 2002-2030. First, it looks at demand growth, its drivers and the regional and sectoral growth patterns. Second, it examines the technology and fuel mix used in electricity generation and how they might evolve during the projection period. Third, it reviews the investment requirements in power generation, transmission and distribution to match the projected electricity demand. Fourth, it looks at the implications for carbon-dioxide emissions arising from fossil fuels used in power plants. The chapter then provides a summary of the implications of the Reference Scenario projections and outlines the challenges facing the electricity sector with regard to ever-growing demand, adequacy of investment, rising emissions and increased dependence on natural gas imports.

Electricity Demand

Drivers of Electricity Demand

Demand for electricity is closely linked to economic growth. Over the past thirty years, the global economy grew by 3.3% per year, on average, and electricity demand grew at 3.6%. Over the projection period, electricity demand is expected to grow at an average annual rate of 2.5%, as the global economy increases at 3.2% per year (Figure 6.1). The world will consume twice as much electricity in 2030 as it does today.

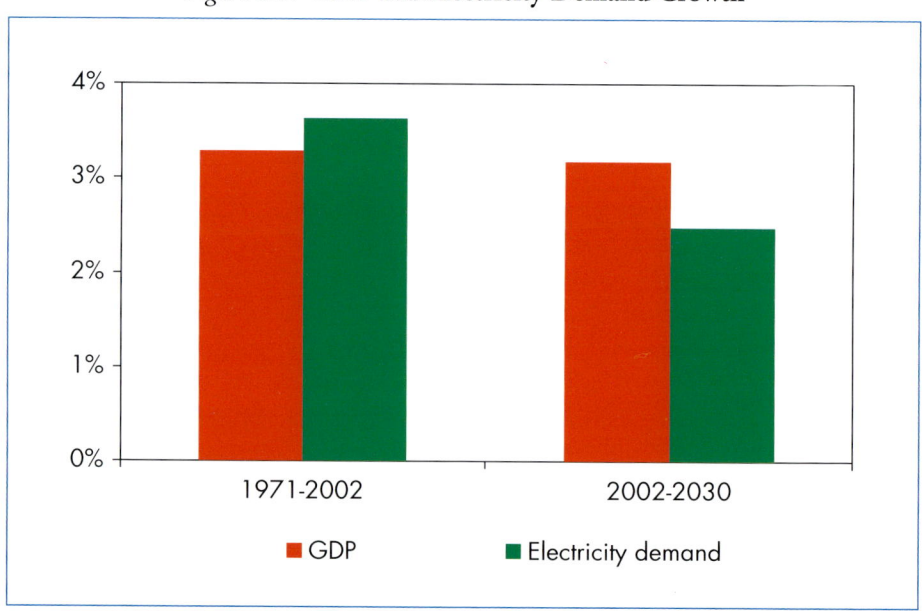

Figure 6.1: **GDP and Electricity Demand Growth**

Developing countries will drive the increase in global demand for electricity, which will rise at about the same rate as their GDP. Developing countries' demand will more than triple by 2030. In the OECD, the pace of growth will be slower, at 1.4% per year. Even so, at the end of the projection period, the 1.3 billion people in the OECD will still be consuming more electricity than the 6.5 billion people in the developing world. Some 1.4 billion in the developing regions will still lack any access to electricity.

Outside the OECD, Asian economies will experience the highest growth in electricity demand. Indonesia's demand is expected to increase by 5.2% per year, while demand in India and China will grow at 4.9% and 4.5%. In 2030, China will generate as much electricity as the United States.

In the transition economies, demand will grow at 2% per year, as these countries are already large consumers of electricity. Moreover, they have the opportunity to use electricity much more efficiently, particularly in industry.

Sectoral Growth

The share of electricity in total final energy consumption will rise from 16% in 2002 to 20% in 2030. The share of electricity will increase in industry, in households and in the services sectors in all regions. The increase will be more substantial in developing countries (Table 6.1).

The largest sectoral increase will be in residential electricity consumption, at 119%, followed by the services sector (97%) and industry (86%). Industry is likely to remain the largest final consumer of electricity throughout the projection period (Figure 6.2).

Table 6.1: **Electricity's Share of Energy Demand by Sector** (%)

	OECD		Transition economies		Developing countries	
	2002	2030	2002	2030	2002	2030
Total final consumption	20	22	13	15	12	20
Industry	25	27	18	22	17	25
Residential	32	38	11	14	8	20
Services	48	57	24	25	31	47

Power Generation

World electricity generation is projected to rise from 16 074 TWh in 2002 to 31 657 TWh in 2030, growing at an average rate of 2.5% per year. The largest increase will be in China, which will raise production by 3 898 TWh from now to 2030, a quarter of the world's projected increase.

Figure 6.2: **Sectoral Growth in Electricity Demand**

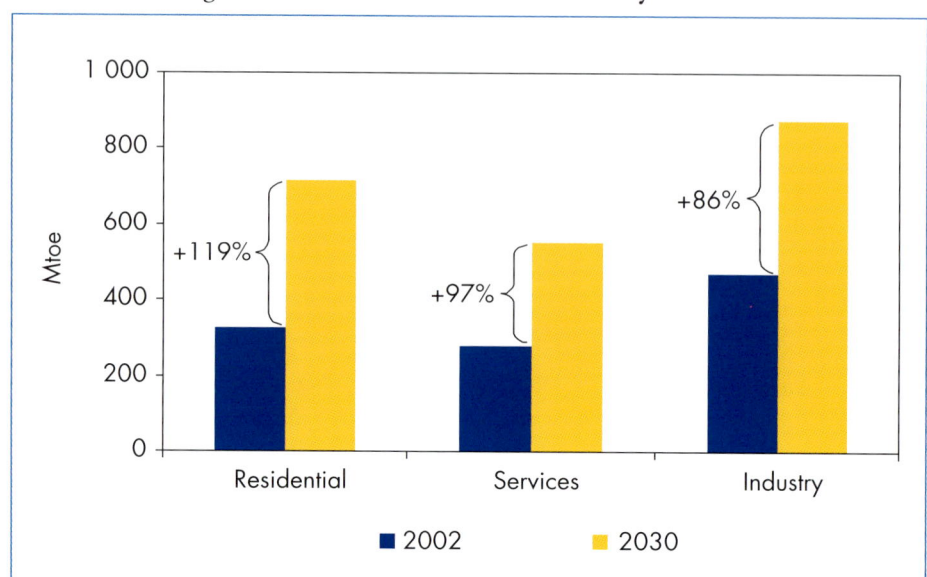

Table 6.2 shows the world's largest electricity markets in 2002. The United States is by far the largest market and could stay at the top of the list in 2030, but China will be nearly as large by that time. India is likely to be third on this list in 2030, with a market about a third of that of China.

Table 6.2: **Top-Ten Electricity Markets in 2002**

	Electricity generation (TWh)	**Share in world** (%)
United States	3 993	25
China	1 675	10
Japan	1 088	7
Russia	889	6
Canada	601	4
Germany	567	4
India	598	4
France	555	3
United Kingdom	384	2
Brazil	345	2

Choice of New Plant

Between now and 2030, power companies around the world will build thousands of new power plants to meet rising electricity demand. They will have to choose among a number of technologies and fuels: coal, gas, nuclear or one of several renewables. These decisions will be based on economic evaluations of the various options, taking into account the relevant risks of each technology.[1]

Indicative generating costs of four key options (gas, coal, nuclear and wind) are shown in Figure 6.3. These estimates are based on current technologies. The fuel component of gas and coal plants show low and high values, reflecting prices in different markets and likely future price increases. While there is no significant variation in *total* costs, their *composition* varies widely.

- Combined-cycle gas-turbine (CCGT) plants have the lowest capital-cost component but the highest variable costs. They are, therefore, quite sensitive to changes in natural gas prices, which in this example range between $3 and $4.5 per MBtu.
- Coal plants have relatively high capital cost. Fuel costs account for a smaller percentage of their total costs, and they can be quite low in coal-producing regions. Coal prices tend to be somewhat more stable than gas prices.
- Nuclear plants have high investment requirements but very low running costs.
- The generating cost of electricity from wind turbines depends on wind speed. The costs shown here are for good sites in Europe and North America. Wind turbines generally have high transmission costs. The need for backup capacity also tends to increase costs.[2]

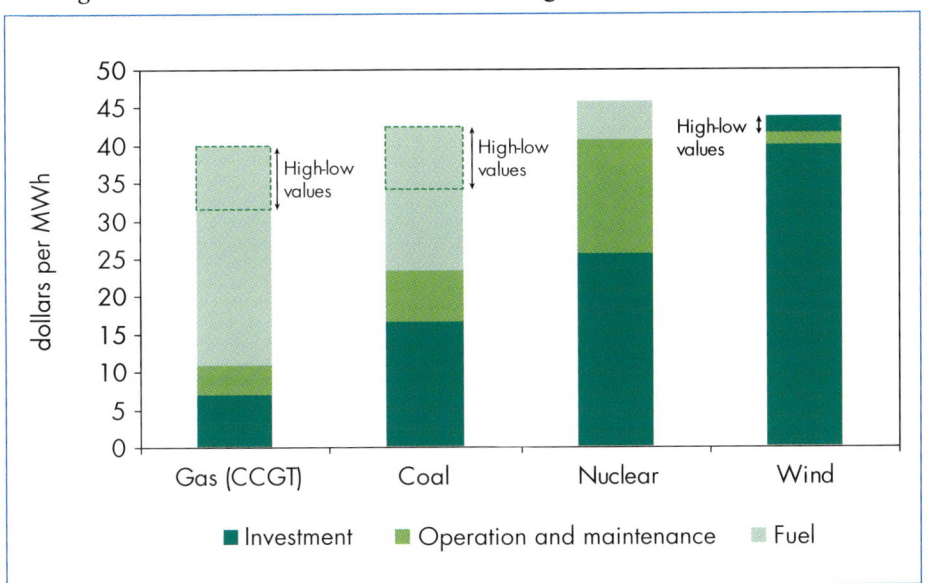

Figure 6.3: **Indicative Mid-Term Generating Costs of New Power Plants**

1. See IEA (2003a) for a discussion of the risks associated with each power-generation technology.
2. These extra costs are discussed in Box 7.1.

Future environmental restrictions on carbon dioxide emissions and pollutants such as sulphur dioxide and nitrogen oxides could increase the costs of coal, and to a lesser extent, of CCGT plants. They could make nuclear and wind generation more attractive.

For over a decade now, CCGT plants have been the preferred option for new power generation. They have lower capital costs than any other type of baseload plant – half as much as a coal plant, a quarter as much as a nuclear plant. Construction time for a CCGT plant is two to three years; it takes at least twice as long to build a coal-fired or nuclear plant. But rising natural gas prices over the course of the *Outlook* are expected to reduce the attractiveness of this type of generation.

CCGT plants have the lowest carbon dioxide emissions of all fossil-fuel based technologies, because of the low carbon content of natural gas and the high efficiency of the plants themselves (Figure 6.4). This advantage reduces investment risk for gas-fired power plants in countries that plan to limit CO_2 emissions. Natural gas is free of sulphur dioxide, while CCGT technology reduces emissions of nitrogen oxides and particulates.

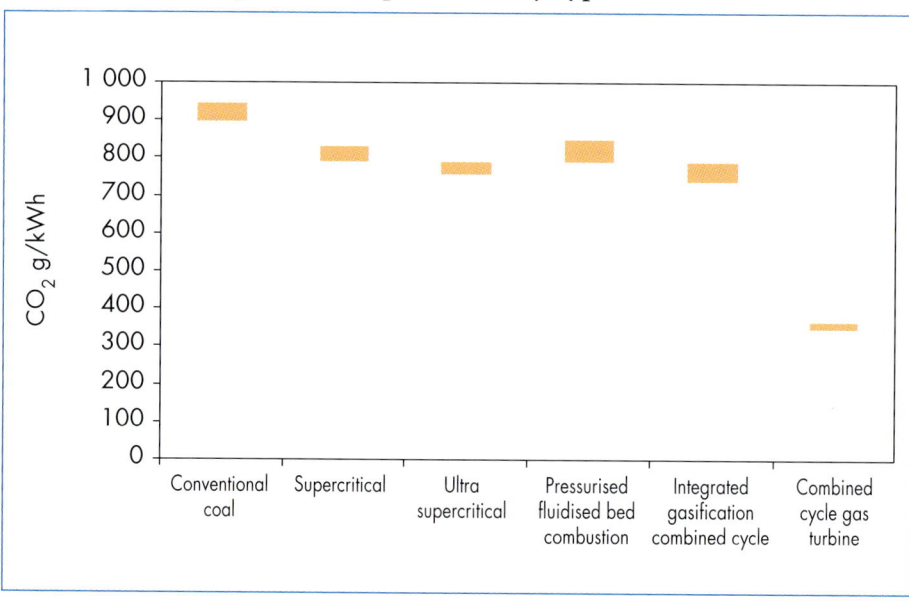

Figure 6.4: **CO_2 Emissions by Type of Plant**

Note: The emissions shown in this chart are based on a range of efficiencies.

The Electricity-Generation Mix

Coal- and gas-fired generation will provide over three-quarters of the world's incremental demand for electricity between now and 2030. Natural gas and renewables[3] will continue to increase their market shares (Table 6.3). Coal will lose some of its market share, although it will remain the predominant fuel. The share of oil, already small, will decline still further. The share of hydropower will fall from 16% now to 13% in 2030. Nuclear power will lose a large part of its market share, which could drop from 17% now to 9% in 2030.

Table 6.3: **Market Shares in Electricity Generation** (%)

	OECD		Transition economies		Developing countries	
	2002	2030	2002	2030	2002	2030
Coal	38	33	22	16	45	47
Oil	6	2	4	2	12	5
Gas	18	29	37	54	17	26
Nuclear	23	15	18	11	2	3
Hydro	13	11	19	15	23	16
Other renewables	3	10	0	2	1	3

Coal-fired power plants provided 39% of global electricity needs in 2002. This share will fall only slightly, to 38% in 2030 (Figure 6.5). Coal-fired generation in OECD countries reached 3 733 TWh in 2002. Very few coal plants are now under construction in OECD countries, where economic factors and the anticipation of future environmental regulation continue to favour gas-based electricity generation. Coal's market share in the OECD, is expected to decline substantially over the projection period, falling from 38% in 2002 to 33% in 2030. The decline could be even sharper if efforts to reduce CO_2 emissions are strengthened.[4]

Nearly 60% of the world's current coal-based electricity production is in OECD countries. Over the projection period, most new coal-fired power plants will be built in developing countries, especially in developing Asia. Coal will remain the dominant fuel in power generation in those countries because of their large coal reserves and coal's low production costs. Developing countries are projected to account for almost 60% of world coal-based electricity in 2030. China and India together will account for 44% of worldwide coal-based electricity generation.

Over the period 2003-2030, nearly 1 400 GW of new coal-fired power capacity will be built worldwide. About two-thirds of these plants will be built in

---

3. Biomass, wind, geothermal, solar, tidal and wave energy.
4. However, the deployment of carbon sequestration and capture technology could help coal maintain a high share in electricity generation.

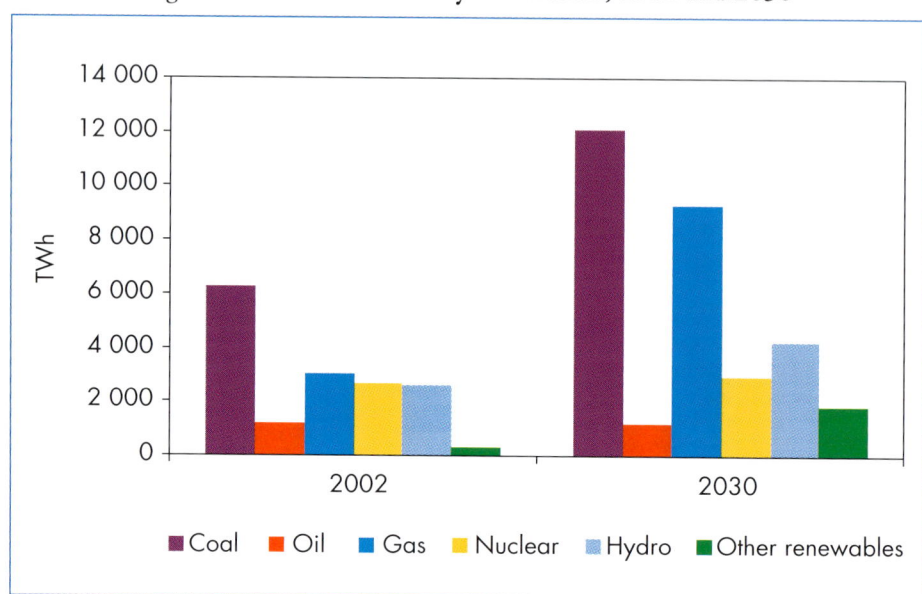

Figure 6.5: **World Electricity Generation, 2002 and 2030**

developing countries. They will be, in general, less efficient than coal plants in OECD countries, because of the technology used, the type of coal burnt, the mediocre maintenance of the plants and their size. In many developing countries the efficiency of coal use is still at the level reached by OECD countries over 50 years ago (IEA Clean Coal Centre, 2002). The average efficiency of coal-fired generation in the OECD was 36% in 2002, compared with just 30% in developing countries. This means that one unit of electricity produced in developing countries emits almost 20% more carbon dioxide than does a unit of electricity produced in an OECD coal plant. The efficiency gap between developed and developing countries will narrow, but not close. In 2030, the average conversion efficiency of coal plants in developing countries will reach 36%, while the OECD will have attained 40%.

Natural gas-based electricity production is expected to triple between now and 2030. Gas's increasing market share continues a trend that began in the late 1980s and early 1990s. In the OECD, the share of gas rose from 10% in 1990 to 18% in 2002 (Figure 6.6). Gas-fired electricity generation will increase everywhere in the OECD, but its prospects in North America are less certain than elsewhere.[5]

In developing countries, the share of gas increased from 11% in 1990 to 17% in 2002. This share is expected to rise to 26% by 2030. Most of the increase will be in Latin America, the Middle East and Africa (Figure 6.7). The

---

5. See Chapter 4 for a discussion of North American natural gas supply prospects.

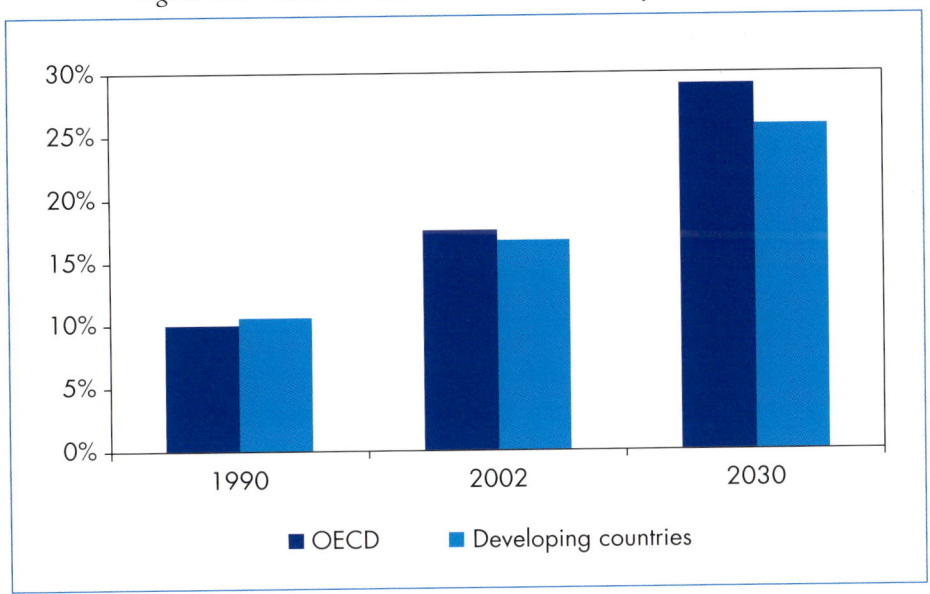

Figure 6.6: **Share of Natural Gas in Electricity Generation**

transition economies will also see a substantial increase in gas-fired electricity generation, mainly in Russia and the Asian part of the transition economies. In the region as a whole, gas accounted for 37% of electricity generation in 2002; it is projected to reach 54% in 2030.

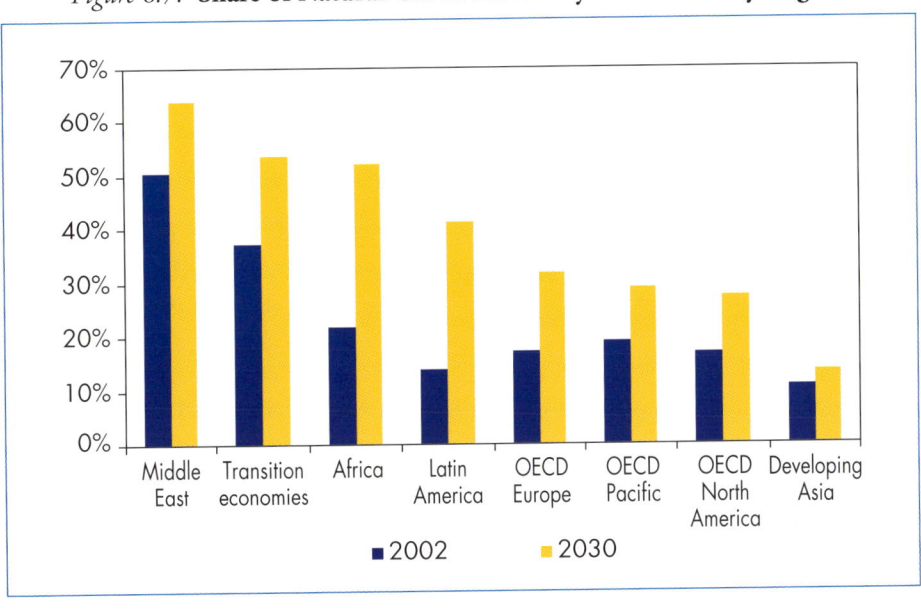

Figure 6.7: **Share of Natural Gas in Electricity Generation by Region**

Chapter 6 - Electricity Market Outlook

This relatively recent surge in gas-fired electricity generation and the projected rise of its future share are largely due to the use of CCGT plants, which will remain the preferred option for new power generation because of their economic and environmental advantages.

CCGT plants will be used to meet base and mid-load demand. Natural gas will also provide for the bulk of peak-load requirements, via simple-cycle gas-turbine technology. Gas turbines will also be used increasingly in decentralised electricity generation, where a gas grid is available. Fuel cells using hydrogen from reformed natural gas are expected to emerge as a new source of power generation, especially after 2020. They will produce a little more than 1% of total electricity output in 2030. Higher natural gas prices in the second half of the projection period will probably make coal-fired generation more attractive for new plants. Volatility is inherent in natural gas prices and is well known to investors. It does not seem to have affected their preference for gas-fired generation so far. As a result of the growing interdependence of gas and electricity, gas price volatility is expected to be increasingly reflected in electricity prices. This could have a considerable impact on industrial and commercial customers.

Oil-fired electricity generation accounted for 7% of world power production in 2002. This share is a third of what it was thirty years ago, because many countries reduced oil use in power generation after the first oil shock. The share of oil will continue to diminish in the future, falling to 4% in 2030. Future oil-fired generation will be concentrated in distributed-generation applications in industry and in remote areas.

There are 31 countries in the world operating *nuclear* power plants. These plants had a capacity of 359 GW in 2002, when they produced 2 654 TWh of electricity. Over 85% of nuclear electricity is produced in 17 countries that are members of the OECD (Box 6.1). In the Reference Scenario, world nuclear capacity is projected to increase slightly, reaching 376 GW in 2030. New nuclear plants with combined capacity of 150 GW are expected to be built around the world, the largest number in OECD Europe (Figure 6.9). However, most of this capacity will replace older reactors in France, the only OECD country so far that anticipates such a large-scale replacement of its nuclear base before 2030.[6] Three-quarters of existing nuclear capacity in OECD Europe is expected to be retired by 2030, because reactors will have reached the end of their life or because governments have adopted policies to phase out nuclear power. Nuclear capacity will increase in a number of Asian countries, notably in China, South Korea, Japan and India. Nuclear will increase its share in electricity generation in all four countries.

---

6. Many of France's existing reactors are expected to be replaced with new ones after they reach 40 years of operating lifetime.

Nuclear capacity in North America is expected to increase in the near term and then fall back to its current level by 2030. Most existing power plants are expected still to be in operation in 2030. Over a hundred nuclear units in the United States have already implemented power level increases (also known as "uprates") or have applied for them.[7] These increases are expected to add the equivalent of a few new plants to US nuclear capacity.

> *Box 6.1:* **Nuclear Policy in OECD Countries**
>
> Nuclear power contributes to the electricity-supply mix of 17 out of the 30 OECD countries. It also contributes indirectly to the power supply of some countries that do not have nuclear plants but import power from countries that do.
>
> These nuclear plants had a combined capacity of 302 GW and produced 2 276 TWh of electricity in 2002. The share of nuclear in total electricity generation varies significantly from one country to another, from 79% in France to 4% in the Netherlands (Figure 6.8).
>
> Three countries have policies in place to phase out nuclear power. The German government and the country's electricity industry agreed in June 2000 to a phase-out of existing nuclear stations after about 32 years of operation. Belgium plans to shut down its nuclear stations after 40 years of operation. Sweden's parliament has voted to phase out nuclear power, but the timing is not yet clear. The Slovak Republic is expected to shut down two reactors, considered unsafe, following an agreement with the European Union. In Spain, the new government has talked about phasing out nuclear power but there is no firm decision yet.
>
> Four countries, France, Finland, Japan and Korea, plan to increase their use of nuclear power. In France the National Assembly has endorsed nuclear power as a priority and France plans to put a demonstration European Pressurised Reactor (EPR) in place in about a decade. EPRs will replace existing reactors, which are to be shut down after 40 years of operation. Finland's fifth nuclear reactor, of EPR design, is expected to be completed by 2010. Japan and Korea are continuing to build nuclear plants, but Japan recently scaled down its nuclear programme significantly.
>
> There are no nuclear plants under construction, nor are there specific restrictions, in the remaining OECD countries. In Switzerland, a moratorium on construction of new nuclear plants has expired. The Dutch government considers that the Borssele plant, which was to have been shut down in 2004, should stay open until the end of its economic life. The future role of nuclear power is being keenly debated in a number of countries, including the United States, Canada, the Czech Republic, Turkey and the United Kingdom.

---

7. US Nuclear Regulatory Commission, *Fact Sheet on Power Uprates for Nuclear Plants* (www.nrc.gov).

Figure 6.8: **Share of Nuclear Power in Electricity Generation in OECD Countries, 2002**

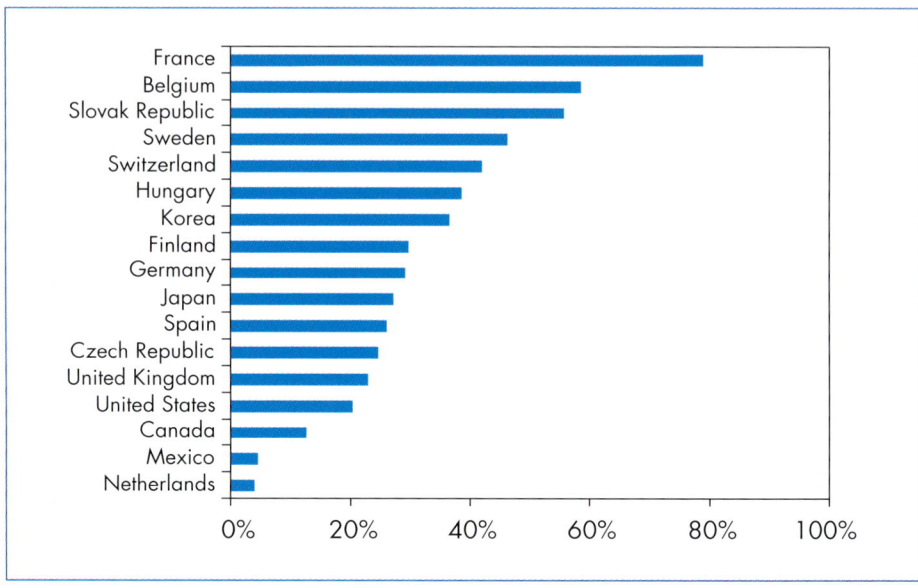

Nuclear capacity will increase in Russia. It will decline in the rest of the transition economies. Several new reactors will be built in Russia, but most existing nuclear plants are expected to be decommissioned by 2030. Lithuania and Bulgaria have agreed with their European Union partners to shut down some of their older reactors in the near future.

Figure 6.9: **Nuclear Plant Capacity Additions by Region, 2003-2030** (GW)

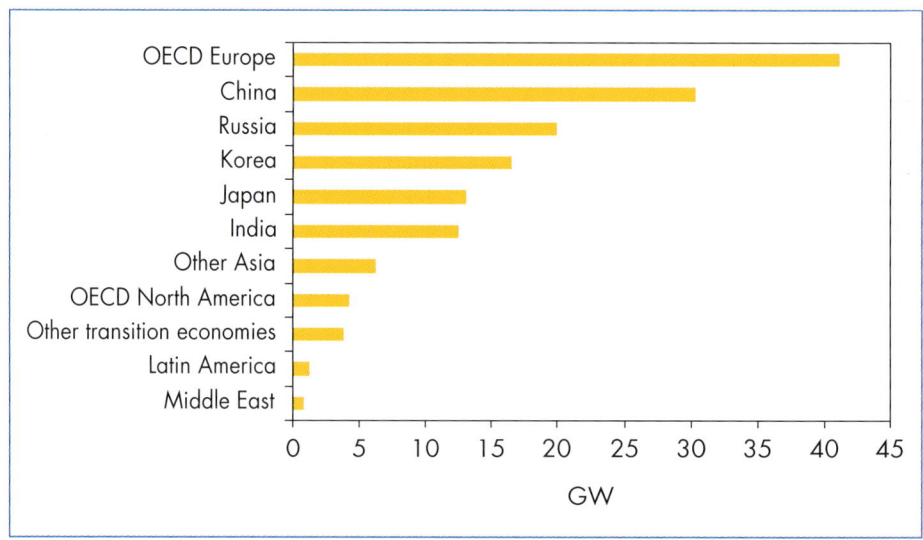

Electricity from *renewable energy* sources amounted to 2 927 TWh in 2002, or 18% of world electricity production, some 96% of that from hydropower and biomass power plants.[8] Over the *Outlook* period, the share of hydropower is likely to decline, though its use will increase in absolute terms. All other renewable sources will gain share. Overall, renewables will account for 19% of world electricity generation in 2030.

Hydropower could increase most in developing countries where its remaining potential is still high. However, there is much discussion about the environmental and social effects of building large dams and such issues could adversely affect the future of hydropower. Growth of hydro-electricity in the OECD will be limited by the lack of available sites and by environmental regulations. Some OECD countries provide incentives for the development of small hydropower plants. Globally, electricity generation from hydropower is expected to increase from 2 610 TWh in 2002 to 4 248 TWh in 2030, but its share will fall from 16% in 2002 to 13% in 2030.

Non-hydro renewable sources will substantially increase their contribution to electricity generation, growing nearly sixfold between now and 2030. Their contribution to electricity generation will increase from 2% in 2002 to 6% in 2030. This increase will be largely driven by government action in OECD countries to reduce CO_2 emissions and dependence on fossil fuels. Several developing countries are also adopting policies to increase the use of renewables. The share of non-hydro renewables in OECD countries is expected to increase from less than 3% in 2002 to 10% in 2030. In the developing countries, the share of non-hydro renewables is projected to increase from 1% to 3%.

Our projected increase in renewables is based on current policies *only*. Many OECD countries and, more and more, countries in the developing world, already have ambitious targets to increase electricity production from renewable energy sources. If they adopt even stronger policies to promote renewables, then a much higher contribution from renewables can be expected by 2030, particularly if such efforts are combined with measures to reduce the growth of electricity demand.

Among these sources, *wind power* will see the biggest increase in market share. Wind accounted for just 0.3% of global electricity supply in 2002, but the figure is expected to be ten times higher in 2030, unless environmental opposition to wind farms slows the pace. Wind power is projected to overtake *biomass* as the largest source of non-hydro renewable electricity generation by the middle of the next decade. Nonetheless, electricity generation from biomass will triple between now and 2030. *Geothermal power* will grow at the same rate as biomass. *Solar, tidal* and *wave* energy will make more substantial contributions towards the end of the projection period.

8. Biomass includes waste.

Figure 6.10: **Renewables in World Electricity Generation**

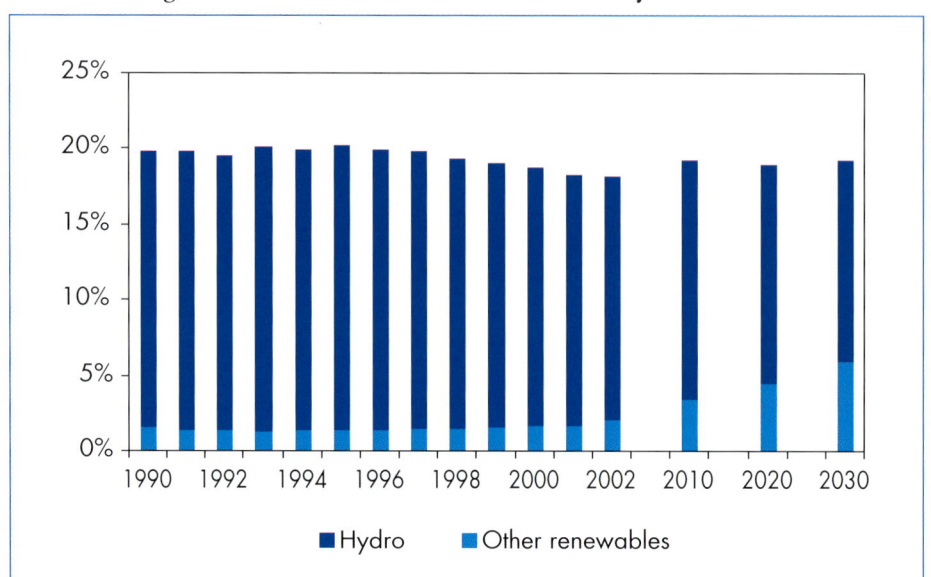

Table 6.4: **Electricity Generation from Renewable Energy Sources**

	2002		2030	
	Electricity generation	Share in total renewables	Electricity generation	Share in total renewables
	TWh	(%)	TWh	(%)
Hydropower	2 610	89	4 248	69
Biomass	207	7	627	10
Wind	52	2	929	15
Geothermal	57	2	167	3
Solar	1	0	119	2
Tide/wave	1	0	35	1
Total	**2 927**	**100**	**6 126**	**100**

Technology Outlook

Nearly two-thirds of the electricity produced in 2002 was based on fossil fuels. This share is expected to rise to over 70% in 2030. Advanced technologies over the projection period are expected to improve the efficiency of fossil fuels and diminish their polluting effects.

Today, most electricity is produced in conventional steam boilers – mostly coal-fired – with average efficiencies ranging between 30% and 42%, and increasingly in CCGT plants with efficiencies that typically exceed 50% and often reach 55% or more.[9] Figure 6.11 shows the fossil-fuel-based technologies that are expected to become available during the *Outlook* period and their efficiencies. In general, new and more efficient technologies will be deployed first in OECD countries, with developing countries following.

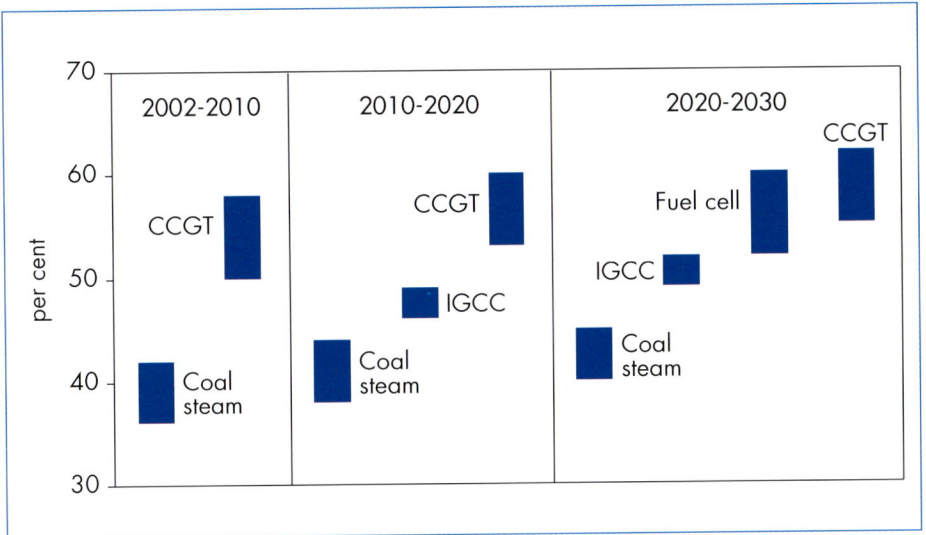

Figure 6.11: **Commercial Availability and Efficiency Improvements of Key Technologies, 2002-2030**

Note: IGCC = Integrated gasification combined cycle.
CCGT= Combined-cycle gas turbine.

In the first decade of the *Outlook*, CCGT technology is expected to dominate new plant construction. Few new coal plants are likely to be built in OECD countries, but those few will be fairly efficient. A substantial number of coal-fired power plants will be built in developing Asia, particularly in China and India. Their efficiency is expected to be higher than that of existing plants in those countries, but lower than in the OECD. In the 2010s, new power plant construction is expected to be based initially on CCGT plants, but as natural gas prices rise, more coal plants will be built. The efficiency of CCGT and coal plants will continue to improve. Later in this decade, IGCC (integrated-gasification combined-cycle) technology is expected to begin its deployment.

In the third decade, the efficiency of coal plants is expected to increase still further, and IGCC will gain market share. CCGT efficiency will stabilise at around 62%. Fuel cells will begin to enter the market, principally in decentralised

9. A number of recently built state-of the-art power plants have achieved even higher efficiencies.

applications. The fuel cells that are expected to achieve commercial viability first will involve the reforming of natural gas inside the cell. Their efficiency will reach 60% by 2030.

New nuclear plants are expected to be based on existing technologies, mostly pressurised water reactor and its variants. Renewable-energy technologies are expected to show progress over time, with improved wind turbines and biomass gasification. The share of decentralised generation is also expected to increase.

The technology improvements described in this section assume continued investment and progress in energy research and development. However, the move towards liberalised electricity markets has caused R&D budgets to decline in many cases and has shifted the focus of research toward short-term achievements. Under these conditions, governments may need to play a greater role in stimulating long-term R&D.

Impact on Fuel Markets

The projected increase in coal- and gas-fired electricity generation implies that fuel supplies to power stations will increase substantially over the *Outlook* period (Figure 6.12). Oil consumption in power stations is very small and is projected to be even smaller by 2030, but world coal and gas supply will be increasingly driven by demand from power stations.

Coal consumption in power generation will increase from 1 641 Mtoe in 2002 to 2 815 Mtoe in 2030. The increase will be modest in the OECD, but quite substantial in developing countries, where coal deliveries to power stations will increase by over a thousand Mtoe between 2002 and 2030. Together, power stations in India and China will consume about 1 300 Mtoe in 2030, more than a third of global coal consumption and more than the total coal consumption in OECD countries in that year.

Coal and gas markets will become more dependent on the electricity sector. Nearly 70% of the world's coal production now goes to power stations. This share will rise to 78% in 2030. Future coal markets will be more sensitive than ever to power sector policies, and in particular on policies to reduce carbon dioxide emissions.

World natural gas consumption in power stations will increase from 796 Mtoe in 2002 to 1 932 Mtoe in 2030. Natural gas markets will become more dependent on power markets, although less so than coal.[10] The share of natural gas production that goes to power stations will increase from 36% in 2002 to 47% in 2030. As there will also be a higher share of gas in electricity generation, interdependence between electricity and gas markets will grow.

10. IEA (2004) analyses the impact of greater use of gas on the OECD power sector.

Figure 6.12: **Fuel Requirements in Power Plants**

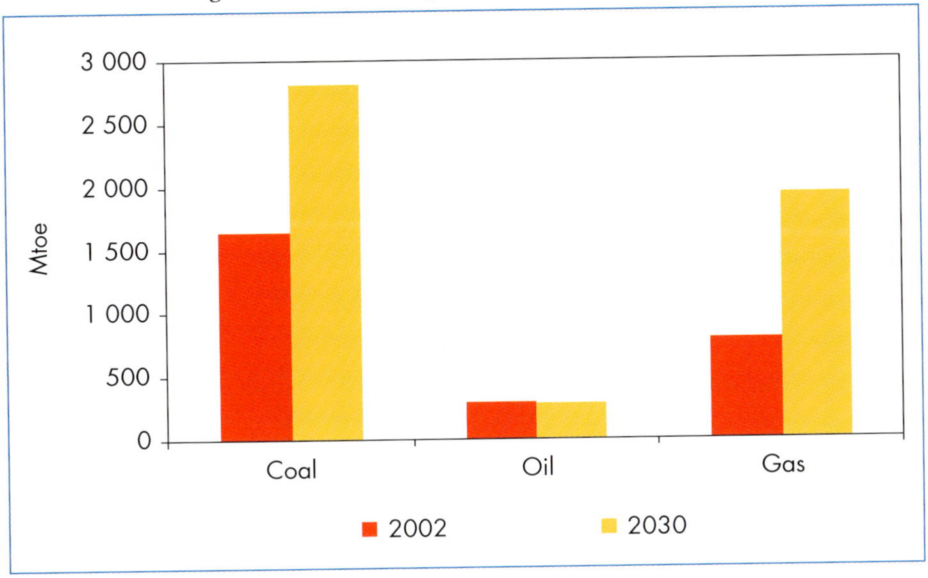

Capacity Requirements and Investment Outlook[11]

New power plants with combined capacity of 4 800 GW are expected to be built worldwide over the period 2003-2030. Half of these new power plants will be in developing countries (Table 6.5). OECD countries will need nearly 2 000 GW. More than a third of this new capacity will be built to replace ageing power plants in the region. Most existing coal-fired capacity will have to be replaced by 2030 (Figure 6.13). Over a third of existing nuclear plants in the OECD are expected to be shut down before 2030, either because they become too old or because of government policies to phase out nuclear power. The transition economies will have to build some 370 GW, with half of this capacity replacing ageing nuclear and fossil-based plants.

Figure 6.14 shows the capacity requirements by region. China will need the largest increases. New capacity requirements will also be substantial in OECD North America and OECD Europe. About 8% of this new capacity is now under construction and another 21% in planning. The largest as-yet-unplanned capacity additions will be in OECD North America and OECD Europe. Africa, Latin America (excluding Brazil) and Indonesia have very little capacity being built. These three regions could fall short of meeting local demand if they fail to attract sufficient investment to speed up construction. China will need to accelerate the pace of construction of new power plants if it is to avoid a repetition of recent electricity shortages. India will also need to accelerate capacity additions to meet increasing demand and to improve electrification rates.

11. A detailed analysis of electricity sector investment can be found in IEA (2003a).

Chapter 6 - Electricity Market Outlook

Table 6.5: **New Electricity Generating Capacity and Investment by Region, 2003-2030**

	Capacity additions (GW)	Investment in electricity sector ($ billion)			
		Generation	Transmission	Distribution	Total
OECD Europe	801	842	125	433	1 399
OECD North America	842	910	273	643	1 827
United States and Canada	*758*	*840*	*240*	*568*	*1 648*
OECD Pacific	332	416	100	199	714
OECD Asia	*275*	*346*	*73*	*150*	*569*
OECD	**1 975**	**2 167**	**498**	**1 276**	**3 940**
Russia	*154*	*138*	*26*	*92*	*256*
Transition economies	**372**	**287**	**79**	**287**	**653**
China	860	883	378	802	2 063
East Asia	391	364	133	302	798
Indonesia	*77*	*69*	*29*	*67*	*166*
South Asia	349	306	155	340	801
India	*272*	*256*	*132*	*289*	*678*
Latin America	373	317	122	269	708
Brazil	*114*	*125*	*46*	*102*	*273*
Middle East	195	118	48	107	272
Africa	269	165	127	271	563
Developing countries	**2 437**	**2 153**	**962**	**2 090**	**5 205**
World	**4 784**	**4 607**	**1 539**	**3 652**	**9 798**
European Union	*766*	*788*	*121*	*423*	*1 332*

Some 40% of new capacity will be gas-fired (Figure 6.15). Coal-fired capacity additions will account for about 30% and renewables for nearly 25%.

The projected capacity requirements will cost over $4 trillion. Renewables will require the largest investment, about $1.6 trillion between 2003 and 2030. Coal-based power plants will need a third of total investment and gas-fired power plants a little more than a fifth. Total power-sector investment over the next three decades, including generation, transmission and distribution, will be almost $10 trillion.[12] Transmission and distribution networks will need over $5 trillion, of which two-thirds will go for distribution networks. China will have the largest investment requirements, exceeding

---

12. Investment needs have been calculated using the methodology outlined in the *World Energy Investment Outlook* (IEA, 2003b) and on the basis of *WEO-2004* demand-supply projections and recent market developments.

Figure 6.13: **Impact of Plant Age on OECD Capacity Requirements** (GW)

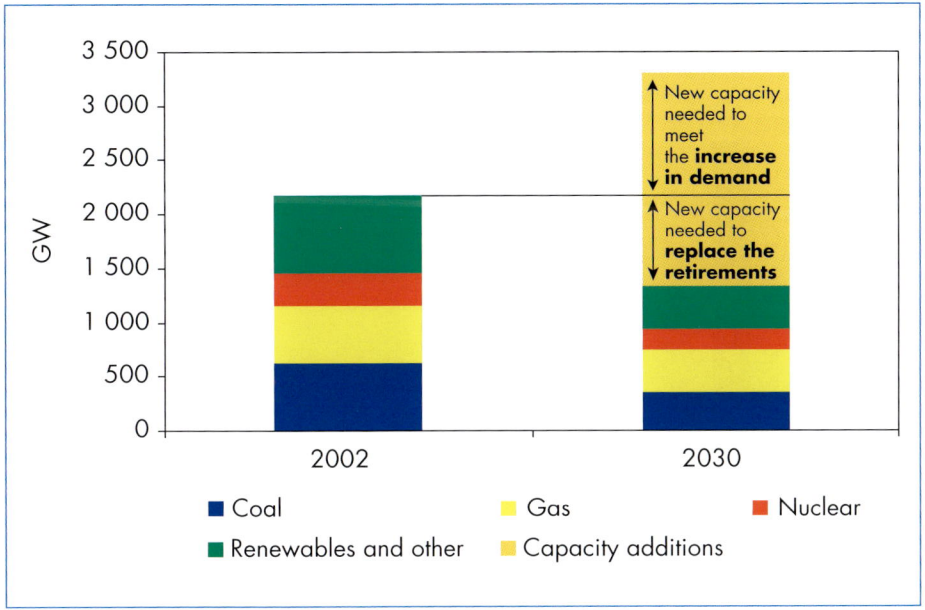

$2 trillion. Investment needs will also be very large in OECD North America and Europe (Figure 6.16). Attracting this investment in a timely manner may not be easy.

Figure 6.14: **Capacity Requirements by Region** (GW)

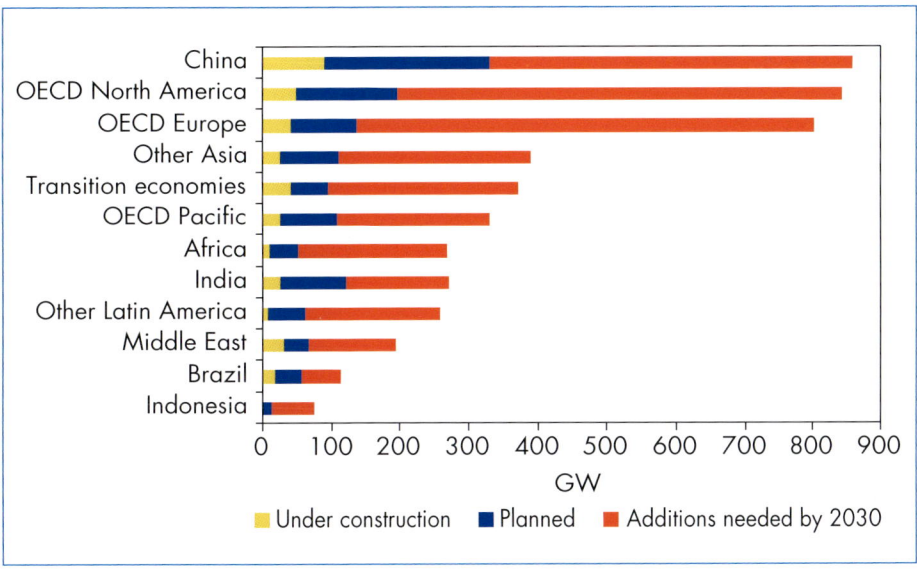

Source: IEA analysis. Data for plants under construction and planning are from Platts (2003).

Chapter 6 - Electricity Market Outlook

Figure 6.15: **World Power Generating Capacity Additions and Investment, 2003-2030**

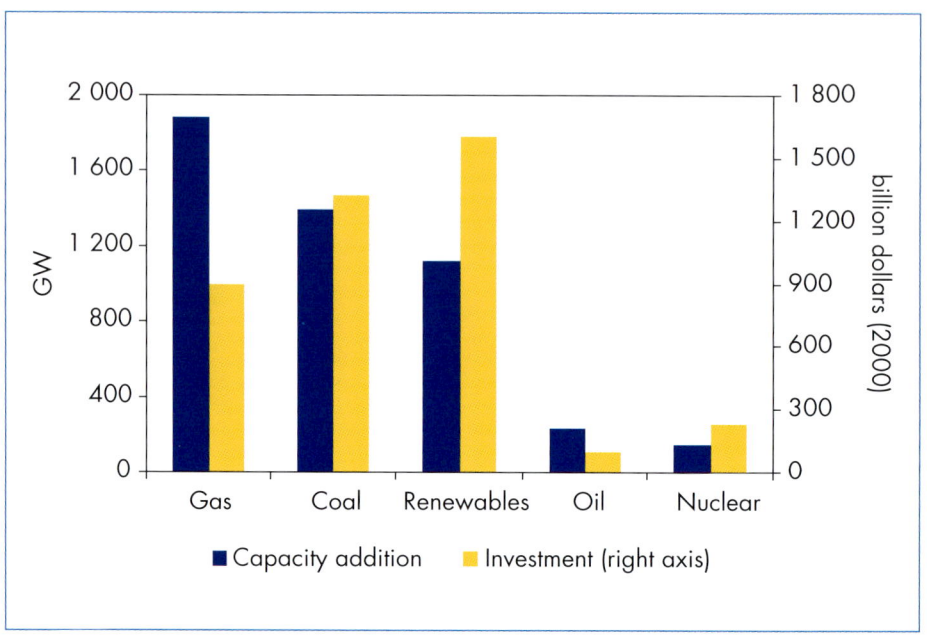

Figure 6.16: **Investment Requirements in Electricity Generation, Transmission and Distribution by Region**

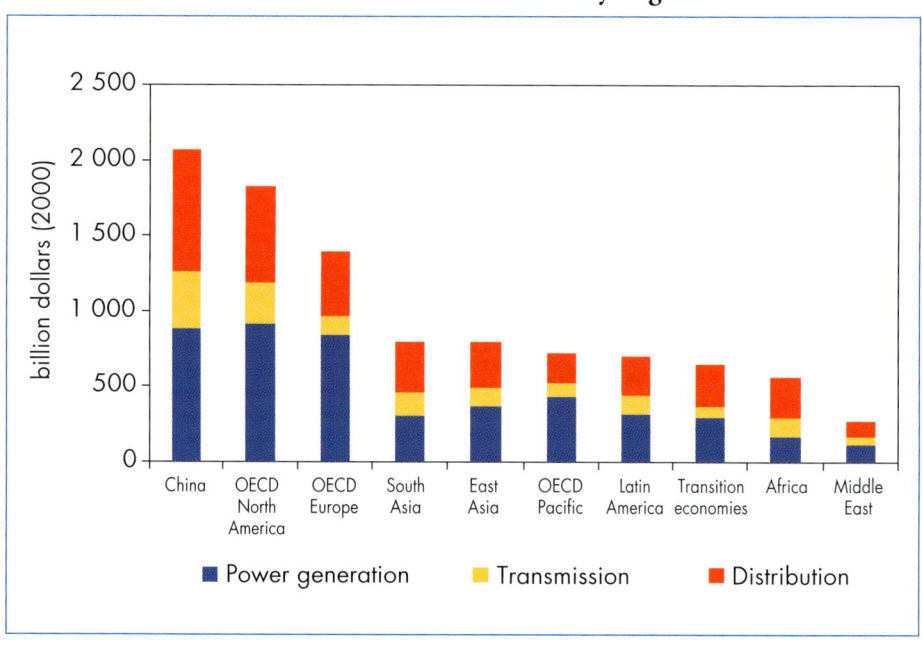

CO_2 Emissions

World CO_2 emissions from power plants are projected to increase by 2.1% per year over the period 2002-2030. Emissions will grow at a lower rate than total electricity generation because of improved generating efficiency, a decline in the share of coal and an increasing share of gas. But the part of power generation in global energy-related CO_2 emissions will increase from 40% in 2002 to 44% in 2030, because of the growing share of electricity in overall energy consumption.

CO_2 emissions from power stations in developing countries will increase nearly threefold from 2002 to 2030. Power plants in developing countries released 3 354 million tonnes of CO_2 into the atmosphere in 2002, about a third less than power plants in OECD countries. In 2030, power sector emissions in developing countries will be 44% higher than in the OECD (Figure 6.17).

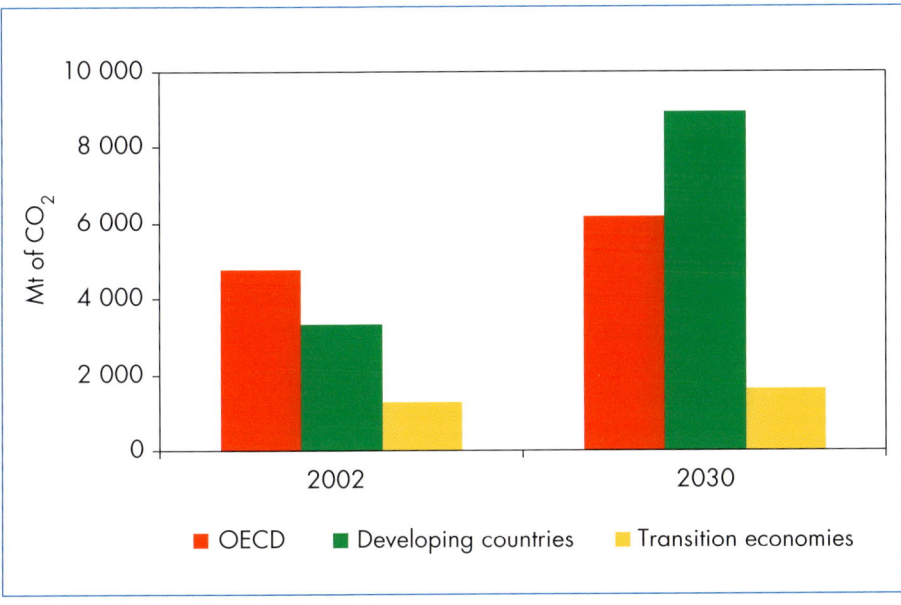

Figure 6.17: **Power-Sector CO_2 Emissions by Region**

Coal plants in developing countries will produce more CO_2 in 2030 than the entire power sector in OECD countries (Figure 6.18). This figure underlines the urgent need to improve efficiency in existing as well as new coal plants, particularly in China and India. There is also a fairly large potential for improving the efficiency of existing power stations.[13]

13. IEA Clean Coal Centre (2003) contains a detailed analysis of the potential to improve the efficiency of existing coal plants in developing countries.

Figure 6.18: **Power-Sector CO_2 Emissions of Coal, Oil and Gas-Fired Power Plants**

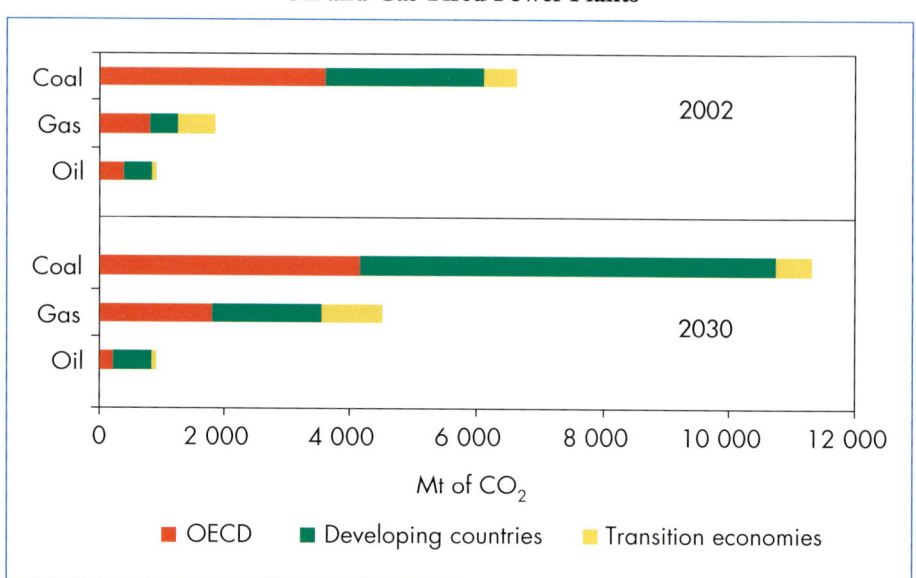

Electricity Markets and the Status of Reforms

Over the past few years, electricity markets around the world have faced serious difficulties. The United States has excess generating capacity, insufficient supplies of natural gas and inadequate transmission networks. The European power sector is finding it hard to become more competitive and faces serious uncertainties about future environmental regulations. European electricity prices are rising and critical investment decisions must be made soon. The blackouts in North America and Europe in the summer of 2003 and in Greece and Spain in the summer of 2004 highlighted the importance of improving transmission and distribution networks (Box 6.2). All these problems have tended to slow the pace of market reform, despite the obvious benefits that liberalisation has brought in many markets, including greater economic efficiency and flexibility, increased competitiveness and, in certain cases, lower prices to consumers.

Several developing countries have suffered power shortages for varying reasons, including breakneck demand growth in China, insufficient investment in power generation in Indonesia, very dry weather in Brazil and natural-gas shortages in Argentina.

In many countries outside the OECD, market reforms have failed to produce the desired results. Recent problems in OECD markets have rattled governments in some countries and, in some cases, turned them against reform altogether. Some countries that had planned reforms may now choose to defer them (IEA, 2003b).

Europe

The European Union's Electricity Market Directive entered into force on 1 July 2004, replacing an earlier directive adopted in 1996. Only the Netherlands and Slovenia, however, had fully implemented the directive by that date. Key provisions include: immediate free choice of supplier for business customers, with free choice for households to follow in 2007; the legal unbundling of transmission system operators; and the appointment of national regulators.

In December 2003, the European Commission announced a proposal for a directive to safeguard security of electricity supply and infrastructure investment. The directive states that further steps towards a competitive internal electricity market are vital to Europe's strategy for security of energy supply.

France has decided to change the status of the utility Electricité de France from a public enterprise to a corporation, with 30% of its shares to be traded freely. In Italy, the electricity exchange Gestore del Mercato Elettrico opened for spot trade on 1 April 2004. The Spanish and Portuguese governments signed an agreement on 19 January 2004 to form a common Iberian electricity market, MIBEL. The agreement was provisionally implemented on 22 April 2004 and will eventually lead to a common electricity exchange.

North America

In the United States, the creation of Regional Transmission Organisations (RTOs) and the growth of existing RTOs progressed during the past year. The formation of RTOs and independent system operators is an important element in the Federal Energy Regulatory Commission's plans for a reformed electricity sector. In Canada, the government of Ontario proposed the Electricity Restructuring Act in June 2004. In Mexico, market reforms have been stalled since 2002.

Pacific

In Australia, the Ministerial Council of Energy agreed on a new reform package in December 2003. Its main provision was to concentrate many regulatory responsibilities in one body instead of leaving them to regulators in the states and territories. In New Zealand, the government announced in May 2003 that measures will be taken to ensure that electricity demand will be met in the case of a so-called "one in sixty year" drought. In Korea, the unbundling process ground to a halt in June 2004; the third phase of a four-phase restructuring plan that was to have been initiated in 2004 did not go into effect. It included plans to divide the distribution assets of the Korean Electricity Power Corporation into separate distribution companies that were to be privatised over time. It also included the full privatisation of previously unbundled generation assets.

Box 6.2: **Major Blackouts in OECD Countries in 2003**

Three severe supply disruptions involving failures of transmission services struck North America and Europe during 2003:

- *North-Eastern United States and Ontario, Canada:* The largest supply disruption in North American history struck at about 4 p.m. on 14 August 2003, affecting eight US states and the Canadian province of Ontario. About 61 800 MW of electricity load was lost and fifty million people were disconnected. While most services in the United States were restored within two days, in some areas it took up to four days. Much of Ontario operated under power restrictions for over a week.
- *Sweden and Denmark:* The Nordic transmission system experienced its worst disruption in 20 years at 12.35 p.m. on 23 September 2003. Southern Sweden lost about 4 700 MW of supply, while Denmark lost about 1 850 MW. Four million people were disconnected, including many in Copenhagen. Transmission services in southern Sweden were restored within an hour, with complete services restored within a few hours.
- *Italy:* The worst supply disruption since World War II struck Italy at 3.28 a.m. on 28 September 2003 following a loss of about 6 400 MW imported into Italy from its northern bordering countries. The incident cascaded into a total loss of around 25 000 MW. An area of over 277 000 square kilometres was affected – most of Italy with the exception of Sardinia. Nearly 56 million people were disconnected, with services restored within 24 hours.

The US-Canada Power System Outage Task Force determined that the North American failure was due to lack of adherence to industry standards, deficiencies in corporate policies and inadequate management of reactive power and voltage (US-Canada Power System Outage Task Force, 2004). First Energy, a utility in Ohio, was singled out as having violated basic reliability procedures. The Task Force also described some standards and processes of the North American Electric Reliability Council as inadequate because they did not give sufficiently clear direction to industry on preventive measures needed to maintain reliability.

A report on the Scandinavian blackout by Elkraft System, the network operator in eastern Denmark, indicated that the disruption was caused by the simultaneous occurrence of mechanical faults at three different points in the southern Sweden power system (Elkraft System, 2003). The report concluded that, given the present design of the power system, a power failure could not have been prevented in the very unusual circumstances.

Three separate reports on the Italian blackout have been released- by the Swiss Federal Office of Energy, by the Italian regulator jointly with the French regulator and by UCTE (Union for the Co-ordination of Transmission of Electricity), the association of transmission system operators in continental Europe (Swiss Federal Office of Energy, 2003, AEEG and CRE, 2004 and UCTE, 2004). The Swiss report indicated that the causes of the outage were unresolved conflicts between the involved countries and companies on the one hand, and technical requirements of the existing transnational electricity system on the other. The joint report by the Italian and French regulators blamed inappropriate technical measures and bad communication by the operators of the Swiss transmission grid following the initial failure of two Swiss power lines. The UCTE report analysed the event from a more technical perspective. It described in detail the sequence which started from the failure of the Lukmanier line in Switzerland and ended with the isolation of the Italian grid from the European network and its subsequent collapse.

China

The Chinese government has embarked on a long-term reform of its electricity sector aimed at lowering prices and improving efficiency by introducing competition among generators. Little progress has been made in the past few years, however, as the government has concentrated on addressing power shortages around the country. The North-East continues to experiment with competitive pricing, but less than 10% of the region's demand is met by electricity sold at market prices. Eastern China has a similar pilot programme, but significant shortages have prevented true competition from emerging there, too.

India

In India, the Electricity Act of 2003 consolidates and replaces a number of previous laws. Key measures include reduced licensing restrictions for power projects based on fossil fuels and open access to transmission. The act imposes an obligation on states to establish regulatory commissions which would set retail tariffs on the basis of full costs and would promote competition. It requires that any subsidies on electricity retail sales be paid out of state budgets rather than through cross-subsidisation.

Brazil

In March 2004, the Brazilian government approved a new plan for the electricity sector, the New Electricity Model, which aims to strengthen supply security, increase competition, and rationalise regulation in order to attract greater investment. Implementation of the plan began in July/August 2004 with the passing of three law decrees. The New Electricity Model provides for auctions based on the lowest-tariff criteria. It creates two contracting systems to work simultaneously, one regulated (ACR) and another free (ACL).

It allows for long-term bilateral contracts and introduces a System Security Reserve. It removes licensing power from the regulator and returns it to the Ministry of Mines and Energy. It creates an Energy Research Enterprise (EPE), a Power Energy Trading Chamber (CCEE) to replace the current Wholesale Market (MAE) and an Electricity Sector Monitoring Committee (CMSE) to monitor supply conditions and recommend preventive actions to restore security of supply.

Russia[14]

Electricity-market reform progressed in Russia in 2003 with the adoption of six new laws in March and April 2003 and a plan for the restructuring of UES (Unified Energy System), the main power company. The laws establish basic rules governing liberalised markets and the remaining state-controlled monopolies, while the plan covers primarily the restructuring of industry assets to create a more competitive market structure during its transition period.

Regional Trends

United States and Canada

Electricity generation in the United States and Canada is projected to grow at 1.3% per year over the period 2002-2030. Most electricity now comes from coal-fired power stations, at 47% in 2002, followed by nuclear, at 19%, natural gas, at 16% and hydro, at 13%. Oil and renewables (other than hydro) account for a little over 2% each.

Over the past decade, the United States saw a significant increase in gas-fired generation. In the United States, following market liberalisation, many new gas-fired power plants were built (Figure 6.19). Between 1997 and mid-2004, 194 GW out of the 202 GW of new capacity built in the United States was gas-fired. On average, the United States now has excess capacity, although some areas need new capacity. Power companies have announced plans to build another 50 GW of gas-fired capacity in the near future.

On the other hand, supply constraints in the United States have caused natural gas prices to rise substantially since 2000 (Figure 6.20). The power sector has responded by cutting its use of natural gas. Consequently, the share of natural gas in electricity generation declined in 2003, and this decline continued in the beginning of 2004. High natural gas prices combined with relatively low electricity prices (largely resulting from excess capacity) have resulted in low utilisation rates of new gas-fired capacity.

In the future, the United States, like the entire North American region, will have to rely increasingly on natural gas imports. Gas-fired generation is expected to

---

14. See Chapter 9 for a more detailed discussion of Russia's electricity market reforms.

Figure 6.19: **US Capacity Additions since 1997**

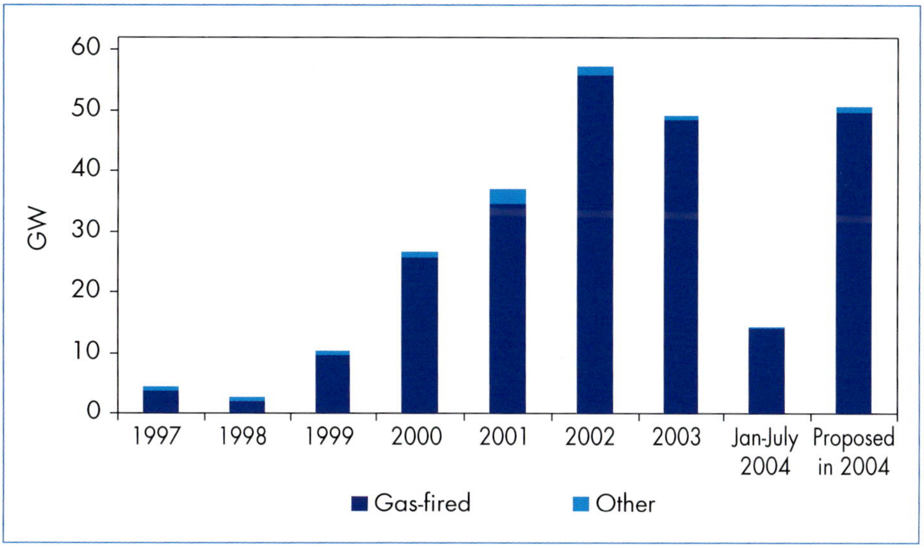

Source: United States Energy Information Administration.

continue to increase, but at a more modest rate than shown in previous *Outlooks*. The share of natural gas is projected to reach 25% by 2030.[15] Natural gas will also be used to produce hydrogen in fuel cells after 2020. This new source of electricity could account for nearly 2% of total electricity generation in 2030.

The need for new capacity in the long term will be filled by coal-fired power plants. Coal-fired electricity generation is projected to increase by 1% per year, reaching 2 861 TWh in 2030. Coal's share, however, will decline from 47% in 2002 to 43% in 2030.

Nuclear capacity in the region is expected to increase in the medium term, as some shut-down reactors in Canada are brought back to operation and because of uprates planned in the United States.[16] After 2020, some older reactors could be retired, bringing nuclear capacity to 109 GW by 2030, the same level as in 2002. No nuclear power plants are projected in the region, as gas and coal plants are expected to be more economical. The United States Department of Energy (US DOE) has put in place a programme to identify sites for new nuclear power plants, to develop advanced nuclear plant technologies, and to demonstrate new regulatory processes. The DOE expects this programme will lead private companies to decide by next year to order new nuclear power plants by 2010. Three applications from industry consortia have been submitted so far.

15. This compares with the 32% projected in *WEO-2002*.
16. A power uprate is an increase in the power level of a power plant. Some uprates can increase power by as much as 20%.

Figure 6.20: **US Natural Gas Prices and Share of Gas in Electricity Generation**

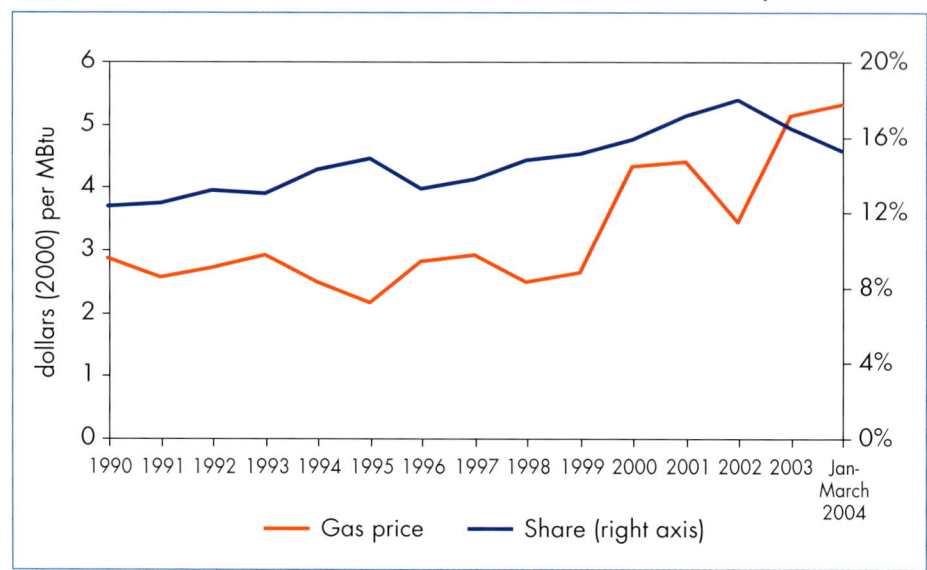

Source: United States Energy Information Administration.

Construction of new hydropower plants is expected to be limited. All new development will be in Canada. The share of hydropower in electricity generation is projected to decline from 13% in 2002 to 10% in 2030. Other renewables will increase their share as a result of government intervention. In the United States, about a third of the fifty states have adopted renewables portfolio standards (RPS).[17] California's RPS is one of the most ambitious, calling for an increase in renewables use of one percentage point per year beginning in 2003 and reaching a total of at least 20% by 2017. Non-hydro renewables are projected to increase their share in the region from 2% in 2002 to 7% in 2030.

European Union

European electricity generation is projected to increase at an average annual rate of 1.3% between 2002 and 2030. The European Union will see a pronounced increase in natural gas and renewables. Gas-fired electricity generation will increase from 521 TWh in 2002 to 1 458 TWh in 2030. The share of gas will double, accounting for over a third of the total in 2030.

Electricity reserve margins were high when liberalisation of the electricity sector started, but have since declined (Figure 6.21). The power-supply situation now appears to be tightening, although some countries still enjoy fairly high

17. Database of State Incentives for Renewable Energy (www.dsireusa.org).

reserve margins. New capacity is being continually added almost everywhere, but in many cases the capacity additions have not kept up with the increase in the load. There is also a degree of market concentration, which can inhibit competition. At the beginning of 2003, seven companies controlled more than half of Europe's generating capacity.

Figure 6.21: **Growth in Installed Generating Capacity, Peak Load and Available Capacity in the European Union**

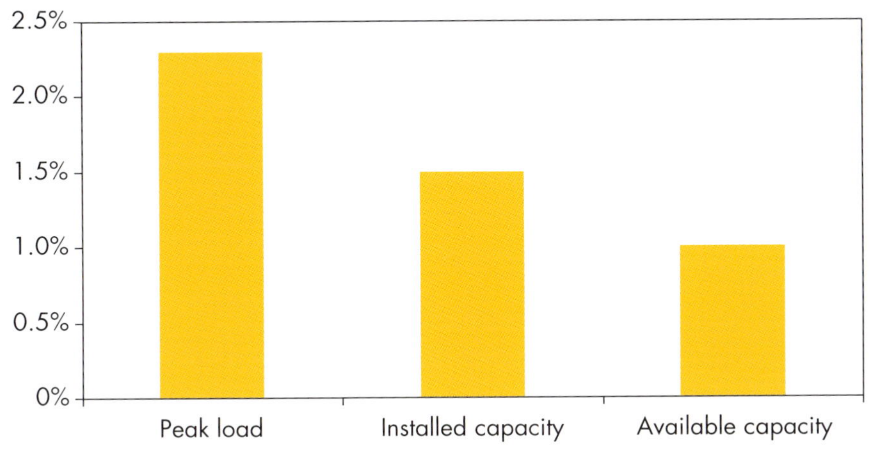

Nuclear power is now the largest source of electricity generation in the region, supplying nearly a third of electricity demand. More than three-quarters of existing nuclear capacity will be retired by 2030 because many reactors will reach the end of their life and because some countries in the region have policies to phase out nuclear. About 40 GW of new capacity will be built in the period to 2030, mostly to replace old reactors in France. Nuclear capacity in the European Union is expected to decline from 133 GW in 2002 to 71 GW in 2030. The share of nuclear power in electricity generation will fall to 13%.

Coal-fired electricity generation is projected to increase at a modest 0.6% per year. Many of the region's coal-fired power plants are expected to be retired. New coal plants are expected to be built in the region in the second half of the projection period when natural gas prices rise, making coal competitive. New coal plants will be based on much more efficient technology than the current stock.

Renewables will see a substantial increase. Their share in Europe's electricity generation will double, from 13% now to 26% in 2030. Wind power is projected to increase from 36 TWh in 2002 to 480 TWh in 2030, reaching 11% of total electricity generation. This sizeable increase will require a redesign of networks and the raising of substantial financial resources to develop wind energy.[18]

OECD Pacific

OECD Asia

Electricity generation in Japan and Korea is expected to increase by 1.3% per year in the period to 2030. Japan generates three times as much electricity as Korea, but Japan's share is expected to drop by 2030 because of saturation of demand and decreasing population. Natural gas (primarily LNG) and nuclear energy will increase their shares significantly. Coal-fired electricity generation will increase at just 0.5% per year and its share will fall from 30% in 2002 to less than a quarter by 2030.

Both Japan and Korea have nuclear programmes. Japan recently revised its long-term nuclear targets downward although it still expects nuclear to play a significant role in meeting future energy demand. Nuclear capacity in the region could increase from 59 GW in 2002 to 87 GW in 2030. Recent problems in Japan's nuclear operations, including an accident in the summer of 2004 that cost the life of four workers, have undermined public confidence in nuclear power and may hinder future development. Siting new nuclear plants is a problem in both countries.

OECD Oceania

Electricity generation in Australia and New Zealand is projected to increase at an annual rate of 1.6% per year. Australia's share is 85%. Coal-fired generation will see its share decline to the benefit of gas. Coal now accounts for nearly 80% of Australia's electricity generation because the country has low-cost reserves of both hard coal and lignite. The share of gas-fired generation will increase from 14% in 2002 to 25% in 2030.

China

China's electricity production is expected to climb at 4.4% per year in the period 2002-2030, more than tripling. The projected growth rate is, nonetheless, much lower than rates recorded over the recent past. In the 1990s, electricity generation grew by an average rate of 8% per year. In 2003, electricity generation increased by 16% (Lawrence Berkeley National Laboratory, 2004) and data for the first half of 2004 indicate an increase of 18%.[19] Following a period of excess capacity in the late 1990s, installed

---

18. See Box 7.1 for a discussion of grid-related wind costs.
19. Power in Asia, *"China: No letup in demand growth"*, 3 August 2004.

capacity increased at a slower pace than did electricity generation, although new construction now appears to have increased (Figure 6.22). In 2002, 2003 and 2004 demand often outstripped supply, resulting in power rationing, and brownouts or blackouts in many areas. A dry year in 2003 cut hydropower production. Insufficient transmission capacity exacerbated the situation. In light of these problems, China needs urgently to create an attractive investment framework both for generation and for networks. China also needs to improve its energy efficiency, particularly in industry, in order to manage electricity-demand growth.

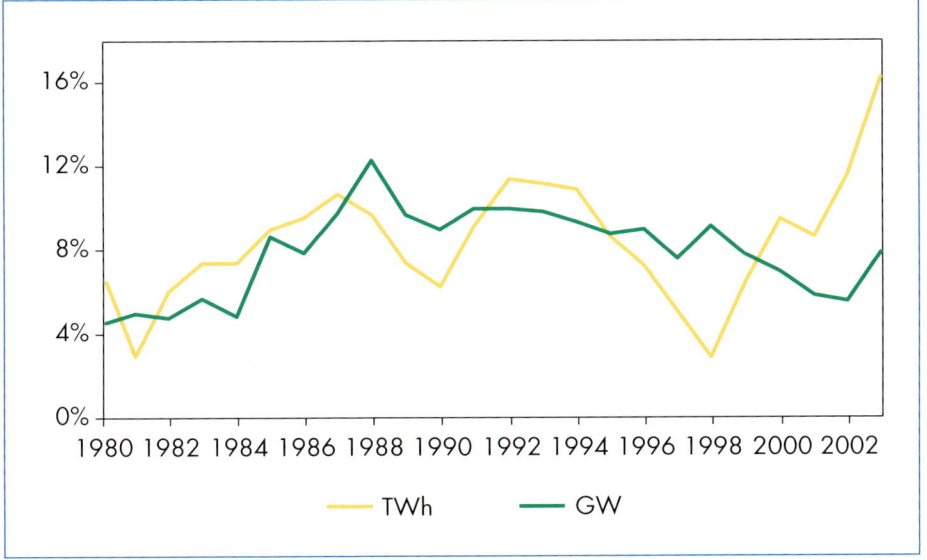

Figure 6.22: **Annual Growth Rates of Electricity Generation and Installed Capacity in China, 1980-2002**

The Chinese electricity-generation mix will continue to be dominated by coal, although coal's share is projected to decline from 77% in 2002 to 72% in 2030. Similarly, hydropower – the second largest source of electricity – will increase, but its share will decline. Natural gas, nuclear and non-hydro renewables will gain market share. China plans to increase the use of liquefied natural gas in coastal areas to diversify supplies and to reduce pollution from coal-fired plants. The share of gas in electricity generation is projected to increase from 1% to 6%. China is well endowed with renewable energy resources, besides hydropower. They could account for 3% of electricity generation in 2030.

China had six nuclear reactors in commercial operation at the end of 2002 (Table 6.6). Two more reactors started operation in 2003 and one in 2004 bringing total capacity to 6.6 GW. Two more reactors are under construction at Tianwan, with one of them expected to begin commercial operation in 2004. There are plans to build more nuclear power plants in the future. Indeed, the programme appears to have accelerated recently with the approval of two new projects in July 2004. China's nuclear capacity is projected to rise to 35 GW by 2030 to supply 5% of the country's electricity.

Table 6.6: **Nuclear Power Reactors in China, 2004**

Units	Net capacity (MW)	Start-up date
Daya Bay 1, 2	2 × 944	1994
Qinshan 1	279	1994
Qinshan 2, 3	2 × 610	2002, 2004
Lingao 1, 2	2 × 935	2002, 2003
Qinshan 4, 5	2 × 665	2002, 2003
Total	**6 587**	

Sources: World Nuclear Association (2004) and CEA (2003).

India

Electricity generation in India is projected to increase at 4.4 % per year between 2002 and 2030. Over 80% of India's electricity now comes from coal and hydropower plants. The share of coal in total generation will drop from over 70% now to 64% by 2030, while that of hydropower is likely to increase up to 2010 and then fall back, as the most economic sites are exhausted, returning to its present share of 11% in 2030. Nuclear power is expected to account for over 5% of Indian electricity supply in 2030, compared with 3% now.

Several new power projects fuelled by imported LNG could be built where coal is expensive to transport. Gas-based electricity generation is expected to increase fivefold and to account for 16% of the total, compared with 10% now.

India is actively promoting renewable energy, particularly to provide power to remote areas that cannot be connected through the grid.[20] India's wind-power capacity, which reached more than 2 GW in 2003, is among the highest in the world. It is projected to reach 7 GW in 2030. In total, non-hydro renewables are projected to account for 2.1% of electricity generation in 2030, compared with 0.7% in 2002.

The Indian electricity sector faces enormous challenges in providing reliable service and meeting rising demand. Demand exceeds supply, particularly at

20. See also Chapter 10.

periods of peak usage. There are frequent blackouts and brownouts. India has one of the highest rates of transmission and distribution losses in the world, mostly due to theft and unmetered consumption. The uncertain pace of electricity market reforms is a major drag on India's electricity supply prospects.

The Indian private sector has reacted positively to the new business opportunities arising from the Electricity Act 2003. The act recognises transmission and trading as separate activities and permits private participation in both. A first transmission licence has been awarded by the Central Electricity Regulatory Commission to a public-private joint-venture in which the private sector partner is the majority shareholder. Several additional licences have been issued and even more applications received. The regulatory commission has also issued five power trading licences and is reviewing several more applications. It has published a five-year tariff order stipulating a flat 14% return on equity for very large private generation projects. It has also set down the terms and conditions of interstate trading and set in motion open access in interstate transmission.

Brazil

Brazil's electricity production is projected to increase at 3.1% per year over the period to 2030. Its heavy dependence on hydropower, which accounted for 83% of electricity production in 2002, is likely to be reduced in the future to the benefit of natural gas.

Construction of hydro plants will gradually slow down as hydro sites that can be tapped economically are exhausted. Environmental considerations may also have an impact on hydro expansion, since much of the remaining potential is in the Amazon. The share of hydropower will drop to 65% by 2030. New hydro development is likely to remain a government responsibility since private investors are showing no interest in building new hydropower plants.

The Brazilian government hopes to attract private investment in gas-fired power plants. But how much gas-fired capacity will be built is uncertain and will depend on the cost of natural gas, on the development of gas infrastructure, on tariffs charged and on contracts for the supply of natural gas. If the development of a gas market proceeds well, gas-fired electricity generation could reach 22% of the total in 2030.

CHAPTER 7

RENEWABLE ENERGY OUTLOOK

HIGHLIGHTS

- Renewable-energy consumption will increase from about 1 400 Mtoe in 2002 to over 2 200 Mtoe in 2030, a rise of almost 60%. The aggregate share of renewables in total energy consumption will remain largely unchanged. Traditional biomass now accounts for 7% of world energy demand, but its share will fall as incomes and urbanisation increase. The share of hydroelectricity will remain stable. The shares of other renewables will increase.
- The power sector will lead the increase in renewable-energy consumption. The share of non-hydro renewables in electricity generation will triple, from 2% in 2002 to 6% in 2030. Most of the increase will be in wind and biomass. The largest increases in renewables will be in OECD Europe, driven by strong government support.
- Wind power is projected to be the second-largest source of renewable electricity after hydroelectricity in 2030, but siting land-based wind turbines is becoming more of a challenge in some areas. Problems such as intermittency, low reliability and difficulties in connecting wind-driven generators to the grid can increase the cost of wind power by anything from $5 to $15 per MWh.
- The investment needed to develop all renewables-based power generation is expected to reach about $1.6 trillion. This is nearly 40% of power-generation investment over the projection period. Of this, a trillion dollars will go to developing non-hydro renewables. Their investment prospects will depend crucially on the effectiveness of government measures to promote renewables.
- The use of commercial biomass, solar water heaters and geothermal heat in industry and buildings will increase. Most of the growth in biomass is likely to come from combined heat-and-power installations in industry. Solar energy consumption for heating water is expected to rise from 4 Mtoe in 2002 to about 35 Mtoe by 2030.
- The share of biofuels in total transport consumption was only 0.4% in 2002. Biofuel consumption is expected to increase more than fourfold by 2030, reaching 36 Mtoe. Government policies are in place to spur biofuel consumption in several countries, especially in the United States, in the European Union, in India and Brazil.

This chapter analyses global trends in renewable energy supply. First, it looks at renewables in total energy consumption. It then analyses the prospects for renewables in electricity generation and summarises the investment requirements. Finally, it examines the role of renewables in industry, buildings and transport.

Renewable Energy Demand[1]

Renewable energy accounted for 14% of the world's total primary energy demand in 2002 (Table 7.1). Biomass[2] is by far the largest renewable energy source. Over two-thirds of biomass is used for cooking and heating in developing countries.[3] Much of these uses is unsustainable. Hydropower is the second-largest renewable source, while solar, geothermal, wind, tide and wave energy each accounts for only a small part of global energy demand.

Table 7.1: **World Renewable Energy Consumption**

	2002		2030	
	Renewables use (Mtoe)	Share of total demand	Renewables use (Mtoe)	Share of total demand
Biomass	1 119	11%	1 605	10%
of which: traditional biomass	765	7%	907	6%
Hydro	224	2%	365	2%
Other renewables	55	1%	256	2%
Total	**1 398**	**14%**	**2 226**	**14%**

Renewable energy consumption is expected to increase from 1 398 Mtoe in 2002 to 2 226 Mtoe in 2030. The share of renewables in total energy consumption is expected to remain largely unchanged. The share of hydropower will remain stable at around 2% of total energy consumption. The share of traditional biomass will fall, but this decrease will be offset by an increase in the share of other renewables.

---

1. Supply and cost issues are discussed extensively in IEA (2001).
2. Biomass includes renewable waste (2.4% in 2002) and a small fraction of non-renewable waste (0.4% in 2002).
3. This type of biomass is referred to as "traditional" throughout this *Outlook*. All biomass used in OECD countries and transition economies, as well as biomass used in industry, in transport and in power generation in developing countries, is termed "commercial".

Three-quarters of renewable energy is now consumed in developing countries, principally in the form of traditional biomass and hydropower (Figure 7.1). OECD countries accounted for 23%, while the transition economies used just 3%. Developing countries will remain the largest consumers of renewables in the future, but their share in global renewables use will fall to about two-thirds by 2030. This is because traditional biomass will lose market share, as it is replaced by other fuels, and because renewable energy use is expected to increase substantially in OECD countries.

Traditional biomass use in total final consumption is expected to increase in developing countries over the projection period, but at a modest rate, so its share will fall. Higher per capita incomes and increased urbanisation will promote its replacement by fossil fuels. The biomass that continues to be used will be consumed in a more efficient and sustainable way. The main form of biomass in many developing countries is firewood, supplies of which are already becoming scarce. Traditional biomass is discussed in Chapter 10.

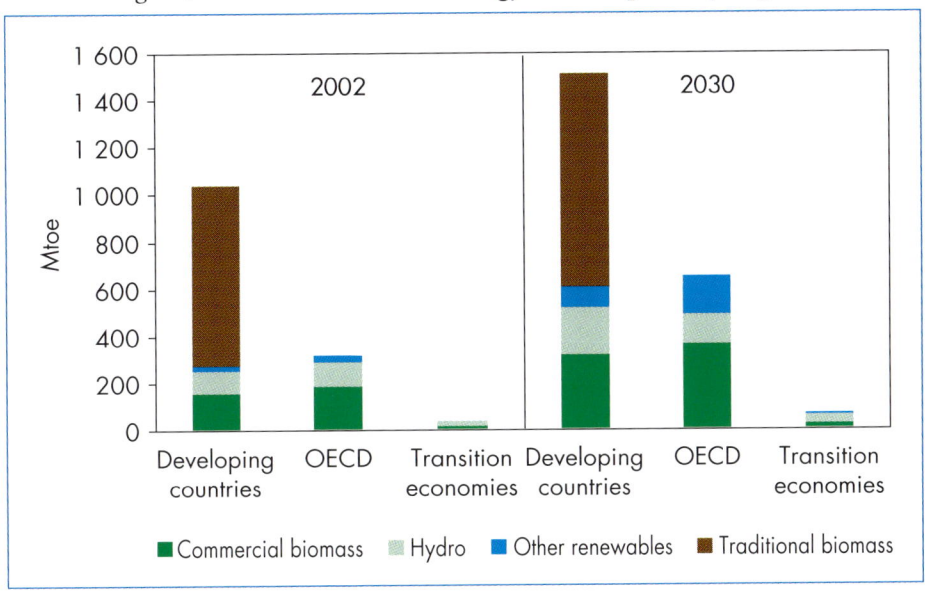

Figure 7.1: **World Renewable Energy Consumption by Region**

While heating and cooking will remain the principal use for renewables, the power sector will lead the increase in renewable energy consumption between 2002 and 2030. The power sector accounted for just a quarter of global renewable energy consumption in 2002. Its share is projected to rise to 38% by 2030. Biofuels in transportation now account for less than 1% of renewables use. They will triple their share by 2030.

Renewables in Power Generation

Electricity generation from renewable-energy sources amounted to 2 927 TWh in 2002. Worldwide, 18% of electricity demand was met by renewables, including hydroelectricity. The share of renewables varies between countries and regions. A few countries meet almost 100% of their electricity needs with renewables, mostly hydropower. In Paraguay, Iceland, Nepal, Congo, Mozambique, the Democratic Republic of Congo, Uruguay and Norway, over 99.5% of electricity generation is currently based on renewables. On a regional basis, Latin America has the highest share of renewables in power generation because of its extensive use of hydropower (Figure 7.2).

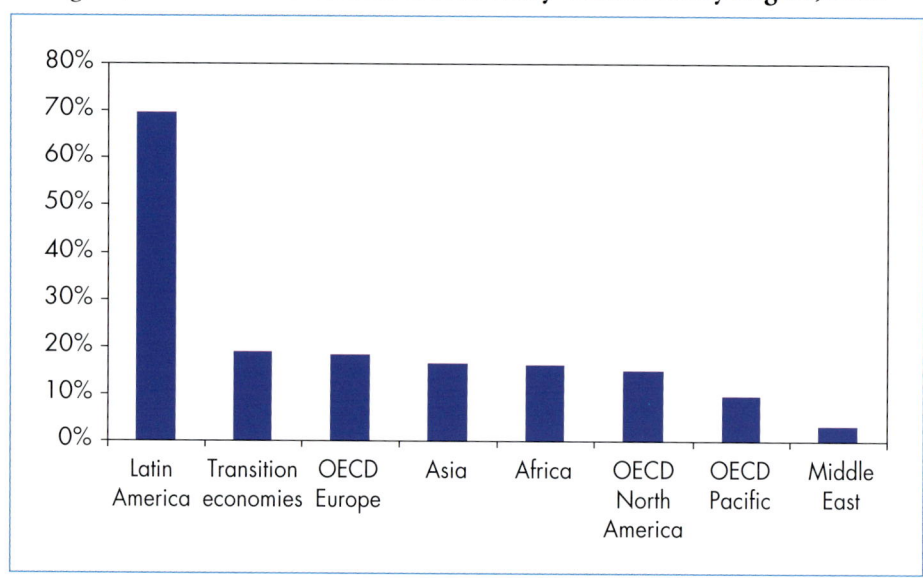

Figure 7.2: **Share of Renewables in Electricity Generation by Region, 2002**

The share of renewables in electricity generation is projected to increase slightly, from 18% in 2002 to 19% in 2030. The share of hydropower will fall, but non-hydro renewables will see their share triple, from 2% in 2002 to 6% in 2030. The largest increase will be in OECD Europe, driven by strong government support and ambitious official targets. The share of non-hydro renewables in Europe's power production is projected to rise from 3% now to 16% in 2030 (Figure 7.3). The other OECD regions will also see significant increases.

Non-hydro renewable energies have yet to gain a significant share in the fuel mixes of OECD countries. Over the past five years, only six countries - Denmark, Iceland, Finland, Spain, Germany and the Netherlands – increased

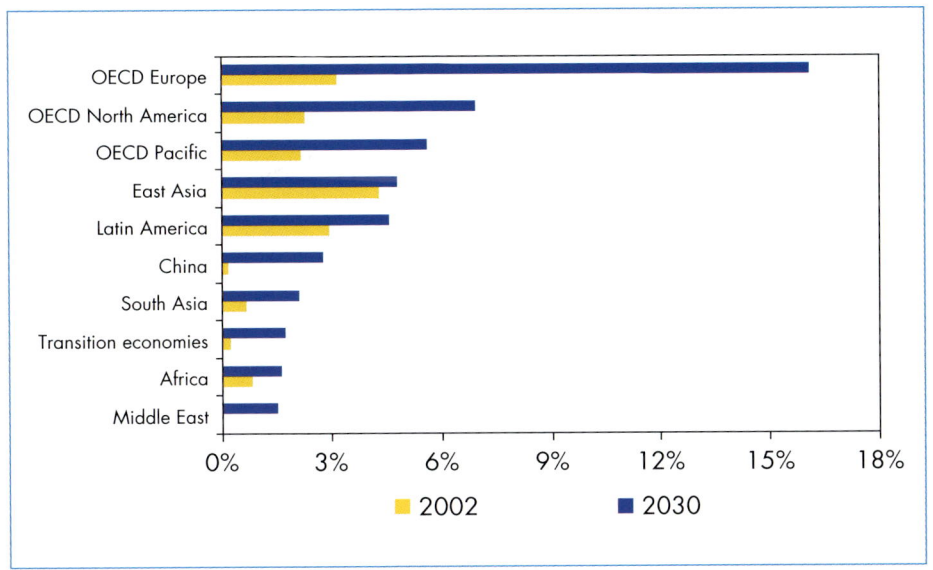

Figure 7.3: **Shares of Non-Hydro Renewables in Power Generation in 2002 and 2030**

the share of non-hydro renewables by more than three percentage points (Figure 7.4). The increase was less than two points in most other countries, while a few saw a decrease in the share of renewables. Across the OECD, the share of non-hydro renewables increased just marginally, from 2% in 1997 to 2.6% in 2002. In absolute terms, the United States was the largest market for non-hydro renewables in 2002, followed by Japan and Germany (Figure 7.5). Production in the United States was slightly less than in the European Union as a whole.

Over the projection period, small yet significant increases are expected in developing countries, some of which are actively promoting renewables for electricity generation. Large-scale development of renewables may be too costly for many of these countries, but the large increases in OECD will help eventually to bring the cost of renewables down, making them more affordable for developing countries too. The cause of renewables may also benefit from international efforts to reduce greenhouse gas emissions, such as the clean development mechanism under the Kyoto Protocol, and from international efforts to increase general access to electricity.

World electricity generation from renewables is projected to double between 2002 and 2030. Hydropower will increase by over 60%, while non-hydro renewables will increase sixfold (Figure 7.6). Most new hydropower will be in developing countries. In the OECD, most of the promising sites have already been utilised. Only a small fraction of the long-term potential of other

Figure 7.4: **Share of Non-Hydro Renewables in Electricity Generation in OECD Countries, 1997 and 2002**

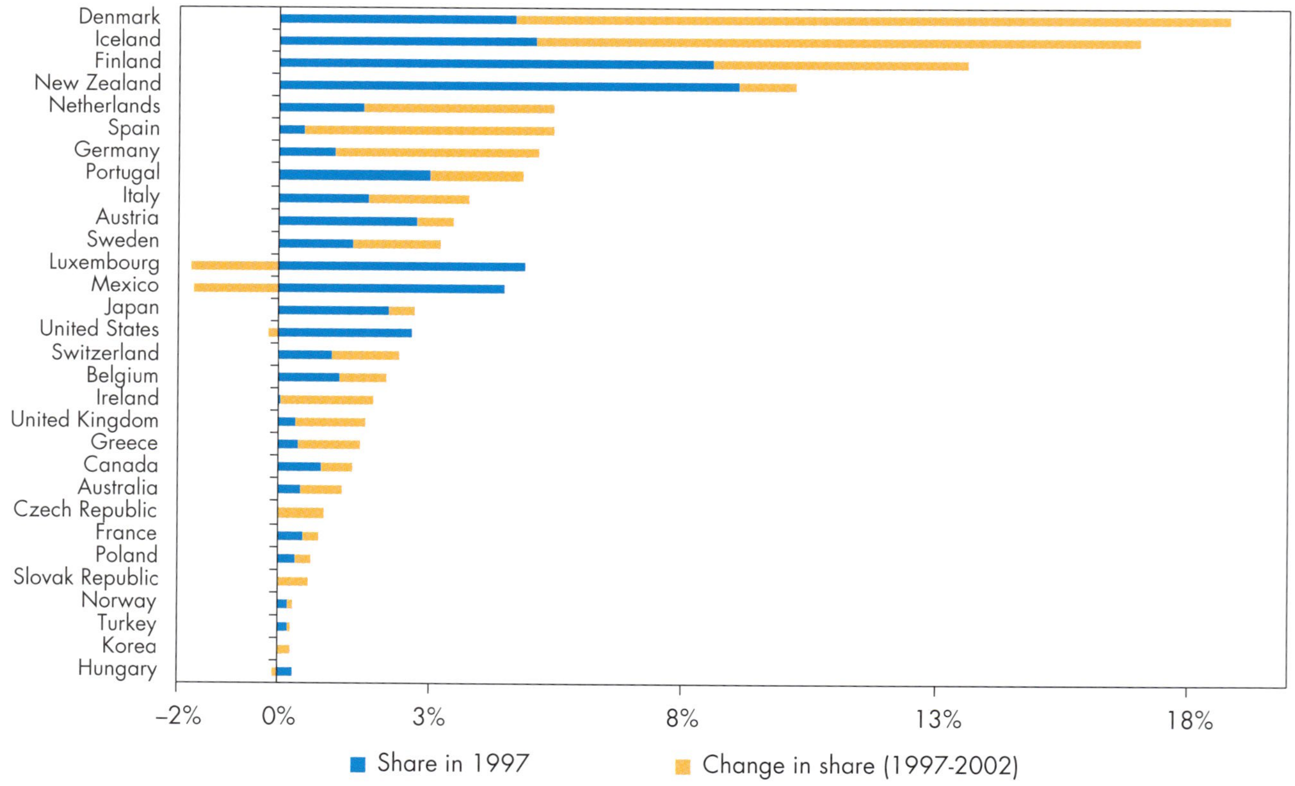

Figure 7.5: **Electricity Generation from Non-Hydro Renewables in OECD Countries, 2002**

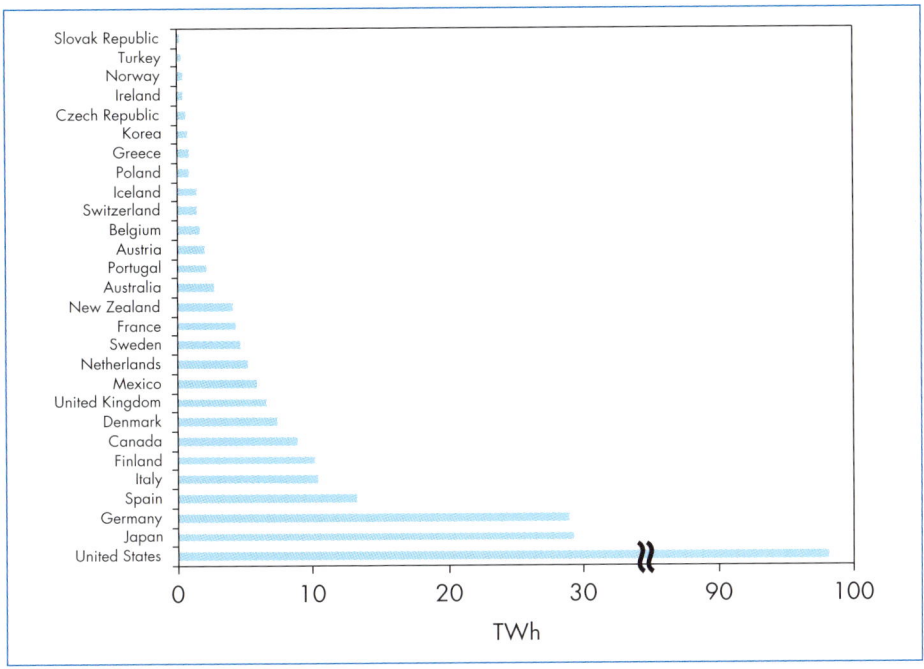

Note: Hungary and Luxembourg are not shown in this chart because their consumption is very small (less than 0.1 TWh).

Figure 7.6: **World Electricity Generation from Non-Hydro Renewable Energy Sources**

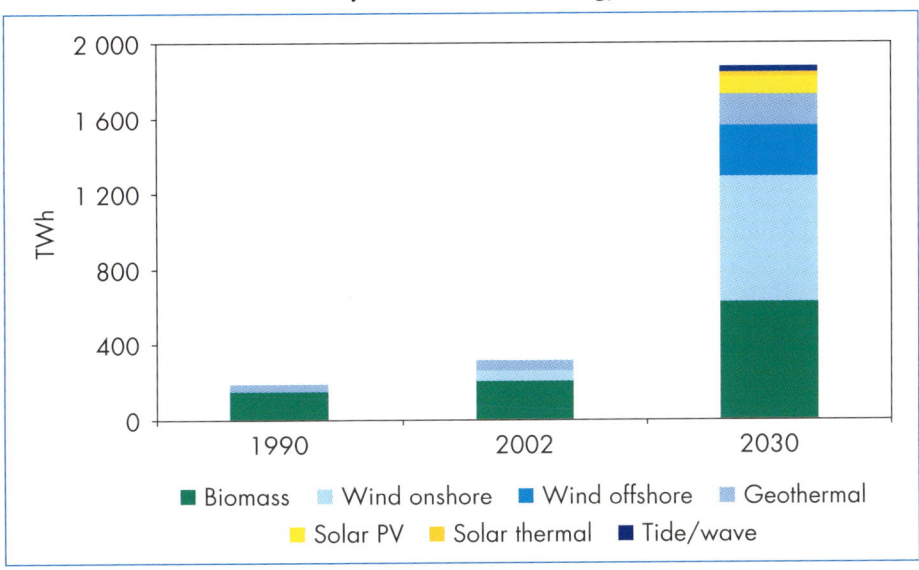

Chapter 7 - Renewable Energy Outlook

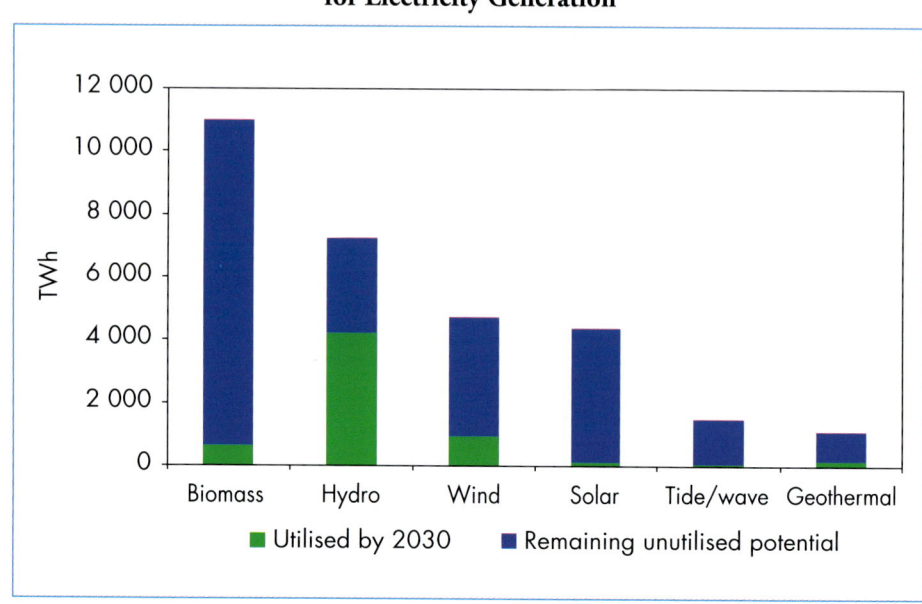

Figure 7.7: **World Long-Term Renewable-Energy Potential for Electricity Generation**

Source: IEA analysis. See Annex C for an explanation of how this potential has been estimated.

renewables will have been exploited by 2030 (Figure 7.7). While hydropower will remain the largest source of electricity generation by renewables in 2030, the contribution of wind and biomass will also become quite substantial.

Cost Developments

The capital costs of renewables are assumed to go on declining in the future (Figure 7.8). The rate of decline will depend on the rate at which they are deployed and on the maturity of each technology. The fastest rate of decline will come in the capital cost of photovoltaics, the most capital-intensive of the renewable energy technologies considered here. Substantial decreases are also expected in the capital cost of offshore wind, solar thermal and tidal and wave technologies. The capital cost of hydro is expected to remain broadly unchanged, since the technology is well known and already mature.

The electricity-generating costs of renewables depend on the capital cost of the technology and on the quality of the resource – strong winds or abundant sunshine, for example. Figure 7.9 shows ranges of generating costs for 2002 and 2030. While the generating costs of renewables will decline generally as a

Figure 7.8: **Capital Costs of Renewable Energy Technologies, 2002 and 2030**

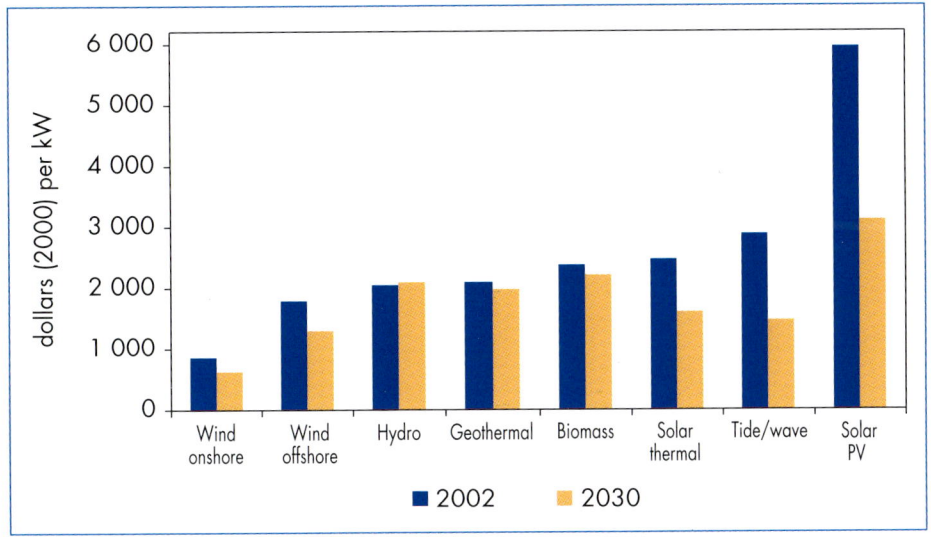

Figure 7.9: **Electricity-Generating Costs of Renewable Energy Technologies, 2002 and 2030**

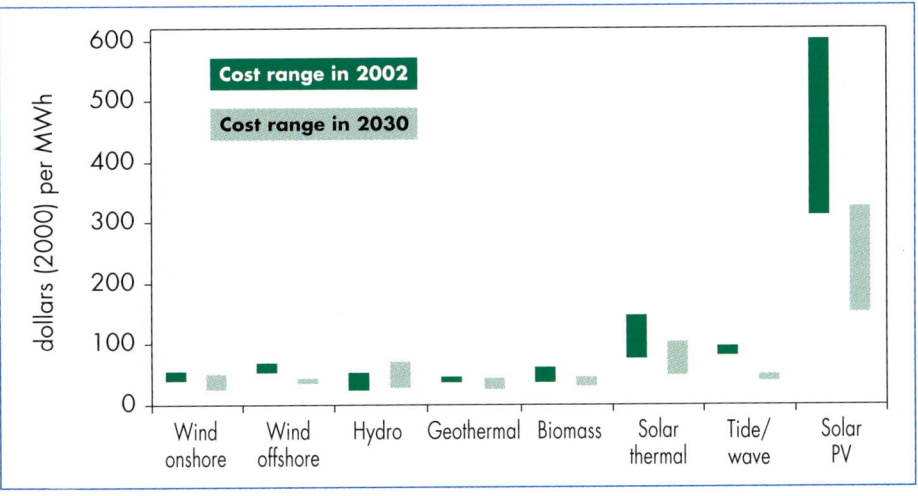

result of falling capital costs, some renewables will become more expensive in some areas because the best sites for them will already have been exploited. This will be the case of hydropower in almost all regions, and of onshore wind in OECD Europe.

Chapter 7 - Renewable Energy Outlook

Outlook by Source

Hydropower

Hydropower use is expected to increase from 2 610 TWh in 2002 to 4 248 TWh in 2030. China and Latin America will account for 60% of the increase worldwide. China will have more installed hydropower capacity than any other country in the world in 2030, having overtaken the United States. However, hydro's current market share will decrease in all regions. In Latin America, hydro now accounts for two-thirds of electricity generation, but this will fall to 46% in 2030.

Biomass

Biomass is the second-largest source of renewables-based electricity generation after hydro. It is often used in combined heat and power production. Most biomass-based electricity is produced in OECD regions, accounting for between 1% and 3% of electricity generation. In Finland, it was 14% in 2002. Electricity generation from biomass is less widespread in developing countries, although it is fairly important in some of them, particularly in Latin America, where bagasse from sugar production is the most copious source of commercial biomass. Biomass power plants supplied 3% of electricity in Brazil in 2002, the same share as in Austria.

Over the next three decades, world biomass-fuelled electricity production is expected to triple worldwide. Biomass will fuel 2% of global electricity production in 2030, up from a little more than 1% now. The most significant increase will be in OECD Europe, where biomass is projected to reach 4% of electricity generation in 2030, up from less than 2% now.

Most future growth in biomass will come as a result of government policies to promote renewables or to increase the use of combined heat and power. Biomass will be used mostly for the production of electricity and heat in decentralised applications in industry, or district heating. A small percentage of it is likely to be used in co-firing with coal, as a way to reduce CO_2 emissions from coal-fired power plants.

Wind Power

Electricity from wind farms will register the greatest increase in market share over the *Outlook* period. Wind's share in total electricity generation is expected to grow from 0.3% in 2002 (52 TWh) to 3% in 2030 (929 TWh). This increase will make wind the second-largest source of renewable electricity after hydro. Wind-based electricity will increase substantially in OECD Europe, where it is projected to meet over 10% of the region's electricity needs by 2030, compared with 1% now. OECD North America will be the second largest wind producer in 2030. But siting land-based wind turbines is becoming more of a challenge in some areas and could impact the development of wind farms.

There has been a considerable increase in wind power in Europe over the past decade. OECD Europe accounted for two-thirds of global wind power in 2002. The increase, driven by government incentives, has been the largest in Denmark, Spain and Germany. These three countries together produced 58% of the world's wind power in 2002 and 82% of OECD Europe's. The success of these countries in developing wind power is largely based on feed-in tariff systems which offer high buy-back rates and guarantee a market for wind-farm output.[4] Wind power has also increased in the United States. The country had the second-largest wind-based electricity production in 2002 behind Germany, but the American share of wind in total electricity was still just 0.3%.

Most existing wind farms are onshore. A substantial contribution is expected from *offshore* wind farms by 2030. Over 40% of wind power in OECD Europe could be offshore by 2030 when nearly 80% of the world's offshore wind power would be concentrated in Europe.

The generating costs of wind power are, on average, higher than those of fossil fuels, ranging from about $45 per MWh at good sites to $55 per MWh at moderate sites. While the generating costs in good sites are quite close to the cost of conventional technologies, additional costs to cope with intermittency and grid integration can increase the generating cost of wind substantially (Box 7.1).

Box 7.1: **Economics of Wind-Power Integration**

Wind power already has significant shares of the electricity-generating capacity in several OECD countries. The main technical features that distinguish wind power from traditional generating capacity are its intermittency, its low reliability and problems involved in connecting it to the grid. The extra costs of integrating wind power include:

– *Backup capacity and operational costs.* Alternative generating capacity must be available to supply when there is no wind. Because of the extreme difficulty of predicting wind patterns more than 36 hours in advance, the provision of a steady supply of power is complex and costly. Access to alternative flexible resources is necessary. Flexibility has a cost. Assessed cost range: $5 to $10 per MWh.

– *Grid costs.* Wind turbines are often connected to the grid at low-voltage levels, a practice that may save grid losses but adds to the complexity of system control and operation. Offshore wind farms extend the transmission system to new territories and this adds costs. Assessed cost range: $2.5 to $4 per MWh.

---

4. A feed-in tariff is the price per unit of electricity that a power company has to pay to purchase renewable electricity from private generators. The tariff rate is fixed by the government at a level high enough to encourage electricity generation from renewables.

The total range of extra costs seems to be on the order of $5 to $15 per MWh (Figure 7.10). Actual costs of any given wind farm will depend on the specific system, the share of wind power and the organisation of the market for electricity and ancillary systems.[5]

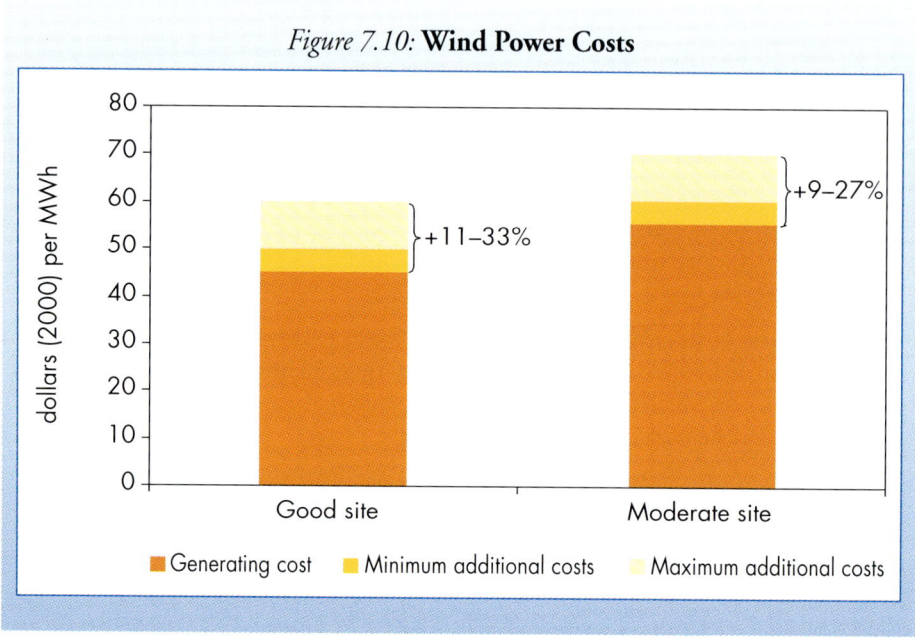

Figure 7.10: **Wind Power Costs**

The rising significance of wind power has coincided with a general reform of the electricity sector in OECD countries. Until recently, most electricity systems were highly centralised. With the reform of the sector, more decisions are now taken locally and under pressure from competition. Business risks and economies of scale are now perceived differently.

Solar Power

Electricity generation from solar power is expected to reach 119 TWh in 2030. Over 80% of it will be from photovoltaics (PV), while the rest will be produced in solar thermal power plants.

Photovoltaic technology has the highest investment and generating costs of all commercially deployed renewable energy sources. Average generating costs range between $350 and $600 per MWh, compared to around $40 per MWh or less for gas. The range of costs is wide principally because of differences in the amount of sunshine available in different regions (also known as

---

5. A more detailed analysis of the economics of wind power is given in IEA (2004a).

"insolation"). Consequently, PV will not compete with other technologies for large-scale centralised electricity generation, unless there are dramatic cost decreases achieved through technology improvements.

Most current PV power is decentralised in buildings and this is expected to remain its main use throughout the projection period. Electricity generation from PV is economically attractive in areas with abundant sunshine and high electricity prices. PV power is most valuable when maximum PV production coincides with peak electricity demand. In remote areas in the OECD and in developing countries, PV can be a cost-effective option.

Solar thermal power is projected to reach 21 TWh by 2030, produced almost exclusively in OECD countries. Electricity generation from solar thermal power now costs between $85 and $135 per MWh, which is two to three times higher than the cost of conventional energy sources. The economics of generation by solar thermal power will improve over the projection period, but it will not become cost-competitive on a large scale before 2030. Toward the end of the *Outlook* period, solar generating costs are expected to fall to around $55 per MWh in sunny areas.

Solar technologies are suitable for large-scale electricity generation. There are a number of options available, at different stages of development. The most promising technologies are the parabolic trough, the central receiver and the parabolic dish. The parabolic trough is already commercially available, while the parabolic-dish and central-receiver technologies are still at the demonstration stage. They could eventually achieve higher conversion efficiencies and lower capital costs than parabolic-trough technology.

Geothermal Power

Electricity production from geothermal energy reached 57 TWh in 2002. The United States is the largest producer of geothermal electricity in the world, at 15 TWh in 2002, followed by the Philippines with 10 TWh and Indonesia with 6 TWh. While geothermal power is now used in only seventeen countries, it has a high share in electricity generation in some of them (Figure 7.11).

Geothermal power is projected to triple to 167 TWh by 2030, while its share in the global electricity generation mix will improve marginally. About 40% of the projected growth will be in North America.

Tide and Wave

Tide and wave energy is still in its infancy. Only France, Canada, China, Russia and Norway now operate tidal power stations, with an estimated overall capacity of 325 MW. There is also a commercial wave power plant in the United Kingdom and a few demonstration projects around the world. Tidal and wave power could reach 35 TWh in 2030, compared with just 1 TWh now.

Figure 7.11: **Share of Geothermal Power in Total Electricity Generation, 2002**

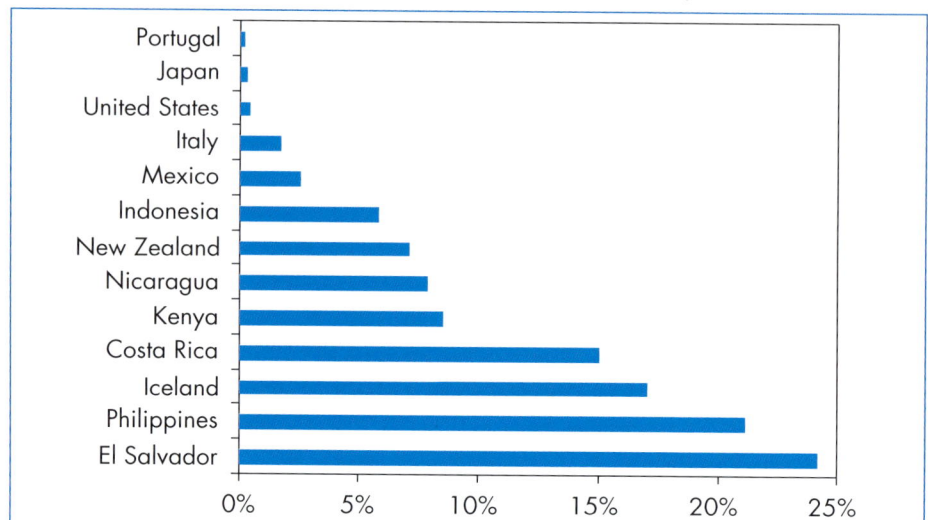

Capacity and Investment Outlook[6]

Over the next thirty years, nearly a quarter of new power-generating capacity will be based on renewable energy. The development of renewables-based power generation is expected to cost about $1.6 trillion (in year-2000 dollars), nearly 40% of power generation investment over the period 2003-2030. The share of renewables in power generation investment as a whole is higher than their share of newly-built capacity because renewables require very high initial investment. Hydropower will account for about half of the projected 1 113 GW of new renewable capacity. The cost will be on the order of $640 billion. Figure 7.12 shows that the largest capacity increases after hydro will be of onshore wind farms.

Government policies will be vitally important in the future of non-hydro renewables. Under present conditions, renewables are, on average, more expensive than fossil fuels. Their share is projected to increase from 2% now to 6% in 2030. This increase implies that 12% of the capacity additions and 22% of the investment in power generation between 2003 and 2030 must be for non-hydro renewables.

---

6. Investment needs have been calculated using the methodology outlined in the *World Energy Investment Outlook* (IEA, 2003) and on the basis of *WEO-2004* demand-supply projections and recent market developments.

Figure 7.12: **Renewables Capacity Additions, 2003-2030**

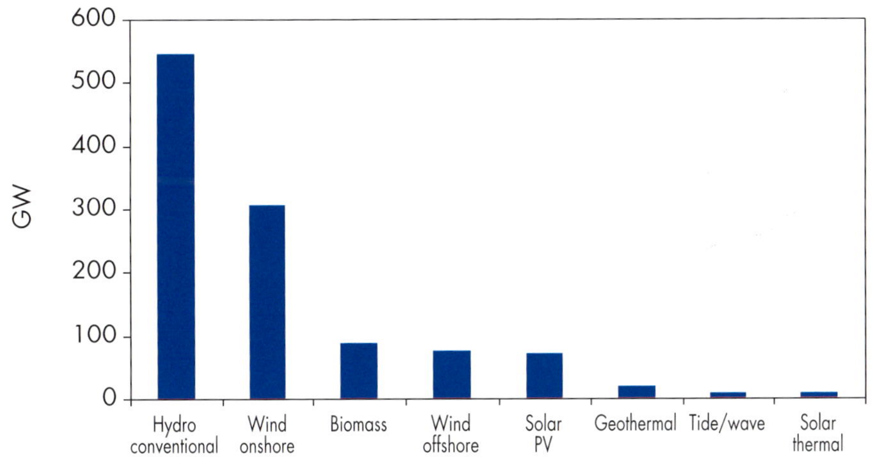

Renewables in Industry and Buildings

Industry and buildings consumed about 1 000 Mtoe of renewable energy in 2002, including 729 Mtoe of traditional biomass in developing countries. Most of the remainder was commercial biomass used in boilers. Solar heat amounted to 4 Mtoe and geothermal to 3 Mtoe.[7] Commercial biomass use reached 225 Mtoe in 2002 and is projected to rise to 338 Mtoe in 2030. Most of the increment will be in the OECD.

In OECD countries, about half of commercial biomass is consumed by industry and half in buildings. Biomass is projected to increase its share in OECD industrial energy demand from 6% in 2002 to 9% in 2030. Most of the growth will come from combined heat and power installations in industrial facilities. The growth of biomass use in buildings will be more modest.

In developing countries industrial consumption of biomass is considered commercial. In Brazil, it accounted for 36% of industrial energy demand in 2002; in Africa 29% and in India 22%. These high shares are expected to decrease in the future, as industries in developing countries switch to other forms of energy.

Many countries use solar water heaters, although only in a few, notably Israel and Cyprus, do such heaters provide a substantial part of the energy needed for water heating. Solar energy is also used for space and swimming-pool heating,

---

7. Not all countries report solar heat data to the IEA. China, for example, does not report any data although it is a large user of solar heat.

Chapter 7 - Renewable Energy Outlook

Figure 7.13: **Projected Share of Solar in Energy Consumption for Hot Water in the OECD**

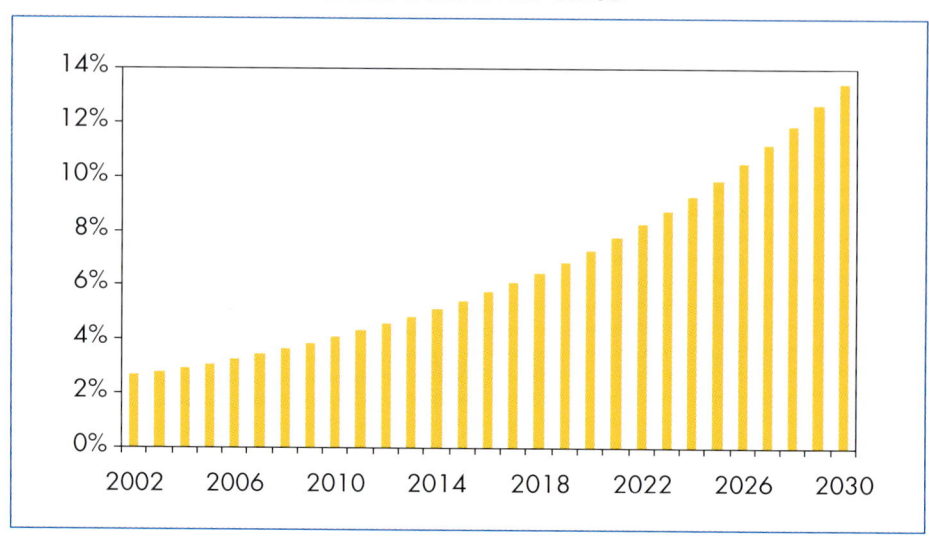

but to a very small extent. Over 90% of solar heat is used in the residential sector, which is expected to remain the largest consumer to 2030. Solar-heat use is expected to rise to about 35 Mtoe by 2030. OECD North America and China will be the largest users.

From 15% to 20% of residential energy consumption in the OECD is used for heating water. So the potential for solar water heaters is large, particularly in sunny areas. Many OECD countries offer incentives to promote the use of solar heating. A much larger share of energy consumption for hot water could be solar in 2030 (Figure 7.13). The potential in developing countries is also quite large, since the solar-water heaters now used in these countries are often manufactured locally and cheaply.

Geothermal-heat use is projected to reach almost 5 Mtoe in 2030. Geothermal now makes a significant contribution to end-use energy demand only in Iceland, where it met 45% of demand for energy in buildings in 2002.

Biofuels

Biofuels are transportation fuels, either in liquid form such as fuel ethanol or biodiesel, or in gaseous form such as biogas or hydrogen, derived from agricultural sources. Cereals, grains, sugar cane and other starches can be fermented to produce ethanol, which can either be used as a motor fuel in pure form or blended into gasoline. Oil-seed crops – rapeseed, soybean and sunflower seed – can be converted into methyl esters, a liquid fuel which can be blended with conventional diesel fuel or burnt as pure biodiesel.

In OECD countries, most fuel ethanol is produced from the starch component of grain crops. Corn is mainly used in the United States and wheat, sugar beets and barley in Europe. In conventional grain-to-ethanol processes, only the starchy part of the crop plant is used, the corn kernels or the whole wheat kernel. These starchy products represent a fairly small percentage of the total plant mass, and there are considerable fibrous residues in the form of seed husks and stalks.

Today, there is virtually no commercial production of ethanol from cellulosic biomass, but there is substantial research going on in this area in OECD countries, particularly the United States and Canada. There are several potentially important benefits from commercial cellulosic ethanol process:

- The possibility of using a much wider array of potential feedstocks, including waste materials, grasses and trees;
- A much greater reduction in "well-to-wheel" CO_2 per litre of fuel, due to nearly completely biomass-powered systems.[8]

If ethanol can be produced from cellulose, using biomass as the fuel to drive the conversion process, net greenhouse-gas emissions can be cut to near zero on a "well-to-wheel" basis. This is also true for a variety of other processes being developed to convert biomass to biofuels. Such advanced processes can produce much more environment-friendly biofuels than current processes relying on fossil energy to drive the conversion of starches to ethanol or oil seeds to biodiesel.[9]

Biofuels production in OECD countries is still relatively expensive, about two or three times the cost of gasoline. Production in some developing countries, like Brazil, is much cheaper, indeed not much more expensive than the cost of gasoline.

Global biofuel consumption was 8 Mtoe in 2002, of which Brazil accounted for 70% and the United States for 23%.[10] The share of biofuels in total transport consumption was only 0.4%. In the Reference Scenario, biofuels consumption is expected to more than quadruple by 2030, reaching 36 Mtoe. Government policies are in place to spur biofuels consumption in several countries, especially in the United States, in the European Union, in India and in Brazil.

---

8. "Well-to-wheel" refers to the complete chain of fuel production and use, including feedstock production, transport to the refinery, conversion to final fuel, transport to refuelling stations, and final vehicle tail-pipe emissions.
9. See IEA (2004b) for a detailed discussion of these technologies.
10. Based on IEA data. Other sources suggest that the United States and Brazil have similar shares. See, for example, IEA (2004b).

In the United States, ethanol is used as an octane enhancer in gasoline, with a 1.5% market share. The market share of biodiesel is negligible. The Energy Policy Act of 2003 requires the production and use of 3.1 billion gallons of fuels from renewable sources in 2005, increasing to 5 billion by 2012. Both ethanol and biodiesel production are eligible for tax credits.

In 2003, the European Union set an indicative target for its member countries to replace 5.75% of the gasoline and diesel they use in transport with biofuels before the year 2010. But the target is not mandatory and no specific measures are indicated on how to reach it.

The greatest successes of biofuels have been scored in Brazil, where ethanol based on sugar cane is blended with gasoline in a proportion that varies between 20 and 25%. In 2003, "flex-fuel" cars were launched in the Brazilian market and they now represent 18% of new car sales. These vehicles can run on gasoline, ethanol or any blend of the two. In addition to ethanol, the Brazilian government is also aiming at blending 5% of biodiesel with ordinary diesel by 2010.

In India, the Planning Commission has proposed a programme to produce ethanol to be blended with gasoline, and biodiesel to be blended with high-speed diesel. In order for India to replace 5% of the oil it uses in cars and trucks with biofuels, 2.3 million hectares of land need to be dedicated to biofuel plantations. A demonstration project of 0.4 million hectares will test the feasibility of the target (TERI, 2003).

CHAPTER 8

REGIONAL OUTLOOKS

HIGHLIGHTS

- Primary energy demand in OECD countries is projected to grow by 0.9% per year over the projection period. It will be almost a third higher in 2030 than it is today. The shares of natural gas and non-hydro renewables will increase at the expense of coal, oil and nuclear. The OECD's share of global energy use will continue to fall, from 52% in 2002 to 43% in 2030.

- Among OECD regions, North America and Oceania will experience the fastest growth in energy demand. OECD Asia demand will grow slightly less quickly, with robust growth in Korea balancing sluggish demand in Japan. OECD Europe will see the lowest rate of demand growth.

- Total primary energy demand in the developing countries as a whole is projected to rise by 2.6% per year over 2002-2030. Developing countries will account for about two-thirds of the increase in world energy demand. Their share in world energy demand will rise from 37% today to nearly half in 2030.

- China will be responsible for 21% of the increase in world energy demand to 2030. Coal will continue to be the dominant fuel in China, but the shares of oil, natural gas and nuclear energy in the primary fuel mix will grow. By 2030, Chinese oil imports will equal the imports of the United States today. China will account for 26% of the world's incremental carbon dioxide emissions from now to 2030.

- India's primary energy demand will increase by 2.3%, reaching 1 026 Mtoe by 2030. Biomass and waste, the main fuels in the primary energy mix today, will be increasingly displaced by coal and oil.

- Brazil's energy demand will grow at an annual average rate of 2.5% from now to 2030. Oil and renewables are expected to remain the key fuels in its energy mix. Gas will make major inroads in power generation, particularly towards the end of the projection period.

- The amount of energy that each person consumes will continue to vary widely across regions. Even in 2030, per capita energy use in Africa and South Asia will be less than 15% of that in the OECD. The transition to modern fuels is expected to continue in developing countries, but Africa and large parts of Asia will remain heavily dependent on biomass.

This chapter summarises energy trends by region bringing together the market trends and prospects for each fuel discussed in the preceding chapters. Focus is given to underlying market conditions and the general energy-policy framework, as well as the results of the projections for energy demand and supply as a whole. OECD regional trends are discussed first, followed by developing countries and the transition economies.

OECD Regions and the EU

Overview

The OECD's primary energy demand is projected to grow by 0.9% per year over the projection period. Demand will be almost a third higher in 2030 than today. There will be a marked shift in the primary fuel mix, mainly because of fuel switching in the power sector. The use of natural gas will increase strongly, at 1.6% per year, and its share in primary demand will rise from 22% to 26% (Figure 8.1). The shares of coal and nuclear power will continue to decline. The share of oil will also decline marginally from 41% to 39%. The share of non-hydro renewables will almost double to more than 7%.

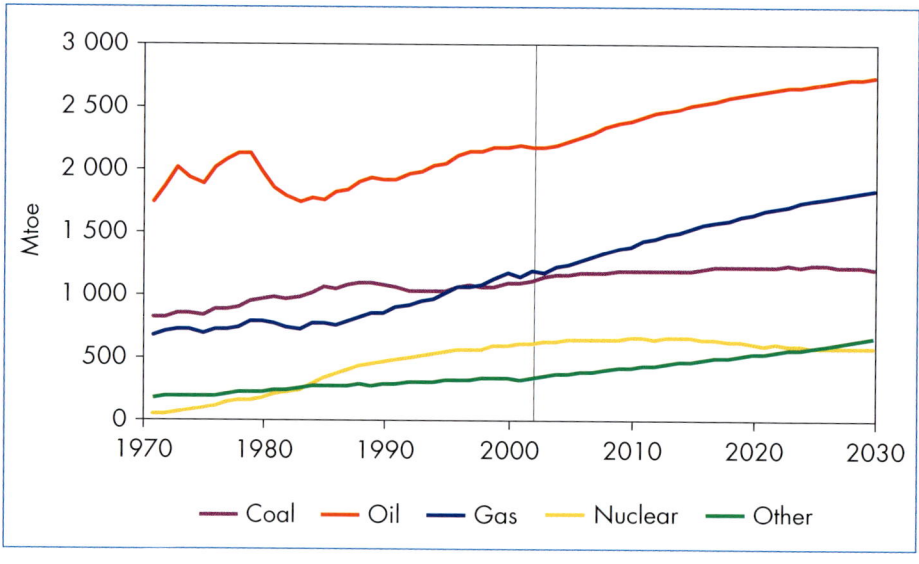

Figure 8.1: **Primary Energy Demand in OECD Countries**

The OECD's share of global energy use will continue to fall, from 52% in 2002 to 43% in 2030, as demand in developing countries grows rapidly. The dependence of OECD countries on energy imports as a whole will rise, from 28% to 39% over the projection period. Most of the increase in imports will be oil and gas.

OECD North America will see the fastest growth in energy demand, partly driven by strong growth of energy use in Mexico. North America will still be by far the largest OECD market in 2030 (Table 8.1). Energy demand in OECD Asia will grow slightly less quickly, with robust demand in Korea balancing sluggish growth in Japan. OECD Europe will see the lowest rate of demand growth, averaging 0.7% per year over 2002-2030.

Table 8.1: **Primary Energy Demand in the OECD** (Mtoe)

	1971	2002	2010	2030	2002-2030*
North America	1 779	2 698	3 035	3 634	1.1%
United States and Canada	*1 735*	*2 540*	*2 840*	*3 316*	*1.0%*
Mexico	*43*	*157*	*195*	*318*	*2.5%*
Europe	1 263	1 795	1 964	2 187	0.7%
Pacific	346	852	971	1 132	1.0%
Asia	*286*	*721*	*824*	*957*	*1.0%*
Oceania	*59*	*131*	*147*	*176*	*1.0%*
Total	**3 387**	**5 346**	**5 970**	**6 953**	**0.9%**
European Union	*1 211*	*1 690*	*1 848*	*2 048*	*0.7%*

* Average annual growth rate.

OECD North America[1]

Current Trends and Key Assumptions

Primary energy demand in North America has levelled off since the start of the current decade. An economic slowdown was the main reason for the dip in consumption in 2001, but demand has hardly increased since then, despite a marked upturn in economic activity (Figure 8.2). Demand increased by only 1.6% in 2002 and, according to preliminary data, was virtually flat in 2003. Much higher energy prices have choked demand in the industrial sector, largely offsetting growth in other sectors and resulting in a fairly strong decoupling of energy demand and gross domestic product (GDP).

Following a slowdown in 2000 and 2001, the North American economy has rebounded strongly. The GDP of the United States, which makes up 85% of the region's economy, grew by 3% in 2003. Strong consumer demand and a revival in investment are driving growth. The economies of Canada and Mexico grew at a more modest pace. The combined GDP of the United States, Canada and Mexico is assumed in this *Outlook* to increase at an average annual rate of 2.4% over the projection period. This compares to 3.1%

---

1. Canada, Mexico and the United States.

Figure 8.2: **GDP and Primary Energy Demand Growth in North America**

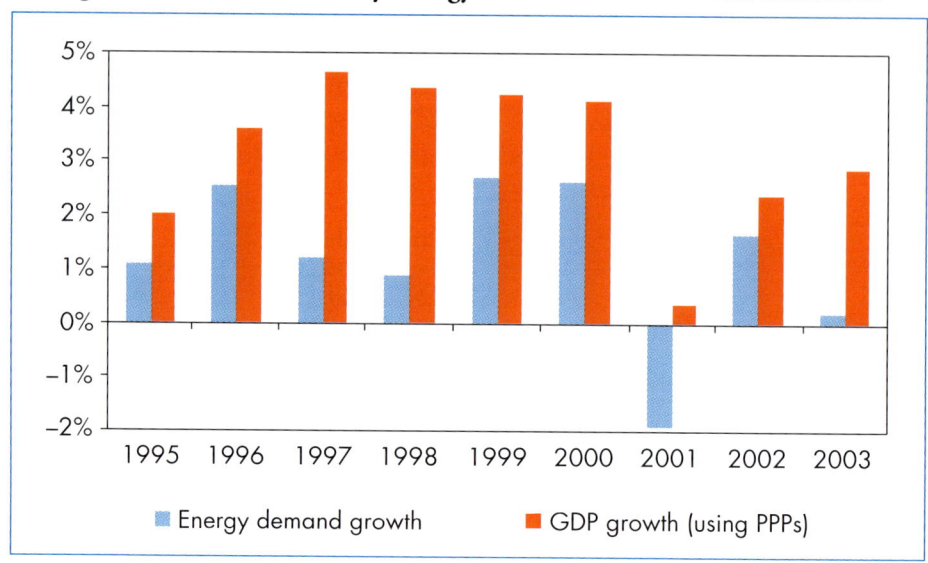

between 1971 and 2002. GDP will grow quickly in the current decade, at 3.1%, but will slow to 1.9% between 2020 and 2030. A gradual slowing of population growth, from 1.3% over the past three decades to 0.9% over the projection period, will contribute to lower GDP growth.

A surge in international oil prices since 1999 has resulted in big increases in gasoline prices at the pump. Gasoline prices in North America are much more sensitive to higher international prices than are prices in most other OECD regions because fuel taxes are much lower. The prices of other oil products and natural gas have increased even more than gasoline in percentage terms. Gas prices have been driven higher by the price of oil and by supply constraints. Coal prices were hardly affected by these developments until late 2003, but a sudden jump in international coal prices, exacerbated by production problems in the United States, has driven US coal prices up in recent months.

The retail prices of oil products are assumed to fall back from recent highs, and then remain flat in real terms through to 2010. Pre-tax prices are assumed then to rise steadily till 2030, following higher crude oil prices. Gas prices are assumed to follow oil prices in the longer term, but to fall less rapidly in the short term because of delays in bringing new import capacity on stream. In the longer term, electricity prices will increase broadly in line with gas prices, but slightly less rapidly. Improved efficiency in power generation, resulting partly from the restructuring of the electricity industry, will cushion the increase in input fuel prices. Full retail competition in electricity is assumed to be introduced gradually across the United States and Canada.

The US administration is continuing to try to implement the recommendations of the 2001 National Energy Policy Report, some of which require Congressional approval. In April 2003, the House of Representatives passed a wide-ranging energy bill, but the Senate passed a different bill in July 2003. At the time of writing, Congress had not yet agreed on a single bill for the President to sign. As a result, none of the measures included in these bills is yet in force. Measures include opening up more federal lands to oil and gas drilling, and establishing a minimum requirement for renewable energy use. The bills also provide for tax incentives, standards and voluntary programmes to decrease energy intensity in various end-use sectors. Since they are not yet effective, the Reference Scenario projections presented here do not take the bills' provisions into account. The possible effects of those measures are analysed in the Alternative Policy Scenario.

Results of the Projections

Total primary energy use in North America is projected to rise by 1.1% per year from 2002 to 2030, slower than the 1.4% rate of 1971-2002 (Table 8.2). Oil will remain by far the most used fuel, continuing to account for around 40% of total primary energy consumption over the projection period. The share of coal will continue to decline, from 21% in 2002 to 18% in 2030. The share of gas will increase from 24% in 2002 to 26% in 2030. The share of non-hydro renewables will rise even more quickly, overtaking nuclear power just before 2030. Nuclear output is projected to increase over the next two decades, but then fall away. By 2030, it will be at the same level as in 2003, but its share will have dropped to 6% from 9% at present. Primary energy intensity, expressed as primary energy demand divided by GDP, will decline by 1.3% per year, about the same rate as between 1990 and 2002.

Table 8.2: **Primary Energy Demand in North America** (Mtoe)

	1971	2002	2010	2030	2002-2030*
Coal	297	578	633	672	0.5%
Oil	826	1 079	1 218	1 478	1.1%
Gas	557	651	743	946	1.3%
Nuclear	12	232	245	233	0.0%
Hydro	38	52	58	62	0.6%
Biomass and waste	49	90	108	165	2.2%
Other renewables	1	16	30	77	5.9%
Total	**1 779**	**2 698**	**3 035**	**3 634**	**1.1%**

* Average annual growth rate.

The biggest increase in energy demand will be for oil, coming almost entirely from the transport sector. Gas use will also increase substantially, though significantly less than was projected in *WEO-2002*, partly because prices are assumed to be higher. Most of the increase in gas use will be for power generation. The consumption of non-hydro renewables will increase much more than in the past three decades. Most of this increase is expected to come from wind power. The increase in non-hydro renewables supply will be bigger than projected in *WEO-2002*, because gas is expected to become less competitive in power generation, especially towards the end of the projection period.

The sectoral breakdown of primary energy use will change little over the projection period. The transport sector will increase its share, from 26% in 2002 to 29% in 2030. Transport fuel demand will continue to rise more or less in line with rising personal incomes and industrial production, which will boost freight. A recent tightening of US fuel-efficiency standards for light trucks, including sport utility vehicles, is expected to have only a marginal impact on overall fuel consumption. The new standards require that new light trucks have a minimum average fuel economy of 21 miles per gallon (mpg) for model year 2005, 21.6 mpg for 2006 and 22.2 mpg for 2007 and beyond. The old standard was 20.7 mpg. The shares of industry, households and services in overall energy use will decline slightly, while that of the power sector will remain at around 40%.

Final energy consumption is projected to rise slightly faster than primary demand, at 1.2% per year from 2002 to 2030. The difference is due to a continuing improvement in the thermal efficiency of power generation. Electricity will account for a growing share of final energy use, while the share of gas will decline. Oil's share of total final energy demand will edge up from 53% at present to 54% in 2030, thanks to continuing strong growth in transport.

Indigenous energy production will not keep pace with the projected increase in demand. As a result, net imports – all in the form of oil and gas – will meet an increasing share of North American energy needs (Figure 8.3). By 2030, more than a half of oil supply and nearly one-fifth of gas supply will come from abroad. Net imports will account for 27% of total energy consumption in 2030, compared with 14% in 2002.

Energy-related CO_2 emissions will rise at the same pace over the projection period as over the past three decades. Emissions will reach 8 596 million tonnes (Mt) in 2030, up from 6 480 Mt in 2002 – an average annual increase of 1%. By 2010, emissions will be almost a third higher than in 1990. Power generation will remain the biggest CO_2-emitting sector in 2030. Its share will remain steady at 40%, despite a modest decline in emissions per unit of electricity produced as a result of an increase of gas and renewables in the generating fuel mix. The transport sector will be the largest contributor to

Figure 8.3: **Net Imports of Oil and Gas as Share of Primary Demand in North America**

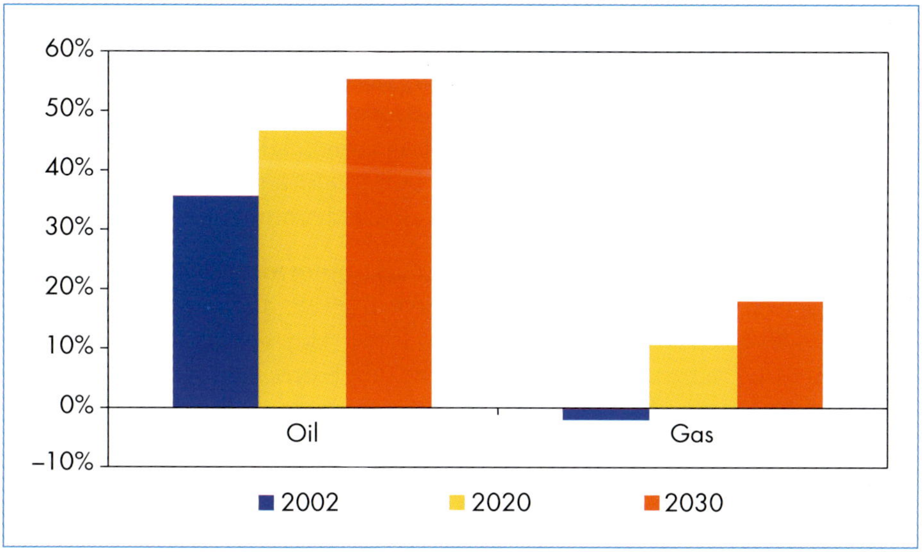

increased CO_2 emissions over the projection period, due to rapid growth in transport and the sector's continuing dependence on oil. The carbon intensity of the North American economy is projected to fall by 1.4% per year over 2002-2030. This is due almost entirely to the projected drop in energy intensity.

European Union[2]

Current Trends and Key Assumptions

EU primary energy demand has continued to edge higher since 2000. Demand dipped in 2002, largely because of mild winter weather, but rebounded by almost 2% in 2003, according to preliminary data. Recently, some power producers have switched back to coal from natural gas, which has proven to be much more expensive because of its *de facto* indexation to oil. Oil remains the predominant energy source, although its share in total primary energy use has declined since the 1970s.

---

2. Projections for the enlarged Union are presented for the first time in this *Outlook*. Ten countries – the Czech Republic, Cyprus, Estonia, Hungary, Latvia, Lithuania, Malta, Poland, Slovenia and the Slovak Republic – joined the European Union in May 2004, increasing the membership to 25. Most EU members belong to the OECD, with the exception of Cyprus, Estonia, Latvia, Lithuania, Malta and Slovenia. These countries are included in the transition economies grouping. Iceland, Norway, Switzerland and Turkey are members of the OECD but not of the European Union. See Annex E for detailed definitions of the regions.

Economic growth in the European Union remains sluggish in major countries, though the pace of recovery has picked up in some countries, notably Spain and the United Kingdom. Overall, EU's GDP grew by only 1.1% in 2003. The economies of the main euro-zone countries – France, Germany and Italy – continue to lag. Unemployment remains stubbornly high, averaging nearly 9% across the region in August 2004, and consumer demand is weak. A sharp appreciation in the euro against most leading currencies has depressed exports and undermined industrial production but it has also shielded European consumers from some of the pain of high dollar-denominated oil prices.

Near-term economic prospects are nonetheless improving. Growth is expected to average over 2% in 2004 and above 2.6% in 2005. Over the period 2002-2010, growth is assumed to average 2.3%. It then slows to 2.1% from 2010 to 2020 and to 1.7% from 2020 to 2030. The differences in growth rates among countries are expected to shrink with the macroeconomic convergence that should result from economic and monetary integration. European population is assumed to remain broadly unchanged over the projection period, rising very gradually through to the mid-2010s and falling back very slowly thereafter. As a result, GDP per capita will be 75% higher by 2030 than in 2002.

New EU directives on electricity and gas and a regulation on cross-border electricity exchanges entered into force in July 2004. Under the new directives, all energy users will have the right to choose their electricity and gas supplier by July 2007. The directives also require the legal unbundling of network activities from generation and supply, and the setting-up of a regulator with well-defined functions. Network operators must publish tariffs for third parties wishing to make use of their grids and their public-service obligations are reinforced, especially for the most vulnerable customers. The regulation establishes common rules for cross-border trade in electricity. These moves, together with national initiatives, are expected to accelerate the opening of energy markets to competition, a trend which will ultimately lead to more rational investment in infrastructure and, in some cases, to lower costs to customers. This will partly offset the effect of other factors that are expected to drive prices up during the second and third decades of the projection period.

In 2003, the European Union adopted an emissions-trading directive that requires all member states to set limits on CO_2 emissions from energy-intensive plants by allocating them emission allowances. Trading will start in January 2005. It is expected that more than 12 000 plants will fall under the scope of the directive. Companies that do not use all their allowances will be able to sell the balance to those that fail to keep their emissions within the allocated allowances. In this way, emissions will be cut where it is cheapest to cut them. The emissions-trading directive required member states to submit national allocation plans, establishing an overall limit on total annual emissions, by

31 March 2004. National plans are not yet in force because the European Commission is still reviewing them. As a result, the trading scheme has been taken into account in the Alternative Policy Scenario but not in the Reference Scenario. Most of the plans submitted involve aggregate allowances close to actual emissions, which will keep the market price of allowances down. Most countries will, therefore, need to rely on additional measures if their national emissions-reduction targets under the Kyoto Protocol are to be met.

Results of the Projections

The European Union's primary energy demand is projected to grow by 0.7% per year over 2002-2030 – the lowest growth rate of any WEO region and well below the 1.1% rate of 1971-2002 (Table 8.3). The pattern of energy use will change considerably (Figure 8.4). Consumption of coal will drop by 10%, and coal's share in total primary energy use will decline from 18% in 2002 to 13% in 2030. The share of nuclear power is also expected to fall sharply, especially during the second half of the *Outlook* period. At 7% in 2030, it will be less than half that in 2002. The share of gas will increase, from 23% now to 32% in 2030, mainly because almost all new fossil-fuel power plants will be gas-fired. The share of non-hydro renewables will also rise sharply, overtaking nuclear power early in the 2020s. Most renewables will be used to generate electricity. Primary energy intensity is projected to fall by 1.3% per year over 2002-2030, about the same as in the last three decades, due to energy-efficiency gains and a continuing structural shift of the EU economy away from energy-intensive activities.

Table 8.3: **Primary Energy Demand in the European Union** (Mtoe)

	1971	2002	2010	2030	2002-2030*
Coal	426	303	307	274	-0.4%
Oil	633	648	687	743	0.5%
Gas	93	389	468	649	1.8%
Nuclear	13	251	251	146	-1.9%
Hydro	20	26	30	33	0.8%
Biomass and waste	25	65	84	147	3.0%
Other renewables	2	8	21	57	7.2%
Total	1 211	1 690	1 848	2 048	0.7%

* Average annual rate of growth.

Final demand will grow by 0.9% per year, sightly higher than primary demand. Annual growth of final oil demand, at 0.7% from now to 2030, will be higher than the 0.4% of the past three decades. In the transport sector, where oil use is increasingly concentrated, the pace of demand growth will slow. Saturation effects and major improvements in the fuel efficiency of new cars and trucks will largely cancel out the effect of rising incomes on personal mobility and freight. Alternative transport fuels, including natural gas, will also displace oil-based fuels. Electricity will take a growing share of final energy use. The 1.4% per annum increase in electricity consumption will come mainly from homes and offices. Most of the 0.9% yearly increase in gas use will come from two areas: industry and space and water heating for households.

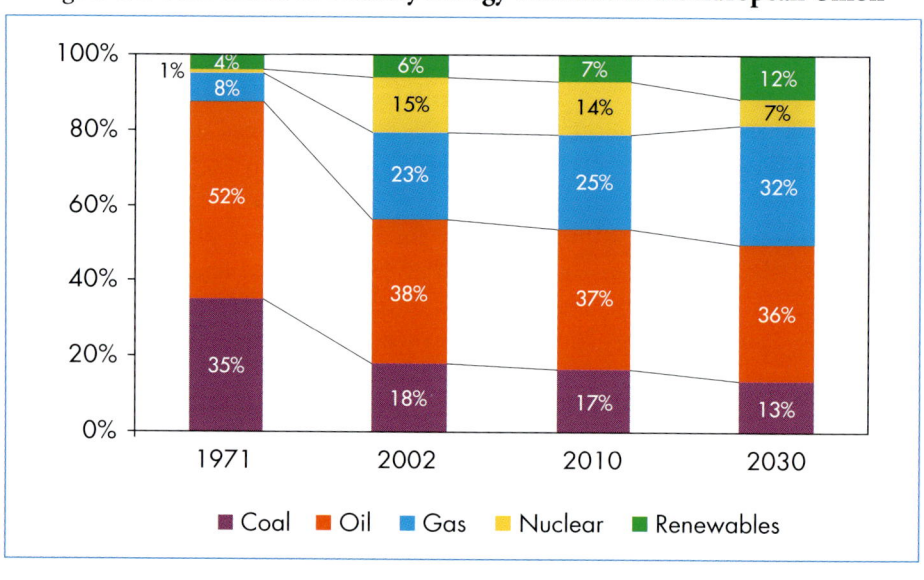

Figure 8.4: **Fuel Shares in Primary Energy Demand in the European Union**

The European Union's imports of fossil fuels will increase substantially as indigenous production dwindles and demand edges up (Figure 8.5). Imports already meet 76% of EU primary oil demand, and this share will grow to 94% by 2030. Production from the North Sea, the main source of indigenous supply, has already peaked. Its decline – led by production from the United Kingdom – is expected to accelerate over the coming years. Total EU oil production is projected to fall from 3.2 mb/d in 2002 to 2.2 mb/d in 2010 and to 1 mb/d in 2030. The increase in Europe's gas imports will also be pronounced. The gap between production and demand will continue to widen, from about 230 bcm in 2002 to 640 bcm in 2030. This implies an increase in EU gas-import dependence from 49% to 81%. A growing share of EU gas imports will be shipped as liquefied natural gas (LNG). Coal imports will also grow, mainly due to further closures of unprofitable mines in those countries which still have them: the Czech Republic, Germany, Greece, Poland, Spain and the United Kingdom.

Figure 8.5: **Fossil Fuel Net Imports in the European Union**

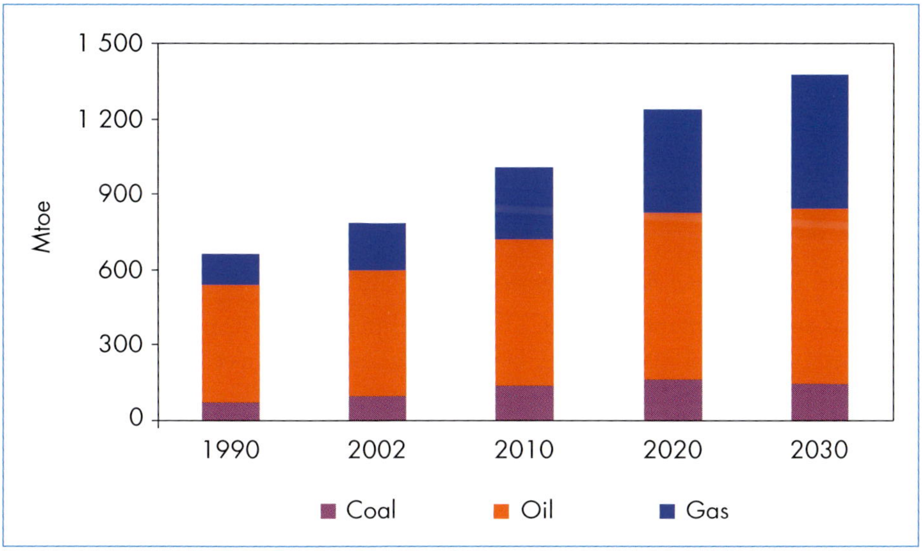

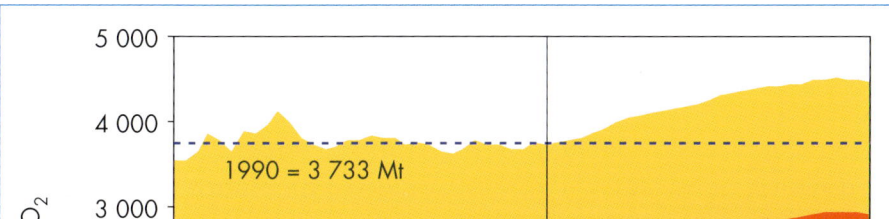

Energy-related CO_2 emissions in the European Union will rise as fast as primary energy demand. By 2030, emissions will reach 4 488 Mt, 20% above the 2002 level (Figure 8.6). Power generation will remain the single biggest CO_2-emitting sector in 2030. Its share will increase from 35% in 2002 to 37% in 2030. The share of transport jumps from 24% to 28% over the same period.

Figure 8.6: **Energy-Related CO_2 Emissions in the European Union**

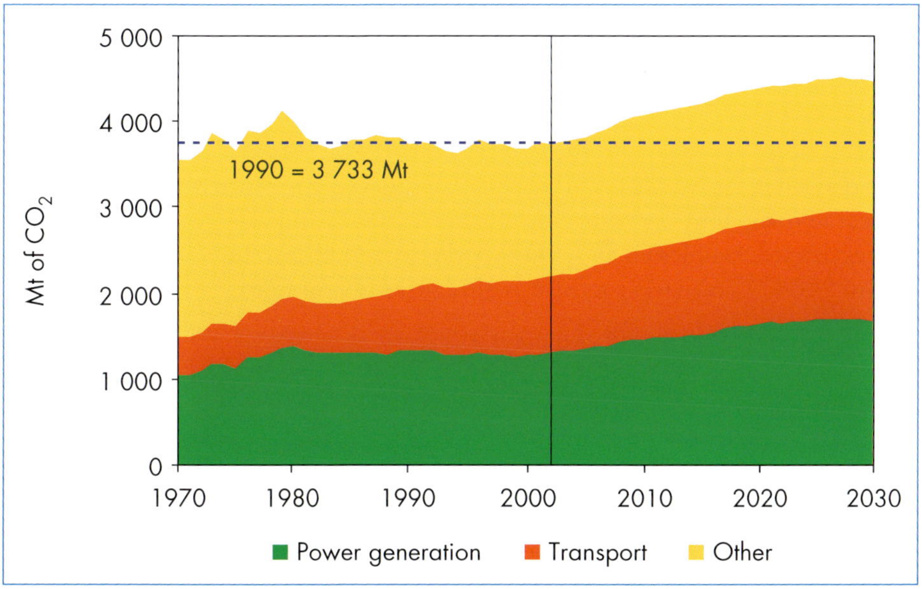

The Reference Scenario projections show the European Union as failing to meet its commitment under the Kyoto Protocol to cut greenhouse-gas emissions to 8% below their 1990 level by the period 2008-2012. The Europeans could only do so if additional EU measures were imposed or if they bought emission credits from non-EU countries. Europe's energy-related CO_2 emissions are expected to be 9% higher in 2010 than in 1990.

OECD Asia[3]

Current Trends and Key Assumptions

Primary energy demand in OECD Asia declined by 0.2% in 2001, and then increased by a tepid 1.3% in 2002, largely explained by Japan's economic recession. The region's economy is the most oil dependent in the OECD, with oil accounting for 50% of the primary fuel mix. The region imported nearly 100 bcm of natural gas in 2002, or two-thirds of global LNG trade. Japan and Korea are the world's largest and second largest coal importers. Nuclear energy is important to the region, meeting 15% of its primary energy needs. Japan accounts for 72% of energy use in OECD Asia, but its share will decline to 62% in 2030.

The economy of OECD Asia is expected to grow by 1.9% per year over the projection period, just half as fast as between 1971 and 2002 and the slowest among WEO regions. Large demographic changes are in store. Japan's population will start declining in about five years, and Korea's will follow around 2025. The share of population aged 65 and over will increase rapidly. These changes will have important but unpredictable impacts on economic growth and energy demand.

Both Japan and Korea have been undertaking regulatory reforms in the energy sector, particularly in gas and electricity. In Japan, 40% of the electricity market is now opened and high-voltage consumers can choose their suppliers. Full market opening is envisaged in 2007, but that date could slip. Gas market liberalisation has also reached nearly 45%. In Korea, unbundling and privatisation of Korea Electric Power Corporation is the centrepiece of electricity market reform. The corporation was split into six companies in 2001. Further moves towards wholesale and retail market competition, however, ground to a halt in 2004. Gas market reform is also currently on hold.

Under the Kyoto Protocol, Japan is committed to reduce its greenhouse gas emissions to 6% below their 1990 level by the period 2008-2012. In 2002, the Japanese government issued a "New Guideline for Measures to Prevent Global Warming" which lays out the country's strategy, including energy-conservation measures, to achieve this target. The Korean government recently expanded its Energy Efficiency Standards and Labelling Programme.

3. Japan and Korea.

Results of the Projections

Total primary energy demand in OECD Asia is projected to rise by 1% per year from 2002 to 2030 (Table 8.4). The annual growth rate of demand will fall to just 0.5% in the 2020s, compared to 1.7% from now to 2010. This is primarily due to the penetration of more energy-efficient technologies and the decline in population. Primary energy intensity will fall by 0.9% per year over the projection period, much faster than the past three decades' annual decline of 0.5%.

Table 8.4: **Primary Energy Demand in OECD Asia** (Mtoe)

	1971	2002	2010	2030	2002-2030*
Coal	62	146	158	168	0.5%
Oil	211	357	378	395	0.4%
Gas	3	88	120	174	2.5%
Nuclear	2	108	139	172	1.7%
Hydro	7	7	8	9	0.9%
Biomass and waste	0	11	15	24	2.8%
Other renewables	0	4	6	14	4.6%
Total	286	721	824	957	**1.0%**

* Average annual rate of growth.

Among fossil fuels, gas will grow strongly. The power sector will account for 70% of the increase in gas demand. The share of gas in the fuel mix will increase from 12% to 18%, at the expense of coal and oil. Oil's share will decline to 41% by 2030. Nuclear energy is projected to increase by 1.7% per year and its share in the fuel mix to rise from 15% to 18%. The projected growth rate of non-hydro renewables is highest, underpinned by government policy measures, especially in Japan.

Final energy consumption is projected to rise at 0.9% per year from 2002 to 2030. The shares of electricity, gas and renewables will increase at the expense of coal and oil. Industry will remain the largest end-use sector in 2030, but structural changes and ongoing energy conservation will limit the growth of its energy use to 0.7% per year, much lower than in other sectors. Demand for transport fuels will increase by 1% per year. Energy use in the residential and services sectors will grow at 1.1% per year. Electricity will become the most important fuel in these sectors in 2030, replacing oil. More appliances will be used in the next ten year or so, especially in Korean households.

The region's projected 11% increase in primary oil demand will be met by imports. Heavy reliance on oil from the Middle East is already a serious concern for both the Japanese and Korean governments. Oil from Russia has been considered as an option to diversify supply sources. For the past several months, Japan and China have been vying over the destination of an eventual oil pipeline from Angarsk in Siberia. While China wanted the pipeline to go to its largest domestic oil field in Daqing, Japan has managed to get it routed to the port of Nakhodka, where oil can be shipped to Japan, Korea, China and even to the United States. Korea has also been trying to diversity its sources of oil imports. The Korea National Oil Corporation has become involved in overseas exploration and production projects in recent years.

Most of gas imports to Japan and Korea come from countries in the Asia-Pacific region such as Indonesia, Malaysia and Australia. Qatar and other Middle East countries currently serve as swing suppliers. Uncertainty about future growth of LNG imports into China and the United States complicates the task of projecting demand-supply balance of LNG in the Asia-Pacific market. Diversification of gas supply is also a key policy issue in OECD Asia. Russia will become an important supplier to this region, when the Sakhalin II project starts LNG exports to Japan in 2007. In the long term, piped gas from Russia could also offer a good option. Korea is keen to develop a pipeline network from the Kovykta field in the Irkutsk region of Russia to South Korea via China and the Korean Peninsula. A pipeline from Sakhalin I is also being considered.

Electricity demand in OECD Asia will grow by 1.3% per year between 2002 and 2030. Generating capacity needs to increase from 336 GW in 2002 to 503 GW in 2030. We assume that nuclear capacity will rise from 59 GW today to 70 GW in 2010 and then to 87 GW in 2030. This is lower than what was projected in *WEO-2002* and in previous government plans. The reasons are lower-than-expected electricity demand growth in Japan and difficulties encountered in finding acceptable nuclear sites in Korea. Nonetheless, the electricity generation mix is set to shift towards nuclear and natural gas (Figure 8.7).

Figure 8.7: **Changes in Electricity Generation by Fuel in OECD Asia**

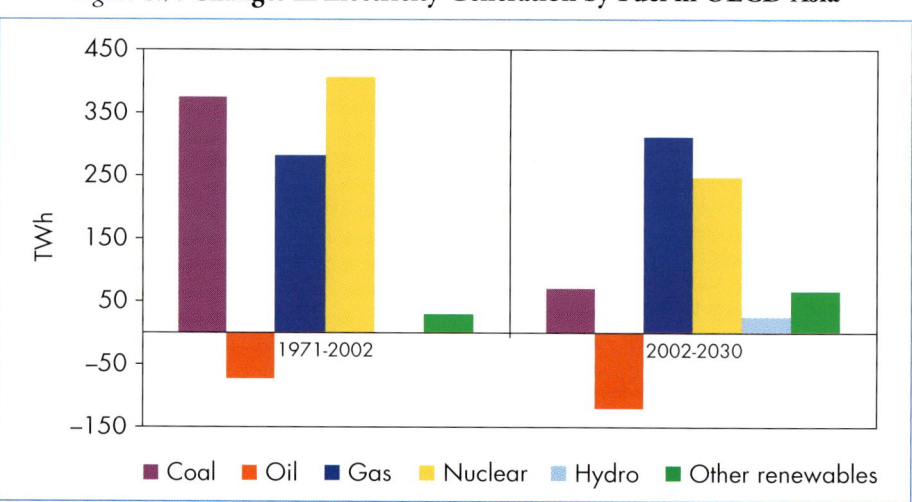

Energy-related CO_2 emissions will rise at 0.7% per year over the projection period – much less rapidly than their 2.4% annual growth in the past three decades. The emissions will increase more slowly than projected primary energy demand, which is set to grow by 1% per year. If existing policy measures and technologies are unchanged, CO_2 emissions from Japan and Korea are projected to stabilise towards the end of the projection period. Power generation will remain the biggest carbon-emitting sector in 2030 and the largest contributor to increased emissions over the projection period, followed by the transport sector (Figure 8.8).

Figure 8.8: **Increase in Energy-Related CO_2 Emissions by Sector in OECD Asia**

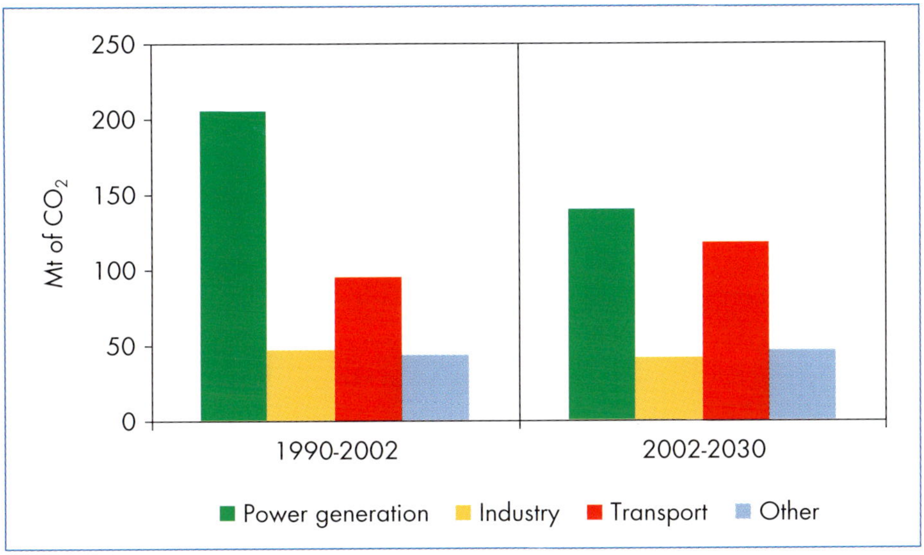

OECD Oceania[4]

Current Trends and Key Assumptions

Australia accounts for 86% of the economy and energy market of the two-country region. In 2002, Australia exported 104 Mt of coking coal, 99 Mt of steam coal, and 10 bcm of natural gas, mainly to Asia-Pacific markets. New Zealand also exports some coal, while domestic production covers the country's entire gas needs. Seven per cent of the region's oil demand is currently met by imports.

The economy of OECD Oceania performed strongly in 2003. Employment continued to grow and the unemployment rates declined to historical lows in both countries. In this *Outlook*, the GDP of OECD Oceania is assumed to

4. Australia and New Zealand.

continue its steady growth in this decade, at an annual rate of 3.1%, slightly down from 3.4% between 1990 and 2002; it will grow at 2.3% per year over the entire projection period.

A national review of the electricity transmission network has been undertaken in Australia to encourage new investment and to improve consistency, transparency and economic efficiency. A review of national gas-market issues, including infrastructure development, is also in preparation. In the wake of power shortages, New Zealand has decided to establish an Electricity Commission to replace the self-regulatory arrangements of the New Zealand Electricity Market. New Zealand ratified the Kyoto Protocol in 2002. Australia has not yet ratified the Protocol, but recognises a commitment to mitigate its greenhouse-gas emissions.

Results of the Projections

Total primary energy demand in OECD Oceania will grow by 1.0% per year from 2002 to 2030 (Table 8.5). Underpinned by steady demand growth in the transport sector, oil will replace coal as the dominant fuel and will account for 33% of the primary fuel mix by 2030. Consumption of other renewables will grow most strongly of all energy sources, at 3.8%, followed by gas. The share of non-hydro renewables in the primary fuel mix will increase from 8% in 2002 to 11% in 2030. The region's energy intensity will decline by 1.2% per year over the period to 2030.

Table 8.5: **Primary Energy Demand in OECD Oceania** (Mtoe)

	1971	2002	2010	2030	2002-2030*
Coal	22	50	52	53	0.2%
Oil	29	41	48	57	1.2%
Gas	2	26	31	42	1.8%
Nuclear	0	0	0	0	–
Hydro	2	3	4	4	0.3%
Biomass and waste	4	9	9	13	1.4%
Other renewables	1	2	3	6	3.8%
Total	59	131	147	176	1.0%

* Average annual growth rate.

Total final consumption will grow by 1.3% per year over the projection period. This is just half the rate of the past three decades and is due to saturation of vehicles and appliances as well as to energy-efficiency measures. Oil will continue to dominate final energy consumption and will account for 47% of the fuel mix in 2030, but its share will slip slightly against gas and electricity. Energy use in the services sector is projected to continue to grow fairly strongly at 1.9% per year, with 1.2% annual growth in the industry and transport sectors, respectively, and 1.3% in the residential sector.

While primary coal demand in OECD Oceania will grow quite modestly, at 0.2% per year, the region's exports will continue to expand strongly. Production, centred in the Australian states of Queensland and New South Wales, is projected to increase from 346 Mt in 2002 to 515 Mt in 2030. Although Australia has one of the world's most competitive coal industries, it will face increasing competition from China, Indonesia, Colombia and Venezuela.

Driven by rapid growth in LNG exports, natural gas production in the region, including gas-to-liquids production, is projected to increase from 41 bcm in 2002 to 98 bcm in 2030. Nearly all this incremental gas production will come from Australia, as gas fields in New Zealand are almost fully depleted. Japan will continue to be the largest importer of Australian LNG, but other markets are likely to emerge in developing Asia and possibly the United States. Exports from Australia's gas-rich North West Shelf to China's first LNG terminal, at Guangdong, are planned to begin in 2005.

Electricity demand in OECD Oceania will rise by 1.6% per year between 2002 and 2030. To meet this demand and to replace retiring plants, power generating capacity of 57 GW needs to be added. Gas-fired power plants will account for 35% of this addition and another 21% will be plants fuelled by non-hydro renewables, especially solar and wind power. Most of new capacity of coal plants will be built to replace ageing plants.

OECD Oceania's energy-related CO_2 emissions will increase by 0.8% per year from 2002 to 2030. In 2010, they will be 11% above the 2002 level. Emissions from power generation will continue to be largest, but because of the increasing use of renewables and a strong shift from coal to gas in the generation mix, the annual increase of the sector's emissions will be only 0.6%. The transport sector will account for a whopping 39% of the increase in total CO_2 emissions over the projection period.

Developing Countries[5]

Total primary energy demand in the developing countries as a whole is projected to rise by 2.6% per year over the period 2002-2030. Although natural gas, nuclear power and other renewables are expected to grow faster in percentage terms, oil will remain the single most important source of energy in 2030, followed by coal (Figure 8.9). Almost half the projected increase in primary demand in the developing world will come from power generation. Primary energy intensity will continue to decline by 1.6% per year in the developing countries as a whole.

5. China, East Asia, South Asia, Latin America, Africa and the Middle East.

Figure 8.9: **Primary Energy Demand in Developing Countries**

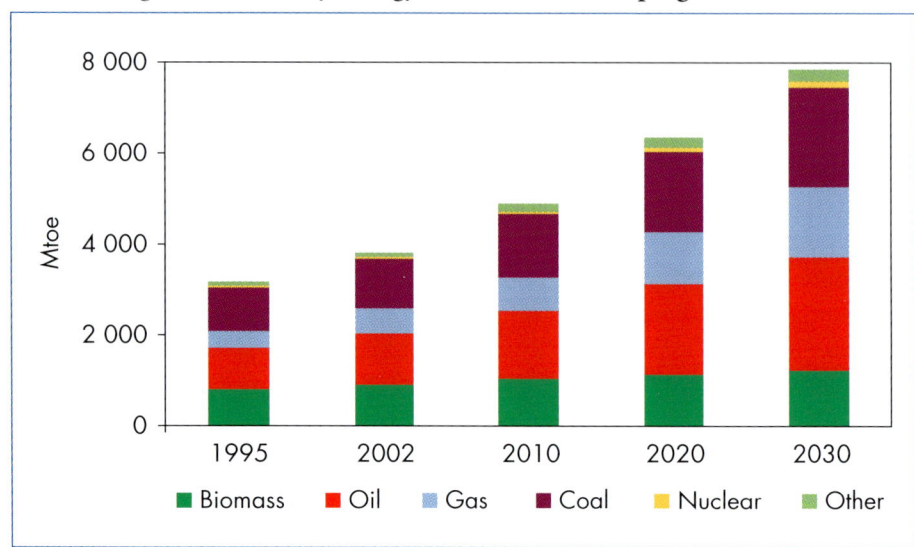

The share of modern fuels in total final consumption, as opposed to traditional biomass and waste, is expected to increase substantially, from around 72% in 2002 to 82% in 2030. Electricity consumption will grow at 4.2% per year, as rising income is accompanied by rapid penetration of appliances. Oil products, which made up 34% of the final fuel mix in 2002, will account for 42% in 2030, owing to booming transport demand and to a shift away from biomass and waste for cooking and heating. The share of natural gas in final consumption is expected to increase, while that of coal is expected to drop by nearly half. Traditional biomass and waste will remain an important source of energy in the household sector, mainly in poor countries in Africa, Asia and Latin America.

The transition to modern fuels in the household sector is expected to continue in all the main developing regions. Total household energy use is projected to increase from 1 017 Mtoe in 2002 to 1 632 Mtoe in 2030. The share of electricity is expected to increase substantially, from 8% to 20%. The share of gas will double, to 8%, while that of oil will rise from 10% to 15%. Coal's share will drop by more than half, from 5% to 2%. Biomass use is projected to grow at 0.6% per year from now to 2030, though its share in household energy consumption will fall from 72% to 53% (Figure 8.10).

Large differences in the pattern of household energy use among regions will remain in 2030. The Middle East will continue to rely least on biomass and most on electricity, oil and gas, with each of the three accounting for around a third of final household consumption in 2030. Africa and large parts of Asia will remain heavily dependent on biomass, using relatively little electricity – as many households will still lack access to it or be unable to afford it (see Chapter 10).

Figure 8.10: **Household Energy Consumption in Developing Countries**

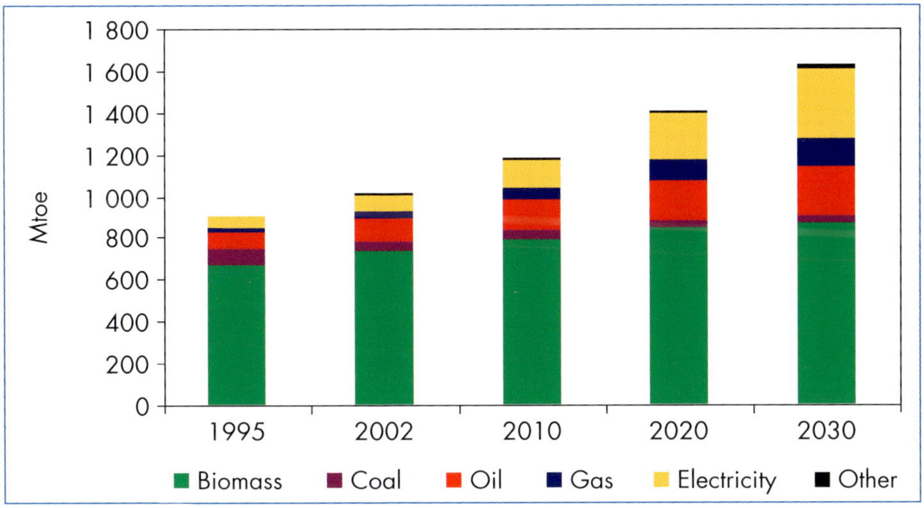

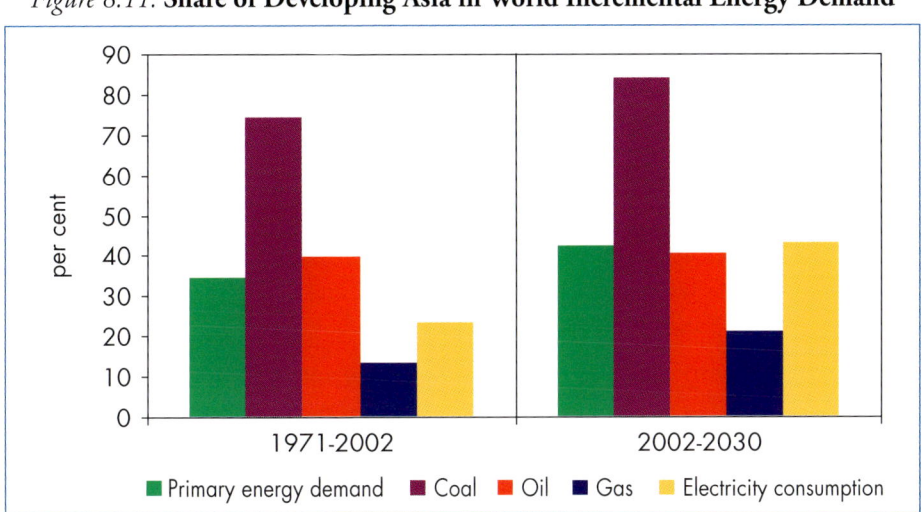

Developing Asia[6]

Developing Asia achieved annual GDP growth of 6.8% in 1990-2002, and rising income led to a surge in demand for energy. The annual growth rate of primary oil demand was 5.5% and electricity consumption leapt by 7.3% per year during the same period. By 2002, the region accounted for about a quarter of global GDP and of total primary energy demand. The region is projected to account for 42% of the increase in world primary energy demand between 2002 and 2030 (Figure 8.11). Its share in the global energy market will reach nearly a third in 2030.

Figure 8.11: **Share of Developing Asia in World Incremental Energy Demand**

6. China, East Asia and South Asia.

Although the region has the most dynamic economy and energy market in the world, there are wide variations among countries in their stage of development. Per capita income and per capita primary energy consumption in Singapore and Chinese Taipei are higher than in some OECD countries. China, the world's fastest-growing economy, is becoming a key strategic buyer in international energy markets, and India has started along the same path. But poverty and lack of access to modern energy are still widespread, particularly in China, India and other South Asian countries, and in several transition countries, like Vietnam. The region contains major energy producers such as China (coal), Indonesia (coal, oil and gas), Malaysia (oil and gas) and Brunei (gas). These countries export energy not only within the region but also to some OECD countries, particularly Japan and Korea.

The future of the economies and energy markets in the region will be determined by what happens in China and India. But events in other countries, such as Indonesia's becoming a net oil importer, will resonate as well. Energy demand growth will outpace increases in indigenous energy supply, a fact that will affect international energy prices, trade and investment flows.

Developing Asia will play an increasingly visible role in global oil markets. In 2030, its oil demand of 30 mb/d will exceed the 28mb/d of the United States and Canada combined and will account for 26% of the world total. Oil production in China is expected to start declining very soon. Output elsewhere in the region, including Indonesia, will be more or less flat over the projection period. As a result, almost all incremental oil demand in developing Asia will have to be met by imports from other regions, mainly from the Middle East. The region's oil import dependence is projected to nearly double from 43% in 2002 to 78% in 2030, intensifying supply security concerns not only to the region itself but also in Japan and Korea.

The region will contribute 21% of world incremental gas demand from now to 2030. Natural gas markets in both China and India will grow most in the current decade. With producers like Indonesia, Malaysia, Pakistan and Thailand, intra-regional gas trade will further develop, while imports from Australia, the Middle East and probably Russia will also grow. The region will remain the world's largest coal market, and the locus of around 80% of growth in world coal demand over the projection period.

Electricity consumption in developing Asia will grow robustly, providing 43% of the worldwide increase. Nonetheless, access to electricity will remain scarce in poor areas, especially in South Asia. About 1 600 GW of new generating capacity needs to be added, at a cost of $1 450 billion (in year-2000 dollars). Attracting capital at reasonable costs and in a timely fashion is a major challenge to the region's electricity sector.

Dependence on traditional biomass and waste is still high in India and other South Asian countries and even in some South-East Asian countries such as Indonesia. It accounted for 76% of the region's residential energy use in 2002. Assumed income growth will lead many households to switch to oil, but without strong government intervention, traditional biomass and waste use in the residential sector will not start to decline in volume before 2015.

China

Current Trends and Key Assumptions

China, the world's most populous country, is the second largest economy and the second largest consumer of primary energy after the United States. It currently accounts for 12% of global GDP and primary energy demand.

China's oil consumption increased from 5.2 mb/d in 2002 to 5.7 mb/d in 2003, accounting for more than 18% of global oil-demand growth. In the process, China became the second largest oil consumer, surpassing Japan. Oil demand is set to exceed 6 mb/d this year. Surging Chinese demand is one of the factors behind the very high oil prices since late 2003. With marginal growth in domestic production over the last several years, China's oil-import dependence, which was 23% in 1998, reached 37% in 2003.

Electricity demand grew by 11% in 2002 and sky-rocketed by more than 15% in 2003, strongly outpacing economic growth. The trend continued through the first half of 2004. Electricity shortages occurred in late 2002 and have intensified since then. Demand is estimated to exceed capacity by 20 to 30 GW.

China is the world's largest coal producer, with nearly 12% of total proven reserves. Even so, the domestic coal market has experienced shortages of supply in recent years. Coal demand grew by 11% in 2002, led by 13% growth in demand from power-generation and steady growth in industry. Persistent bottlenecks in the railway system contributed to the tightening of the coal market. This in turn led to a shift in China's coal policy, away from favouring exports to focusing on the domestic market.

China's GDP growth in 2003 reached 9% despite the epidemic of severe acute respiratory syndrome (SARS). The main driver of economic growth was investment, especially in industries such as iron and steel, cement and automobiles. In the first half of 2004, GDP expanded by a further 10%. Signs of economic overheating emerged in the form of rising raw material prices, rapid growth in oil demand and shortages of coal and electricity. Concerns about possible over-investment prompted the government to tighten monetary policy and impose restrictions on investments in those industries. The *Outlook* assumes that China's GDP will grow at 6.4% per year through the current decade and 5% per year over the full projection period. It will be close to that of OECD North America by 2030.

Price liberalisation is assumed to continue in China, as are the elimination of trade restrictions on energy products and the opening of the energy sector to foreign companies. By 2010, end-use prices are assumed to reflect the economic cost of supply and to follow trends in international energy prices.

The Chinese government has pledged to boost energy supply. The Tenth Five-Year Plan for the period of 2001-2005 puts energy conservation near the top of China's energy policy agenda.[7] The Reference Scenario incorporates policies and measures taken under the Plan for Energy Conservation and Comprehensive Resources Utilisation attached to the Tenth Five-Year Plan. These include energy-efficiency standards for appliances and a scheme to reduce average energy intensity by 20% in key industries such as cement. China is about to introduce more stringent fuel-economy standards for new vehicles than those in force in the United States.

Results of the Projections

China's total primary energy demand is projected to grow by 2.6% per year from 2002 to 2030 (Table 8.6). Coal will continue to dominate the fuel mix, especially in power generation, but its share in overall consumption will drop slightly, from 57% today to 53% in 2030. Primary oil consumption will increase by 3.4% per year, driven by transport demand. Strong demand for natural gas will come from power generation and the residential sector. Gas demand will grow by more than 5% annually, and the share of gas in the fuel mix will double to 6% by 2030. Nuclear supply is projected to increase tenfold. The country's energy intensity will decline by 2.3% per year over the projection period.

Table 8.6: **Primary Energy Demand in China** (Mtoe)

	1971	2002	2010	2030	2002-2030*
Coal	192	713	904	1 354	2.3%
Oil	43	247	375	636	3.4%
Gas	3	36	59	158	5.4%
Nuclear	0	7	21	73	9.0%
Hydro	3	25	33	63	3.4%
Biomass and waste	164	216	227	236	0.3%
Other renewables	0	0	5	20	–
Total	**405**	**1 242**	**1 622**	**2 539**	**2.6%**

* Average annual growth rate.

7. Objectives of the plan include diversification of the energy mix, security of energy supply, improvement of energy efficiency and environmental protection. It has been reported that energy conservation is the top priority of the Energy Plan to 2020, which is under preparation.

China's presence in global energy markets will further increase with major consequences for energy prices and trade. Between 2002 and 2030, the country's share in world primary energy demand will increase from 12% to 16% (Figure 8.12). Its share in world oil demand will reach 11% in 2030, compared to 7% today, and it will continue to be the world's largest coal market. This means that China will account for 21% of world growth in primary energy demand, 53% of incremental coal demand and 19% of incremental oil demand over the projection period.

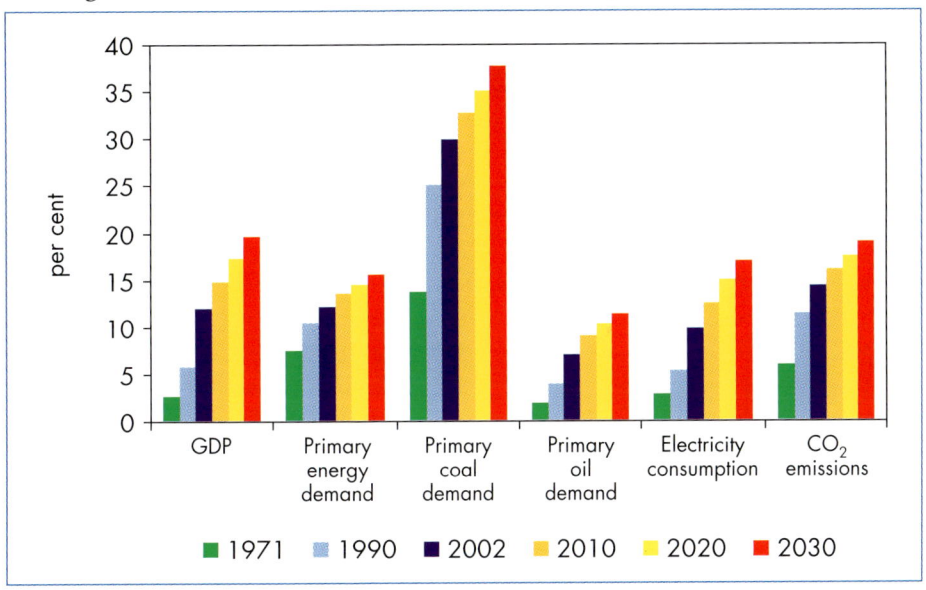

Figure 8.12: **China's Share in the Global Economy and Energy Markets**

Total final energy consumption is projected to grow at 2.2% per year. Coal and traditional biomass will be increasingly replaced by oil, gas and electricity.

The *Outlook* projects that total Chinese vehicle ownership will reach more than 90 per 1 000 persons in 2030. This implies a total vehicle stock of over 130 million by the end of the projection period, compared to 220 million in the United States today (Figure 8.13).[8] Despite the introduction of fuel-consumption standards, energy demand for road transport will climb by 6.5% per year during this decade. Energy consumption in transport as a whole will grow by 4.6% per year over the projection period, and its share in total final consumption will nearly double, from 11% today to 20% in 2030.

---

8. As one of its measures to promote a soft landing of the overheating economy, the Chinese government has imposed restrictions on automobile loans. This could slow the growth of vehicle ownership at least in the short term.

Figure 8.13: **Vehicle Stock and Oil Demand for Road Transport in China**

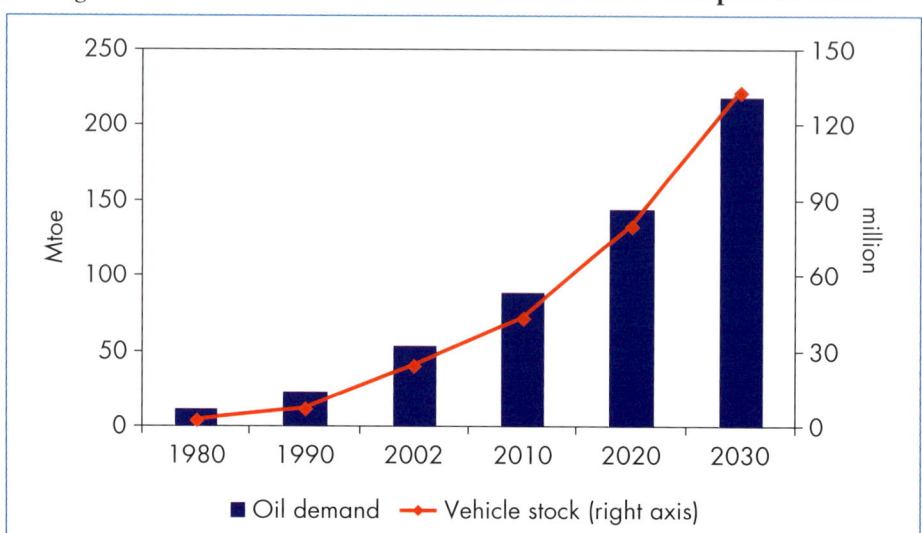

Primary oil demand will reach 13.3 mb/d in 2030. In 2030, China is expected to import almost 10 mb/d, as much as the United States imports today. China's import dependence will increase to 74% (Figure 8.14). To cope with possible supply disruptions, the Chinese government has set up strategic oil reserves. It plans to have reserves corresponding to 90 days of net imports within 15 years.

Figure 8.14: **Oil Balance in China**

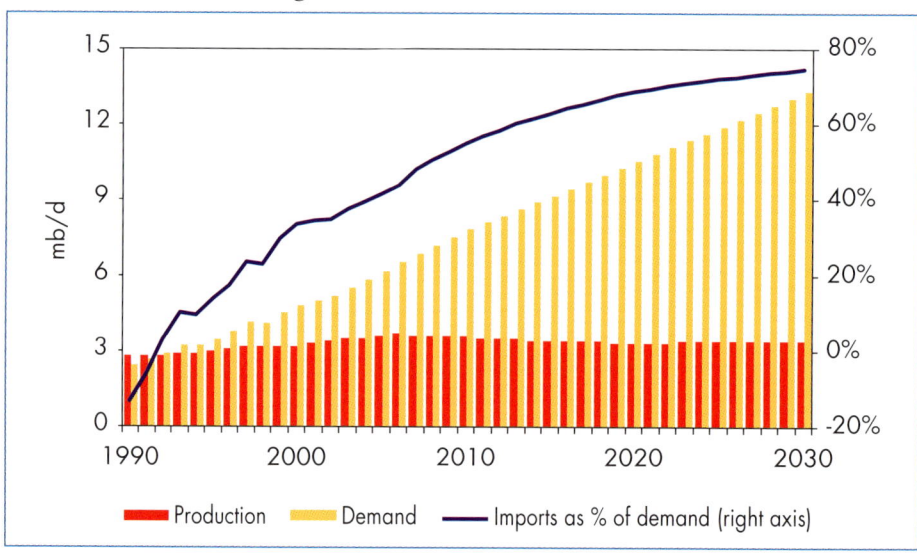

About half of China's oil imports come from the Middle East, mainly Saudi Arabia, Oman and Iran. In order to secure future oil supplies, Chinese national oil companies have been investing in exploration and production in Russia, the Caspian countries, Indonesia, Africa and Latin America. They are also actively involved in energy diplomacy with countries like Algeria, Gabon, Egypt and Mongolia.

Primary Chinese gas demand is projected to increase from 36 bcm in 2002 to 59 bcm in 2010 and to 157 bcm in 2030. Gas will then account for 6% of the primary fuel mix. The government has set a target of 50 bcm by 2005. Although there is good potential for gas growth in China, barriers are yet to be overcome.[9] Downstream infrastructure is the weakest link in the gas supply chain. The lack of sound legal and regulatory frameworks is another problem. In the longer term, the price competitiveness of gas against coal in power generation remains very uncertain. Nonetheless, projected growth in demand for gas will outpace the expected increase in China's domestic gas production from 36 bcm in 2002 to 115 bcm in 2030. As a result, China will need to increase its gas imports. Australia will start sending LNG to the Guangdong terminal in 2005. Later, Indonesia will supply the Fujian terminal. In the longer term, Russia could become a major supplier of piped gas, most likely from the Kovykta field in Siberia. In the very long term, pipelines from Central Asian transition economies such as Kazakhstan are also possible.

China's primary coal demand will grow at an average rate of 2.3% between 2002 and 2030. Most of new demand will come from the power generation sector. The use of coal by industry will grow at only 0.7% per year and consumption will decline in other end-use sectors. Although China could continue to be the world's largest coal producer in the foreseeable future, it is not sure that China can and will make the major investments needed both in mining and transport networks. In order to increase its production from 1 398 Mt in 2002 to 2 490 Mt in 2030, the Chinese coal sector would have to invest $130 billion.

China's electricity market, which is the second largest after the United States, will grow by 4.5% per year on average, with 6% annual growth in the current decade. More than 50% of incremental electricity demand will come from the residential and services sectors, followed by industry with 43%. Electricity generation will reach 5 573 TWh in 2030 from 1 675 TWh in 2002 (Table 8.7). Per capita electricity consumption will almost triple from 1 300 kWh in 2002 to 3 860 kWh in 2030, but it will still be less than half the average in OECD countries. The share of gas and nuclear power in electricity generation will increase from 1% to 6% and from 2% to 5%, respectively, by 2030, at the expense of coal.

---

9. See IEA (2002b) for detailed discussion about the key policy issues for Chinese gas market development.

Table 8.7: **Electricity-Generation Mix in China** (TWh)

	1971	2002	2010	2020	2030
Coal	98	1 293	2 030	2 910	4 035
Oil	16	50	59	65	53
Gas	0	17	55	196	315
Nuclear	0	25	82	180	280
Hydro	30	288	383	578	734
Biomass and waste	0	2	31	58	84
Other renewables	0	0	13	31	72
Total	**144**	**1 675**	**2 653**	**4 018**	**5 573**

At the end of 2002, China had 360 GW of generating capacity. In the wake of power shortages in 2003, the government approved the construction in about 35 GW of capacity during 2003 and another 40 GW during 2004 (CERA, 2004). Development of transmission networks from the western part of the country to the demand centres in the coastal zones is ongoing. But electricity shortages are likely to continue in the short term. From now to 2030, China will have to add new generating capacity of 860 GW. This capacity requirement is largest in all WEO regions. The capacity of coal-fired power plants will increase threefold to 776 GW. Gas-fired capacity will surge from 8 GW in 2002 to 111 GW in 2030, and nuclear-powered capacity is projected to reach 35 GW. Among renewable energy sources, wind power is the most promising, followed by bioenergy. The capacity of non-hydro renewables is assumed to increase from 2 GW today to 38 GW in 2030.

China's investment requirement for electricity from now to 2030 is $2 trillion, the largest in the world. The Chinese power sector could run short of capital because of weakness in the country's still-developing financial markets, limited inflows of international capital and inadequacies in the generating companies' own management practices (IEA, 2003).

China is the second largest contributor to global energy-related CO_2 emissions after the United States. The annual growth rate of Chinese emissions is projected to be 2.8%, compared to energy demand growth rate of 2.6% and GDP growth of 5%. China's share in global emissions will increase from 14% to 19% over the projection period. China will account for 26% of new global emissions between 2002 and 2030, exceeding the increase in emissions from all the OECD countries combined. While the largest absolute amount of emissions will come from the power generation sector, the transport sector will see the fastest growth of emissions.

India

Current Trends and Key Assumptions

India's primary energy demand has grown over the last thirty years at an average of 3.6% a year. Including traditional fuels, it now accounts for 5% of total world demand. Coal is the dominant commercial fuel in India, meeting half of commercial primary energy demand and a third of total energy demand. Power generation accounts for about 73% of India's coal consumption, followed by industry. Coal resources are plentiful but of low quality.

In 2003, India's oil demand reached 2.6 mb/d, or about 22% of the country's total energy demand, some 70% of which is met by imports. India's indigenous oil production has stagnated in recent years. Demand for natural gas, mainly from the power sector, has been rising faster than that for any other fuel, from barely 12 bcm in 1990 to 28 bcm in 2002. Some recent gas discoveries hold promise. Biomass and waste represent 39% of India's current energy demand and nearly 85% of the energy used in the residential sector. Some 595 million people – 60% of the Indian population – depend on traditional biomass for cooking and heating.

India's GDP is assumed to increase at an average of 4.7% per year over the full projection period. The growth will be fastest in the current decade, at 5.6%, and then will slow, dipping to 4% from 2020 to 2030. India's population will exceed 1.4 billion by 2030. Despite increasing urbanisation, the rural population will still represent nearly 60% of total population.

Results of the Projections

India's primary energy supply is projected to rise by 2.3% per year between 2002 and 2030 (Figure 8.15) – well below the assumed GDP growth rate of 4.7% over the same period and also below the 3.6% rate for energy supply growth between 1971 and 2002. Energy demand will decelerate gradually in line with an assumed slowdown in economic and population growth after 2010. Biomass and waste, the main fuels in the primary energy mix in 2002, will be increasingly displaced by coal and oil. Natural gas use will increase rapidly, but from a low base, so its share in total primary energy supply will reach only 9% in 2030 compared to 4% in 2002. A small number of new nuclear plants will be built, leading to a tripled increase in the nuclear share. Hydropower will increase strongly up to 2020, but its growth will then slow as most good sites are exhausted. India's energy intensity will decline by 2.3% per year, faster than the 2% a year by which it fell between 1990 and 2002.

Electricity output, which increased at 6.2% per year over the past decade, is projected to grow at 4.4% per annum over the projection period. The

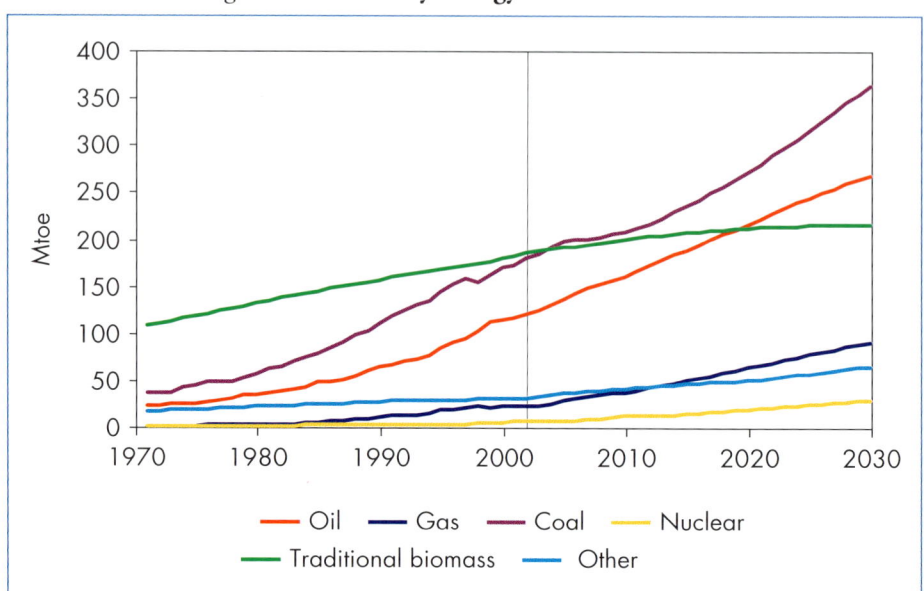

Figure 8.15: **Primary Energy Demand in India**

electricity industry faces enormous challenges in providing reliable service and meeting rising demand (TERI, 2003). Transmission losses, theft, unmetered connections and heavily subsidised prices will continue to hamper profitability. These problems will affect the quality of grid-based electricity supply and slow the rate of electrification. The investment cost of meeting the projected increase in generating capacity, transmission and distribution is estimated at around $680 billion from now to 2030.

Total final energy consumption will increase at 2.1% per annum from 2002 to 2030. The share of biomass and waste will decrease from 54% in 2002 to 36% in 2030. The absolute amount of biomass used, however, will continue to rise by 0.6% a year. The average consumption of energy per capita will increase by more than 30% over the *Outlook* period to reach 479 kgoe per capita in 2030 (Figure 8.16), well below the developing countries' average of 810 kgoe per capita.

The residential sector will account for 48% of total final consumption in 2030. The switch away from traditional biomass will favour increased use of liquefied petroleum gas (LPG), kerosene, gas and electricity. Oil consumption in the residential sector is set to more than double by 2030. Gas consumption will rise fivefold to 2.7 Mtoe in 2030. Electricity's share in residential consumption will reach 15% in 2030, reflecting an increase in electrification from 44% of population in 2002 to 68% in 2030. Without the introduction of major new government initiatives, India is unlikely to achieve its target of full electrification by 2012. Coal use will drop both in share and in absolute level. Biomass will still account for two-thirds of residential consumption in 2030.

Figure 8.16: **Total Final Energy Consumption per Capita in India**

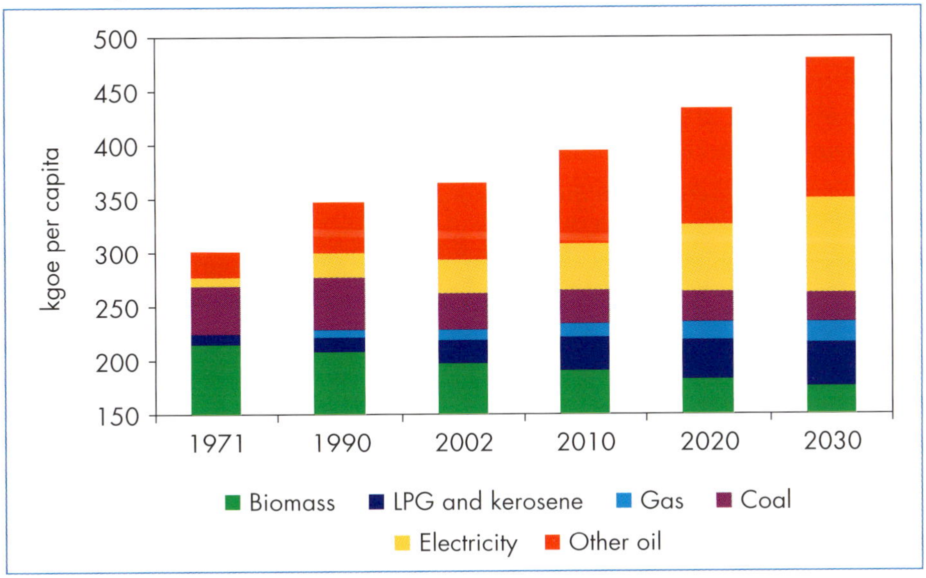

[Stacked bar chart showing kgoe per capita from 1971 to 2030, with categories: Biomass, LPG and kerosene, Gas, Coal, Electricity, Other oil. Values rise from about 300 in 1971 to about 485 in 2030.]

Oil consumption in the transport sector will continue to rise, driven by an increase in passenger-vehicle ownership. The number of diesel and gasoline vehicles will nearly double between 2002 and 2030. The share of the vehicle fleet fuelled by compressed natural gas and biofuels will continue to grow.

Industry's share of India's total final consumption will increase from 27% in 2002 to around 32% in 2030. Oil will still be the dominant fuel in industry, accounting for 43% of consumption in 2030. The share of coal and renewable energies will decline, while that of oil, electricity and gas will increase. By 2030, gas and electricity will represent 27% of total industry consumption.

Our projections imply that Indian CO_2 emissions will more than double by 2030, reaching 2 254 Mt. The annual growth rate of emissions over the projection period will be 2.9%, compared to 2.3% growth in primary energy demand and 4.7% growth of GDP. This increase will be driven by robust energy demand and the continued dominance of coal in the energy mix for power generation. The power sector will remain the biggest source of CO_2 emissions, contributing 61% of the total by 2030. Local pollution is an even greater concern, as more than 420 thousand people die each year from pollution-induced illnesses.

Indonesia

Indonesia has been and will continue to be a key energy producer. Currently, it is the world's largest exporter of LNG, and the third largest hard coal exporter after Australia and China.

With macroeconomic stability restored after several years of political turmoil, Indonesia returned to the international debt markets in early 2004 for the first time in eight years. GDP growth improved to 4.1% in 2003, and faster growth is expected in the short term. This *Outlook* assumes annual economic growth of nearly 4% from 2002 to 2030, though this projection is subject to political uncertainties, especially the impact of this year's presidential election.

Total primary energy demand in Indonesia is projected to grow by 2.7% annually from 2002 to 2030. With growth of 3.8% per year in transport demand, oil will continue to dominate the fuel mix, accounting for 38% of total demand in 2030. Coal demand is expected to grow most strongly by 4.6% per year, as the power generation sector will increasingly rely on it. Because the country is a huge archipelago, it is difficult and costly to deliver modern fuels, especially electricity, to a large part of the rural population. So, demand for traditional biomass and waste will continue to grow, but at a much slower pace than in the past three decades. As a result, Indonesia's dependence on fossil fuels will increase from 69% in 2002 to 82% in 2030.

Indonesia's oil production has been declining since the mid-1990s and was running at 1.3 mb/d in 2002. That production figure is expected to increase to 1.6 mb/d in the current decade, as the Cepu field in Java starts operations. Production of non-conventional oil is expected to increase gradually towards the end of projection period. With strong inland demand, however, Indonesia will become a net oil importer towards the end of this decade (Figure 8.17).

Figure 8.17: **Oil Balance in Indonesia**

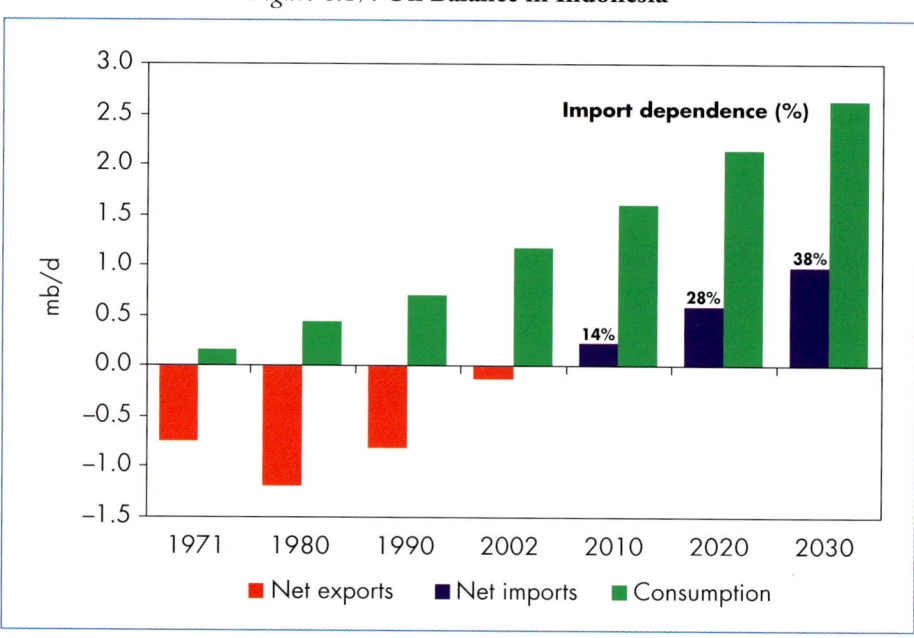

Gas production, on the other hand, is expected to grow rapidly, from 76 bcm in 2002 to 166 bcm in 2030. Gas exports will rise by more than 80% between 2002 and 2030, but they will face increasing competition from Australia, the Middle East and eventually Russia. Proven coal reserves are more than ample to meet projected domestic demand and provide a surplus to export for several decades to come.

Indonesia needs almost to triple its power generating capacity to 109 GW to meet the projected electricity demand growth of 5.2% per year over the projection period. The generation fuel-mix is expected to shift strongly to coal, whose share will increase to 55% in 2030 from 40% in 2002. Gas will gain three percentage points, fuelling 25% of electricity generation in 2030.

Finding investment finance is a major challenge to the Indonesian energy sector. More than $300 billion will be required to meet domestic demand and to expand exports. Under the Oil and Gas Law, the monopoly held by state-owned Pertamina has been ended and some of its functions transferred to two newly-created agencies. But the unclear division of responsibilities among these three entities has generated confusion among private companies. The electricity sector still suffers from the consequences of the Asian economic crisis. Perusahaan Listrik Negara, the state-owned utility, is heavily indebted, and the sector remains highly centralised. There are worries about the security of energy production facilities. The poor investment climate, with inadequate legal and regulatory frameworks, could continue to undermine investors' interest in the energy sector.

Latin America[10]

Latin America's economy rebounded in 2003 from a downturn in the previous year. GDP growth for 2003 is estimated at 1.8%. Economic expansion is expected to continue during 2004. With strong economic policies now in place, bolstered by a currency devaluation in Argentina, confidence and activity in Brazil and Argentina have been recovering strongly. But economic recovery in Venezuela and other Andean countries have been hampered by political uncertainties and social tensions. GDP growth of 3.2% a year is assumed for Latin America over the projection period, stronger than that of the past three decades. Average per capita income will almost double over the period, reaching $12 000 in 2030.

Three countries, Brazil, Argentina and Venezuela, accounted for two-thirds of the region's energy consumption in 2002. Venezuela has the largest oil reserves outside the Middle East. Venezuela also has substantial gas reserves, as do Bolivia, Argentina and Trinidad and Tobago. The region's energy demand is

---

10. See Annex E for a list of countries in the Latin American region.

projected to grow less rapidly than its economy, at 2.6% per year over the projection period, thanks to improvements in energy efficiency.

Oil demand is expected to grow strongly at 2.3% per year, spurred by increased use of oil in the transport sector. Transport, the fastest growing end-use sector, is expected to account for 73% of the growth in oil demand, which will reach 8.4 mb/d by 2030. Oil production is set to increase even faster, so net exports will increase from 2.7 mb/d today to 3.3 mb/d in 2030. Venezuela will remain the largest oil producer in the region, while Brazil will become a net exporter in a few years.

Gas is expected to grow fastest among fossil fuels, at 4.4% a year. Gas will gain 12 percentage points in the fuel mix over the projection period, accounting for 31% of energy consumption in 2030. The power sector will drive this growth. Gas-fired power plants will account for nearly half of power generating capacity in Latin America, at 255 GW. Over the projection period, the vast gas resources in Venezuela, Bolivia and Argentina will be tapped and Latin America will become a net exporter of gas, especially in the form of LNG. Coal consumption is expected to remain marginal in the Latin American energy mix, accounting for a mere 4% in 2030.

Electricity demand will grow strongly, at 3.6% per year, driven by increasing prosperity and rural electrification. The region already relies heavily on renewable energy sources, largely hydroelectric power, to meet its electricity needs. In view of their very heavy dependence on hydro, which can be crippled in drought years, many countries in the region are planning to diversify their electric-power fuel mix. Hydro is expected to grow by 2.1% per year. At that rate, it will lose market share in electricity generation in favour of gas.

Brazil

Current Trends and Key Assumptions

Brazil is Latin America's largest energy consumer, accounting for 40% of the region's consumption in 2002. Its primary energy mix is dominated by oil (47%), hydropower (13%) and other renewable energies (25%). Despite its large resources, Brazil consumes more energy than it produces. In 2002, gas production was 9 bcm for a demand of 13 bcm, while oil production was 1.5 mb/d, and demand was 1.8 mb/d.

With an estimated 175 million inhabitants, Brazil has the largest population in Latin America and the fifth largest in the world. This *Outlook* assumes that the population will increase by 0.8% per year, reaching 220 million by 2030. Brazil's economy is the largest in Latin America. But its economic growth slowed during 2001 and 2002 to less than 2%, reflecting the downturn in major world markets. GDP *declined* by 0.2% in 2003, the first year of negative growth since 1992. Growth has since rebounded strongly and is expected to reach 3.3% in 2004. This *Outlook* assumes that the Brazilian economy will grow by around 3% a year through to 2030.

Brazil is one of the world's largest users of hydroelectricity, which currently accounts for more than 80% of its power production. During 2001 and 2002, Brazil suffered an electricity crisis due to a drought that left hydroelectric dams low on water. This led to mandatory electricity rationing for a period of nine months, as well as to electricity price increases. In August 2004, in a bid to avoid a repetition of such events, the Brazilian government introduced a new regulatory framework aimed at luring investment into the power sector.

Biomass meets a quarter of Brazil's primary energy demand. But, unlike many other developing countries, Brazil uses biomass for more than cooking and heating. Both the industry and transport sectors make extensive use of biomass.

Petrobras, Brazil's state oil company, hopes to end oil imports by 2006. Achieving this target will depend largely on the company's success in expanding the capacity of high-cost deepwater and ultra-deepwater oil fields in areas such as the Campos basin north of Rio de Janeiro. In view of the very large amount of money needed for these projects, foreign investment is being encouraged. Recently, however, oil majors have shown less interest in Brazil, owing to disappointing results from exploration drilling. Some government policies, such as requirements for domestic participation and the imposition of onerous procurement conditions, have also undermined Brazil's attractiveness to foreign oil companies.

In the last decade, Brazil's energy sector underwent profound regulatory and structural changes. Electricity generation and distribution have been opened up to private capital. The monopoly of the state-owned company Petrobras in oil and gas exploration and production concessions has ended. There are now two transnational gas pipelines and several electricity transmission lines linking Brazil with neighbouring countries. Many others are under construction or in the planning stage.

This *Outlook* assumes that energy prices in Brazil will become more market-oriented as reforms proceed and that, as a consequence, prices of all energy products will converge towards international price trends.

Results of the Projections

The Reference Scenario projects an average annual growth rate of 2.5% in Brazil's primary energy demand from 2002 to 2030 (Table 8.8). Demand will grow slightly more rapidly in the period up to 2020, then slow to 2.3% in the third decade. By comparison, demand grew by 3.2% per year from 1971 to 2002. Brazil's energy intensity will gradually decline, by 0.5% per year, as the structure of its economy progressively approaches that of OECD countries today.

Oil will remain the dominant fuel in Brazil's energy mix. Its share of total primary energy supply will remain broadly unchanged at around 46% throughout the

Table 8.8: **Primary Energy Demand in Brazil** (Mtoe)

	1971	2002	2010	2030	2002-2030*
Coal	2	13	14	22	1.9%
Oil	28	88	109	172	2.4%
Gas	0	12	18	59	5.8%
Nuclear	0	4	4	6	2.0%
Hydro	4	25	31	45	2.2%
Biomass and waste	35	46	51	65	1.2%
Other renewables	0	0	0	2	42.9%
Total	**70**	**188**	**228**	**372**	**2.5%**

* Average annual growth rate.

projection period. Natural gas use will increase most rapidly, at an annual rate of 5.8%, mainly as a result of surging demand for power generation. Gas's share of the total primary energy supply will rise from 6% in 2002 to 16% in 2030. Although hydroelectricity will grow rapidly, its share in the energy mix will remain steady around 13%. Nuclear energy supplies will also retain a constant share, at 2%. Although the use of biomass will increase in absolute terms, its share of the primary energy mix will decline from 25% in 2002 to 18% in 2030.

Oil, half of which is used in transport, will extend its position as the most important end-use fuel. Oil demand in the transport sector is projected to increase to 98 Mtoe from 43 Mtoe over the projection period. Brazil is expected to become self-sufficient in oil before 2010 (Figure 8.18). Gas import dependence will rise

Figure 8.18: **Oil Balance in Brazil**

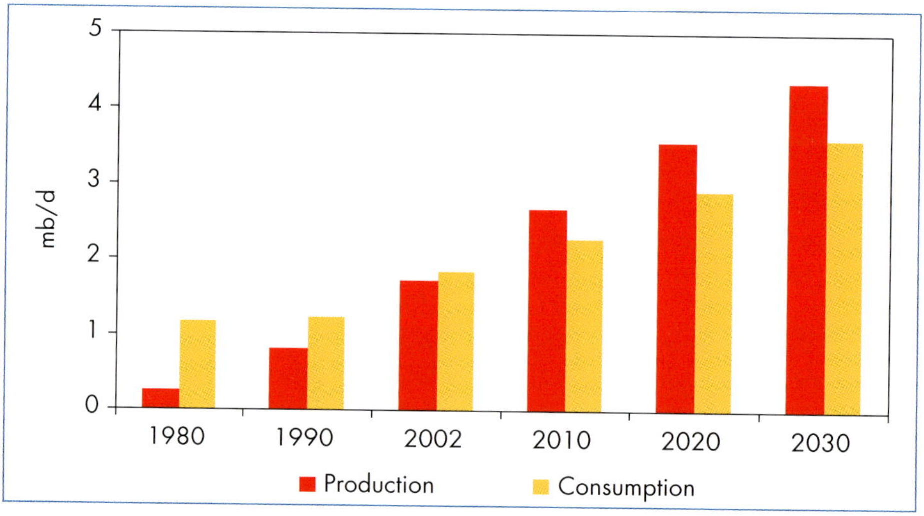

rapidly in the first decade of the *Outlook* period, but once Brazil taps its vast gas resources, that trend will be reversed and the country will end the projection period as a net gas exporter.

Electricity demand will grow by 2.9% per year until 2030. Electricity rationing in the years 2001-2002 produced short-term energy savings, especially in industry, and experience gained at that time is expected to contribute to long-term savings. Electricity demand is expected to surpass pre-crisis levels in 2004. In 2002, electricity generation in Brazil totalled 345 TWh. Generation is projected to reach 808 TWh in 2030, after growth of 3.1% per year (Table 8.9)

Table 8.9: **Electricity Generation Mix in Brazil** (TWh)

	1990	2002	2010	2020	2030
Coal	5	8	8	14	23
Oil	6	13	14	17	12
Gas	0	13	21	77	179
Nuclear	2	14	15	24	24
Hydro	207	285	362	450	529
Biomass and waste	4	11	17	23	28
Other renewables	0	0	3	6	14
Total	223	345	441	611	808

Renewables will retain a sizeable, yet declining, share of total final energy consumption. Increased amounts of biomass will be used in the power sector. The Brazilian National Development Bank is financing additional power generation from bagasse, the residue from sugar-cane processing. Biofuels are expected to increase their share of Brazil's road transport fuel demand from 13% in 2002 to 16% in 2030. This trend will be bolstered by strong growth in sales of flex-fuel vehicles, which can run on gasoline or ethanol or a mixture of both. Flex-fuel vehicles are proving popular because they offer motorists a chance to save on fuel costs. They also alleviate concerns about air pollution and about the early depletion of Brazil's oil reserves. Another contributing factor to the growth in biofuels is a programme to add 2% of biodiesel to diesel fuel commencing in late 2004.

The investment required to expand Brazil's energy system from now to 2030 is projected to be nearly $450 billion (in year-2000 dollars). The Brazilian public sector alone will not be able to provide that much financing. Private investment will only be forthcoming, however, if Brazil's regulatory regime becomes more transparent and consistent.

Energy-related CO_2 emissions from Brazil are expected to reach 665 Mt by 2030, up from 302 Mt in 2002. Brazil's energy system is one of the least carbon-intensive in the world, because of the wide use of hydropower and active government encouragement of biomass fuels.

Middle East

Primary energy demand in the Middle East is projected to increase at an average annual rate of 2.5% until 2030. This strong growth will be fuelled by population increases of 1.9% per year and by solid economic growth of 3% per year. Energy demand growth will slow over the period from an annual rate of 3.2% in the first decade to 1.5% in the third decade. Expected growth in energy demand is well below that experienced in the region over the last three decades, during which consumption increased eightfold. Considerable scope exists for energy savings throughout the region.

As in the past, the Middle East's fuel mix will consist almost exclusively of oil and gas (Table 8.10). By 2030 these two fuels will meet 96% of all energy needs. In 2003, the region's domestic oil demand stood at 4.4 mb/d. This figure is expected to increase to 7.8 mb/d in 2030 in line with increased demand for transport fuels, which are priced well below international levels in most Middle Eastern countries. Gas demand is expected to double, from 219 bcm in 2002 to 470 bcm in 2030. Demand for gas will come mainly from the power sector and from industry, where it is used primarily as a petrochemical feedstock and to fuel water-desalination plants. Demand for renewable fuels is expected to grow rapidly, from a very low base. Renewables and coal together will represent less than 4% of the region's total primary energy supply by 2030.

Table 8.10: **Primary Energy Demand in the Middle East** (Mtoe)

	1971	2002	2010	2030	2002-2030*
Coal	0	8	9	14	2.3%
Oil	38	206	257	374	2.1%
Gas	11	189	250	405	2.8%
Nuclear	0	0	2	2	–
Hydro	0	1	2	3	3.1%
Biomass and waste	1	2	2	7	5.0%
Other renewables	0	1	1	3	5.5%
Total	**51**	**407**	**524**	**809**	**2.5%**

* Average annual growth rate.

The Middle East is expected to experience a large demographic change over the projection period. Its population will age significantly, with the age group from infancy to 19 dropping from 48% now to 39% in 2030 (Table 8.11). To meet the needs of the region's rapidly growing population, electricity demand is set to grow at 3.2% per year through to 2030. Over 60% of incremental generating capacity will be gas-fired, as countries seek to make more oil available for export and for domestic consumption in the transport sector. Very large investments in electricity and gas infrastructure will be needed. They may be hard to mobilise, however, given that electricity prices in the region tend to be well below economic levels, making it difficult for generators to recoup their costs.

Table 8.11: **Population Structure in the Middle East** (% of total)

Age groups	2002	2010	2020	2030
0-19	48	45	42	39
20-60	46	49	50	51
60+	6	6	8	10

Source: UN (2003).

The Middle East supplies about 28% of the world's oil needs and about 10% of its gas. It also possesses about two-thirds of the world's proven oil reserves and around 40% of proven natural gas reserves. By 2030, the region's share of global oil supply will increase to over 40% and it will be the largest exporter of natural gas. The Middle East's role as a major oil and gas exporter is analysed in detail in Chapters 3 and 4.

Africa

In 2002, Africa had the lowest GDP per capita of all *WEO* regions and energy use per capita of just 0.6 toe. Economic activity, even in the poorest parts of sub-Saharan Africa, has been resilient in the recent past despite the global economic slowdown. Economic growth in 2004 is expected to increase still further. While this *Outlook* assumes that Africa's economy will grow at 3.8% per year between 2002 and 2030, there will be sharp differences among countries.

The region's primary energy demand has almost tripled over the last three decades to reach 534 Mtoe in 2002. Even so, Africa still accounted for barely 5% of world energy demand. Only 36% of the African population has electricity. More than 80% of its rural population has none. Traditional biomass has dominated the African energy scene for the past thirty years, accounting for 39% of total energy demand in 2002. The inefficient ways in which biomass has been used have resulted in an energy intensity higher than the world average despite Africa's low rates of electrification, industrialisation and vehicle ownership.

Africa produced 10% of world crude oil in 2002. Its net exports reached more than 70% of its production. Also, it accounted for around 5% of world coal production. But the continent includes very different energy entities. North Africa is a major oil and gas producer and has good electricity infrastructure with electrification rates of 94% in 2002. It uses little traditional biomass – less than 2.2% of overall energy demand in 2002. West Africa is a major oil producer and a heavy consumer of traditional biomass. Almost all the continent's coal production comes from South Africa. Overall, sub-Saharan Africa has very poor energy infrastructure, with less than 24% of the population enjoying electricity and over 630 million people relying on traditional biomass for cooking and heating.

Over the *Outlook* period, total primary energy demand in Africa is projected to increase by 2.6% per year on average. Oil demand will nearly triple and natural gas demand will rise fourfold between 2002 and 2030. The share of natural gas in the energy mix will increase from 11% in 2002 to 21% in 2030, at the expense of coal, the share of which will drop from 17% to 13%.

The transport sector will continue to grow at 3.5% per year on average, slightly higher than the 3.4% of the past three decades. Growth in transport will drive oil demand growth of 3.4% per year. A switch away from traditional biomass in all end-use sectors will result in increased use of other fuels. In the residential sector, oil use, mainly in the form of LPG and kerosene, will increase by 4.4% per year over the projection period. Despite this substantial shift from traditional to modern energy services, traditional biomass will remain the dominant fuel.

Electrification rates will improve significantly over the next three decades to reach 58% in 2030 for Africa as a whole and 51% for sub-Saharan Africa. Extensive electrification will boost electricity consumption in the industry, residential and services sectors. Electricity consumption will grow at 4.4% per year until 2030. Nonetheless, the population lacking electricity will reach 586 million people by 2030, an increase of 51 million people from 2002 – equivalent to the combined population of Spain and Portugal.[11]

The investment needed to meet the projected increase in inland energy demand and exports to other regions will be of $ 1.2 trillion over the projection period, of which half would go to the electricity sector alone. The challenge is clearly immense, especially for the poorest African countries.

Africa's CO_2 emissions will more than double over the *Outlook* period to reach 1 861 Mt in 2030. Still, African emissions will average only 1.3 tonnes of CO_2 per capita, just a tenth of the OECD average in 2030. The most pressing environmental issues are those arising from inefficient use of biomass and from local vehicle pollution.

---

11. See Chapter 10 for further discussion on energy and development issues.

Transition Economies[12]

Primary energy demand in the transition economies, excluding Russia, is projected to grow by 1.4% per year over the period 2002-2030. Energy demand growth will slow over the period from an annual rate of 1.9% in the first decade to just 1% in the third decade. Gas will extend its position as the dominant fuel in the region. Its share in total primary energy use will rise from 43% in 2002 to 48% in 2030, as most new power generators will be gas-fired. The share of oil is also expected to increase, from 23% in 2002 to 27% in 2030, driven by strong demand for transportation fuels. Although coal use will increase slowly in absolute terms, its share of the energy mix will fall from 21% in 2002 to 16% in 2030. Nuclear's share will also decline, as plant retirements will greatly outweigh additions, particularly during the 2020s.

The robust economic growth experienced in the region over the past few years is expected to continue. GDP growth is assumed to average 4.8% from now to 2010. Growth then slows to 3% from 2020 to 2030, as the economies in the region mature. Growth rates in the region's Central Asian countries are expected to be slightly higher than those in its European countries but this will depend largely on the Central Asian countries' ability to develop their rich energy resources. The population of transition economies is assumed to remain broadly unchanged at around 200 million.

Although the region as a whole is a net natural gas importer, relying primarily on supplies from Russia, there are a number of significant gas producers in the Caspian region: Azerbaijan, Kazakhstan, Turkmenistan and Uzbekistan. Expansion of production will depend largely on the development of sufficient export pipeline capacity. Russia's Gazprom has refused to grant national gas companies in the Caspian region the affordable access to its transmission system which would allow them to sell directly to buyers in Europe. A new pipeline system is being built to carry gas from the Shah Deniz field in Azerbaijan to Turkey via Georgia, thereby bypassing Russia. A long-distance line from Turkmenistan to Turkey could be built towards the end of the projection period. However, this will depend on Turkmenistan's success in proving up more reserves and on geopolitical developments in the region. A proposed pipeline from Turkmenistan to China, which could pick up gas from Kazakhstan and Uzbekistan, is also assumed to proceed in the last decade of the projection period, but considerable uncertainty surrounds the project.

The transition economies are set to become large-scale oil exporters. Oil production currently just meets demand, but it will increase rapidly, and there

---

12. See Annex E for the list of countries. This section does not include Russia. Chapter 9 has an in-depth analysis of Russia.

will be net exports of almost 2 mb/d by 2010, again from Caspian Basin countries. Oil exports will remain around that level for the remainder of the *Outlook* period, though this will depend on the construction of new export routes. We expect that the Caspian Pipeline Consortium's line from Kazakhstan to the Black Sea will expand from 0.6 mb/d to over 1 mb/d by the end of the current decade. We also foresee completion of the 900 kb/d Baku-Tblisi-Ceyhan pipeline, which will bring Azeri oil across Georgia to the deepwater port of Ceyhan on the Mediterranean coast of Turkey. Shipments North via the Russian Transneft system and increased Caspian-Iranian oil swaps may also be possible. In the longer term, a pipeline to China could be built. In Kazakhstan, there is potential for production growth at the two main existing fields, Tengiz and Karachaganak.

CHAPTER 9

RUSSIA – AN IN-DEPTH STUDY

HIGHLIGHTS

- Russia will continue to play a central role in global energy supply and trade over the *Outlook* period, with major implications for world supply security. The Russian energy sector has undergone a dramatic transformation in recent years. It has been the principal driver of the country's economic recovery since the late 1990s.
- The prospects for Russian oil production are very uncertain. This *Outlook* projects production to continue to increase, though more slowly than in recent years, from 8.5 mb/d in 2003 to 10.4 mb/d in 2010 and 10.8 mb/d in 2030. Most of the increase in the short to medium term will be available for export. But the share of Russian exports in world trade will fall back after 2010.
- Russia will still be the world's biggest gas exporter in 2030. But output from the country's old super-giant fields is declining, and huge investments in greenfield projects will be needed to replace them. Gazprom will rely more on imports from Central Asia, allowing it to put off development of its own reserves. The prospects for gas production from independent producers will depend on their gaining access to Gazprom's network.
- The Russian economy's dependence on the oil and gas sectors has grown in recent years, owing to rising prices and production. It now approaches that of some OPEC countries. Russia's long-term economic prospects hinge on improving the competitiveness and diversity of other manufacturing sectors and of internationally traded services.
- Developing Russia's huge energy resources will call for investment of more than $900 billion from now to 2030. A stable and predictable business regime and market reforms will be critical to the prospects for financing this investment. If gas-sector reform is delayed, worries about the security of future supply will increase. Large amounts of foreign capital are unlikely to become available for projects that are not aimed at export markets.
- Russian energy demand will continue to recover steadily from the lows reached at the end of the 1990s. Because of price reform, changes in the structure of the economy and investment in more efficient technology, energy demand will grow much more slowly than GDP. Russia's energy-related carbon dioxide emissions will rise, but will still be 27% below their 1990 level in 2010 and 11% lower in 2030.

This chapter analyses the outlook for Russia's energy sector to 2030. The first section summarises the importance of energy to Russia's economy and reviews recent trends in energy supply and demand and structural changes in the sector. This is followed by a description of the macroeconomic and policy assumptions underlying the projections. Projections of overall energy demand are presented first, followed by the projections of supply and demand by fuel. The final section assesses the implications of these projections for Russia's position in global energy supply, investment needs and the environment.

Energy Market Overview

Russia is exceptionally well-endowed with energy resources and the energy sector plays an increasingly central role in the Russian economy. Russia holds the world's largest proven natural gas reserves, the second-largest coal reserves and the seventh-largest oil reserves (BP, 2004). It is the world's biggest producer and exporter of natural gas, providing close to a quarter of OECD Europe's total gas needs. It is also the second-largest oil producer and a major exporter of oil to Europe and increasingly to Asia. Although domestic demand is well below what it was at the end of the Soviet era, Russia is still the third-largest energy consumer in the world, after the United States and China. Russia's energy production is low relative to its reserves of coal and gas, but is high for oil. Russia accounted for almost 10% of world primary energy production in 2002 (Figure 9.1). Energy is by far the largest industrial sector in Russia. Oil and gas alone contribute around a quarter of gross domestic product (World Bank, 2004).

Production and exports of oil plunged in the early 1990s as a result of the economic dislocation caused by the collapse of the Soviet Union. But they have rebounded strongly since 1999 in the wake of higher world prices. Gas production declined through much of the 1990s, mainly because of falling domestic consumption. But gas exports – entirely to Europe and to other former Soviet Union countries – have held up much better. Net exports amounted to 169 bcm in 2002, well below the 1999 level of close to 200 bcm. Coal and electricity production also fell substantially in the 1990s, but have recovered since 2000.

Domestic energy demand is recovering steadily in line with the strong economic rebound since 1998, but the pace of demand growth is much slower than past trends would have suggested. From 1999 to 2002, Russian GDP (adjusted for purchasing power parity) grew on average by 6.2% per year, while energy demand grew by only 0.8%. Throughout much of the 1990s, the rate of *decline* in energy demand was much closer to that of GDP (Figure 9.2). Final energy consumption fell sharply in all end-use sectors, most dramatically in industry. Industrial consumption in 2002 was less than two-thirds of that in 1992.

Figure 9.1: **Share of Russia in World Energy, 2002**

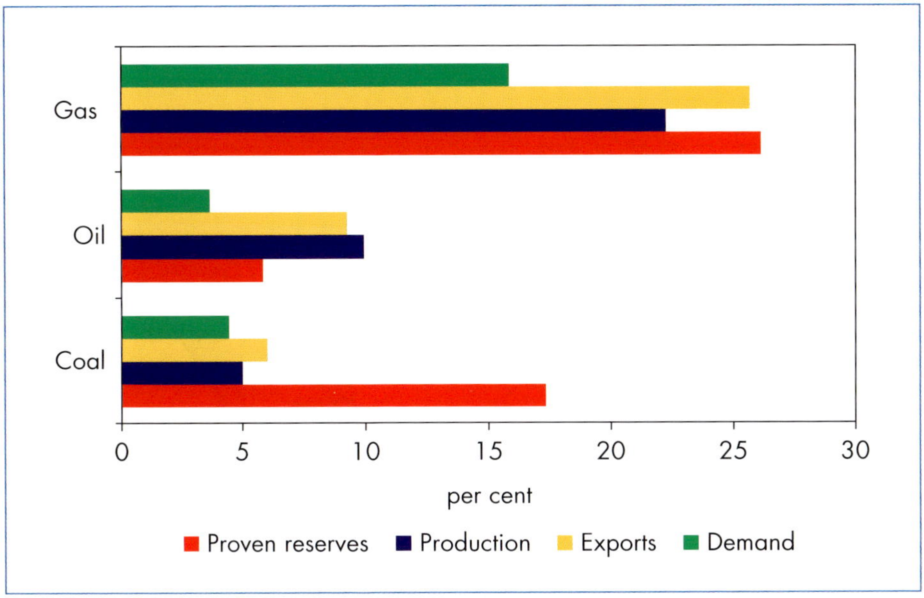

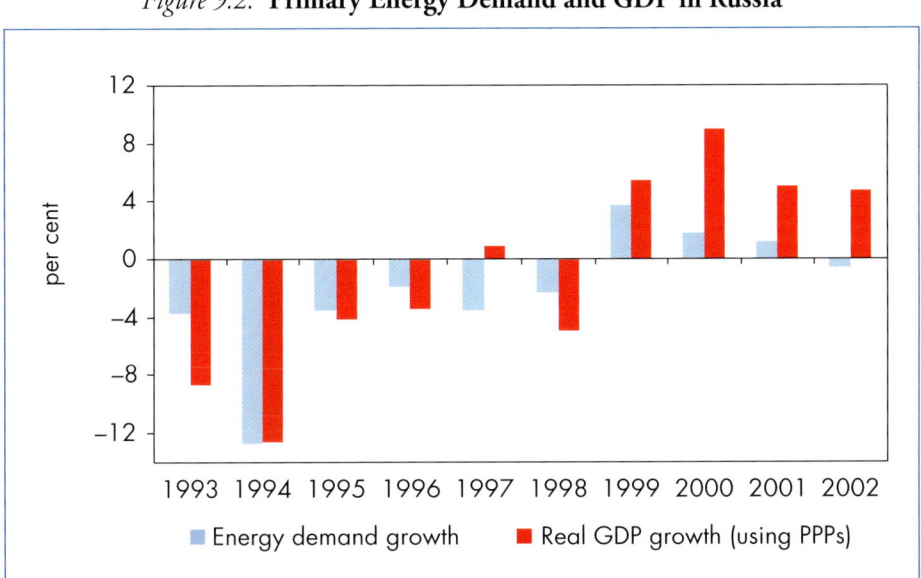

The Russian fuel mix has changed markedly over the past decade (Figure 9.3). The share of natural gas in total primary demand jumped from 47% in 1992 to 53% in 2002, mainly at the expense of oil. The share of oil dropped from 28% to 21% over the same period, primarily due to a drop in the use of fuel

Figure 9.2: **Primary Energy Demand and GDP in Russia**

Chapter 9 - Russia – an In-depth Study

oil in power generation. Hydroelectricity and nuclear power have seen their shares increase. The share of gas in final consumption has also risen, while that of coal has declined slightly.

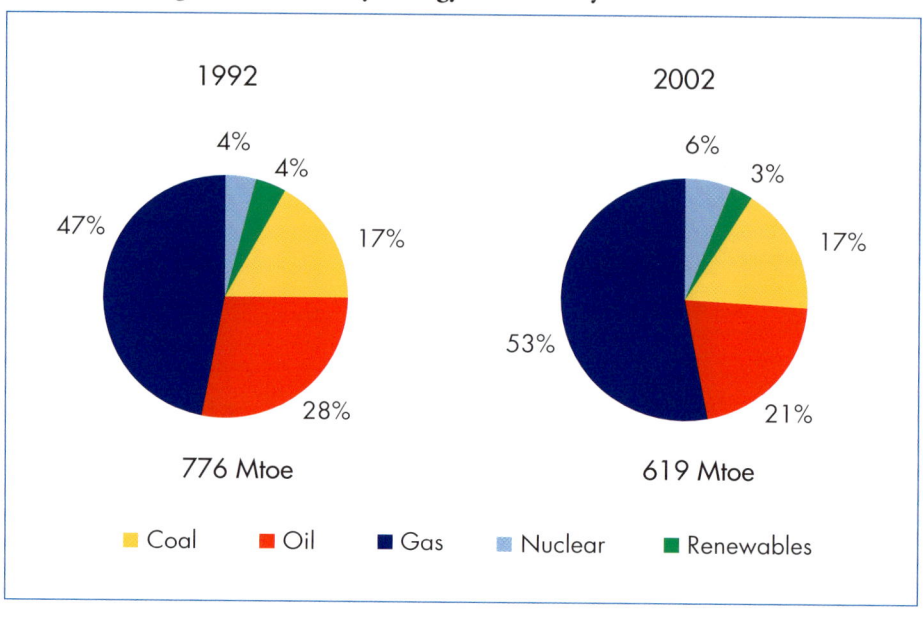

Figure 9.3: **Primary Energy Demand by Fuel in Russia**

In the initial restructuring of the energy industry after the collapse of the Soviet Union, commercial disciplines were introduced in state-owned enterprises, and many of them were partially or completely privatised. The oil industry was reorganised into separate companies, including some vertically-integrated businesses, and largely sold off to private interests. Mergers and acquisitions have since led to consolidation and the emergence of a small number of large companies. The state has retained majority shareholdings in some energy companies, including Rosneft, an oil-production company, the oil-pipeline monopolies, Transneft and Transnefteprodukt, and Unified Energy System (UES), the national power company. It also holds large minority stakes in the giant gas company, Gazprom, which it controls, and in three oil companies. The government plans further sales of state energy holdings, notably in UES.

Macroeconomic Context

Russia has made considerable progress in its transition to a market economy, though many legacies of the old centrally-planned system are still evident and many serious problems remain. Major institutional, regulatory and legal reforms still need to be implemented. Property rights need to be guaranteed and efficient markets established. The restructuring of key sectors of the economy, including energy, has yet to be completed.

Table 9.1: **Key Economic and Energy Indicators for Russia and the World**

	2002		1992-2002 (%)*	
	Russia	World	Russia	World
GDP ($ billion in year-2000 dollars, PPP)	1 064	47 658	−1.1	3.3
GDP per capita ($ in year-2000 dollars, PPP)	7 390	7 714	−0.8	1.9
Population (million)	144	6 178	−0.3	1.4
Total primary energy supply (Mtoe)	619	10 200	−2.2	1.5
TPES/GDP**	0.58	0.21	−1.2	−1.7
Energy production/TPES	1.67	1.01	1.5	−0.1
TPES per capita (toe)	4.30	1.65	−1.9	0.2
CO_2 emissions (million tonnes)	1 488	23 116	−2.3	1.3
CO_2 emissions per capita (tonnes)	10.3	3.7	−2.0	−0.1

* Average annual growth rate.
** Toe per thousand dollars of GDP in PPP terms at 2000 prices.

Russia's economic performance since the financial crisis of August 1998 has been phenomenal. GDP growth peaked at 10% in 2000, and has averaged almost 6% since then. Inflation has been contained, with consumer prices rising by 12% in 2003, compared to 84% in 1998. Unemployment has fallen steadily, down to a little over 8% of the workforce by the end of 2003 (OECD, 2004a). A jump in oil-export earnings has created large trade surpluses, amounting to $36 billion in 2002 – or 9% of GDP – and an estimated $49 billion in 2003. The general government budget, which ran a deficit of over 5% of GDP in 1988, has been in surplus since 2000. Rising private consumption has helped to sustain economic growth. Investment has finally started to pick up, growing by 13% in 2003, while political stability over the past five years has boosted investor and consumer confidence. However, growing concerns about arbitrary state intervention threaten to undermine that confidence.

Surging oil production and exports, buoyed by high oil prices, have played a big part in this economic recovery. According to a 2004 report by the OECD, oil and gas contributed more than half of the increase in Russia's industrial production in 2002 and 2003 (Figure 9.4).[1] The oil sector led the Russian

---

1. Official data significantly understate the size of the hydrocarbon sector relative to the rest of the Russian economy, because they do not take into account transfer-pricing effects. The OECD analysis was based on World Bank estimates of the relative weight of different sectors in the Russian economy. See OECD (2004a) and World Bank (2004).

export boom between 2001 and 2003. Net oil exports increased by a third while exports of metals and machinery grew by much less. The big increase in oil- and gas-export earnings worked its way through the economy, stimulating household consumption, commercial activity and fixed investment.

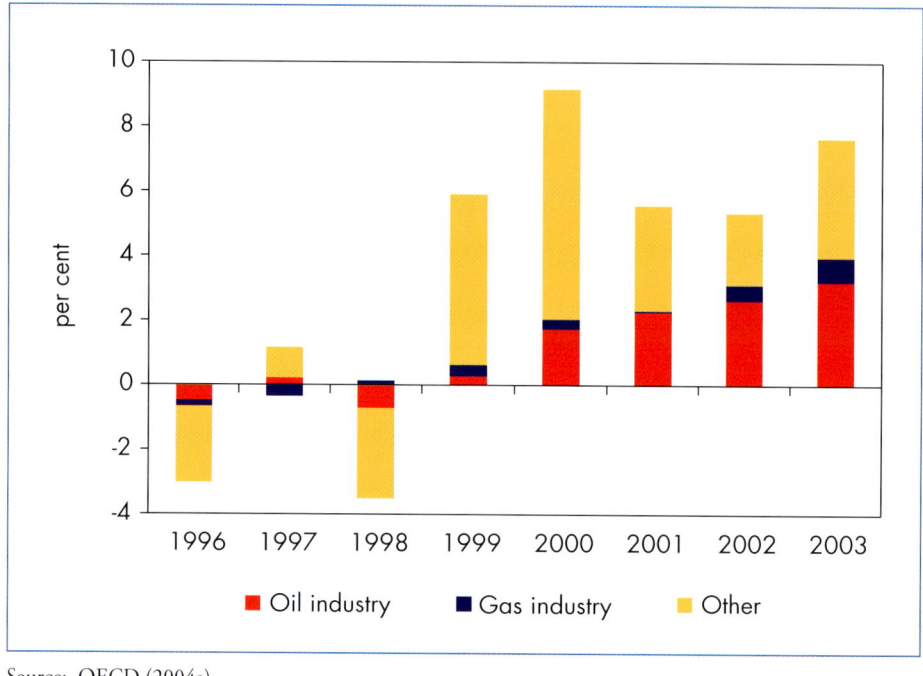

Figure 9.4: **Growth in Industrial Production in Russia**

Source: OECD (2004a).

The share of oil and gas production in the Russian economy has grown sharply in recent years. It now approaches that of some OPEC countries (Figure 9.5). Sustaining high economic growth in the short term will depend on continuing strong energy exports. In the longer term, growth prospects will hinge on Russia's ability to improve the competitiveness and diversity of its manufacturing sector and achieve more broad-based economic development.[2] This would also help to reduce the country's excessive dependence on oil and gas exports. But bringing about such an economic transformation will not be easy. Higher energy exports tend to raise the rouble's exchange rate, making it harder for other industries to export. Investment in the Russian economy has lagged, especially in the manufacturing sector. Gross capital formation has

2. See Concluding Statement of the IMF mission to the Russian Federation, 24 June 2004 (available at www.imf.org/external/np/ms/2004/062404.htm).

trailed that of the OECD and of several developing regions. Many of Russia's production facilities are physically and technologically obsolete. More far-reaching economic policy reforms are needed to tackle these problems.

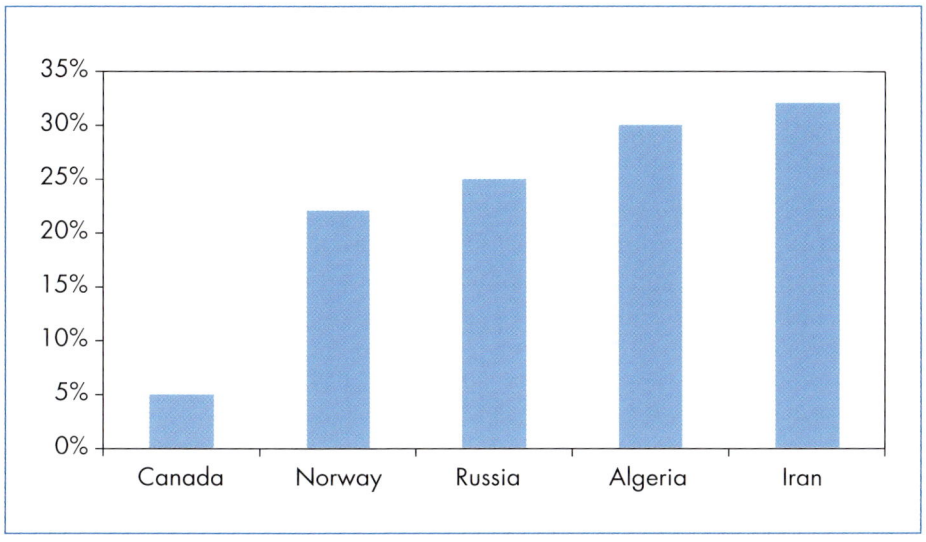

Figure 9.5: **Contribution of Oil and Gas Sectors to GDP in Selected Countries, 2002**

Note: Data on Russia are for 2000.
Sources: OECD (2004b) and World Bank (2004).

Fiscal and monetary policies will remain critical to stabilising the Russian macro-economy. Conservative fiscal policies and an expansion of the stabilisation fund, a mechanism to absorb windfall government revenues and to smooth spending, could reduce the economy's vulnerability to volatile international oil prices. Current plans, however, call for a fund equivalent to less than 4% of GDP, an amount that will allow it to play only a limited role. In the longer term, growth will depend on further improvements in the business climate, which will require institutional, regulatory and legal reforms. Major reforms have already been launched, but many still need to be completed. These include banking and competition reforms to make markets work better, reform of the social-welfare and pensions systems and further rationalisation of the tax system. Improvements are needed in the quality of state institutions to ensure that new commercial and civil laws are applied and that corruption is effectively policed. These reforms would strengthen confidence in the enforcement of commercial law, would deter capital flight and would create a more favourable climate for investment.

Ensuring that all Russians benefit from economic growth is another important challenge facing the government. The strong growth of the last few years has boosted wages, income and employment, and these factors have underpinned a surge in household spending and increased savings. Survey data confirm that ordinary Russians consider themselves better off now than in the late 1990s (OECD, 2004a). But real wages have not increased at the same pace in all sectors, and a new class of "working poor" has emerged, made up largely of young adults, families with children and single parents. Some non-economic indicators of human health and welfare also show little improvement. Life expectancy has yet to recover from a sharp fall in the 1990s. Deaths and illnesses from diseases commonly associated with poverty remain stubbornly high.

The government forecasts GDP growth of 6.9% for 2004 and average growth of 6.2% per year over 2000-2007. These rates are below the 7.3% rate of 2003. Moreover, President Vladimir Putin set a target in 2003 of doubling GDP within ten years, a feat which would require an average annual growth rate of over 7%. But the Reference Scenario projections presented here assume lower oil and gas prices in the second half of the current decade. If our assumption proves accurate, it will be very hard for Russia to meet its President's target. The Russian economy has never grown by more than 5.5% in a year when there was no increase in oil prices (World Bank, 2004).

Our projections assume that Russia's GDP[3] will grow on average by 4.0% per year from 2004 to the end of the current decade, and by 2.8% per year in the 2020s (Table 9.2). Real per capita incomes will more than triple over 2002-2030, from $7 390 to over $23 000 in year-2000 dollars. Industrial production is assumed to grow more slowly, as the share of service activities in the economy increases.

Table 9.2: **Macroeconomic and Demographic Assumptions for Russia**
(annual average rate of change, %)

	1992-2002	2002-2010	2010-2020	2020-2030
GDP	−1.1	4.4	3.4	2.8
Population	−0.3	−0.6	−0.6	−0.7
GDP per capita	−0.8	5.1	4.1	3.6

One reason for the assumed slowdown in economic growth in the medium to long term is the expected ageing and contraction of the Russian population. The population has already fallen from 149 million at the start of the 1990s to

3. In purchasing power parity terms.

145 million in 2001. According to the United Nations Population Division, it will drop to 137 million by 2010 and to 120 million by 2030 – a cumulative fall of over 17% (UNPD, 2003). This will lead to a substantial fall in the size of the workforce and a contraction of the productive potential of the Russian economy.

Energy Policy Developments

Restructuring and liberalisation of the electricity and gas sectors are progressing at varying rates. Reforms are most advanced in the electricity sector. Several important steps were taken in 2003 toward restructuring the vertically-integrated monopoly utility, UES, creating competitive markets in wholesale and retail supply and revamping the regulation of transmission and distribution. The pace and direction of gas-sector reform, which will have important implications for the evolution of the electricity sector, are less certain. Discussions about how to restructure Gazprom, the dominant gas producer and monopoly transmission company, have been going on for several years. President Putin has ruled out breaking up Gazprom, but he has acknowledged the need for more transparency and tariff reform. Still, it remains very uncertain how quickly cost-reflective pricing of domestic gas sales can be achieved. There are also differences of opinion about how "cost-reflectivity" should be defined.

Reform of electricity and gas pricing is a critical element of energy-market restructuring. The utilities' ability to modernise, to replace obsolete infrastructure and to expand exports is severely hampered by consumer prices that fail by a wide margin to cover long-run marginal costs. Domestic energy prices rose much less than average producer prices in the 1990s, depriving the energy sector of investment funds and encouraging continued waste. Consumers had little incentive to conserve energy or use it more efficiently. Cross-subsidies remain enormous: the average electricity price paid by industry in 2002 was 45% higher than that paid by households. In OECD Europe, average prices charged to industry are around 50% *lower* than those paid by households.

The government recognises the need to raise electricity and gas tariffs, but is concerned by the impact this will have on poor households, on the competitiveness of Russian industry and on economic growth in the near term. In early 2004, the government established a new mechanism for setting electricity tariffs in 2005-2007. Under this scheme, they will hardly be increased in real terms. They are, however, expected to rise substantially once market reforms have been fully implemented. Cross-subsidies are also expected to be eradicated gradually.

The Russian government's *Energy Strategy* document envisages a continuing rapid increase in gas tariffs (Box 9.1 and Table 9.3). Russian gas prices increased by about 70% on average in real local currency terms between the beginning of 2000 and the first quarter of 2004, including an 18% rise at the start of 2004. In August 2004, the government approved a tariff increase of

Box 9.1: **Russian Energy Strategy**

The *Energy Strategy of Russia for the Period to 2020*, adopted by the Russian parliament in August 2003, sets out the government's strategic thinking about the evolution of the energy sector and provides a framework for future policy and regulatory actions. It forecasts trends in energy production and demand and identifies the main challenges facing the sector. These include: mobilising investments in production and export capacity; restructuring the gas, electricity and coal industries; limiting the social impact of energy-price rises; and improving energy efficiency.

The *Strategy* envisages large overall increases in fossil fuel and electricity output and it calls for a cut in the share of gas in the power-generation fuel mix in favour of coal, hydro and nuclear power. But the policies and measures necessary to achieve these outcomes are not specified in detail. The document simply acknowledges the need for reform in the areas of pricing and taxation, and the regulation of natural monopolies. It also calls for measures to promote energy efficiency and conservation. The *Strategy* adopts a multi-scenario approach to the outlook for demand and supply (Table 9.3).

23% for 2005. It plans to raise tariffs by a further 11% in 2006 and 8% in 2007. The percentage increases after 2005 are only slightly higher than expected inflation, so prices will scarcely increase in real terms. In May 2004,

Table 9.3: **Russian *Energy Strategy* and *WEO-2004* Projections to 2020**

	2002	2020 Energy Strategy (high and low scenarios)	2020 WEO-2004
Energy sector			
Primary energy demand (Mtoe)	619	794 – 881	802
Oil sector			
Production (Mt)	383	450 – 520	531
Exports of crude and products (Mt)	248	305 – 350	351
Gas sector			
Production (bcm)	584	680 – 730	801
Exports (bcm)	169*	275 – 280	249*
Power sector			
Electricity generation (TWh)	889	1 215 – 1 365	1 200

* Net exports.
Sources: Government of the Russian Federation (2003); IEA databases (2002 data).

Russia agreed with the European Union to raise its domestic gas tariffs in return for EU backing of Russia's entry into the World Trade Organization. The Union had argued that below-cost domestic tariffs represent a hidden trade subsidy. The Russian government promised to raise average gas prices to industry from the current $27 per thousand cubic metres to between $37 and $42 in 2006 and between $49 and $57 in 2010, about the same levels as foreseen in the *Energy Strategy*.

Our projections assume that average domestic gas prices will reach cost-reflective levels, equivalent to around $50 per thousand cm by 2014 and $60 in 2020 in real year-2000 dollars. After 2020, prices are assumed to rise in line with international prices (Figure 9.6).[4] Cross-subsidies between industrial and household consumers are assumed to be removed by 2020. The gradual removal of gas subsidies will temper the increase in electricity tariffs, as gas will remain the leading fuel for power generation. We assume that wholesale and retail competition will gradually develop and that retail electricity prices will be entirely deregulated by around 2015. By then, electricity prices will be fully determined by market forces. As a result, they are expected to treble in real terms from 2003 to 2030. The large share of gas in the fuel mix in power generation means that lingering downside distortions in gas pricing, especially

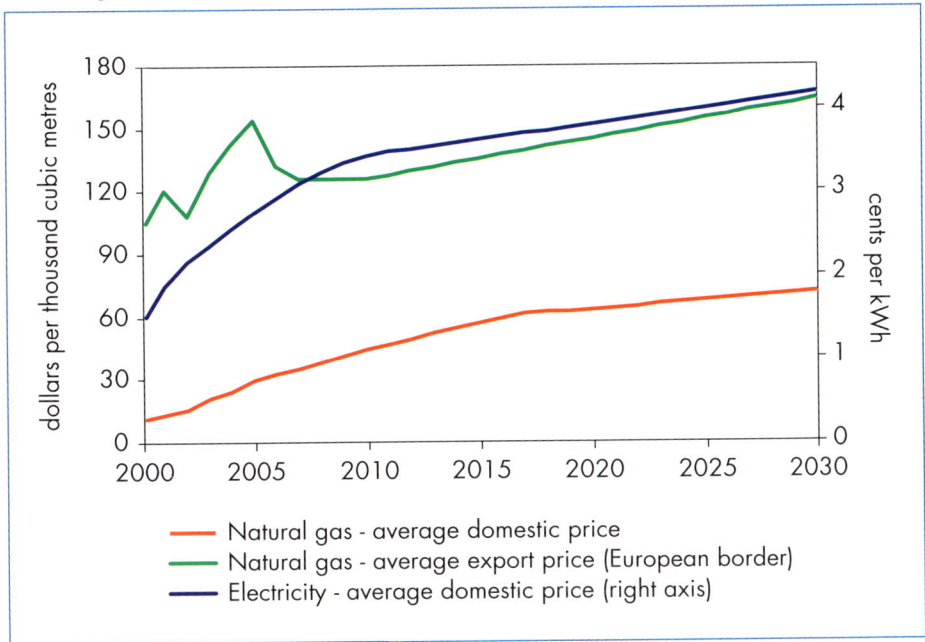

Figure 9.6: **Real Average Electricity and Natural Gas Prices in Russia**

Note: Prices are national averages for the household and industrial sectors.

4. Rouble prices will increase at the same rate as the exchange rate is assumed to be constant.

Chapter 9 - Russia – an In-depth Study

during the transition to a liberalised electricity market, will tend to skew investment decisions in favour of gas-fired capacity.

In 1998, the Russian government agreed in principle to co-operate with OPEC on production and export policies aimed at stabilising world-market prices at that time. In practice, this involved imposing some volume restrictions on oil exports by pipeline, but those limits were largely circumvented by oil shipment by truck and rail. They were quickly lifted once prices rebounded. The government has not officially indicated whether it would limit output again in the future, though President Putin declared in September 2003 that Russia *would* co-operate with OPEC in the event of a severe weakening of prices. The projections in this *Outlook* assume no intervention that would materially affect production, but such a move cannot be ruled out.

The *Energy Strategy* acknowledges the critical importance of energy efficiency and conservation in improving the competitiveness of the Russian economy and in freeing up supplies for export. A fall of 11% in the primary energy intensity of the Russian economy in 1992-2002 resulted primarily from the decline in heavy industry. It did not reflect any significant improvement in the technical efficiency of energy equipment and appliances. Higher prices are expected to provide the main stimulus for energy-efficiency improvements in the long term. Other efficiency and conservation measures, most of which come under the federal *Energy Efficient Economy* programme of 2001, are expected to contribute to energy savings and emissions reductions.

The potential for energy savings is undoubtedly huge, especially in industry, power generation and buildings (IEA, 2002a). But the slow replacement of capital stock will limit how quickly efficiency gains can be achieved. The installation of meters and thermostats for gas and heat supplies is critically important to conserving energy and promoting the deployment of more efficient boilers and heat plants. All regional energy-efficiency laws include requirements for compulsory metering, but the pace of meter installation varies widely. Other measures that are expected to be implemented include efficiency standards, energy labelling, building codes and energy audits in industry. But these measures are unlikely to have much impact on consumption so long as end-user prices remain subsidised. The effects of additional policies to improve energy efficiency are analysed in the World Alternative Policy Scenario (Chapter 11).

Energy Demand Outlook

Overview

Russia's primary energy demand is projected to grow at an average rate of 1.3% per year from 2002 to 2030 (Table 9.4). Growth is expected to be most rapid in the current decade, at 1.7% per year, and then to slow to 1% per year in the

2020s as the pace of economic expansion slackens. These projections are slightly lower than in the last edition of the *Outlook*, mainly because higher domestic prices, which dampen demand in buildings and industry, are assumed in this analysis.

Table 9.4: **Total Primary Energy Demand in Russia** (Mtoe)

	1992	2002	2010	2020	2030	2002-2030*
Coal	132	107	118	125	117	0.3%
Oil	221	128	149	171	199	1.6%
Gas	364	326	371	433	489	1.5%
Nuclear	32	37	45	47	48	0.9%
Hydro	15	14	16	17	17	0.7%
Other	12	7	9	10	15	2.7%
Total	**776**	**619**	**708**	**802**	**885**	**1.3%**

* Average annual growth rate.

Demand for oil is expected to grow faster than that for any other major fuel, at 1.6% per year. Oil's share of primary demand will increase by 1 percentage point, to 22%, in 2030 (Figure 9.7). This will reverse the trend of the last decade, when oil demand fell by more than 5% per year following a collapse in

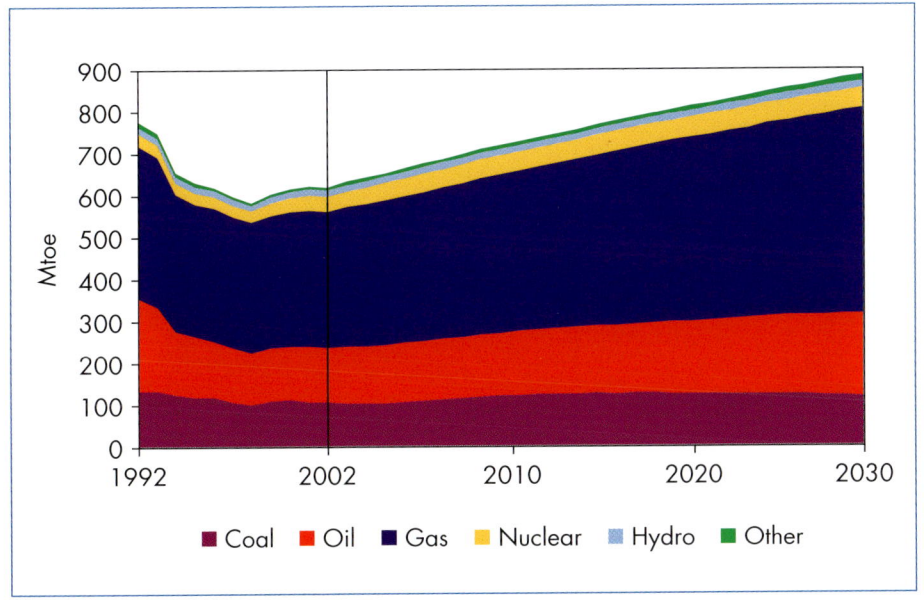

Figure 9.7: **Fuel Shares in Primary Energy Demand in Russia**

industrial consumption. Yet oil demand in 2030 will still be below its 1992 level. As in other countries, transport will account for most of the growth in oil demand. Natural gas, demand for which is expected to rise by an average 1.5% per year, will also see its share increase, from 53% to 55%. This increase will arise from the increased use of gas in power generation and the underlying growth in demand for electricity. In stark contrast to the official Russian vision, our *Outlook* sees demand for coal *declining* slightly, so that its share of total energy demand will drop from 17% in 2002 to 13% in 2030. The power sector will dominate coal demand even more in the future than it does today. We projected the share of nuclear power to rise only a little in the coming decade or so. It will decline gradually thereafter, on our assumption that only limited financing can be found for building new reactors to replace those that will have to be retired before 2030. The Russian *Energy Strategy* projects a substantial *increase* in nuclear power.

The fuel mix in *final* demand is expected to change much more dramatically than in primary demand over the projection period (Figure 9.8). Final demand also grows slightly less rapidly, reaching exactly its 1992 level in 2030. The share of district heat is projected to slump from 32% at present – by far the highest share in any large country – to under a quarter by 2030. District heat's share already fell by three percentage points over the ten years to 2002. Oil, gas and electricity will account for most of the decline in heat's share of final energy use.

Figure 9.8: **Total Final Consumption by Fuel in Russia**

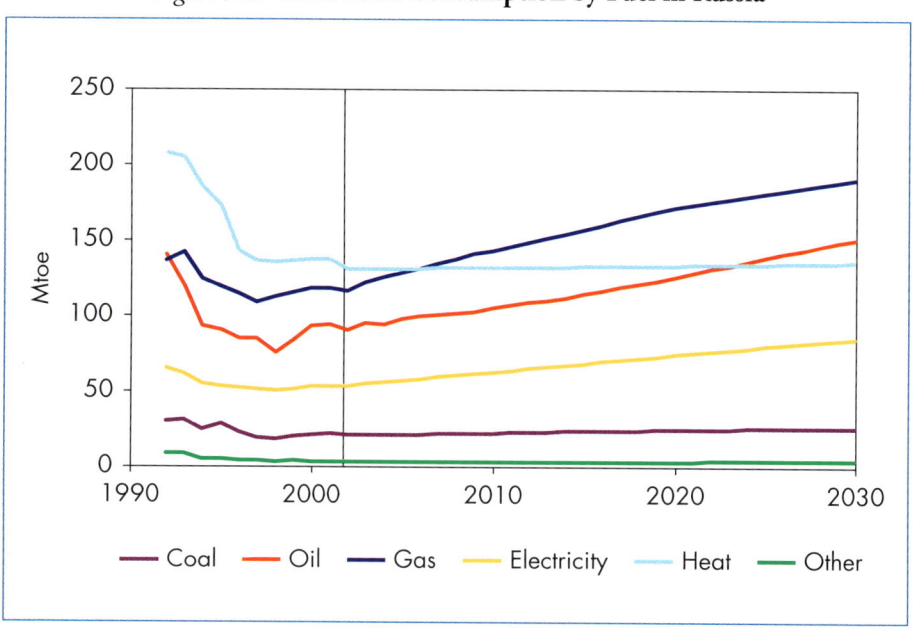

Primary and final energy demand will grow much less rapidly than GDP, resulting in a substantial fall in energy intensity (Figure 9.9). Primary energy intensity will fall by almost half from now to 2030, but will still be about two-and-a-half times higher than the OECD average. Per capita demand will increase in line with rising household incomes and economic activity.

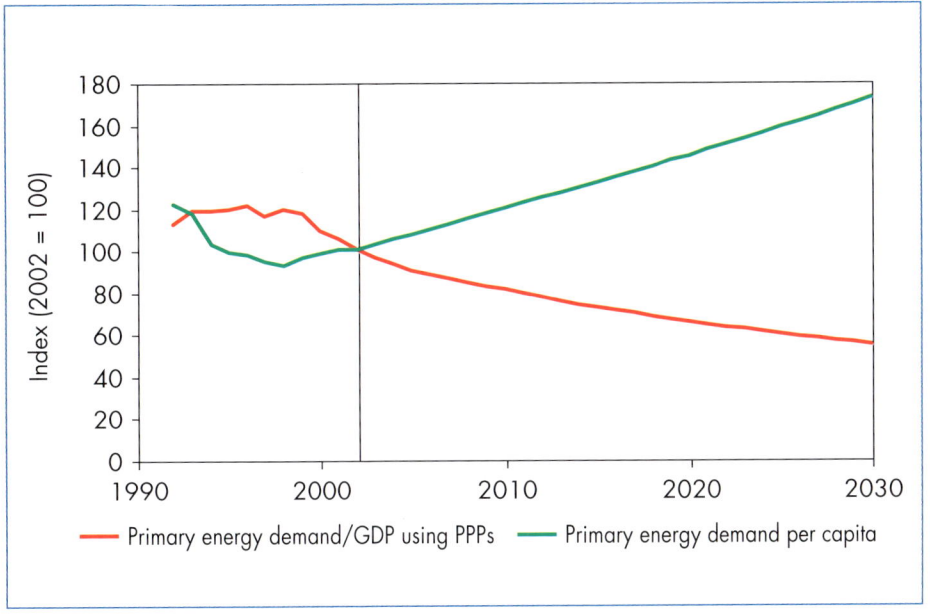

Figure 9.9: **Energy Intensity Indicators in Russia**

Sectoral Trends

The share of the power sector in total primary demand will drop sharply, from 56% to 51% over the projection period. Relatively brisk growth in final demand for electricity will be more than offset by large improvements in the thermal efficiency of power and heat plants as old, inefficient stations are retired.

Among end-use sectors, *transport* demand will grow most briskly, by 2.1% per year. Its share of total final consumption is projected to rise from 20% in 2002 to 25% in 2030. This compares with an increase from 29% to 32% over the same period in the European Union. Transport will account for about 37% of incremental final energy demand and 77% of incremental final oil demand over the period. Private car ownership, at 119 vehicles per 1 000 people, is very low compared with Poland's 259 and Germany's 542. Rising incomes will lead to increased car ownership and driving, as well as to more freight. This will offset any improvement in vehicle fuel efficiency. Air travel will also expand quickly, accounting for a growing share of total transport demand. Energy demand in the transport sector is much more closely correlated with GDP than demand in other end-use sectors (Figure 9.10).

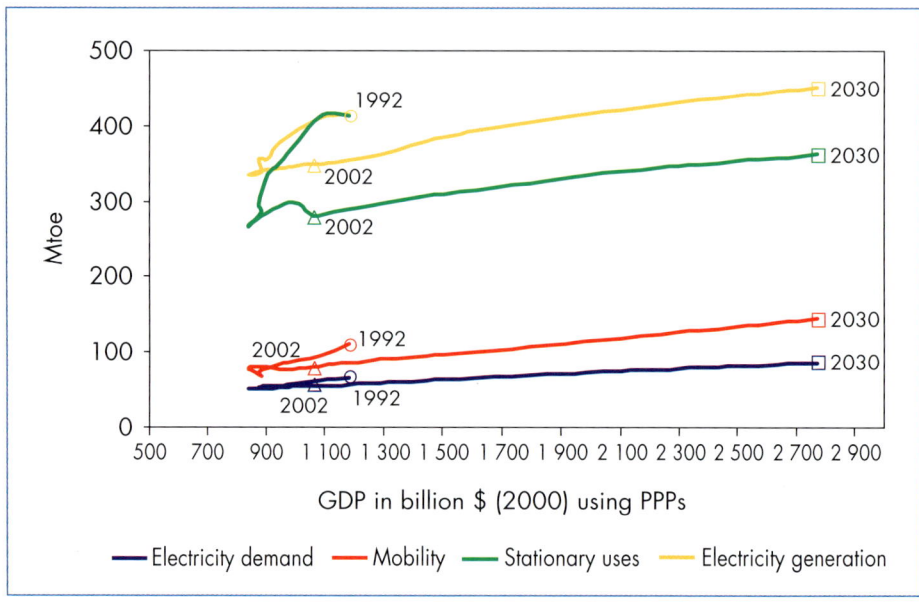

Figure 9.10: **Energy Services and GDP in Russia**

Industrial demand, which fell precipitously in the 1990s with the closure of inefficient, highly energy-intensive manufacturing plants, is expected to recover slowly. There is still plenty of scope for reducing energy intensity and improving the efficiency of energy use in specific end-uses in Russian industry. In most industrial sectors, the amount of energy used per dollar of value added is considerably higher than in the largest OECD countries, although energy intensity has fallen significantly since the early 1990s as the most inefficient plants have closed (Figure 9.11).

Higher energy *intensity* in Russian industry is in large part due to its relatively low level of energy *efficiency*, which in turn is due to outdated technology and waste. Over the years, low energy prices gave little incentive to firms to use more efficient technology and stamp out waste. In the iron and steel industry, for example, one tonne of output required 0.31 toe of energy input in 2002, compared with 0.17 toe in the United States, 0.12 toe in Germany and 0.10 toe in Japan (Figure 9.12). Russia's industrial energy intensity has, nonetheless, fallen by more than 10% since 1998.

Total final energy use in industry is projected to rise by more than 40% over 2002-2030, or 1.2% per year. This increase will be less than that of industrial

Figure 9.11: **Energy Intensity of Industrial Production in Selected Sectors and Countries, 2000***

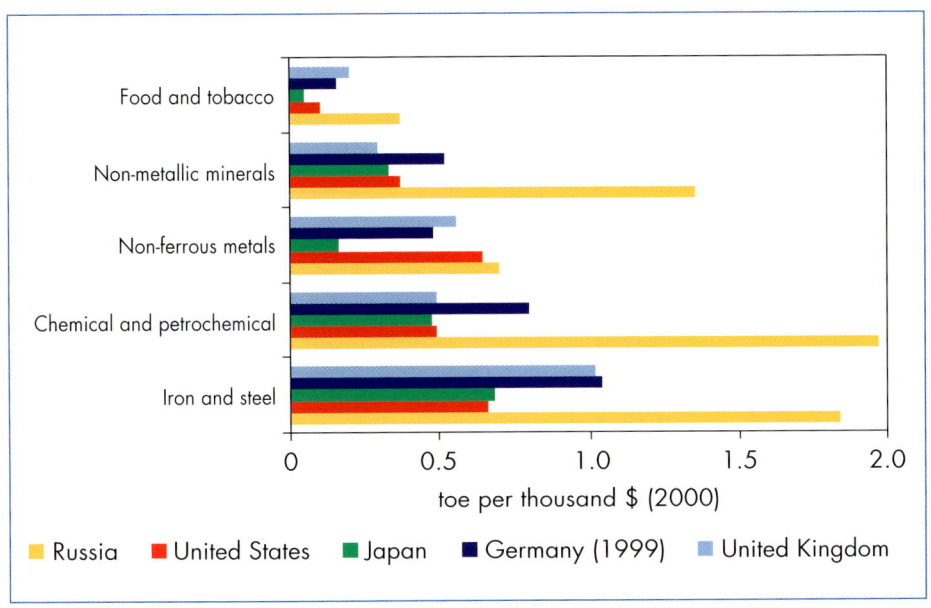

* Energy consumption in toe per thousand dollars of value added, adjusted for PPP.
Sources: IEA analysis; UNIDO database.

Figure 9.12: **Energy Intensity of Iron and Steel Production in Selected Countries***

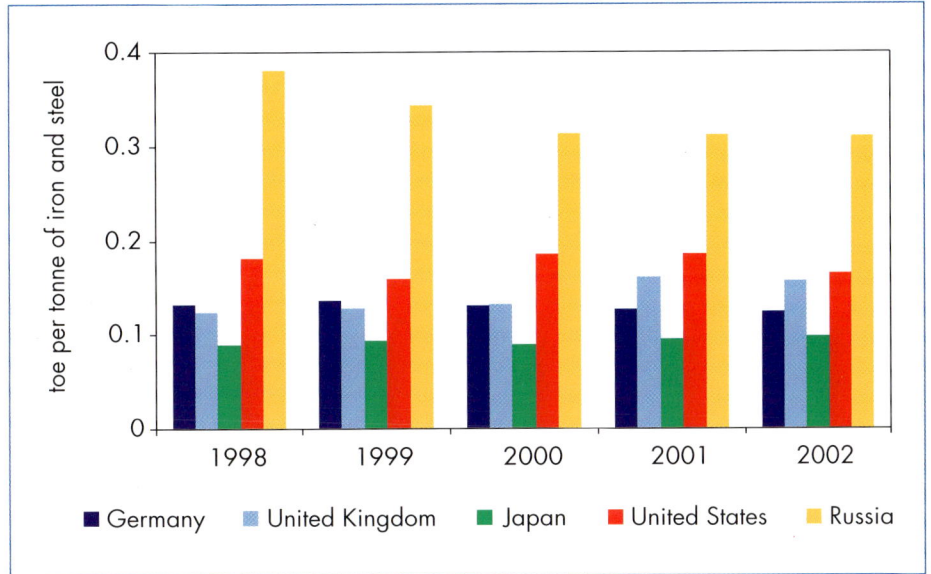

* Energy consumption per tonne of iron and steel production.
Source: IEA analysis; UNIDO database.

Chapter 9 - Russia – an In-depth Study

output, thanks largely to continuing improvements in the efficiency of producing direct process heat and in eliminating waste. A gradual shift toward lighter manufacturing will also contribute to an overall decline of about a half in the energy intensity of industry between 2002 and 2030. The share of gas and electricity in final consumption in industry will rise, at the expense of heat. Industry's use of district heat is expected to grow only very slowly, as manufacturers build more on-site co-generation plants and boilers.

The decline in the population, the gradual replacement of inefficient gas-fired and district-heating systems and better insulation of apartments and offices will limit the growth of energy use in the *residential and services* sectors. However, the addition of new buildings, particularly in the commercial sector, will offset these factors. Families and businesses are expanding their space requirements as the economy grows. Demand in these two sectors is projected to grow by 1% per annum from 2002 to 2030. Increases in natural gas and heat tariffs will have little impact on demand in the short term, as many households still have no means of adjusting their heat supply. Metering remains the exception rather than the norm. But the gradual replacement of the building stock and collective heating systems will provide an opportunity for more choice and efficiency. Most households are expected to be metered by 2030 and thermostatic controls are expected to become more widespread. In addition, improved insulation of existing buildings is expected to reduce heating needs. As a result, the share of district heat in total energy use in the residential and services sectors is expected to decline from 48% in 2002 to 42% in 2030. Natural gas supplied directly to buildings to fuel collective and individual boilers will win most of this market share.

Oil Supply Outlook

The fall-back in oil prices assumed in our Reference Scenario is expected to dampen investment in the upstream industry and growth in production capacity. The current tax system will accentuate the impact of lower prices on investment. In the short to medium term, inadequate export infrastructure will also limit production growth. Oil production is, nonetheless, expected to continue to grow, albeit at a slower pace than in the last five years, from 8.5 mb/d in 2003 to 10.4 mb/d by 2010. Production is then expected to rise much more slowly in the next decade, reaching 10.8 mb/d in 2030 (Figure 9.13). Domestic demand will increase less rapidly than production in the first decade or so. Net exports of crude oil and refined products are, therefore, projected to rise further, from 5.6 mb/d in 2003 to 7.3 mb/d in 2010. Exports will start to decline gradually soon after 2010, as domestic demand outstrips the increase in production. Production and export trends will remain highly sensitive to oil prices, costs and taxes.

Figure 9.13: **Russian Oil Balance**

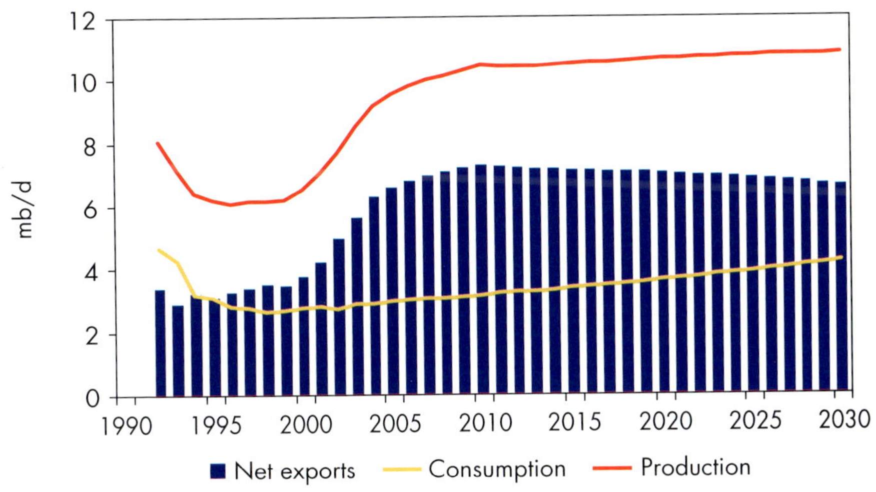

Resources

At the end of 2003, Russia had 69 billion barrels of proven oil reserves, or 6% of the world total (BP, 2004). Over 70% of Russian reserves and a similar share of current crude oil production are in West Siberia. The rest of the country's reserves are in the Volga-Urals region (14%), Timan-Pechora (7%), East Siberia (4%) and the Far East (3%). Figure 9.14 shows the location of these basins and existing oil pipelines. Russian proven reserves can sustain production at current rates for 22 years. The internationally audited[5] proven reserves of the six largest companies operating in Russia – Yukos, Lukoil, TNK-BP, Surgutneftegaz, Sibneft and Tatneft – amount to 62 billion barrels.

Ultimately recoverable resources are much larger. DeGoyler and MacNaughton, a leading auditor of Russian oil reserves, estimates proven, probable and possible (3P) reserves at 150 billion barrels (Brunswick UBS, 2004). A 2000 study by the US Geological Survey estimates undiscovered resources of oil and natural gas liquids that are expected to be economically recoverable at 115 billion barrels, or 12% of total world undiscovered resources. IHS Energy (formerly Petroconsultants) put Russia's resource potential at 140 billion barrels at the end of 2001.

---

5. In response to concerns among foreign investors about the reliability of Russian reserves data, most leading Russian oil companies have had their estimates audited by independent auditors using consistent and internationally accepted methodologies. As a result, their reserves data were revised down, but are now generally considered to be of good quality.

Figure 9.14: Russian Oil Basins and Pipelines

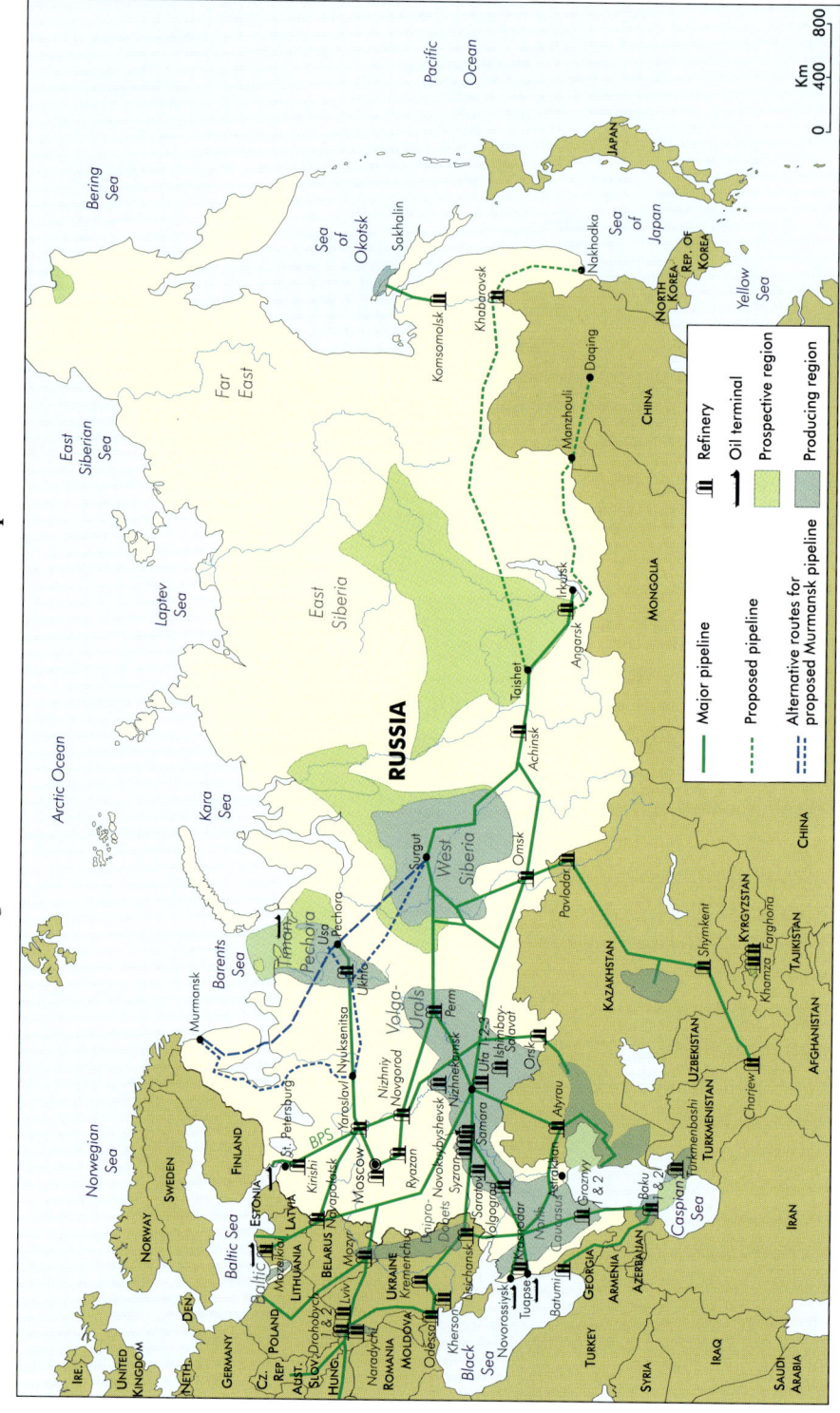

To profit from high prices, Russian oil companies have tended to focus their attention on raising production from working fields rather than replacing proven reserves through exploration. The tax system has also favoured short-term investments over long-term ones. According to official Russian data, only 60% of reserves were replaced in 2003 and 70% in 2002.

Crude Oil Production

The last five years have seen a dramatic turnaround in Russian oil production. Causes include higher prices, the 1998 rouble devaluation, a surge in investment and the adoption of more modern technology and management practices. Production almost halved between 1987 and 1996, reaching a low of 6.1 mb/d, largely as a result of lower investment by domestic companies after the break-up of the Soviet Union. Production began to recover in 1999, reaching an average of 8.5 mb/d in 2003 and over 9.3 mb/d in August 2004. Output is expected to average 9.2 mb/d in 2004, an increase of 8% over 2003 and of 50% over 1998 but equal to the 1991 level.

There is enormous uncertainty about whether the recent pace of production growth can be sustained. Official and private projections – including those of past *WEOs* – have consistently underestimated the strength of the rebound in Russian oil-production growth. Much of this growth has come from rehabilitating and stimulating existing wells to enhance the recovery of reserves. Yukos and Sibneft have relied mainly on boosting well productivity to increase output. But drilling of development wells has also picked up. The total number of wells in operation is now close to the number in 1990, despite a number of recent well closures. Average well productivity has rebounded, from 51 barrels per day in 1996 to 66 b/d in mid-2003, though it remains far lower than in most other producing countries. But the water cut – the share of water mixed in with the oil extracted – has reportedly resumed its long-term upward trend, after falling back for a short period. The water cut averaged 82% in 2000, compared with 76% in 1990 and 70% in 1986.

The introduction of advanced production technologies and modern management practices has helped to raise productivity and boost output. Higher prices have led to strong cash flows, which have helped finance a surge in investment and made possible partnerships with international oil and oil-service companies. At $7.7 billion, total capital expenditure in the upstream oil industry in 2003 was more than three times higher than in 1999. Most investment is going to West Siberia, much of it into boosting output at already operating fields.

The pace of production growth is expected to slow in the near term as most low-cost opportunities to boost output – the "low-hanging fruit" – have now been exploited. Capacity additions will increasingly need to come from new greenfield developments in West Siberia, including some large fields that were overlooked during the Soviet era because of poor technology. Later, attention may shift to less mature basins such as Timan-Pechora and to frontier areas such as East Siberia, the Pechora Sea, the Russian sector of the Caspian Sea and the Far East. Development and production costs for these projects are likely to be considerably higher than for existing brownfield projects in West Siberia, because of a lack of infrastructure and more difficult geological and operating conditions. The average investment needed per barrel of capacity stands at around $13 000, which is higher than in most other parts of the world (IEA, 2003a).

Recent tax changes, including the outlawing of schemes to minimise tax liabilities, coupled with rising costs, mean that most oil companies make little additional profit when oil prices rise above $25 a barrel. An increase in export duties, which took effect in August 2004, and further changes in the tax code due for the start of 2005 will increase the oil companies' tax burden. But the increase will be smaller than it would have been had the reforms proposed before the April 2004 presidential election been adopted. The main change is an increase in export duties on crude oil and refined products, which are linked to crude oil prices on the international market. Overall, the changes could increase tax revenues by 6% to 7% at a Brent crude price of around $25/barrel in 2005 (UFG, 2004). The system will remain largely revenue-based. This will simplify the calculation of taxes but will reduce incentives to invest in capital-intensive long-term projects. Following a change in policy in 2003, production-sharing agreements (PSAs) are no longer expected to play a significant role in Russia, except for three that had already gone ahead and perhaps some very large new projects.

Stricter enforcement of existing tax rules could increase revenues even more than the changes in the effective rates. The government's Economic Expert Group estimates that the oil companies have underpaid their taxes, especially corporate profit taxes, since a new tax code came into effect in 2002. The government has claimed $4.1 billion in tax arrears for 2001 and fines from Yukos, the leading Russian oil company. Its former chief executive officer has been imprisoned and put on trial for tax fraud and other offences.

Changes in the upstream licensing regime are also on the cards, but details had not been agreed as of mid-2004. The new Russian minister of natural resources has indicated that he will give priority to clarifying existing regulations and amending licence terms rather than simply revoking licences which have not been strictly adhered to. The oil companies had feared a wholesale reallocation of licences. A new law addressing the way licences are issued is planned, addressing deficiencies in the current system. These include a lack of transparency and excessive scope for government officials to take arbitrary decisions. The new arrangements could give producers an automatic right to develop reserves that they discover. At present, companies do not have the right of first refusal on reserves that they discover, a provision which discourages exploration.

The crude oil pipeline operator Transneft has so far been unable to reach an agreement with producers on the introduction of a "quality-banking" system, which would take account of the different qualities of crude oil fed into its pipelines and compensate producers of higher-value oil. As things stand, there is no financial benefit for producing better-quality crude oil. Almost all production is blended in-pipe.

Refining Capacity and Production

The Russian oil-refining industry continues to suffer from deep-seated structural problems: notably large overcapacity in primary distillation; a lack of secondary processing and upgrading capacity; and poorly located plants. All these are legacies of Soviet central planning. Total annual crude oil distillation capacity amounts to about 270 million tonnes, or 5.5 mb/d – 6% of the world total and almost double current domestic demand. Because of lack of investment, thermal cracking capacity is equal to only 7% of primary distillation capacity, a much lower proportion than in refineries in Western Europe. As a result, the light-to-medium product yield averages only about 60% - well below the average for the rest of the world. Many refineries are located in oil-producing regions, a situation which leads to large regional supply imbalances and high transportation costs.

Refinery output has recovered since 1998, partly as a result of increased export demand. In 2002, output was 16% higher than five years earlier. Higher international prices have made it profitable for many refineries to export part of their output. Rising domestic demand for gasoline and other light products will call for large investments in upgrading capacity, averaging around $700 million per year (in year-2000 dollars) through to 2030.

Export Prospects

Russian exports of crude oil and refined products are limited by a lack of transportation and terminal capacity. Net exports of crude oil, which amounted to 3.5 mb/d, or just under two-thirds of total Russian oil exports in 2003, are mostly shipped by pipeline – either cross-border or to sea terminals. The crude oil pipeline system – built during the Soviet era and controlled entirely by Transneft, a state monopoly – is far from fully utilised. Large parts of the system, including some lines in Latvia and Ukraine, are no longer used because patterns of supply have shifted. But there are still bottlenecks along some export routes. Congestion is particularly acute at the Black Sea ports, Novorossisk and Tuapse, in part because of restrictions on tanker traffic through the Turkish Straits. Export capacity via the Baltic Sea was increased in 2003 with an expansion of the Baltic Pipeline System (BPS) and the commissioning of a new terminal at Primorsk. Crude oil exports by rail have risen sharply as a result of pipeline bottlenecks. Most product exports are already transported by train, barge or road-truck. Net product exports, mostly gas oil and heavy fuel oil, jumped from 1.2 mb/d in 2000 to an estimated 2.1 mb/d in 2003, of which two-thirds were transported by rail.

New infrastructure will have to be built if there is to be further growth in oil exports. In the near term, the following projects will provide additional capacity:

- The second stage of the Baltic Pipeline System expansion was completed in early 2004. It will allow a further increase in exports via Primorsk, from 840 kb/d to around 1 mb/d. Transneft is waiting for government approval to further increase capacity to as much as 1.24 mb/d.
- Lukoil's Vysotsk terminal near Saint Petersburg started operating in June 2004. Initial capacity for crude and products will be about 100 kb/d (5 Mt/year), rising to 240 kb/d (12 Mt/year) in 2005.
- The Klaipeda product terminal in Lithuania will be able to handle 50 kb/d (2.5 Mt/year) of crude oil once an imminent storage upgrade is completed.
- A 180-kb/d pipeline from Odessa to Brody near the Ukraine-Poland border, built in 2002 to export Caspian crude to Europe, has been reversed and will be used to export Russian crude via the Black Sea. First exports are expected in October 2004.
- The Adria pipeline, which runs from the Adriatic Coast in Croatia to the Druzhba line in Ukraine, is being reversed to allow Russian crude oil to be exported via the Mediterranean. In February 2004, the Ukrainian parliament approved the project, which is expected to be completed by late 2004 or early 2005. The line's capacity will be 100 kb/d (5 Mt/year), but that figure could be raised to as much as 300 kb/d (15 Mt/year) at a later stage.

There are a number of other major pipeline projects that have been proposed and are awaiting government approval. Transneft has completed a feasibility study, financed by a consortium of private companies, on a 1.5 to 3 mb/d (75 to 150 Mt/year) pipeline to a new export terminal near Murmansk on the Barents Sea. This line would allow exports on large tankers, providing access to the US market. The government has indicated that 2008 would be the earliest date for commissioning the line. Rosneft, the only oil-producing company that is still entirely state-owned, recently commissioned a small port at Arkhangelsk, from which oil is shuttled to floating storage facilities moored near Murmansk. This port could be expanded to allow Rosneft to export new production from the Barents Sea.

Question marks remain over two planned projects to export oil from East Siberia, where the current pipeline system ends, to markets in the Far East.

A project sponsored by Yukos to build a line from Angarsk to Daqing in China was given the go-ahead, but has stalled because of the company's financial difficulties. The capacity would be 400 kb/d initially, rising to 600 kb/d by 2010. The line would cost around $4 billion. Another project would involve the construction of a 1.6 mb/d line from Taishet to Nakhodka, a port on the Russian Pacific coast, at a cost of $10 billion. The line may be subsidised by the Japanese government through soft loans. The oil would be exported to international markets, notably Japan. A spur line to Daqing is also an option.

Industry Structure and the Role of the State

There has been a profound shift in relations between the oil industry and the state since 2003. The president and the government are reasserting state control over the sector and taming the power and influence of the so-called "oligarchs" – the handful of rich individuals who own and run the companies that emerged from the controversial privatisations of the 1990s. The tax case against Yukos, which threatens to bankrupt the company, has increased worries among the oligarchs that the government will challenge their rights to assets acquired under the privatisation programme. The government has indicated that private crude oil pipelines will not be permitted and that it intends to keep Transneft, the pipeline monopoly, in state hands. In September 2004, the government announced a plan to merge Rosneft into Gazprom, giving the government a majority stake in the group. The government is proceeding with a sell-off of its residual 7.6% holding in Lukoil. Taxes on producers are being increased and alleged malpractices are being pursued more vigorously.

The Yukos affair has revived fundamental concerns about property rights and the independence of the judicial system, harming the business climate and increasing investment risk. Yet foreign interest in investing in Russia's private oil companies remains high. Several foreign oil companies have expressed interest in acquiring stakes in Russian companies, including Sibneft, Yukos and Lukoil. The outcome of the Yukos affair will have a major impact on how the industry evolves and on the interest of foreign oil companies in investing in it in the near term.

Further industry restructuring is likely. The 2003 merger of Yukos and Sibneft is now being unravelled, but consolidation is still the norm in the rest of the industry. Independent upstream companies continue to be absorbed by the vertically-integrated Russian majors. This trend is supported by a tax system that favours larger companies. Small companies lacking their own refineries are forced to sell their crude oil to the vertically-integrated majors. The share of small producers[6] in total Russian crude oil output has fallen from around 9.5% in 1998 to 6.5% in 2003.

---

6. All companies other than Yukos, Lukoil, TNK-BP, Surgutneftegaz, Sibneft, Tatneft, Rosneft, Slavneft, Bashneft and Gazprom.

Gas Supply Outlook

Russia's huge gas resources are expected to underpin a continued increase in production, to meet a gradual rebound in domestic demand and to provide increased exports to Europe and new markets in the East. We project production to rise from an estimated 608 bcm in 2003 to 655 bcm in 2010 and 898 bcm in 2030 (Figure 9.15). Net exports are expected to rise from 169 bcm in 2002 to 182 bcm in 2010 and 274 bcm in 2030. This projection takes account of increased imports from Central Asia, which will make possible higher exports to Europe. Russia will still be the world's biggest gas exporter at the end of the projection period. Higher production will, however, call for considerable investment in greenfield projects to replace declining output from super-giant fields that have been in production for decades. Securing the necessary financing will depend on market reforms, particularly the elimination of domestic price subsidies, and easier access for independent producers to the national transmission system operated by Gazprom, the dominant gas company (IEA, 2003a).

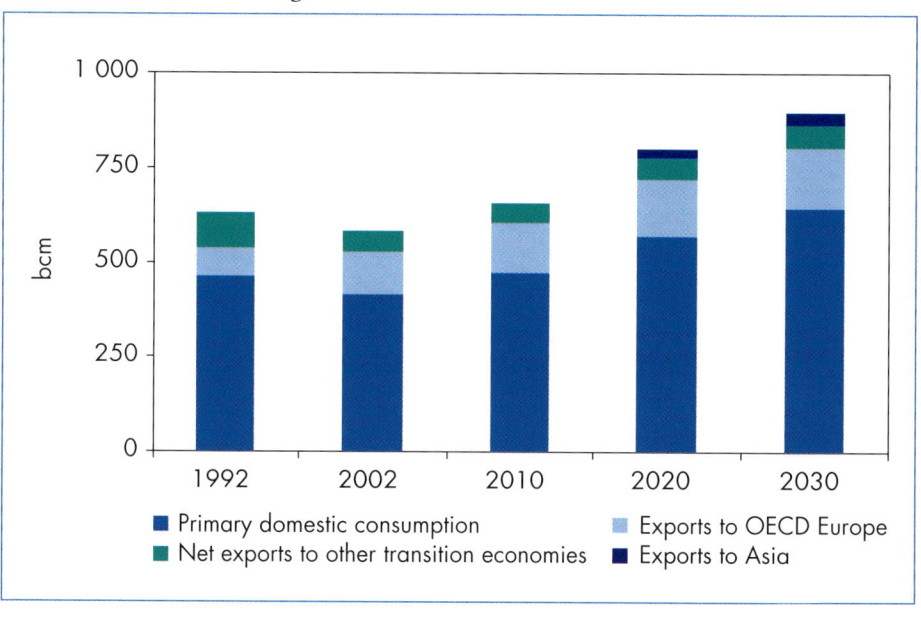

Figure 9.15: **Russian Gas Balance**

Resources and Production Trends

Russia's gas resources are huge. It has 47 trillion cubic metres of proven natural gas reserves, 26% of the world total (Cedigaz, 2004). Gazprom holds the licences to fields holding 55% of these reserves; other producers hold 28%, while 17% are unallocated. Three-quarters of Russian gas reserves – and a

Box 9.2: **Profile of Gazprom**

> Gazprom is the world's largest gas company. It plays a central role in the Russian economy, providing up to a quarter of federal government tax revenues. It accounts for almost 90% of Russian gas production and owns and operates the national network of high-pressure inter-regional gas pipelines, which, at over 150 000 km, is the longest in the world. It is the sole owner of gas storage sites in Russia, operating 22 underground facilities. Gazprom's role in local distribution has risen markedly since the mid-1990s, as it acquired stakes in smaller companies facing financial difficulties. Gazprom has a monopoly on all gas exports outside the Commonwealth of Independent States (CIS) and holds a monopoly on gas processing in Russia, making it the sole buyer of the wet gas produced by Russian oil companies and independent gas producers. Over the years, the company has acquired a vast array of holdings in such sectors as banking, insurance, agriculture, media and construction. It is committed to disposing of many of its non-core assets, but this is proving a slow process.
>
> Though constituted as a joint-stock company, Gazprom operates in many ways as an arm of the state, combining commercial and regulatory functions and maintaining tight control over the sector's infrastructure and over information flows within it. A majority of the shares in the company was sold to private investors in the 1990s. But the state still holds 38% directly and another 16.6% indirectly, giving it majority control of the board.

similar share of current production – are in West Siberia, and most of these are in the Nadym-Pur-Taz region (Figure 9.16). European Russia (including the Barents Sea shelf) accounts for 16% and East Siberia and the Far East together for the remaining 9%. Some 20 giant fields have been discovered, each with more than 500 bcm in reserves, making up three-quarters of total Russian reserves. Only seven of these fields have been brought into production. Reserves are equivalent to about 81 years of production at current rates. In addition to proven reserves, there are an estimated 33 tcm of undiscovered gas resources[7] (USGS, 2000).

Russian gas production fell sharply in the 1990s in response to the collapse in domestic demand following the break-up of the Soviet Union, from a peak of 632 bcm in 1991 to a trough of 561 bcm in 1997. Production has since recovered, largely thanks to rising exports. Output amounted to an estimated 608 bcm in 2003, of which Gazprom produced 540 bcm. The bulk of Russian gas production comes from three super-giant fields in Nadym-Pur-Taz that have

---

7. Mean estimate.

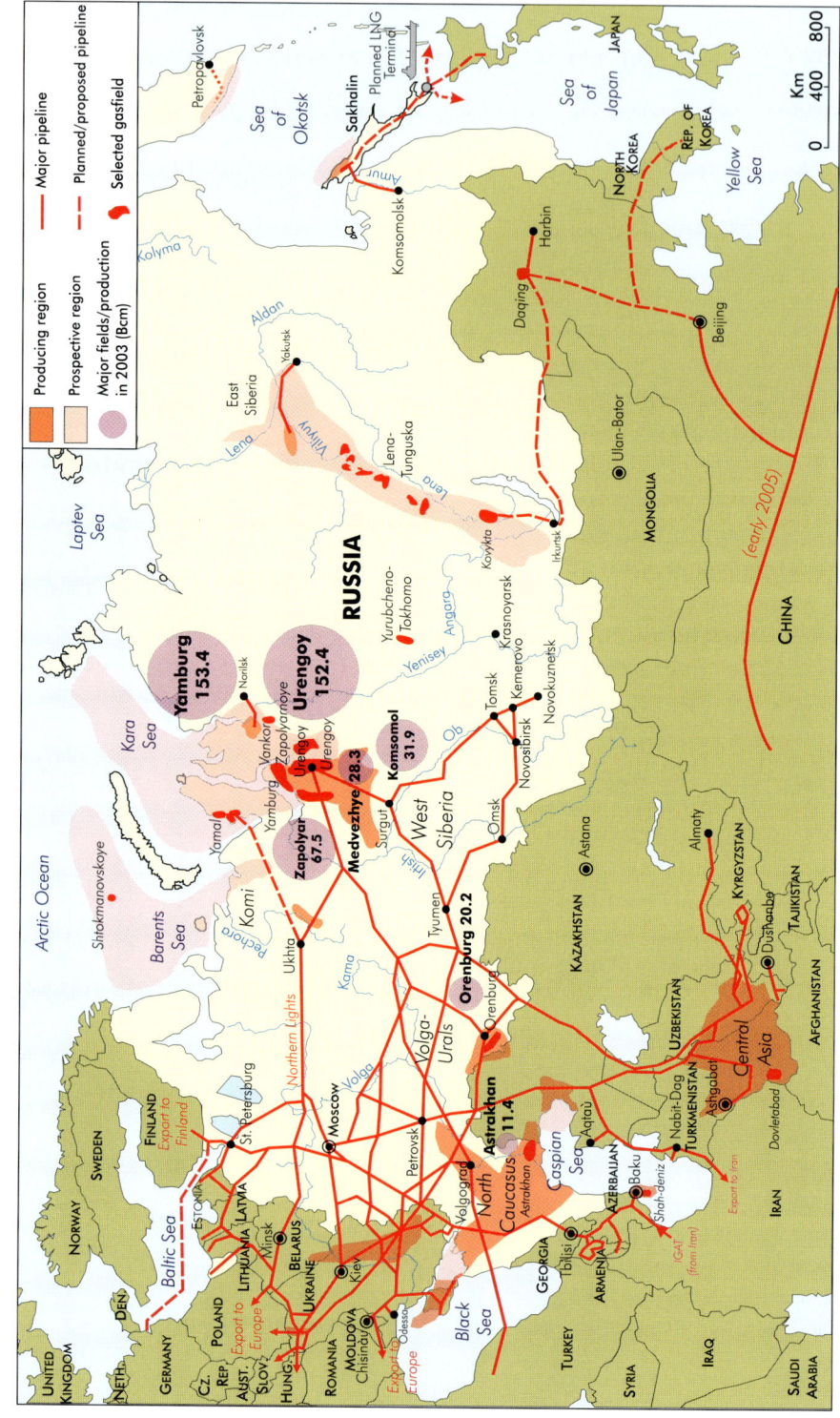

Figure 9.16: Major Gas Reserves and Supply Infrastructure in Russia

been in production for many years and are now in decline: Medvezhye, Yamburg and Urengoye. Rising output from a fourth super-giant field, Zapolyarnoye, which started producing in 2001, is expected to compensate for much of the decline in production at the other super-giant fields over the next few years. Sustainable peak production of around 100 bcm from Zapolyarnoye is expected to be reached in 2008, once a third gas-processing plant is brought on stream.

In the next few years, Gazprom and independent producers will need to bring several new fields on stream in existing producing areas to stem the decline from the three old super-giants. Gazprom expects production from those fields to fall by 7% to 8% per year over the rest of the current decade. If this forecast proves accurate and decline rates remain the same beyond 2010, their combined output would fall from 334 bcm in 2003 to about 200 bcm in 2010 and less than 100 bcm in 2020. Gazprom prefers investing more in new fields rather than trying to sustain production levels at the super-giants – despite the success of past efforts to slow the rate of production decline at the Medvezhye field. The company is giving priority to developing a number of smaller fields in the vicinity of the super-giants which will be able to use spare capacity in the pipeline system running from Nadym-Pur-Taz. These include Pestsovoye, Yen-Yakhinskoye, Yuzhno-Russkoye and shallow-water fields in the Ob-Taz Gulfs. Yuzhno-Russkoye is ear-marked to fill the planned North European Pipeline to Germany and beyond, Gazprom plans to invest over $7 billion in 2004, rising to $9 billion in 2006, most of it in upstream projects.[8]

Contrary to earlier official expectations, there is unlikely to be a need for Gazprom to bring new fields on the Yamal Peninsula into production before the beginning of the next decade and possibly not before the middle of the decade. The timing of these developments will depend partly on Gazprom's imports of Central Asian gas. Yamal's reserves exceed 10 tcm, but the climate and terrain are harsh, and development costs are high (IEA, 2001). New pipelines would need to be built, but costs could be minimised by connecting the fields to the existing pipeline system to the south. The first Yamal fields to be developed will probably be Bovanenkovskoye and Kharasavey, with the potential to produce a total of 150 to 180 bcm a year at plateau. Gazprom is also planning to develop, with foreign partners, the Shtokmanovskoye field in the Barents Sea, which holds 2.5 tcm of proven reserves. Peak production from this field, which may be used for LNG exports, is estimated at 70 bcm a year. But development costs are very high, possibly exceeding $20 billion. We do not expect first gas before 2020, even though production is officially expected to begin in 2010.

---

8. The Ministry of Economic Development and Trade (MEDT) is critical of Gazprom's investment plans. The ministry wants Gazprom to reduce investments and adopt a balanced budget, probably on the grounds that Gazprom's upstream investments are less efficient than those of independent producers.

Oil companies and independent gas producers, who hold around a third of gas reserves, are expected to make a growing contribution to Russian gas production in the coming decades. They already account for an estimated 13%, all of it sold to domestic customers. Several companies are seeking to boost production, much of it associated with oil. Company projections imply that total non-Gazprom output could reach 270 to 290 bcm by 2015 – about a quarter of total Russian gas production (Table 9.5). Such a big increase is, however, unlikely. The *Energy Strategy* projects non-Gazprom production at between 105 and 115 bcm in 2010 and between 140 and 160 bcm in 2020. The prospects for independent production depend critically on transparent and reliable access to Gazprom's gas-processing capacity and transmission system. Large volumes of gas produced by oil companies are still being flared because Gazprom declines to buy it or because the terms of access to processing plants and the network are uneconomic.

Table 9.5: **Russian Gas Production of Non-Gazprom Companies** (bcm)

	2003	Company expectations 2010-2015
Oil companies	**40.4**	**195-215**
Surgutneftegaz	13.9	25
TNK-BP	6.8	20-40
Rosneft	7.1	50
Yukos	5.7	50
Lukoil	4.7	50
Other	2.1	–
Independents*	**35.9**	**75**
Novatek	21.0	52
Nortgaz	5.0	11
PSAs (including Sakhalin)	0.2	12
Other	10.7	–
Total	**76.3**	**270-290**

* Expectations of independents are all for 2010 (based on *Energy Strategy*).
Sources: IEA estimates; company reports; Government of the Russian Federation (2003).

Central Asian gas is expected to play an increasingly important role in meeting Russia's domestic needs, as well as Gazprom's export commitments to Europe. Gazprom has signed deals with Turkmenistan, Kazakhstan and Uzbekistan to import gas. The most important of these deals, with Turkmenistan, provides for annual imports of 5 to 6 bcm in 2004, rising to a plateau of as much as 80 bcm over 2009-2029. The gas, priced at an estimated $29 per thousand cubic metres, will be paid for in cash and in bartered gas equipment and services through to 2006. These arrangements will allow Gazprom to delay the development of its own expensive reserves in Yamal and Arctic regions. They will also reduce Gazprom's need to buy gas from independent Russian producers. Furthermore, they will effectively eliminate Central Asian producers as competitors for sales to Europe and other export markets, since most of their production will go to Russia. However, it is uncertain whether these deals will proceed as planned.

Export Prospects

Russian exports gas exclusively to other CIS countries and Europe. In 2003, Russia exported 119 bcm to OECD Europe. Gazexport, a wholly-owned subsidiary of Gazprom, is the sole exporter to Europe. Other non-Gazprom companies export Russian gas to other CIS countries. Rising gas demand in Europe is expected to remain the primary driver of Russian gas exports over the projection period, although Asia will emerge as an important new market. Exports to the European Union are projected to climb to 137 bcm in 2010 and 155 bcm in 2030. Exports to Asia are expected to reach 30 bcm by 2030. These projections are extremely sensitive to the rate of growth in domestic gas demand.

Increased exports to Europe will require substantial additions to pipeline capacity. Existing capacity can meet projected export needs through to the end of the current decade and probably well beyond in the case of Turkey, where demand is growing much more slowly than was previously envisaged. Completion of the Yamal-Europe Pipeline by 2005 will increase capacity through Belarus and Poland to Germany, but plans for a parallel line have been put on hold. Debottlenecking could also delay the need for new projects. Gazprom and a number of international oil and gas companies are nonetheless considering building a 20-bcm/year North European pipeline, which would run under the Baltic Sea from the Russian coast near Saint Petersburg to the German coast, and possibly on to the Netherlands and the United Kingdom. We do not expect the line, which will cost an estimated $5.7 billion, to be built until after 2010.

Gazprom is also looking into the possibility of developing LNG exports based on reserves on the Yamal peninsula and in the Shtokman field in the Barents Sea. But the costs would be very high, because of the extremely harsh climate. For this reason, we have assumed that no LNG projects other than Sakhalin-2 proceed before 2030.

Gas exports to Asia in the form of LNG from Sakhalin-2 are due to start in 2007. The project, owned by a foreign consortium led by Shell, involves the development of an offshore gas field and the construction of a two-train liquefaction plant with a capacity of 9.6 million tonnes per year. Gas will also come from an adjacent oil and associated-gas field. Total investment will amount to around $9 billion. We expect gas exports from the less advanced Sakhalin-1 project, led by Exxon-Mobil, to start in the 2010s.

Pipeline exports to Asia are expected to begin during the second decade of the projection period. Rusia Petroleum, owned by the TNK-BP joint venture[9], holds a licence to develop gas reserves at the Kovykta field near Irkutsk in East Siberia. It plans to develop the field and build a pipeline to export the gas to China and Korea. According to a feasibility study completed in late 2003, a 34-bcm/year pipeline system, with branches to Dalian and Beijing and an undersea line to South Korea, would cost in the region of $12 billion and upstream development another $5 billion to $6 billion. Gazprom was appointed by the Russian government in 2001 to co-ordinate all East Siberian export projects. It has recently sought greater control over the Kovykta project and has proposed exporting gas from Chayandinskaya, another East Siberian field, if agreement on Gazprom's participation in the Kovykta project with Rusia Petroleum is not forthcoming.

Market Reforms

The domestic gas market in Russia is not a real market at all, but rather a rationing mechanism operated by Gazprom. The company and the government negotiate a gas-supply balance for the country one year ahead, allocating the quantity of gas that Gazprom must supply to domestic consumers at artificially low regulated prices. Since export prices are much higher, Gazprom has every incentive to keep its domestic deliveries as low as possible. Any extra gas that industrial customers or power generators need must be purchased at higher prices, either from independent producers or from Gazprom itself. The government accepts the need for tariff reform and plans to raise prices gradually to full-cost levels. In 2004, the gas-export duty was increased to 30%, a move that the European Union requested in order to reduce the gap between Russian and European gas prices.

Although the principle of third-party access to pipelines is established in law, Gazprom has the ability and the motive to discriminate against other producers. The company is required to grant access only if there is sufficient capacity available in the system. It may refuse access on technical grounds, such as the quality of the proposed gas. But a lack of transparency makes it impossible to assess whether Gazprom is justified in refusing access. Until recently, few non-Gazprom producers in Russia were able to negotiate profitable deals with other buyers and so were obliged to sell their gas directly to Gazprom or flare it. Gazprom claims that the volumes it ships for third parties have increased from

---

9. In 2003, BP announced an equity investment of $6.75 billion in TNK, creating a new company, TNK-BP, Russia's third-largest oil company.

28 bcm in 1998 to an estimated 110-120 bcm in 2004, though this includes gas owned by Gazprom-affiliated companies and gas from Central Asia. Gaining the right to sell directly to end-users and to contract with Gazprom for transportation services on a cost-plus basis would allow independent producers to seek better pricing terms and give them stronger guarantees of future revenues.

The government has been considering for several years how best to deal with these problems. The Ministry of Economic Development and Trade has prepared several sets of reform proposals, including the restructuring and break-up of Gazprom. The most radical plan calls for the unbundling of Gazprom's gas transportation business and central dispatching unit into 100%-owned subsidiaries, the introduction of full wholesale and retail competition and the removal of domestic price controls by 2008-2010. Agreement has not yet been reached, however, partly because of strong resistance from Gazprom itself, which argues that its organisational integrity is critical to the smooth functioning of the nation's gas-supply system. Following his re-election in April 2004, President Putin publicly cautioned against reforming the sector too rapidly and breaking up Gazprom. Nonetheless, Russia has reportedly agreed, as part of its WTO-related deal with the European Union, to take action to make access to the pipeline network easier for third parties. Gazprom itself has also announced that it will unbundle its production, transportation, gas-processing, storage and distribution functions in early 2005 in order to improve operational efficiency and transparency. Gazprom will, however, retain its monopoly over all these functions, as well as over exports to Europe.

Our projections assume that, whatever precise system is eventually adopted, the terms and conditions for non-Gazprom producers will gradually improve. This will strengthen their incentives to invest in new upstream capacity and enable them to compete with Gazprom in supplying domestic markets. In the longer term, the rising cost of developing its own reserves is expected to encourage Gazprom to facilitate third-party access in order to help meet its domestic and export supply obligations.

Coal Supply Outlook

Russian coal production is projected to grow slowly over the projection period, from around 240 Mt in 2002 to 276 Mt in 2030 (Figure 9.17). This is well below the peak of 437 Mt achieved in 1988. Incremental output will go almost entirely to domestic markets, as infrastructure constraints and strong competition from other lower-cost producers will keep exports flat – bucking the strongly rising trend of the past few years. Net exports jumped from 5 Mt in 1998 to an estimated 27 Mt in 2003. Demand from the power sector holds the key to the Russian coal-supply outlook. Stronger government support for coal use in that sector could enhance the prospects for overall coal demand and production.

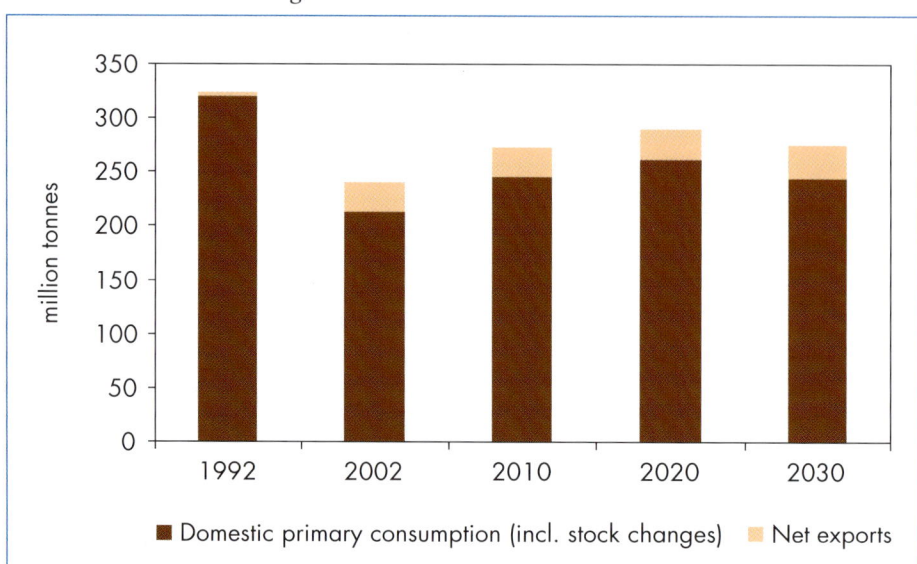

Figure 9.17: **Russian Coal Balance**

Russia has 157 billion tonnes of proven coal reserves, second only to the United States and equal to 16% of total world reserves. The mining industry is in poor shape, because of mismanagement during the Soviet era and under-investment in recent years. Production plunged in the early 1990s in line with demand, resulting in the closure of many unprofitable mines. Output stabilised by the end of the decade, and has started to recover, reaching 241 Mt in 2003, according to preliminary data.

Restructuring of the Russian coal industry began in 1993 with the break-up of the state monopoly, RosUgol. An estimated 86% of coal production now comes from independent private producers. Productivity averaged 1 440 tonnes per miner in 2002, 20% more than in 1998 and 88% more than in 1994. Two-thirds of production comes from opencast mines. In total, 230 mines remain in production.

The prospects for domestic demand for Russian coal will depend on the price of competing fuels – particularly natural gas – in power generation. Our projections show only limited growth in the use of coal and a drop in the share of coal in the power generation fuel mix.[10] Gas will remain competitive against coal in most instances despite expected increases in domestic gas prices. Coal-production growth will be further limited by the large distances between Russia's main reserves and its population centres, industry and ports. Rail-transport costs are high and capacity is fully utilised. Net exports are projected to level off at around 30 Mt per year by the end of the current decade.

10. This is in stark contrast to the official projections in the 2003 *Energy Strategy*.

Power and Heat Sector Outlook

Electricity output is projected to grow at an annual average rate of 1.5% in 2002-2030. Growth will be strongest in the current decade, and will slow thereafter in line with weaker economic growth and electricity demand. Inputs to power generation are expected to grow more slowly as thermal efficiency improves and transmission and distribution losses are reduced. Natural gas will remain the dominant fuel. Heat demand will grow much less rapidly. The implementation of market reforms is needed to put the industry onto a commercial footing so that it can secure investment to replace and refurbish generating plants and transmission and distribution networks.

Capacity Needs and Fuel Mix

Russian power plants generated 889 TWh in 2002, 12% less than in 1992 but 8% up on 1998. Natural gas is the main input fuel in electricity, accounting for 43% of generation in 2002. Hydropower accounts for 18%, coal for 19% and nuclear power 16%. The average capacity factor fell substantially in the 1990s, from 54% in 1992 to 45% in 2000, recovering slightly to 46% in 2002. The fall was most dramatic for fossil-fuel plants, which now have the lowest capacity factors. The shares of nuclear energy and hydropower, which are used to meet baseload demand, increased in the 1990s, as overall generation fell.

The Russian power sector is at a crossroads. Recent decisions on market reforms are paving the way for a radical shake-up of the industry, which will have a major impact on day-to-day decisions on plant dispatch and on fuel choices for new capacity. The introduction of competition into wholesale and retail electricity supply will mean that the economics of inter-fuel competition play the main role in determining electricity prices and the fuel mix of the future.

Natural gas is expected to be the most economic way to produce electricity in Russia, despite the prospect of rising gas prices as domestic subsidies are removed and as international prices increase. Almost 90% of net additions to capacity over the projection period will be gas-fired (Figure 9.18). Most new plants will be combined-cycle gas turbines. The share of gas will grow most rapidly in the east. Other technologies and fuels will, nonetheless, be used at times, because of regional market factors and policy considerations. A small amount of new hydroelectric capacity is expected to be built where conditions are favourable. Hydro capacity is projected to increase from 45 GW in 2002 to 53 GW in 2030. Some new coal plants are also expected to be built close to low-cost mines, but plant retirements will exceed the amount of new capacity built. We project nuclear capacity to increase from 21 GW in 2002 to 26 GW in 2030 (Box 9.3).

The share of natural gas in the electricity generation fuel mix is projected to remain flat through to 2010 and then rise progressively through to the end of

Figure 9.18: **Net Additions to Power-Generation Capacity in Russia**

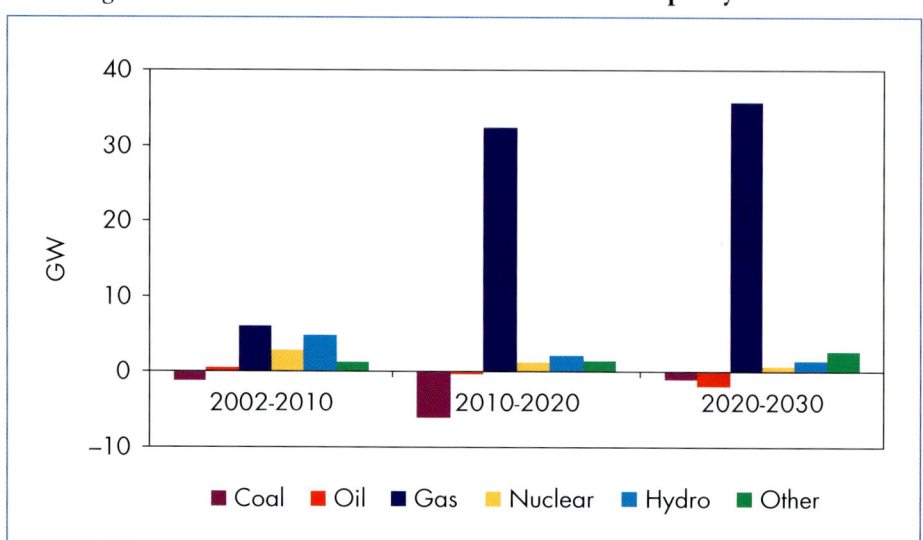

the projection period, reaching 53% in 2030 (Table 9.6). The share of coal will drop sharply, from 19% in 2002 to 15% in 2030. Nuclear power's share of electricity output will rise slightly in the near term, reaching 17% in 2010, but will decline slowly thereafter. By 2030, nuclear power's share will be down to 14%. Hydropower production will increase slowly, but its share will decline. Other renewables will grow rapidly, but their share will remain small as they are starting from a low base.[11]

Table 9.6: **Electricity Generation Fuel Mix in Russia** (TWh)

	1992	2000	2002	2010	2020	2030	2002-2030*
Coal	154	176	170	201	222	206	0.7%
Oil	100	33	27	35	33	28	0.0%
Gas	461	370	385	427	560	720	2.3%
Nuclear	120	131	142	170	179	184	0.9%
Hydro	172	164	162	186	194	200	0.7%
Other	2	3	3	9	13	23	7.7%
Total	**1 008**	**876**	**889**	**1 028**	**1 200**	**1 361**	**1.5%**

* Average annual growth rate.

11. See IEA (2003b) for a more detailed discussion of the potential for increased renewables use in Russia.

Box 9.3: **The Outlook for Nuclear Power in Russia**

Russia has 30 nuclear reactors at ten sites. At the end of 2002, the installed capacity of these reactors was 21 GW. They supplied 16% of the country's electricity in 2002. There are 14 reactors using VVER technology (similar to the pressurised water reactor technology in OECD countries), 15 reactors using RBMK technology and one fast-breeder reactor. Over 90% of these units were built in the 1970s and 1980s. The most recent reactor, a 950 MW VVER unit at the Rostov power plant, was commissioned in 2001. Russian reactors are licensed for 30 years, but their operating lives may be extended by 15 years.

Russia plans to build a lot more nuclear capacity. The country now has a number of incomplete reactors. Their construction started in the 1980s but was suspended following the Chernobyl accident and the break-up of the Soviet Union. In the near term, the government plans to increase nuclear capacity by completing their construction. Four of them – Kalinin-3, Kursk-5, Rostov-2 and Balakovo-5 – are close to being ready. In the longer term, new reactors are expected to be built. These new reactors will probably be based on improved VVER technology. The first ones to be built before 2015 will most likely have a capacity of 1 000 MW – the same as the current generation of reactors. Later, their capacity could rise to 1 500 MW. This *Outlook* assumes that about 20 GW of nuclear capacity will be installed in Russia between 2003 and 2030. Assuming an average lifetime of 45 years, three-quarters of existing nuclear capacity will be shut down by the end of the projection period. As a result, Russia's available nuclear capacity will rise only slightly to 26 GW in 2030.

The availability and performance of Russian nuclear power plants have improved in recent years. The average capacity factor was as low as 56% in 1994 but reached 76% in 2002. We assume that the average capacity factor will increase further, to 82% in 2030. By then, electricity generation from nuclear power is projected to reach 184 TWh – up from 142 TWh in 2002. Nuclear power plants are capital-intensive, requiring large investments. The total cost of the nuclear capacity additions projected here is estimated at about $34 billion (in year-2000 dollars). Russia's ability to raise this capital in domestic or international markets is extremely uncertain.

Heat supply through distribution networks to industrial, commercial and household consumers is an integral part of the electricity industry in Russia. The majority of heat-only plants are fuelled by natural gas. District heat is the main source of energy for household space heating and hot water. Many district-heating systems are poorly maintained and unreliable. Two-thirds of the equipment used to produce heat is over 20 years old and half of that is over

30 (OTAC, 2004). Most municipal heat distribution companies lose money, as tariffs do not cover costs. Raising tariffs to full-cost levels would, however, have only limited immediate impact on household demand, since many housing units are not able to adjust their heat supply. Nor are they billed according to usage, as metering is still rare (IEA, 2004c).

District heat will retain a central role in Russia's energy supply. Its share of final consumption will nonetheless decline with more efficient use. In addition, some consumers will switch to more direct forms of heating. Although most existing housing units and most new ones will still rely on district heat, industrial consumers will increasingly generate their own heat needs on-site. Under the market-reform programme, the government will impose controls on the closure of district-heat plants in order to maintain supplies to buildings that cannot easily be heated with other fuels.

The Russian government is considering a draft heat law. It would require municipalities and regions to prepare heat-supply plans and district heating companies to install heat meters in buildings. It would provide for disconnecting customers who do not pay, but would also make more explicit the companies' obligations to serve customers. Wholesale competition would be encouraged, but heat tariffs to final customers would continue to be set nationally.

Impact of Electricity Market Reforms

In 2003, after many years of debate, Russia embarked on the restructuring of its electricity sector, launching one of the most far-reaching and technically complex reforms of the post-Soviet era.[12] The legal basis for reform is the Decree on Electricity Restructuring, which was adopted in 2001. The decree's approach reflects many of the lessons learnt from similar restructuring programmes that OECD countries have undertaken over the last decade. Restructuring of the Russian electricity sector will involve the break-up of UES and the regional distribution companies (*energos*) into separate production, transmission and distribution businesses. Wholesale and retail supply will be separated from transmission and distribution activities. Competitive wholesale and retail markets will be created and new regulatory arrangements for transmission and distribution will be set up. These reforms are aimed at creating conditions that will encourage both investment in new capacity and greater efficiency of power production and consumption through the operation of market forces.

Six laws were adopted in March and April 2003, and a plan for the restructuring of UES itself has been agreed. The laws establish basic rules governing liberalised markets and the remaining state-controlled monopolies. The

---

12. A detailed review of the issues surrounding electricity market reform can be found in IEA (2004b).

so-called "5+5" plan covers primarily the restructuring of industry assets needed to create a more competitive market structure during the transition period. Under the electricity laws, three specialised entities will handle market and system operation and the transmission infrastructure. Electricity and heat are to become freely tradable commodities, with wholesale and retail markets for electricity and a market for heat. Power prices will be set freely, on the basis of supply and demand, in the competitive segments of power markets. Transmission and distribution tariffs will be set so that they ensure cost recovery and a return on capital.

The generation assets of UES will be spun off into ten wholesale generation companies (*gencos*) organised by plant type: six thermal and four hydroelectric. The national system operator, responsible for dispatch, has already been created as a separate company within UES. The plan also provides for the restructuring of the energos, with the creation of 14 territorial generating companies, up to five inter-regional distribution companies and a larger number of supply companies. Some energo assets will be allocated to the gencos. When complete, the structure of the electricity industry will resemble that in Figure 9.19.

The state will retain 100% ownership of the country's nuclear generating capacity, of the system operator and probably of the hydroelectric gencos. It will also retain a 76% stake in the national transmission company. It will hold 52% (the same as its current UES shareholding) of each of the inter-regional distribution companies, the holding company set up to manage UES stakes in isolated energy systems and other residual UES assets. The government plans to auction off its holdings in the wholesale thermal generating companies and the territorial generators.

The implementation of the restructuring plan is well under way. Actual testing of the wholesale market began in November 2003. Up to 15% of wholesale power production can now be traded at unregulated prices. By September 2004, about 8% of Russian electricity was traded freely. But other aspects of the reform process have encountered delays. In June 2004, Prime Minister Mikhail Fradkov announced a postponement of the sell-off of the thermal generating companies. This is thought to have been prompted by delays in setting up the companies, in making organisational arrangements for the sale and in drafting regulations.

As with all such reforms, many uncertainties remain. They include:

- The pace at which tariffs to captive end-users are raised to full-cost levels, cross-subsidies are eliminated and retail price caps are removed.
- The exact mechanism for reorganising and privatising industry assets and detailed rules and responsibilities of actors in the liberalised market.
- The design of the new regulatory arrangements and how effective the new institutions will turn out to be.

Figure 9.19: **Proposed Structure of the Russian Electricity Industry after Planned Reforms**

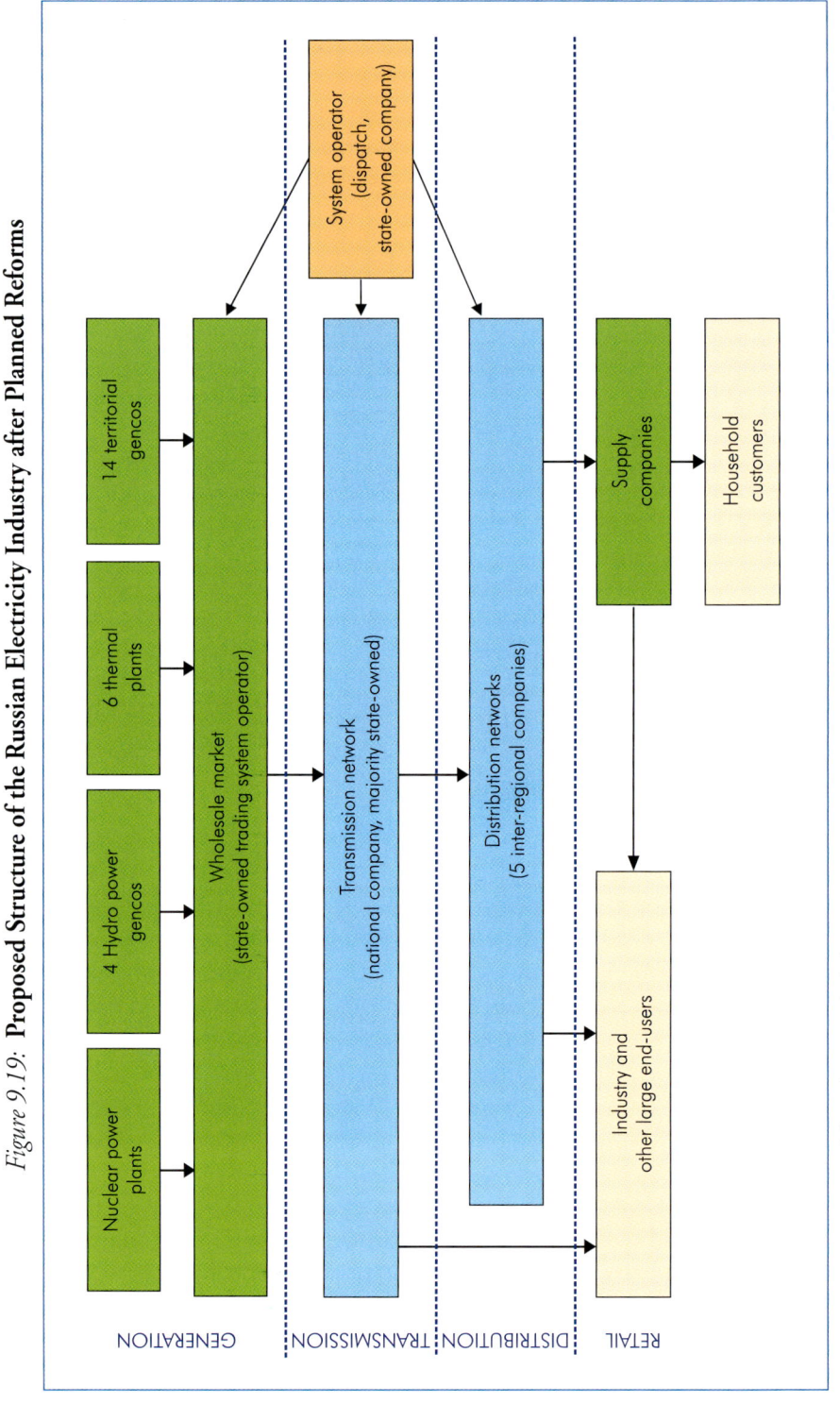

Note: Competitive segments of the industry are shaded in green.

- Arrangements for attracting foreign investment, including the prospects for establishing an investment-guarantee fund and genco management contracts – two proposals currently under consideration.

Successful creation of a robust and competitive electricity market will depend on how the government responds to these and other transitional issues. Inappropriate government actions, such as use of its substantial hydroelectric and nuclear generating assets to manipulate the market, could undermine the market's credibility and investor confidence. Our projections assume that effective competition in wholesale supply develops gradually over the next decade or so. Competition is expected to provide stronger incentives for generating companies to select the most economic technologies and for all operators to improve operating efficiencies.

The pace of reform in the gas sector will strongly affect how the power sector develops. The competitiveness of gas against other generating fuels will depend on how quickly subsidies to domestic gas supplies are removed. There is a pressing need to liberalise the system for allocating gas supplies to domestic customers, including power generators, and to adopt long-term contracts, which do not yet exist in Russia. There is also a need to establish a non-discriminatory regime for third-party access to Gazprom's transmission network in order to eliminate distortions in prices in the wholesale electricity market. Uncertainties surrounding these issues are making investment and management decisions very difficult and are raising the cost of capital.

Implications of the Projections

Russia's Role in the Global Energy Market

Russia will continue to play a central role in global energy supply and trade over the *Outlook* period, with major implications for energy security. Over the rest of the current decade, net exports of oil and gas will increase, both in absolute terms and as a share of world inter-regional trade. In the longer term, however, Russian oil and gas exports will become relatively less important as exports from other regions grow faster (Figure 9.20).

Oil production is projected to continue to rise through to the end of the current decade, albeit at a slower pace than in the last five years. It will outstrip growth in domestic demand. Net oil exports will, therefore, continue to expand, both in absolute terms and as a share of total world supply and inter-regional trade. This assumes that export infrastructure will expand accordingly, but it is not certain that this will happen in the short to medium term. Much of the increase in Russian oil exports will go to Europe, Asia and the Pacific region. If the pipeline to Murmansk is built, oil could be exported to the United States too. This will help to reduce the major oil-consuming

Figure 9.20: **Russian Fossil-Fuel Exports as Share of World Inter-Regional Trade**

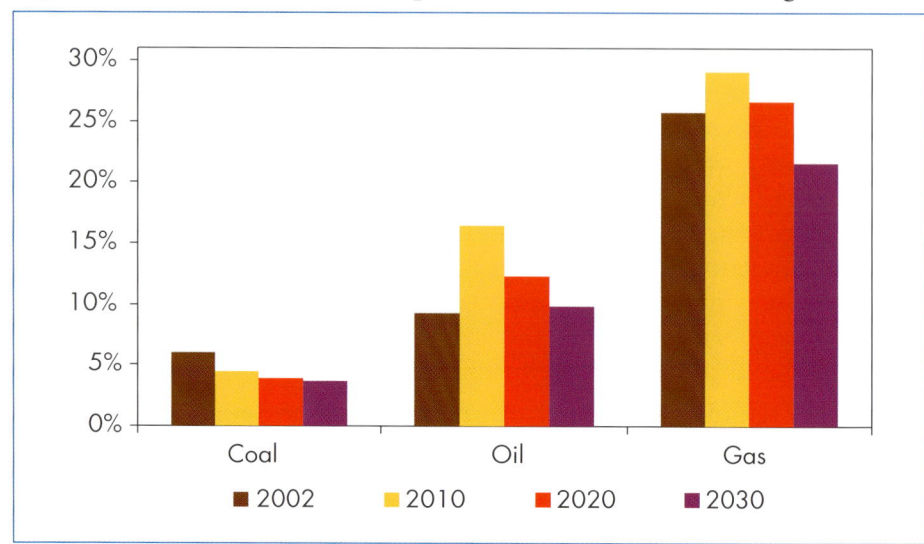

countries' dependence on oil from the Middle East. In the longer term, however, net exports of oil will stagnate as domestic demand grows faster than production.

The growing share of Russian oil exports in global oil trade in the next few years will undoubtedly affect OPEC production and pricing policies over time, with potentially large benefits to oil-importing countries. The recent surge in Russian production has reminded Middle East producers that very high prices tend to stimulate upstream investment in higher-cost regions, undermining their market share. Thus, higher Russian exports will limit OPEC's ability, and will reduce its willingness, to pursue higher prices through production ceilings. Oil-importing countries will benefit in two ways: through lower oil prices and through reduced dependence on oil supplies from the Middle East.

On the other hand, Russia's growing importance in world oil markets will increase the pressure from OPEC on the Russian government to co-operate in its efforts to manage prices. The growing dependence of the Russian economy on oil and gas export revenues increases the probability that Russia will agree to take part in future OPEC-led production cuts. A sharp drop in international prices would increase the likelihood of Russia's making such a move.

Gas exports will also increase further, although Russia's *share* of global inter-regional gas trade will decline steadily over the projection period. Russia will remain the leading supplier of natural gas to Europe, although its share of total European gas supply and imports will decline over the projection period. Russia will also emerge as an important supplier of gas to Asian markets. Buyers' concerns about the security of Russian gas supply will inevitably rise as gas flows increase. There are also concerns about how the liberalisation of downstream markets in Europe and Asia will affect investment in production and transportation infrastructure in Russia. Governments of importing countries in Europe and Asia will seek to improve their energy relations with Russia. This will take the form of more intensive political dialogue and, in some cases, direct involvement in financing or facilitating specific projects. The European Commission, for example, is co-financing a feasibility study of the North European Pipeline. The project is seen as a way of enhancing Europe's supply security, since it would bypass both Belarus and Ukraine. It is unclear whether commercial operators will be prepared to invest in this project. The Energy Charter Treaty, which Russia has signed but not yet ratified, sets out rules on energy trade, investment protection, transit and dispute resolution. Ratification of the treaty, as well as agreement on the Transit Protocol currently being negotiated by Treaty members, could encourage investment in long-distance pipelines.

Russia, for its part, will seek to strengthen its strategic and commercial ties with buyer countries in order to secure long-term outlets for its gas and to extract more value from the supply chain. Gazprom is increasing its direct involvement in downstream European markets. It is buying stakes in European gas-transmission companies and forming joint ventures with foreign companies to distribute and market gas in Europe. Because of the high cost of transporting Russian gas to market, Gazprom is also expected to enter into "swap" agreements with European partners, such as the Netherlands' Gasunie, Germany's Wintershall, Gaz de France and Norway's Norsk Hydro, and to obtain storage capacity in Ukraine and Germany (IEA, 2004a).

Energy Investment Needs and Financing

The amount of investment needed to underpin the projected growth in energy supply in Russia is enormous, totalling $935 billion (in year-2000 dollars) from 2003 to 2030. The oil, gas and electricity sectors will call for roughly equal amounts of funding. Coal investment will be very small by comparison (Figure 9.21). Combined oil and gas investment needs will average $24 billion per year over the projection period.

Projected investment in the energy sector is equal to over 5% of Russia's GDP. Financing this much investment will be difficult, given competing calls on capital, uncertainties about Russia's investment climate and the country's poorly developed domestic capital markets. Russia has a very high savings rate – equal to 37% of GDP in 2000 (World Bank, 2003) – which should in principle provide the major source of private capital. Nonetheless, external financing could account for a significant proportion of total capital flows to the Russian energy sector, especially in the oil and electricity industries.

Figure 9.21: **Cumulative Energy Investment Needs in Russia, 2003-2030**

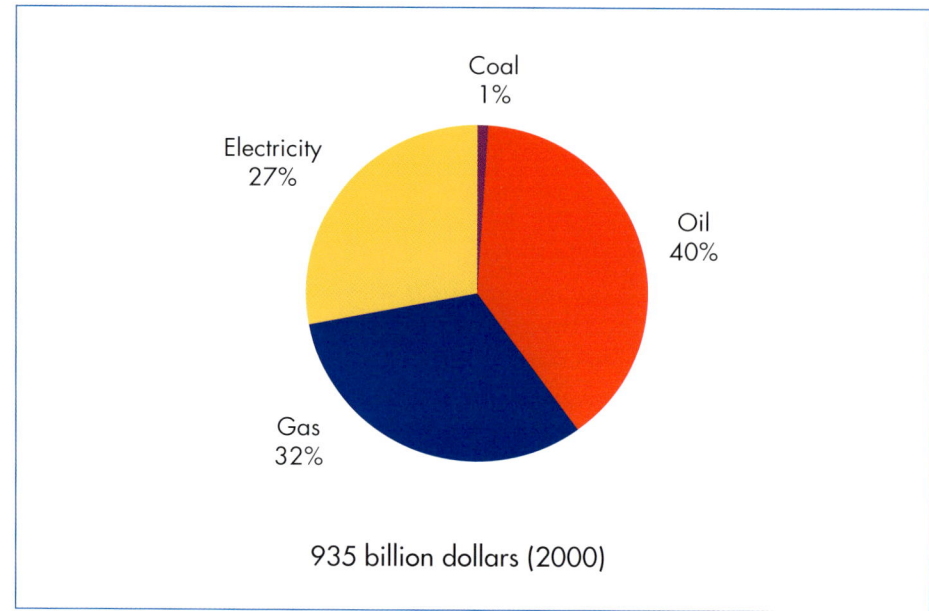

Note: Investment needs have been calculated using the methodology outlined in the *World Energy Investment Outlook* (IEA, 2003a) and on the basis of *WEO-2004* demand and supply projections and recent market developments.

Government policies on the development of oil pipelines, on upstream taxation and on large-scale foreign direct investments are fraught with major uncertainties. Price liberalisation, the removal of subsidies and the establishment of a transparent, non-discriminatory regulatory regime will be the key drivers of investment in the power sector. Apart from demand, the main issues affecting gas-industry investment are price reform and access for independent producers to Gazprom's transmission network.

Environmental Impact

The contraction of Russian industry in the 1990s led to a sharp fall in the country's energy-related emissions of carbon dioxide. Emissions fell by 21% between 1992 and 2002. But the carbon intensity of the Russian economy remains high, because of inefficient technology and waste. Russia's CO_2 emissions per unit of GDP in PPP terms were 211% higher than in the OECD and 205% higher than in developing countries in 2002. Russia's per capita emissions are also among the highest in the world.

Russia's CO_2 emissions are projected to rebound over the projection period, but will still be 11% below their 1990 level in 2030 (Figure 9.22). Power generation will remain the largest contributor to total emissions, although its

Figure 9.22: **Energy-Related CO_2 Emissions in Russia**

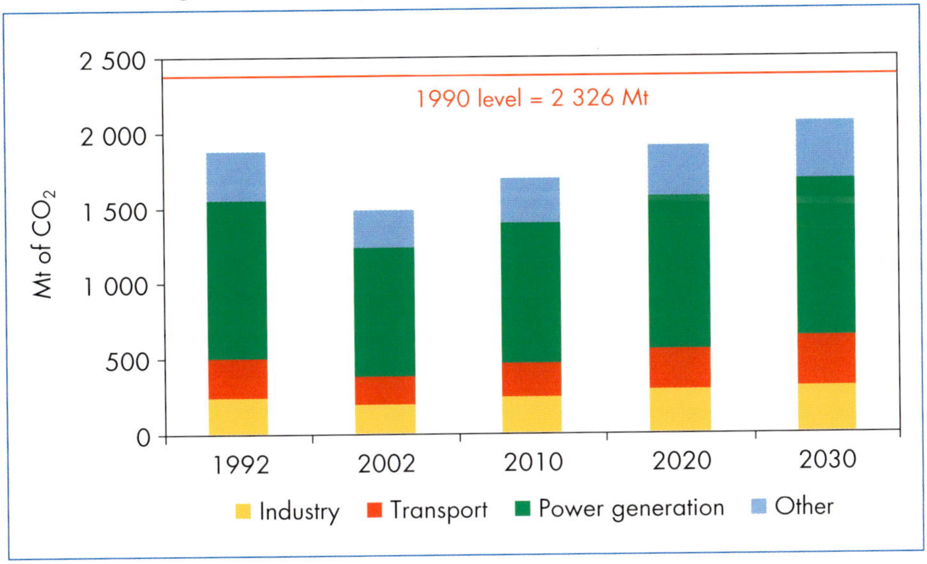

share will fall slightly. The transport and industrial sectors will both account for growing shares of emissions.

Under the Kyoto Protocol, Russia would be committed to limit its average annual greenhouse-gas emissions in the period 2008-2012 to their 1990 level of 2 326 Mt. In fact, energy-related CO_2 emissions in 2008-2012 are expected to be more than 600 Mt, or 28%, lower than in 1990. If Russia ratifies the Protocol, a condition for its coming into effect, the country would be able to sell its surplus emissions under a planned international emissions-trading system. In our Alternative Policy Scenario, Russian emissions fall much faster as more efficient energy technologies are deployed, and are 17% lower in 2030 than in our Reference Scenario.

Until recently, the government's reticence about ratifying the Protocol has stemmed in part from the higher emissions that it expects would occur if Russia were to meet its ambitious goal of doubling GDP between 2003 and 2013. During the series of meetings between Russian and EU officials in May 2004 to discuss Russia's accession to the World Trade Organization, President Putin is reported to have promised to accelerate examination of the ratification issue. As a result, on 30 September 2004, the Russian government approved the Protocol and submitted it to the state Duma for ratification. It remains to be seen whether the Duma will, in fact, ratify it. Ratification of the Protocol could pave the way for more investment in energy savings and conservation in Russia. UES and Gazprom have already started to prepare their greenhouse-gas inventories and to get involved in joint implementation projects that are allowed under the Protocol.[13]

Local and regional pollution remains a major problem in Russia. Airborne emissions of noxious gases, including nitrous oxides, sulphur dioxide and particulates, are among the highest in the world – a legacy of rapid industrialisation in the Soviet era. That has been exacerbated by the low priority given to environmental protection up to now. Pollution from radioactive waste is also a major problem. Although new environmental laws have been adopted since the 1990s, there is a pressing need to step up efforts to tackle air and water pollution. Replacement of unsophisticated energy-burning equipment, elimination of waste and changes in energy-industry practices to minimise its environmental impact could help to alleviate these problems in the long term. But increased energy production and use will inevitably offset some of these improvements.

---

13. See the United Nations Framework Convention on Climate Change (UNFCCC) in-depth review of the third national communication of the Russian Federation, June 2004 (http://unfccc.int/resource/docs/idr/rus03.pdf).

CHAPTER 10

ENERGY AND DEVELOPMENT

HIGHLIGHTS

- Energy is a prerequisite to economic development. The prosperity that economic development brings, in turn, stimulates demand for more and better-quality energy services. Many countries have established a virtuous circle of improvements in energy infrastructure and economic growth. But in the world's poorest countries, the process has barely got off the ground.
- Energy services enable basic human needs, such as food and shelter, to be met. They also contribute to social development by improving education and public health. During the early stages of development, the absolute amount of energy used per capita and the share of modern energy services – especially electricity – are key contributors to human development.
- For the first time, this *Outlook* presents an Energy Development Index – a composite measure of energy use in developing countries and of their progression in the use of modern energy services. The standing of all regions on that index will increase from now till 2030. Yet only a few Middle East and Latin American countries will have reached the stage of energy development that OECD countries had attained three decades ago. Most of Africa and South Asia will remain far behind.
- Almost 1.6 billion people in developing countries did not have access to electricity in their homes in 2002, representing a little over a quarter of world population. Most of the electricity-deprived are in South Asia and sub-Saharan Africa.
- The United Nations' Millennium Development Goals include halving the proportion of the world's people living on less than $1 a day by 2015. In our Reference Scenario, the number of people without electricity in 2015 will be only fractionally smaller than in 2002. It is highly unlikely that the UN poverty-reduction target will be achieved unless access to electricity can be provided to another half-a-billion of the people we expect will still lack it in 2015. This would cost about $200 billion. Meeting the target also implies a need to extend the use of modern cooking and heating fuels to 700 million more people by 2015.

- Governments need to act decisively to accelerate the transition to modern fuels and to break the vicious circle of energy poverty and human under-development in the world's poorest countries. This will entail improving the availability and affordability of commercial energy, particularly in rural areas. The rich industrialised countries have clear long-term economic and security interests in helping developing countries along the energy-development path.

This chapter considers the role of energy in development, focusing on developing countries. It first evaluates the contribution that energy makes to economic development, the energy dimension of sustainability and the relationship between the transition to modern fuels and indicators of human development. It goes on to assess today's patterns of energy use in developing countries using the IEA's newly-created Energy Development Index. The relationship between that index and the UNDP's index of human development is also analysed. It then looks at prospects for development based on the Reference Scenario and EDI projections. It also evaluates the implications of the targeted reduction in poverty by 2015 set by the UN Millennium Development Goals both for electricity access and reliance on traditional biomass. The final section considers the policy implications of this analysis.

The Role of Energy in Development

Energy is deeply implicated in each of the economic, social and environmental dimensions of human development. Energy services provide an essential input to economic activity. They contribute to social development through education and public health, and help meet the basic human need for food and shelter. Modern energy services can improve the environment, for example by reducing the pollution caused by inefficient equipment and processes and by slowing deforestation. But rising energy use can also worsen pollution, and mismanagement of energy resources can harm ecosystems. The relationships between energy use and human development are extremely complex.

The environmental and social dimensions of human development have attracted increased attention in recent years. The United Nations Development Programme defines human development as the creation of an environment in which people can realise their full potential and lead productive, creative lives in line with their needs and interests (UNDP, 2004). In this view, economic growth is only one means – albeit a vitally important one – of extending the range of human choices. UNDP has developed a set of numerical indices designed to measure the stage of human development in individual countries and to facilitate cross-country comparisons (Box 10.1).

> *Box 10.1:* **UNDP Human Development and Poverty Indices**
>
> The United Nations Development Programme has devised five indices of human development, including a summary Human Development Index (HDI), which is applied to all countries, and a Human Poverty Index (HPI-1), specially tailored for developing countries. The indices are updated annually and the results published in the yearly *Human Development Report*. The HDI measures life expectancy at birth; adult literacy and school enrolment; and per capita GDP (adjusted for purchasing power parity). The HPI-1 measures much the same aspects, but uses different indicators: probability at birth of not surviving to age 40; adult literacy; the percentage of the population without access to clean water; and the percentage of children who are underweight for their age. None of the UN indices explicitly takes energy use into account.
>
> Norway, Sweden and Australia headed the HDI rankings for 2002 (UNDP, 2004). The 20 lowest-ranked countries, and 31 of the bottom 35, were all in sub-Saharan Africa. Sierra Leone came last. Among developing countries covered by HPI-1, Barbados, Uruguay and Chile are ranked the most advanced. Again, the sub-Saharan African countries are clustered at the bottom.

The *sustainability* of development can be assessed in economic, environmental and social terms. Energy sustainability requires meeting our energy needs upon which economic development depends, while protecting the environment and improving social conditions. No matter how we define "sustainable" development, most current systems of energy supply and use are clearly not sustainable in economic, environmental or social terms. In practice, sustainable development is about finding acceptable trade-offs between economic, environmental and social goals.

Energy and Economic Growth

Energy alone is not sufficient for creating the conditions for economic growth, but it is certainly necessary. It is impossible to operate a factory, run a shop, grow crops or deliver goods to consumers without using some form of energy. Most studies of the relationship between energy use and economic development have focused on how the latter affects the former. Economic growth almost always leads to increased energy use, at least in the early stages of economic development. Empirical analysis, however, demonstrates the importance of energy in *driving* economic development (Box 10.2).

Box 10.2: **Assessing the Contribution of Energy to Economic Growth**

The neoclassical production function attributes economic growth to increases in the size of the labour force and to the amount of capital available, as well as to increases in "total factor productivity" – a catch-all for any part of growth that is not explained by labour and capital. By explicitly incorporating an energy variable in the production function[1], we have estimated the contribution energy made to the growth of gross domestic product in several countries that grew very rapidly in the 1980s and 1990s. The United States was included in the sample for comparison. The results are summarised in Table 10.1.

In every country studied, except China, the combination of capital, labour and energy contributed more to economic growth than did productivity increases.[2] Energy contributed significantly to economic growth in all countries and was the leading driver of growth in Brazil, Turkey and Korea. Its contribution was smaller in India, China and the United States. Our results suggest that energy plays a bigger role in countries at an intermediate stage of economic development, because industrial production often makes a large contribution to economic growth at this stage. The energy intensity of manufacturing, expressed as the amount of energy used to produce a unit of GDP, is generally much higher than that of other economic activities. As the economy matures, more energy-efficient technology, whose contribution is captured as a part of total factor productivity, kicks in and the amount of energy needed to produce a unit of GDP diminishes. The United States is the clearest example. Recent studies using growth-accounting approaches yield similar results.[3]

The results also reflect government policies and the resource endowment of individual countries. Brazil and Mexico, where energy played the leading role in economic growth, have both industrialised rapidly. In Indonesia, the relatively low importance of energy probably reflects the country's policy of importing sophisticated manufacturing technology via

1. We used the standard Cobb-Douglas formulation: $Y_t = A_t \times (K_t)^{\alpha}(L_t)^{1-\beta}(E_t)^{1-\alpha-\beta}$, where Y is output, K is the stock of capital, L is the labour force, E is primary energy use, A is the economy's total factor productivity and t is the time period. See, for example, Collins and Bosworth (1996).
2. There are doubts about the accuracy of China's official GDP data. Many studies have concluded that official statistics understate GDP and overstate growth rates. This could explain China's very high productivity growth relative to other countries.
3. Ayres and Warr (2003) demonstrate that including energy services measured by useful physical work as a factor of production in the standard production function improves the explanation of the historical growth path of the US economy since 1900. Productivity is only significant as a contributory factor to growth after the 1970s.

foreign direct investment. Korea has depended heavily on the chemical industry as a major engine of growth. Low levels of per capita energy use in India suggest that a lack of available energy may have held back economic growth and development there (per capita GDP growth in India was lower than in most other regions). It follows that development policies need to take into consideration energy-infrastructure needs, especially in the poorest and least industrialised regions.

The complementary relationship between energy use and economic growth is intuitively obvious. Less obvious is the extent to which constraints on the availability of energy and its affordability can affect economic development. Numerous studies have demonstrated that energy, capital and labour can, in principle, be substituted for one another to some degree. An increase in energy-input costs can be compensated by investing more in energy-efficient technology, shifting to less energy-intensive production or using more labour, where it is in surplus supply. In practice, structural economic rigidities and inappropriate government policies can impede the ability of the economy to adjust to changes in energy prices. In many poor countries, under-investment in public utilities, inefficient management, under-pricing and a generally unattractive climate for private investment cause energy shortages and hold back economic growth and development.

Table 10.1: **Contribution of Factors of Production and Productivity to GDP Growth in Selected Countries, 1980-2001**

	Average annual GDP growth (%)	Contribution of factors of production and productivity to GDP growth (% of GDP growth)			
		Energy	Labour	Capital	Total factor productivity
Brazil	2.4	77	20	11	-8
China	9.6	13	7	26	54
India	5.6	15	22	19	43
Indonesia	5.1	19	34	12	35
Korea	7.2	50	11	16	23
Mexico	2.2	30	60	6	4
Turkey	3.7	71	17	15	-3
United States	3.2	11	24	18	47

Sources: IEA analysis based on IEA databases and World Bank (2004).

Energy and Human Development

To understand better the relationship between energy use and human development, it is helpful to analyse the different aspects of energy use. We have identified three key indicators of energy use in developing countries: per capita consumption, the share of modern energy services in total energy use and the share of the population with access to electricity in their homes.

Per Capita Energy Consumption

The absolute amount of energy used by each individual has historically been a key factor in human development during the early stages of the process. There is a very strong link between per capita energy consumption (commercial and non-commercial) and the UN Human Development Index for all countries (Figure 10.1). The link is particularly strong for non-OECD countries with a HDI value of less than 0.8. Very few countries with per capita energy use of less than 2 tonnes of oil equivalent have a HDI score of more than 0.7. Once a country has reached a reasonably high HDI level, variations in its per capita energy use are largely attributable to structural, geographic and climatic factors. For the poorer developing countries, however, the picture is clear: a higher HDI goes hand in hand with increased per capita energy use.

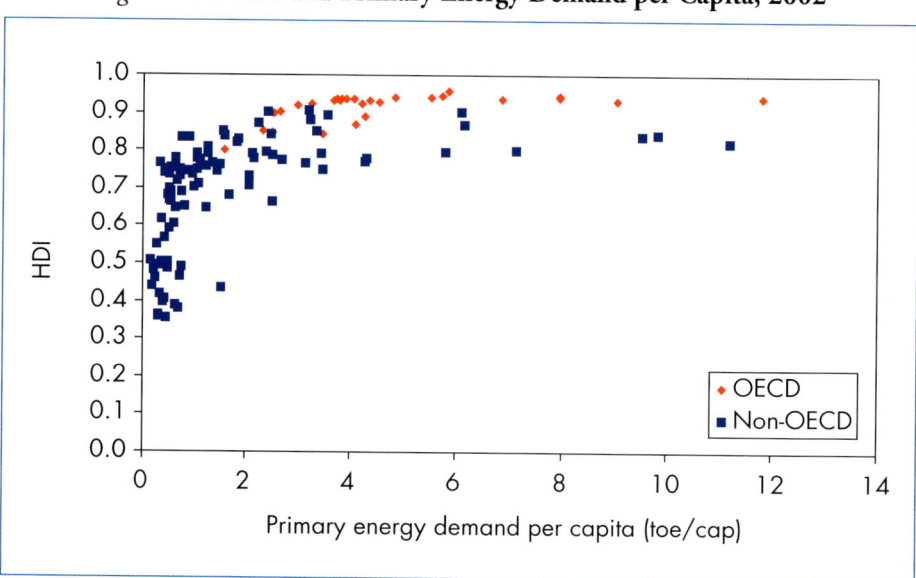

Figure 10.1: **HDI and Primary Energy Demand per Capita, 2002**

Sources: IEA analysis; UNDP (2004).

The link between per capita energy use and human development is much stronger when considering commercial energy alone. Per capita commercial energy demand is ten times greater in the richest developing countries, such as Uruguay and Israel, where less than 5% of the population is classified as poor, than in the poorest countries, such as Nigeria and India, where more than 75% of the population lives on less than $2 a day (Figure 10.2).

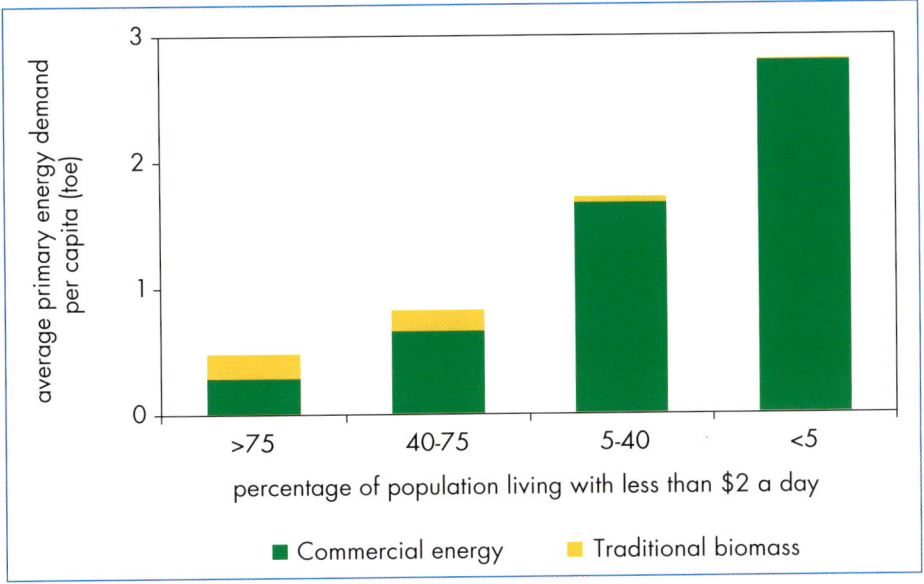

Figure 10.2: **Average Primary Energy Demand per Capita and Population Living on Less than $2 a Day, 2002**

Sources: IEA analysis; World Bank (2004).

The Transition to Modern Energy Services

Access to modern energy services is an indispensable element of sustainable human development. It contributes not only to economic growth and household incomes, but also to the improved quality of life that comes with better education and health services. Without adequate access to modern, commercial energy, poor countries can be trapped in a vicious circle of poverty, social instability and underdevelopment. Increased use of modern energy by households is a key element in the broader process of human development, typically involving industrialisation, urbanisation and increased personal mobility. The facts bear this out: the share of modern energy in overall energy use is strongly correlated with indicators of human development (Table 10.2).

Table 10.2: **Commercial Energy Use and Human Development Indicators, 2002**

Indicator	Commercial energy as share of total energy consumption		
	0-20%	21-40%	41-100%
Average life expectancy at birth (years)	59.8	69.0	69.5
Probability at birth of not surviving to 40 (%)	21.7	9.4	9.1
Gross school enrolment ratio	52.4	65.4	76.9
Children underweight (% of population)	40.9	15.1	11.9
Population without access to improved water (%)	20.9	22.9	12.8
Number of countries in sample	*30*	*7*	*27*
Per cent of total sample population	*42%*	*39%*	*17%*

Note: Indicators are averages weighted by population based on 64 developing countries for which data are available. See the note to Figure 10.5 for the definition of "improved water access".
Sources: IEA analysis; UNDP (2004).

As we stressed in *WEO-2002*, the extensive, and inefficient, use of traditional biomass and waste for energy purposes is both a characteristic of poverty and a cause of its persistence. Traditional fuels include charcoal, wood, straw, agricultural residues and dung. Most of such fuels are not traded commercially. Poor people in rural areas, especially women and children, spend much of their time gathering firewood. This practice generally leads to scarcity and ecological damage in areas of high population density and strong demand for fuelwood. The use of biomass energy can reduce agricultural productivity, because agricultural residues and dung burned in stoves might otherwise be used as fertilizer. Inefficiently burned, biomass can be a major cause of indoor smoke pollution. The World Health Organization estimates that, each year, 1.6 million women and children in developing countries are killed by the fumes from indoor biomass stoves. Over half are in China and India.

As incomes rise, households in developing countries typically switch to modern energy services for cooking, heating, lighting and electric appliances (Table 10.3). How quickly this occurs depends on the affordability of modern energy services, as well as their availability, and on cultural preferences. The process is in most cases a *gradual* one. People generally shift first from traditional fuels to intermediate modern fuels, such as coal and kerosene, and finally to advanced fuels, such as liquefied petroleum gas, natural gas and electricity (Figure 10.3).[4]

---

4. The use of traditional fuels in sustainable and efficient ways may be considered a modern energy service.

But the transition is rarely straight-line. Some households may leap-frog directly to the most advanced fuels if they are available and affordable.[5] Rising incomes also boost demand for personal mobility and, therefore, for transport fuels.

Table 10.3: **Dominant Fuels in Developing Countries by End-Uses**

Sector/end-use	Urban areas		Rural areas	
	Low income	High income	Low income	High income
Households				
Cooking	Wood, charcoal, coal	LPG, kerosene, coal	Wood, residues, dung	Kerosene, biogas, LPG, charcoal
Lighting	Candles, kerosene (or none)	Electricity, LPG	Candles (or none)	Kerosene, LPG, electricity
Space heating	Wood, residues, coal	Wood, coal, kerosene, LPG	Wood, residues, dung (or none)	Wood, coal
Appliances	Batteries (or none)	Electricity	None (or batteries)	Electricity
Agriculture				
Ploughing	–	–	Manual, animal	Diesel, animal
Irrigation	–	–	Manual, animal	Diesel, electricity
Food processing	–	–	Manual, animal	Diesel, electricity
Industry				
Mechanical	Manual, diesel	Diesel, electricity	Manual, animal	Diesel, electricity
Process heat	Wood, charcoal	Coal, charcoal, kerosene	Wood, residues, charcoal	Coal, charcoal, kerosene

Source: World Bank/WLPGA (2002).

5. See IEA (2002) for a detailed discussion of the transition to modern fuels.

Figure 10.3: **Final Energy Consumption per Capita by Fuel and Proportion of People in Poverty in Developing Countries, 2002**

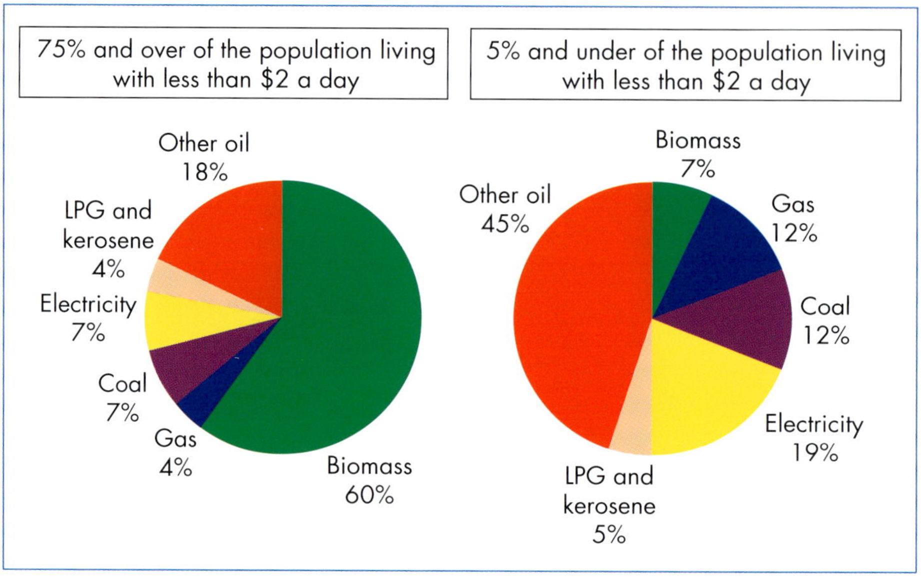

75% and over of the population living with less than $2 a day	5% and under of the population living with less than $2 a day
Other oil 18%	Biomass 7%
LPG and kerosene 4%	Other oil 45%
Electricity 7%	Gas 12%
Coal 7%	Coal 12%
Gas 4%	Electricity 19%
Biomass 60%	LPG and kerosene 5%

Sources: IEA analysis; UNDP (2004).

Kerosene is generally the cheapest fuel for cooking, heating and pumping water, and is the easiest to obtain in developing countries. But it is hazardous as a household fuel. Kerosene stoves are a major cause of fires and source of indoor pollution. Liquefied petroleum gas is a cleaner, safer fuel, but it is poorly distributed in some regions. The cost of using LPG can also be a problem for very poor households, because of the initial cost of the gas cylinder (either the deposit or outright purchase) and the stove, which is usually more expensive than a conventional kerosene stove.

Access to Electricity

Access to electricity is particularly crucial to human development. Figure 10.4 plots per capita electricity consumption against HDI ratings for the largest OECD and non-OECD countries. The correlation is strong and non-linear. The increase in HDI scores is most rapid relative to electricity use at low levels of consumption. Put another way, modest increases in per capita electricity use are associated with much larger improvements in human development. This is because electricity use in poor countries is largely a matter of access. Electricity is, in practice, indispensable for certain activities, such as lighting, refrigeration and the running of household appliances, and cannot easily be replaced by other forms of energy. As we saw with per capita energy use, HDI reaches a plateau when per capita electricity consumption attains a certain level – about 5 000 kWh per year.

Figure 10.4: **HDI and Electricity Consumption per Capita, 2002**

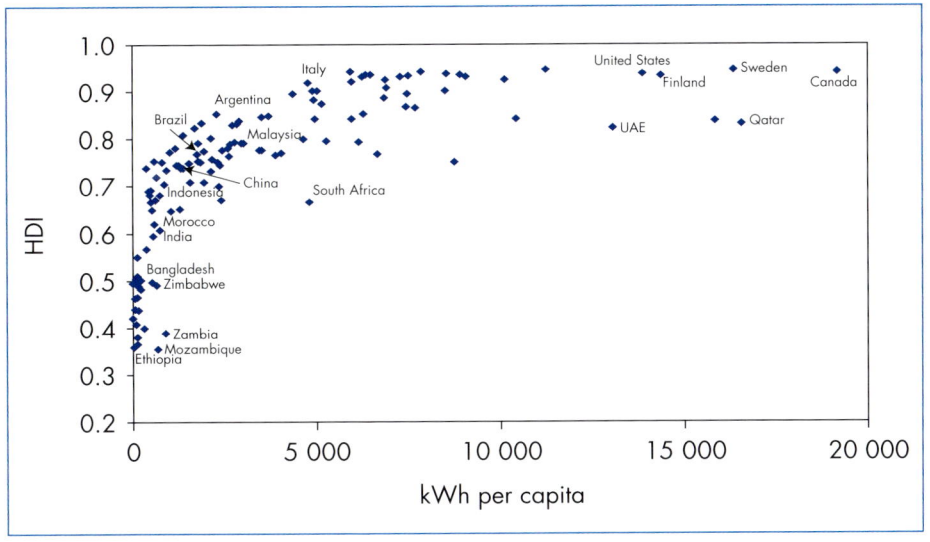

Sources: IEA analysis; UNDP (2004).

Individuals' *access* to electricity gives a better indication of a country's electricity poverty status than do statistics on their average *consumption*. Per capita consumption data can give a distorted impression where a very small minority consumes enormous amounts of electricity, while the majority consumes practically none. Taking the average of the two population segments gives a misleading impression of the prevalence of electricity poverty.

We have updated the database on electrification rates that we built for the *WEO-2002*. We estimate that just over 1.6 billion people[6] in developing countries did not have access to electricity in their homes in 2002, a little over a quarter of the world population. Around two-thirds of the electricity-deprived are in Asia; most the rest are in sub-Saharan Africa. Four out of five people without electricity live in rural areas (Table 10.4). Electrification rates have improved steadily over recent decades, but population increases have offset part of this improvement. As a result, the total number of people without electricity has fallen by fewer than 500 million since 1990. Rapid electrification programmes in China account for most of the progress. Excluding China, the number of people without electricity increased steadily over the past three decades. Detailed country-by-country statistics on electricity access in 2002 can be found in the appendix to this chapter.

6. The figure is slightly lower than that given in *WEO-2002* for 2000, mainly because of new connections.

Table 10.4: **Number of People without Electricity, 2002** (million)

Region	Rural	Urban	Total
Africa	416	118	535
Sub-Saharan Africa	*408*	*117*	*526*
North Africa	*8*	*1*	*9*
Developing Asia	871	148	1 019
East Asia and China	*192*	*29*	*221*
South Asia	*679*	*119*	*798*
Middle East	13	7	14
Latin America	39	1	46
Developing countries	1 339	275	1 615
OECD and transition economies	7	<1	7
World	1 347	275	1 623

Energy Poverty and Human Development Indices

There is an implicit level of energy development that underlies each level of human development. Yet energy development is never identified *per se* by the poverty indices, which focus more on such basic human needs as water, health and education. Energy is a factor in procuring each of these needs, but it is not fully captured by measuring them. The example of access to clean water, a fundamental need, clearly illustrates this point. Among countries which have achieved high levels of access to clean water supplies, defined here as over 70% of the population, access to electricity varies enormously (Figure 10.5). North African, Middle East and Latin American countries have high rates of electricity access, while sub-Saharan African and South Asian countries generally have much lower rates – extremely low in some cases. Such "energy poverty" is not adequately indicated by non-energy indicators. This has important implications for policy-making. Since energy underlies all economic activity, human development may be severely impeded by a lack of energy infrastructure. An index of energy development would, therefore, introduce an important element in understanding the drivers of human development and identifying the policies that can achieve it.

Figure 10.5: **Electricity and Improved Water Access*, 2002**

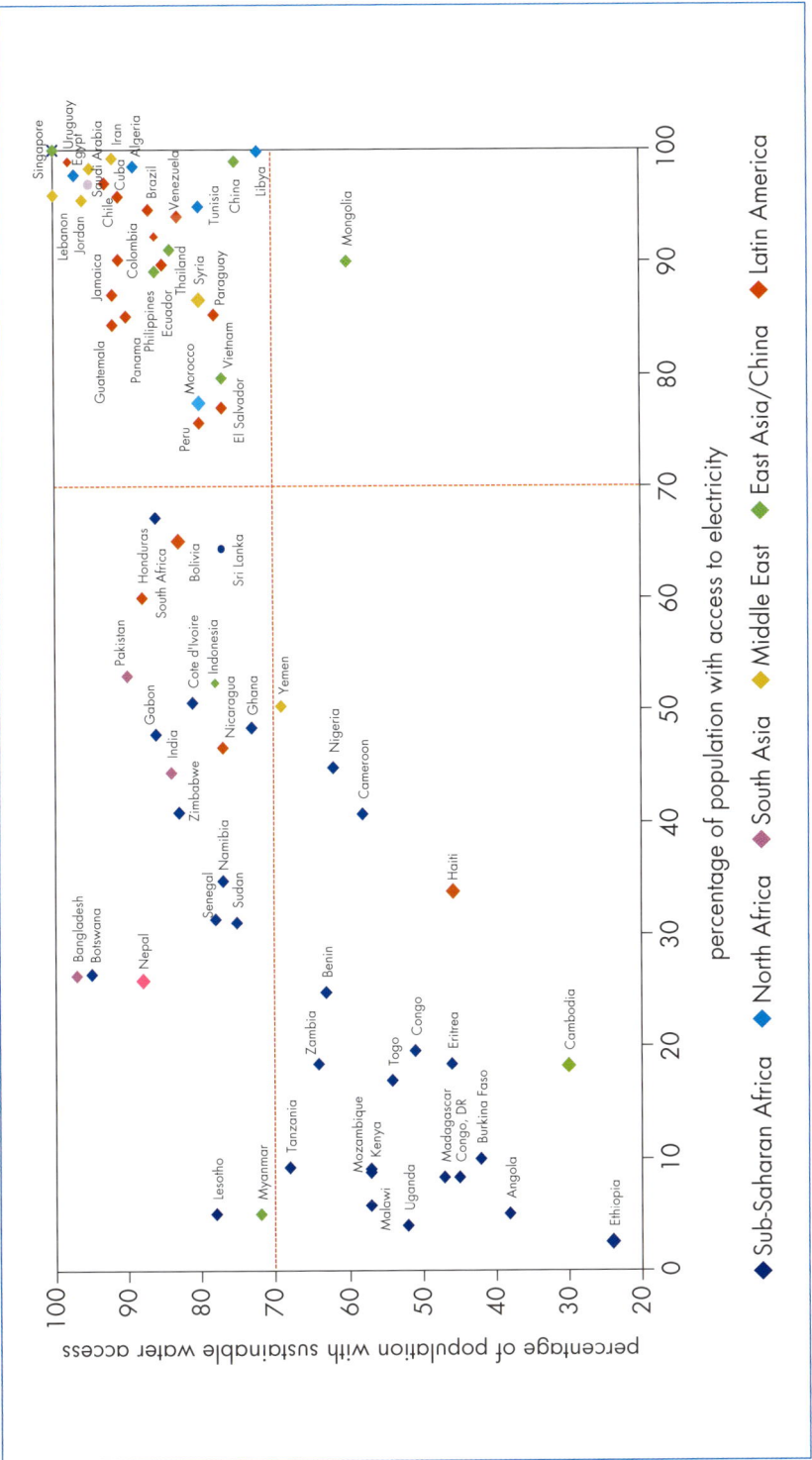

* "Improved water access" is defined by the UN as the share of the population with reasonable access to any of the following types of water supply for drinking: household connections, public standpipes, boreholes, protected dug wells, protected springs and rainwater collection. "Reasonable access" is defined as the availability of at least 20 litres per person per day from a source within one kilometre of the user's dwelling.

Chapter 10 - Energy and Development

The IEA Energy Development Index

To better understand the role that energy plays in human development, the IEA has devised for this *Outlook* an Energy Development Index (EDI). It is intended to be used as a simple composite measure of a country's or region's progress in its transition to modern fuels and of the degree of maturity of its energy end-use. The index seeks to capture the *quality* of energy services as well as their quantity. It is calculated in such a way as to mirror the UNDP's HDI (Box 10.3). *WEO* projections for the developing regions can be used to project future trends in EDI values. The index can, therefore, be used to assess the need for policies to promote the use of modern fuels and to stimulate investment in energy infrastructure in each region.

Box 10.3: **The IEA Energy Development Index**

The EDI is composed of three dimensions:
1. Per capita commercial energy consumption.
2. Share of commercial energy in total final energy use.
3. Share of population with access to electricity.

A separate index is created for each dimension, using the actual maximum and minimum values (known as "goalposts") for the developing countries covered. Performance in each dimension is expressed as a value between 0 and 1, calculated using the following formula:

$$\text{Dimension index} = \frac{\text{actual value} - \text{minimum value}}{\text{maximum value} - \text{minimum value}}$$

The index is then calculated as the arithmetic average of the three values for each country. The goalposts used for calculating the EDI in 2002 are as follows:

Indicator	*Maximum value*	*Minimum value*
Per capita commercial energy use (toe)	9.4 (Bahrain)	0.01 (Togo)
Share of commercial energy use (%)	100 (Israel/Kuwait/ Singapore)	8 (Ethiopia)
Electrification rate (%)	100 (15 countries)	2.6 (Ethiopia)

This is a first effort to produce an index of energy development. We have decided to introduce it here to encourage thinking about the role of energy as a contributory factor in development, rather than simply a consequence.

We have calculated EDI scores for 75 developing countries for which energy data are available, using 2002 data.[7] Figure 10.6 shows selected developing countries according to their EDI rankings. Detailed results are shown in Table 10.5. Ethiopia and Myanmar are the least developed countries in energy terms. The Middle Eastern and medium-income Latin American countries are generally ranked highest, reflecting their high rates of household electrification – often the result of large subsidies – and their limited use of traditional biomass. The sub-Saharan African countries, with uniformly low household incomes and electrification rates, are at the bottom of the rankings.

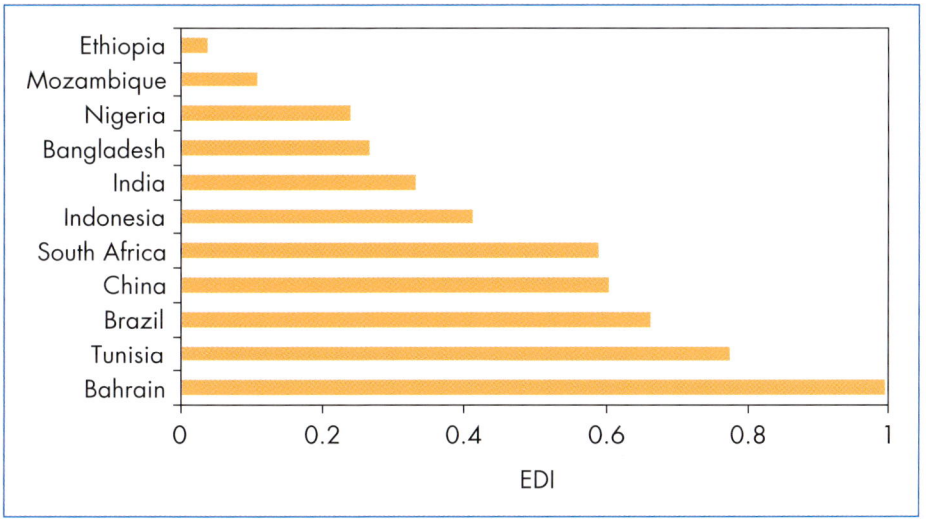

Figure 10.6: **Selected Developing Countries Ranked on the Energy Development Index, 2002**

Figure 10.7 shows that, as expected, there is a strong correlation between the two indices. This correlation is however non-linear, suggesting that the pace of improvement in HDI diminishes as the EDI increases. In other words, the two indices appear to decouple at higher levels of wealth and human development.

Although the rankings on the EDI are broadly similar to those on the HDI, there are notable divergences:

- Oil-producing countries are generally ranked much higher in energy development than in human development, reflecting the abundance and low cost of commercial energy supplies and the large amounts of energy used in

---

7. 2002 is the last year for which detailed energy data and human development indices are available for developing countries.

Table 10.5: **Energy Development Index for Developing Countries, 2002**

Rank	Country	EDI	Commercial energy use per capita index	Traditional biomass use index	Electrification Index	HDI	EDI vs. HDI ranking
1	Bahrain	**0.994**	0.984	1.000	0.999	*0.843*	
2	Kuwait	**0.984**	0.953	1.000	1.000	*0.838*	↑
3	Netherlands Antilles	**0.896**	0.692	1.000	0.995	*..*	
4	Singapore	**0.869**	0.608	1.000	1.000	*0.902*	
5	Brunei	**0.858**	0.609	0.973	0.992	*0.867*	
6	Saudi Arabia	**0.854**	0.577	1.000	0.984	*0.768*	↑
7	Iran	**0.834**	0.552	0.995	0.954	*0.732*	↑
8	Chinese Taipei	**0.801**	0.417	1.000	0.988	*..*	
9	Oman	**0.791**	0.427	1.000	0.946	*0.770*	↑
10	United Arab Emirates	**0.781**	0.369	0.999	0.974	*0.824*	
11	Libya	**0.775**	0.341	0.985	0.998	*0.794*	
12	Israel	**0.773**	0.319	0.999	1.000	*0.908*	↓
13	Tunisia	**0.772**	0.538	0.827	0.950	*0.745*	↑
14	Trinidad and Tobago	**0.728**	0.195	1.000	0.990	*0.801*	
15	Venezuela	**0.716**	0.214	0.994	0.940	*0.778*	
16	Malaysia	**0.711**	0.203	0.958	0.971	*0.793*	
17	Argentina	**0.698**	0.153	0.990	0.950	*0.853*	↓
18	Algeria	**0.693**	0.098	0.996	0.985	*0.704*	↑
19	Jordan	**0.686**	0.104	0.999	0.955	*0.750*	↑
20	Lebanon	**0.683**	0.118	0.972	0.960	*0.758*	↑
21	Cuba	**0.681**	0.122	0.963	0.958	*0.809*	
22	Egypt	**0.679**	0.078	0.983	0.977	*0.653*	↑
23	Iraq	**0.679**	0.044	0.999	0.992	*..*	
24	Thailand	**0.677**	0.211	0.908	0.911	*0.768*	
25	Costa Rica	**0.672**	0.088	0.960	0.970	*0.834*	↓
26	Brazil	**0.662**	0.102	0.938	0.946	*0.775*	↓
27	Syria	**0.657**	0.106	1.000	0.866	*0.710*	↑
28	Chile	**0.652**	0.140	0.846	0.970	*0.839*	↓
29	Jamaica	**0.646**	0.138	0.930	0.870	*0.764*	
30	Uruguay	**0.640**	0.066	0.864	0.990	*0.833*	↓
31	Ecuador	**0.635**	0.067	0.941	0.897	*0.735*	
32	Dominican Republic	**0.617**	0.083	0.845	0.923	*0.738*	
33	Colombia	**0.609**	0.056	0.871	0.902	*0.773*	↓
34	China	**0.603**	0.080	0.738	0.990	*0.745*	
35	Philippines	**0.594**	0.045	0.846	0.891	*0.753*	↓
36	Panama	**0.589**	0.091	0.826	0.851	*0.791*	↓
37	Morocco	**0.589**	0.035	0.956	0.774	*0.620*	↑
38	South Africa	**0.588**	0.226	0.868	0.671	*0.666*	↑
39	Paraguay	**0.541**	0.051	0.718	0.853	*0.751*	↓
40	Bolivia	**0.538**	0.046	0.916	0.651	*0.681*	

Rank	Country	EDI	Commercial energy use per capita index	Traditional biomass use index	Electrification Index	HDI	EDI vs. HDI ranking
41	Peru	0.532	0.037	0.804	0.757	0.752	↓
42	Yemen	0.504	0.022	0.988	0.503	0.482	↑
43	El Salvador	0.489	0.050	0.648	0.769	0.720	↓
44	Guatemala	0.458	0.035	0.496	0.844	0.649	
45	Honduras	0.420	0.033	0.625	0.601	0.672	
46	Namibia	0.414	0.051	0.844	0.347	0.607	
47	Indonesia	0.412	0.054	0.656	0.525	0.692	↓
48	Vietnam	0.409	0.024	0.406	0.797	0.691	↓
49	Sri Lanka	0.409	0.025	0.555	0.645	0.740	↓
50	North Korea	0.407	0.082	0.940	0.200	..	
51	Pakistan	0.387	0.030	0.601	0.530	0.497	↑
52	Gabon	0.333	0.061	0.460	0.479	0.648	
53	India	0.332	0.034	0.519	0.444	0.595	
54	Nicaragua	0.326	0.032	0.482	0.466	0.667	↓
55	Ghana	0.304	0.016	0.412	0.485	0.568	
56	Côte d'Ivoire	0.290	0.014	0.349	0.507	0.399	↑
57	Senegal	0.280	0.016	0.510	0.314	0.437	↑
58	Bangladesh	0.267	0.010	0.528	0.263	0.509	
59	Cameroon	0.253	0.014	0.338	0.407	0.501	
60	Zimbabwe	0.251	0.032	0.311	0.409	0.491	
61	Haiti	0.244	0.009	0.389	0.335	0.463	
62	Nigeria	0.238	0.021	0.246	0.449	0.466	
63	Sudan	0.229	0.013	0.365	0.310	0.505	↓
64	Benin	0.205	0.010	0.357	0.248	0.421	
65	Congo	0.189	0.008	0.364	0.196	0.494	
66	Zambia	0.179	0.018	0.335	0.184	0.389	
67	Togo	0.176	0.001	0.359	0.170	0.495	↓
68	Eritrea	0.165	0.005	0.305	0.184	0.439	
69	Angola	0.149	0.022	0.373	0.050	0.381	
70	Nepal	0.131	0.005	0.129	0.259	0.504	↓
71	Kenya	0.124	0.012	0.271	0.091	0.488	↓
72	DR of Congo	0.118	0.008	0.262	0.083	0.365	
73	Mozambique	0.107	0.009	0.226	0.087	0.354	
74	Myanmar	0.091	0.007	0.217	0.050	0.551	↓
75	Ethiopia	0.037	0.002	0.084	0.026	0.359	

↑ EDI rank is more than 5 ranks higher than HDI   ↓ EDI rank is more than 5 ranks lower than HDI
.. Not available.
Sources: IEA analysis; UNDP (2004).

Figure 10.7: **EDI and HDI in Developing Countries, 2002**

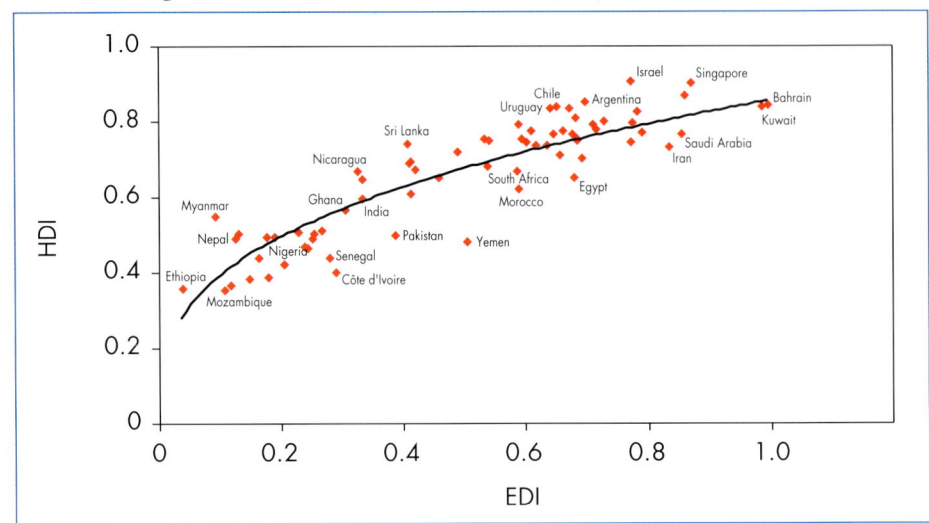

Sources: IEA analysis; UNDP (2004).

energy-production and related industries. Saudi Arabia's EDI, for example, is much higher than Brazil's, but its HDI is lower.
- EDI rankings in Latin America are generally lower than those for human development, despite fairly high electrification rates and low use of traditional biomass. Very low per capita energy consumption accounts for this divergence.
- Most sub-Saharan African countries have low scores on *both* indices, whether they have abundant energy resources or not.

Prospects for Energy Development
EDI Projections to 2030

Based on the Reference Scenario projections described in earlier chapters of this *Outlook*, EDI scores are expected to continue to rise in all developing regions. The index for developing countries as a whole is projected to rise from 0.48 in 2002 to 0.57 in 2030. The biggest increases are expected to occur in Africa and India (Figure 10.8). In 2030, these two regions will, nonetheless, remain the most under-developed in energy terms, and the Middle East and Latin America the most developed. By the end of the projection period, most of the developing regions will remain well below the stage of energy development reached by OECD countries three decades ago. The exception will be the Middle East, which will have reached exactly that level by 2030.

Figure 10.8: **Outlook for Energy Development Index by Region**

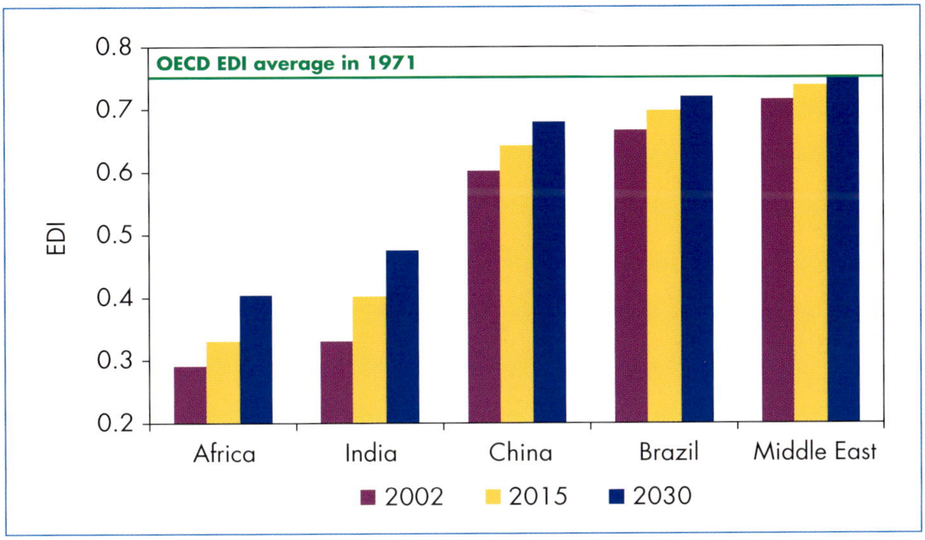

Note: In the absence of historical data for electricity access, the EDI for IEA countries in 1971 is based on an assumed average electrification rate of 90%.

In all regions, each of the three dimensions included in the EDI increases in line with rising incomes. Per capita energy consumption and the share of commercial energy in total final consumption are projected to grow steadily throughout the projection period.[8] Average per capita consumption in developing countries will rise from 0.82 tonnes of oil equivalent in 2002 to 1.2 toe in 2030. The share of commercial energy will rise from 80% to 88% over the same period. The number of people relying on traditional fuels for cooking and heating will, nonetheless, grow, from just under 2.4 billion in 2002 to over 2.6 billion in 2030 (Table 10.6).[9] The share of India and Africa together in the total number of these people will grow from just over half to almost two-thirds. The *proportion* of the population using traditional fuels will remain highest in sub-Saharan Africa.

Electrification rates will also rise over the projection period, from 66% of the population of developing countries in 2002 to 78% in 2030 (Table 10.7).[10] In the Middle East, North Africa, East Asia and Latin America, electrification

---

8. See Chapter 8 for a detailed discussion of these trends.
9. The estimates of the number of people relying on biomass for cooking and heating are based on the assumption that biomass demand per capita in each region is constant over the *Outlook* period at 2002 levels. This is a conservative assumption, establishing a lower limit on the number of people who rely on biomass for cooking and heating. The energy demand projections for biomass take into account technological factors that increase the efficiency of biomass use.
10. Our projections of electrification rates are prepared using the electrification module of the IEA's World Energy Model (described in Annex C). These projections are determined by many factors, including incomes, fuel prices, population growth and technological advances.

Table 10.6: **Population Relying on Traditional Biomass for Cooking and Heating** (millions)

	2002	2015	2030
Africa	646	805	996
South Asia	746	844	883
India	*595*	*665*	*693*
East Asia and China	925	829	693
China	*704*	*618*	*505*
Latin America	79	68	60
Developing countries	**2 398**	**2 549**	**2 634**

Note: Middle East is not included as the numbers are negligible.

Table 10.7: **Electrification Rates by Region** (%)

	2002	2015	2030
Africa	36	44	58
North Africa	*94*	*98*	*99*
Sub-Saharan Africa	*24*	*34*	*51*
South Asia	43	55	66
East Asia and China	88	94	96
Latin America	89	95	96
Middle East	92	96	99
Developing countries	**66**	**72**	**78**

rates will approach 100%. Although rates will improve substantially in sub-Saharan Africa and South Asia, they will remain relatively low. By 2030, half the population of sub-Saharan Africa will still be without electricity.

Despite rising electrification rates, the total *number* of people without electricity will fall only slightly, from 1.6 billion in 2002 to just under 1.4 billion in 2030 (Figure 10.9). In fact, 2 billion more people will gain access to electricity, but this will be largely offset by rising world population. Most of the net fall of 200 million people who will lack electricity will occur after 2015. The number of people without electricity will fall in Asia, but will continue to increase in Africa, peaking at just under 600 million by the end of the 2020s.

These projections are highly dependent on incomes and on electricity-pricing policies, which determine the affordability of electricity. Investment in electricity-supply infrastructure and rates of rural-urban migration are also important factors. Access to electricity will remain easier in urban areas, but the

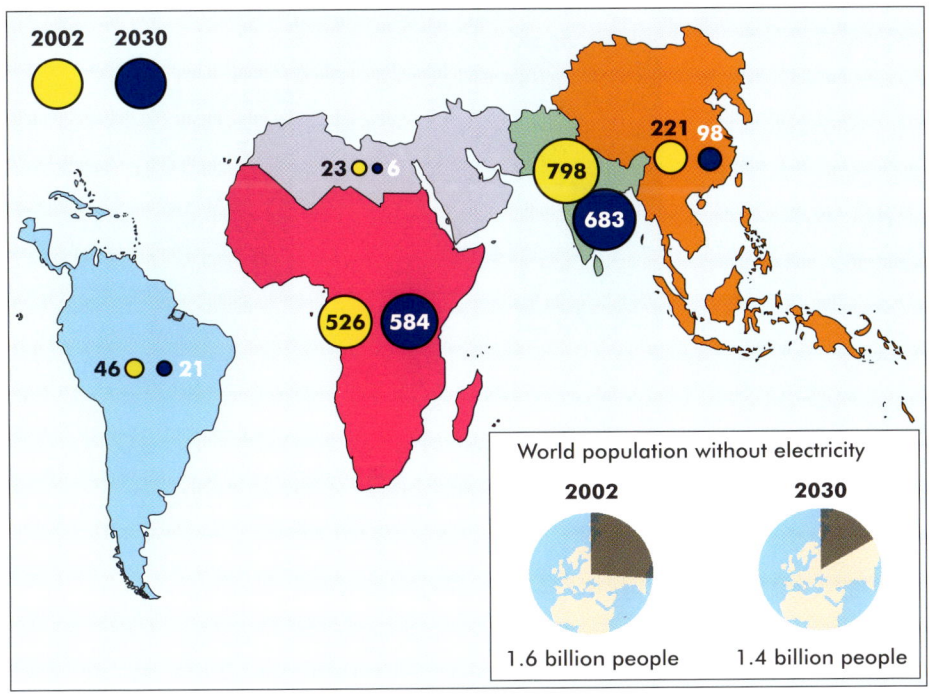

Figure 10.9: **Electricity Deprivation** (million)

absolute number of people without electricity will increase slightly in towns and cities, while it will fall in the countryside with continuing rural-urban migration (Figure 10.10). Detailed projections of electrification rates by urban and rural areas can be found in Table 10.A4 at the end of this chapter.

The EDI projections point to the prospect of considerable advances in human development in all major regions, even though big differences among regions will remain. Expected improvements in living standards in the developing countries depend on heavy investment in energy-supply infrastructure, both to produce for export and for domestic supply. The capital required will represent a sizable proportion of total savings in many regions, especially Africa (IEA, 2003). Other sectors, of course, will also be making large claims on these countries' limited financial resources. Much of the funding will, therefore, need to come from abroad in the form of direct investment and development aid. The latter will need to play an important role in the poorest countries, where the lack of existing infrastructure and a poor commercial environment are major deterrents to inward investment.

Figure 10.10: **World Population without Electricity in Rural and Urban Settings**

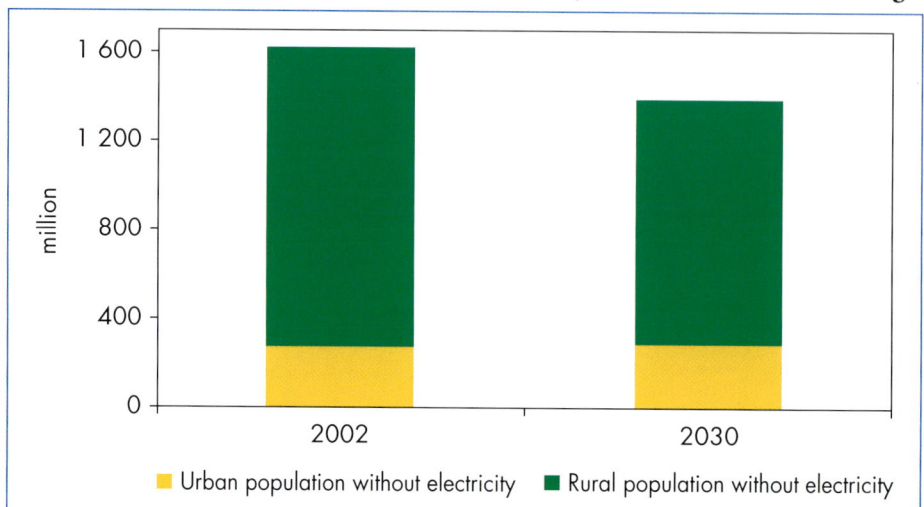

Energy Development and the Millennium Goals

In the year 2000, the United Nations adopted eight "Millennium Development Goals", the first of which was to eradicate extreme poverty (Box 10.4). One of the two targets established to measure progress in achieving this goal was halving the proportion of people living on less than $1 a day by 2015. Because of the strong link between income and access to electricity, meeting this target implies an enormous increase in electrification rates in very poor countries. Put another way, past experience shows that much higher rates of access would normally be expected to accompany the considerable improvement in prosperity that achievement of the poverty-reduction goal would imply. Indeed, expanding electricity access would directly contribute to that objective.

In our Reference Scenario, the overall number of people without electricity in 2015 will still be just under 1.6 billion – practically unchanged from today. This finding suggests that, in the absence of rigorous new policies, the target of halving the proportion of people living on less than $1 a day is very unlikely to be met. We estimate that achieving it would need to be accompanied by a reduction of 600 million in the number of people without electricity, to about 1 billion.[11] Almost all those people would be in South Asia and sub-Saharan Africa (Table 10.8). By 2015, electrification rates will be close to 100% in all other regions. We estimate that the additional investment needed to bring electricity to

---

11. The energy implications of halving poverty in 2015 are projected using regression analysis, applied to each region. The relationships between poverty, energy consumption and electrification rates are based on a cross-country analysis covering 100 countries.

Box 10.4: **The UN Millennium Development Goals**

In September 2000, the member states of the United Nations adopted what they called the "Millennium Declaration". Following consultations with the World Bank, the International Monetary Fund, the OECD and the specialised agencies of the United Nations, the General Assembly recognised eight specific goals as part of the road map for implementing the declaration:

1. Eradicate extreme poverty and hunger.
2. Achieve universal primary education.
3. Promote gender equality and empower women.
4. Reduce child mortality.
5. Improve maternal health.
6. Combat HIV/AIDS, malaria, and other diseases.
7. Ensure environmental sustainability.
8. Develop a global partnership for development.

Yardsticks were established for measuring results and targets for 2015. They concern not just developing countries but also the rich countries that are helping to fund development programmes and the international organisations that are helping countries implement them.

Table 10.8: **Impact of Meeting MDG Poverty-Reduction Target on the Number of People without Electricity and Investment in Developing Countries**

	Population without electricity (million)				Additional cumulative investment, 2003-2015 ($ billion)
	2002	2015 Reference Scenario	2015 MDG Case*	Difference	
Africa	536	601	453	148	46
North Africa	9	3	1	2	1
Sub-Saharan Africa	526	598	452	146	45
South Asia	798	773	417	355	104
East Asia and China	221	127	100	28	22
Latin America	46	27	5	22	28
Middle East	14	9	5	3	3
Total	**1 615**	**1 537**	**981**	**557**	**202**

* Assumes that the Millennium Development Goal of reducing by half the proportion of the population living on less than $1 per day by 2015 is achieved.

these 560 million people would be about $200 billion. This is equal to 10% of the total cumulative investment in the electricity sector in developing countries that we estimate from 2003 to 2015. Three-quarters of this additional finance would be needed in sub-Saharan Africa and South Asia.

In a similar way, the achievement of the Millennium Development Goals would most likely require a substantial reduction in the use of traditional biomass for cooking and heating. The amount of biomass consumption is usually a function of how poor a country is and of the relative availability of commercial and non-commercial fuels. In our Reference Scenario, the number of people relying almost entirely on traditional biomass for cooking and heating will increase slightly from 2.40 billion in 2002 to over 2.55 billion in 2015. Our analysis suggests that if the poverty-reduction target is met, the number would need to be reduced to under 1.85 billion. To accomplish this, governments would need to take new measures to extend the use of modern cooking and heating fuels to more than 700 million people from 2002 to 2015.

Figure 10.11 summarises the implications of meeting the poverty-reduction target for electricity access and traditional biomass use. Increased electricity access and reduced biomass use would also help achieve other Millennium Development Goals (UNDP/UNDESA/WDC, 2004).

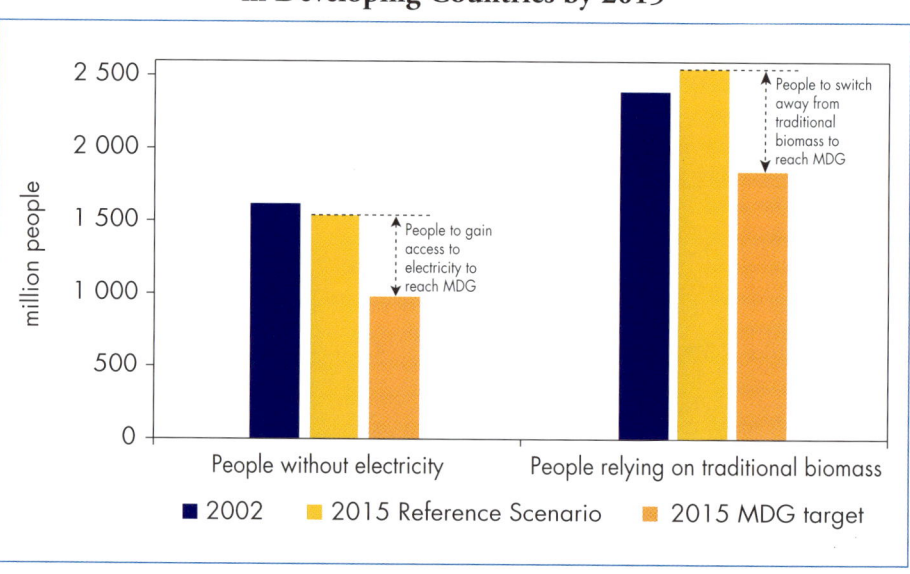

Figure 10.11: **The Energy Implications of Halving Poverty in Developing Countries by 2015**

Policy Implications

The analysis and projections described above provide a compelling argument for decisive action to accelerate the process of energy development in poor countries. Sitting back and waiting for people to become richer as the global economy expands will not be enough. Developing countries are unlikely to see their incomes and living standards increase without a concomitant increase in their use of modern energy services. Energy development is, of course, an effect of economic growth and development. But energy is also a cause. Energy-poverty levels vary widely among the developing countries. Yet, even for the most advanced among them in energy terms, there is much to be done. If the vicious circle of energy poverty and human under-development is to be broken, governments must act to improve the availability and affordability of modern energy services, especially electricity.

Good governance in the energy sector is critical to attracting infrastructure investment. Effective competitive markets give consumers choice and drive down costs. Creating such markets means removing controls on the pricing of petroleum products and other tradeable forms of energy. It means establishing cost-based regulation of energy-network services, and those services must be paid for. Laws and regulations that impede energy trade and investment have to be reformed. And various measures to attract private capital should be considered. From now to 2030, the developing countries as a whole need to secure about $5 trillion in financing for electricity generation, transmission and distribution. Where public funding is limited, private investors will be called upon to provide the lion's share of this capital. Where companies remain state-owned, they should be compelled to compete on an equal footing with private companies.

Public policies aimed at improving both the quantity and quality of energy services need to be backed by broader policies to promote investment, growth and productive employment. These include rural infrastructure development, training and education, and support for micro-credit programmes. More generally, efforts are needed to strengthen the overall legal, institutional and regulatory framework, including the protection of land and property rights. Existing laws and regulations need to be enforced more effectively. In many developing countries, there is a long way to go in applying and respecting the basic principles of good governance.

At the household level, policies need to focus on ways of increasing access to and the affordability of fuels for cooking and heating and of electricity. Policies should also aim to promote more efficient use of all fuels. In practice, a primary objective should be to expand the distribution of petroleum-based fuels and, where available, natural gas. Governments should promote the use of energy-efficient cook-stoves, water pumps and other appliances. And electricity services should be extended to households not yet connected to the grid. Where it is uneconomical to extend the electricity grid to rural areas, the most appropriate solution will often be small-scale generators. In many cases, the cheapest fuel will be diesel or LPG, though renewable energy technologies such as photovoltaic panels and wind power may, in some cases, offer a more suitable option.

As a rule, subsidies to energy services are ineffective, economically inefficient and contrary to good environmental practice. But subsidies may be justified in some cases in order to combat poverty. They should be resorted to under specific conditions: be properly targeted and affordable; deliver quantifiable benefits; be easily administered and not cause large economic distortions; be transparent and limited in duration (IEA, 1999; UNEP, 2004). The way a subsidy is applied is critical to how effective it is and to its cost. Subsidies should normally be restricted to energy services provided through fixed networks: electricity, natural gas or district heat. Subsidies to other forms of energy, such as oil products, can never be properly limited to poor households, because those fuels are freely traded. Policies should target the "poor" very precisely so that the mechanism for subsidising a particular fuel does not allow richer households to benefit from the subsidy.

The case for subsidising electrification in poor developing countries is widely accepted in principle, since the developmental benefits are often judged to exceed the long-run costs involved in providing subsidised electricity. Where high up-front connection charges prevent poor people from gaining access to electricity, "lifeline rates" – special low rates for small users – can be a cost-effective way of making services affordable to poor households. Alternatively, governments can finance part of the connection charge or oblige utilities to spread the cost out over time. The challenge is to ensure that electricity subsidies increase access for the poor at the lowest cost, while ensuring that electricity utilities are still able to make money and to continue to invest. That means limiting the size of subsidies and the number of recipients, and compensating the utility for any loss of revenue. This can be done either through higher charges for other customer categories or direct financial transfers from the government budget.

By improving efficiency and encouraging investment, electricity-sector reforms can speed up the pace of electrification, while improving the quality and lowering the cost of supply. Many developing countries have launched such reforms, but few have implemented them fully. Cost-reflective pricing and effective billing systems are vital to the financial health of electricity companies and to their ability to sustain investment. In countries such as India, theft and unmetered connections deprive the state-owned power companies of money that could be used to upgrade and extend the grid. But it can be very difficult, politically and socially, to raise prices and enforce the payment of bills. Many developing countries have tried to open up their electricity sectors to private capital from domestic and foreign sources. But such investment has slumped since 1997, because of poor returns and uncertainty about possible regulatory developments (IEA, 2003). Official development assistance has also fallen since the mid-1990s (UNDP/UNDESA/WEC, 2004).

Clearly, policy reforms and development priorities must be tailored to each country's situation. In the poorest African and Asian countries at the bottom of the EDI rankings, relying predominantly on private capital to develop energy infrastructure from scratch is unlikely to succeed, because of the risks involved. One way forward for these countries may be to establish public-private partnerships between host-country governments, donors, multi-lateral development banks, non-governmental organisations and private companies.

The rich industrialised countries have obvious long-term economic, political and energy-security interests in helping developing countries along the path to energy development. For, so long as poverty, hunger and disease persist, the poorest regions will remain vulnerable to social and political instability and to humanitarian disasters. The cost of providing assistance to poor countries may turn out to be far less than that of dealing with the instability and insecurity that poverty breeds.

APPENDIX TO CHAPTER 10: ELECTRIFICATION TABLES

Contents

Table 10.A1 – Urban, Rural and Total Electrification Rates by Region, 2002

Tables 10.A2 – Electricity Access in 2002 (country-by-country database)
- Africa
- Developing Asia
- Latin America
- Middle East

Table 10.A3 – Electrification Rate Projections by Region

Table 10.A4 – Projections of Urban and Rural Electrification Rates by Region

Definitions and Approach

Electricity Access

There is no single internationally-accepted definition for electricity access. The definition used here covers electricity access at the household level, that is, the number of people who have electricity in their home. It comprises electricity sold commercially, both on-grid and off-grid. It also includes self-generated electricity for those countries where access to electricity has been assessed through surveys by government or government agencies. The data do not capture unauthorised connections. The main data sources are listed in the tables. Each data point has been validated through a consistency-check process among different data sources and experts. The electrification rates shown in this appendix indicate the number of people with electricity access as a percentage of total population. Rural and urban electrification rates have been collected for most countries. Only the regional averages are shown in this publication. More information on the IEA's work on energy and development is available at http://www.worldenergyoutlook.org/poverty.

Where country data appeared contradictory, outdated or unreliable, the IEA Secretariat made estimates based on cross-country comparisons, earlier surveys, information from other international organisations, annual statistical bulletins, publications and journals. Population and Urban/Rural Breakdown Projections are from *World Population Prospects – The 2002 Revision*, published by the United Nations Population Division.

For the projections of electrification rates to 2030, a detailed model was used for sub-Saharan Africa, India and other South Asia. The electrification module is part of the IEA's World Energy Model.

The projections for African and South Asian regions are quantified using regional regressions based on several determinants: income, population growth, fuel prices, urbanisation rates, poverty levels, present and past electrification rates and electricity and biomass consumption figures. For the other regions where electrification exceeds 85% in 2002, a linear model was used based on gross domestic product and population growth, as well as on past electrification growth rates. The projections of electricity and biomass consumption are based on the World Energy Model described in Annex C.

Biomass Use

Data on biomass consumption are from IEA statistics, *Energy Balances of Non-OECD Countries,* 2004 edition. UN-FAO data are used for information on forest coverage and estimates of biomass supply. Biomass and traditional biomass are defined in Annex E. Projections for both biomass and traditional biomass energy demand by region/country are modelled in the World Energy Model and presented in Annex A.

Abbreviations

ADIAC - Agence d'Information d'Afrique Centrale
AFREPREN - African Energy Policy Research Network
APERC - Asia Pacific Energy Research Centre
AREED - African Rural Energy Enterprise Development
DOE - U.S. Department of Energy
DHS - Demographic and Health Surveys
EEPCo - Ethiopian Electric Power Corporation
ESMAP - Energy Sector Management Assistance Programme
GNESD - Global Network on Energy for Sustainable Development
MEMR - Ministry of Energy and Mineral Resources, Indonesia
OECD - Organisation for Economic Co-operation and Development
OLADE - Latin American Energy Association
UNDP - United Nations Development Programme
USAID - The United States Agency for International Development

Table 10.A1: Urban, Rural and Total Electrification Rates by Region, 2002

	Population	Urban population	Population without electricity	Population with electricity	Electrification rate	Urban electrification rate	Rural electrification rate
	million	million	million	million	%	%	%
North Africa	143	74	9	134	93.6	98.8	87.9
Sub-Saharan Africa	688	242	526	162	23.6	51.5	8.4
Africa	**831**	**316**	**535**	**295**	**35.5**	**62.4**	**19.0**
China and East Asia	1 860	725	221	1 639	88.1	96.0	83.1
South Asia	1 396	390	798	598	42.8	69.4	32.5
Developing Asia	**3 255**	**1 115**	**1 019**	**2 236**	**68.7**	**86.7**	**59.3**
Latin America	428	327	46	382	89.2	97.7	61.4
Middle East	173	114	14	158	91.8	99.1	77.6
Developing countries	**4 687**	**1 872**	**1 615**	**3 072**	**65.5**	**85.3**	**52.4**
Transition economies and OECD	1 492	1 085	7	1 484	99.5	100.0	98.2
World	**6 179**	**2 956**	**1 623**	**4 556**	**73.7**	**90.7**	**58.2**

Table 10.A2: Electricity Access in 2002 - Africa

	Electrification rate %	Population without electricity million	Population with electricity million	Sources
Angola	5.0	12.5	0.7	DOE Country Analysis Brief, AFREPREN (2001)
Benin	24.8	4.9	1.6	ESMAP, DHS (2001)
Botswana	26.4	1.3	0.5	DOE Country Brief on Southern Africa (2004), AFREPREN (2000)
Burkina Faso	10.0	11.4	1.3	OECD (2003), ESMAP (1998/99)
Cameroon	40.7	9.3	6.4	DHS (1998)
Congo	19.6	2.9	0.7	ADIAC
Côte d'Ivoire	50.7	8.1	8.3	DHS (1998/99)
DR Congo	8.3	46.9	4.3	GNESD (2004)
Eritrea	18.4	3.3	0.7	AFREPREN (2001)
Ethiopia	2.6	67.2	1.8	AFREPREN (2001), EEPCo, DHS (2000)
Gabon	47.9	0.7	0.6	ESMAP (2000)
Ghana	48.5	10.5	9.9	AREED, ESMAP (1998), DHS (1998)
Kenya	9.1	28.7	2.9	AFREPREN (2001), ESMAP (1998), DHS (1998)
Lesotho	5.0	1.7	0.1	GNESD (2004)
Madagascar	8.3	15.5	1.4	GNESD (2004)
Malawi	5.8	11.2	0.7	AFREPREN (2001), DHS (2000)
Mauritius	100	0.0	1.2	AFREPREN (2001)
Mozambique	8.7	16.9	1.6	AFREPREN (2000)
Namibia	34.7	1.3	0.7	AFREPREN (2000), DHS (2000)
Nigeria	44.9	66.6	54.3	ESMAP (1999)

Senegal	31.4	6.8	3.1	GNESD (2004), AREED, DHS (1999)
South Africa	67.1	14.7	30.0	AFREPREN (2001), ESMAP (1998)
Sudan	31.0	22.7	10.2	AFREPREN (2000)
Tanzania	9.2	33.0	3.3	AFREPREN (2001), AREED, DHS (1999), Helio International (2002)
Togo	17.0	4.0	0.8	ESMAP (1998)
Uganda	4.0	24.0	1.0	AFREPREN (2001), Ugandan Government, ESMAP (2000/01)
Zambia	18.4	8.7	2.0	AFREPREN (2001), DHS (2001/02)
Zimbabwe	40.9	7.6	5.3	AFREPREN (2001), DHS (1999)
Other Africa	7.0	83.9	6.3	Secretariat estimate
Sub-Saharan Africa	**23.5**	**526.3**	**161.6**	
Algeria	98.5	0.5	30.8	Ministry of Energy and Mining
Egypt	97.7	1.6	64.8	USAID, DHS (2000)
Libya	99.8	0.0	5.4	Secretariat estimate
Morocco	77.4	6.8	23.3	Ministry of Energy and Mines, Office National de l'Electricité
Tunisia	95.0	0.5	9.2	ESI Africa, Institut National de la Statistique
North Africa	**93.6**	**9.3**	**133.6**	

Africa	**36.0**	**535.6**	**295.2**	

Table 10.A2: Electricity Access in 2002 - Developing Asia

	Electrification rate %	Population without electricity million	Population with electricity million	Sources
China	99.0	12.9	1 275.3	Secretariat estimate
Brunei	99.2	0.0	0.3	APERC
Cambodia	18.3	11.3	2.5	GNESD (1998), DHS (2000)
Chinese Taipei	98.8	0.3	22.2	Secretariat estimate
DPR Korea	20.0	18.0	4.5	Secretariat estimate
Indonesia	52.5	100.5	111.2	PLN Statistiks 2002 (2003), MEMR (2002), GNESD (2004), DHS (2002/03)
Malaysia	97.1	0.7	23.3	GNESD (2000)
Mongolia	90.0	0.3	2.3	Helio International (2000)
Myanmar	5.0	46.4	2.4	GNESD (2000)
Philippines	89.1	8.7	69.8	GNESD (2004), DHS (1998)
Singapore	100.0	0.0	4.2	GNESD (2000)
Thailand	91.1	5.5	56.6	GNESD (2004)
Vietnam	79.6	16.3	63.9	GNESD (2001), DHS (2002)
Other Asia	80.0	0.0	0.2	Secretariat estimate
China and East Asia	**88.1**	**221.0**	**1 638.8**	
Afghanistan	2.0	22.5	0.5	World Bank, DOE Country Analysis Brief, UNDP
Bangladesh	26.3	100.5	35.8	GNESD(2000), Bangladesh Power Development Board, USAID, DHS(1999/00)
India	44.4	582.6	465.9	GNESD (2000), DHS (1998/99), Indian Census (2001)
Nepal	25.9	17.9	6.2	GNESD (2000), ESMAP, DHS (2001)
Pakistan	53.0	68.1	76.7	GNESD (2000)
Sri Lanka	65.5	6.5	12.4	GNESD (2001)
South Asia	**42.8**	**798.0**	**597.6**	
Developing Asia	**68.7**	**1 019.0**	**2 236.4**	

Table 10.A2: **Electricity Access in 2002 - Latin America**

	Electrification rate %	Population without electricity million	Population with electricity million	Sources
Argentina	95.0	1.9	36.1	GNESD (2004), OLADE (1998)
Bolivia	65.1	3.0	5.6	DHS (1998), OLADE (2002)
Brazil	94.6	9.5	165.1	OLADE (1999)
Chile	97.0	0.5	15.1	APERC (2001)
Colombia	90.2	4.3	39.3	DHS (2000)
Costa Rica	97.0	0.1	4.0	OLADE (2002)
Cuba	95.8	0.5	10.8	OLADE (2002)
Dominican Republic	92.3	0.7	8.0	DHS (2002), OLADE (2002)
Ecuador	89.7	1.3	11.5	OLADE (2002)
El Salvador	76.9	1.5	4.9	GNESD (2004), OLADE (2001)
Guatemala	84.4	1.9	10.2	ESMAP (1998/99), DHS (1998/99), OLADE (2002)
Haiti	33.5	5.5	2.8	DHS (2000), OLADE (1997)
Honduras	60.1	2.7	4.1	OLADE (2002)
Jamaica	87.0	0.3	2.3	OLADE (2002)
Netherlands Antilles	99.5	0.0	0.2	Secretariat estimate
Nicaragua	46.6	2.8	2.5	OLADE (2002), DHS (2001), Global Environment Facility (2001)
Panama	85.1	0.5	2.6	OLADE (2000)
Paraguay	85.3	0.8	4.9	OLADE (2002)
Peru	75.7	6.5	20.3	OLADE (2002), GNESD (2004), DHS (2000), APERC (2000)
Trinidad and Tobago	99.0	0.0	1.3	OLADE (1997)
Uruguay	99.0	0.0	3.4	OLADE (1997)
Venezuela	94.0	1.5	23.7	OLADE (2002)
Other Latin America	87.0	0.5	3.3	Secretariat estimate
Latin America	**89.2**	**46.3**	**381.7**	

Table 10.A2: **Electricity Access in 2002 - Middle East**

	Electrification rate %	Population without electricity million	Population with electricity million	Sources
Bahrain	99.9	0.0	0.7	Secretariat estimate
Iran	99.2	0.5	64.8	Tavanir, World Energy Council
Iraq	95.4	1.1	23.3	Secretariat estimate based on World Bank
Israel	100.0	0.0	6.5	Israel Electric Corporation (2003)
Jordan	95.5	0.2	4.9	Secretariat estimate based on World Bank
Kuwait	100.0	0.0	2.1	Secretariat estimate
Lebanon	96.0	0.2	4.3	Secretariat estimate based on World Bank
Oman	94.6	0.1	2.4	Secretariat estimate
Qatar	95.6	0.0	0.6	Secretariat estimate
Saudi Arabia	98.4	0.4	21.7	Secretariat estimate
Syria	86.6	2.3	14.7	Secretariat estimate based on UNDP
United Arab Emirates	97.4	0.1	3.0	Secretariat estimate
Yemen	50.3	9.3	9.4	Secretariat estimate based on World Bank
Middle East	**91.8**	**14.2**	**158.4**	

Table 10.A3: Electrification Rate Projections by Region (%)

	2002	2010	2015	2020	2030
North Africa	94	98	98	98	99
Sub-Saharan Africa	24	29	34	39	51
Africa	**36**	**41**	**44**	**49**	**58**
China and East Asia	88	93	94	95	96
South Asia	43	50	55	59	66
Latin America	89	93	95	95	96
Middle East	92	95	96	97	99
Developing countries	**66**	**70**	**72**	**74**	**78**
World	**74**	**77**	**78**	**80**	**83**

Table 10.A4: Projections of Urban and Rural Electrification Rates by Region (%)

	2002		2010		2015		2020		2030	
	Urban	Rural	Urban	Rural	Urban	Rural	Urban	Rural	Urban	Rural
North Africa	99	88	100	96	100	96	100	96	100	97
Sub-Saharan Africa	52	8	55	12	58	16	62	21	70	30
Africa	**62**	**19**	**65**	**23**	**67**	**26**	**69**	**30**	**75**	**38**
China and East Asia	96	83	98	88	100	88	100	88	100	89
South Asia	69	33	73	40	77	44	81	46	88	50
Latin America	98	61	100	68	100	71	100	73	100	76
Middle East	99	78	100	85	100	87	100	90	100	95
Developing countries	**85**	**52**	**88**	**57**	**89**	**58**	**90**	**59**	**92**	**61**
World	**91**	**58**	**92**	**61**	**93**	**62**	**93**	**63**	**94**	**65**

For reference:

Total number of people without electricity (million)	275	1 347	289	1 267	290	1 249	295	1 210	287	1 106

CHAPTER 11

WORLD ALTERNATIVE POLICY SCENARIO

HIGHLIGHTS

- The World Alternative Policy Scenario depicts a more efficient and more environment-friendly energy future than does the Reference Scenario. It demonstrates that policies to address environmental and energy-security concerns that countries are already considering, together with faster deployment of technology, would substantially reduce energy demand and carbon-dioxide emissions.
- Global primary energy demand is about 10% lower in 2030 than in the Reference Scenario. The reduction in demand for fossil fuels is even more pronounced, thanks mainly to policies that promote renewable energy.
- By 2030, oil demand is 12.8 mb/d lower than in the Reference Scenario, an amount equal to the current combined oil production of Saudi Arabia, the United Arab Emirates and Nigeria. Stronger measures to improve fuel economy in OECD countries and faster deployment of more efficient vehicles in non-OECD countries account for almost two-thirds of these savings in 2030. Oil-import dependence in the OECD countries and China diminishes as a result.
- Demand for coal falls more steeply than that for any other fuel. In 2030, it is almost a quarter below the Reference Scenario. The amount saved is roughly equal to the current coal consumption of China and India combined. World natural gas demand is 10% lower. Gas-import needs are 40% lower in OECD North America and 13% lower in Europe. China's imports are higher, because of a switch away from coal.
- Energy-related emissions of carbon dioxide would be reduced by some 6 gigatonnes, or 16%, below the Reference Scenario figure in 2030. This is roughly equal to the combined current emissions of the United States and Canada. OECD emissions peak around 2020, and then start to decline. Almost 60% of the cumulative reduction in CO_2 emissions would take place in non-OECD countries.
- More efficient use of energy in a wide range of applications, including vehicles, electric appliances, lighting and industrial uses, account for almost 60% of the reduction in CO_2 emissions. A shift in the fuel mix for power generation in favour of renewables and nuclear energy power accounts for most of the rest.

- In the Alternative Scenario, larger capital needs on the demand side would be entirely offset by lower investment needs on the supply side – despite a 14% increase in the capital intensity of the electricity sector in the Alternative Scenario. Electricity prices would rise. They would be 12% higher in the European Union, for example. It is uncertain, however, whether all the investment invoked in the Alternative Scenario could actually be financed, especially in developing countries. This is mainly because end-users, who would have to invest more, are likely to find it harder to secure financing than would suppliers, who would need to invest less.

This chapter describes a markedly different energy future from that set out in the preceding chapters. It analyses the effect on global energy markets of energy efficiency and environmental policies beyond those considered in the Reference Scenario. The first section sets out the reasons for preparing the Alternative Policy Scenario, the scope of analysis and our approach. The results are then presented by region and by sector, including trends in overall demand and the fuel mix and their implications for CO_2 emissions, trade and investment needs. Detailed tables are included at the end of the chapter.

Background and Approach

Why an Alternative Scenario?

The Reference Scenario presented so far in this book takes into account all government policies and measures that had been adopted by mid-2004. It does *not* include policy initiatives that might be adopted in the future. Energy markets will very probably evolve in different ways from those depicted in this scenario, because the policy landscape will change.

In the Reference Scenario, global energy use and carbon-dioxide emissions continue to grow rapidly and fossil fuels continue to dominate the energy mix. Almost all OECD countries and an increasing number of developing countries are actively considering a range of new policies and measures to meet the environmental and energy-security concerns sparked by these trends.

The Alternative Scenario analyses how the global energy market could evolve were countries around the world to adopt a set of policies and measures that they are either currently considering or that they might reasonably be expected to implement over the projection period. We examine in detail the effectiveness of those policies in addressing environmental and energy-security concerns, and their implications for supply, trade and investment. We update and extend the detailed analytical work of the last two *World Energy Outlooks*.[1]

For each major region, the Alternative Scenario considers policies and measures to reduce air pollution and greenhouse-gas emissions, and to enhance energy security. Measures to improve energy efficiency and increase the use of renewables are among the main instruments. The basic assumptions about macroeconomic conditions and population are the same as in the Reference Scenario. But energy prices change, because of the new equilibrium between supply and demand.

The investment requirements for energy-supply infrastructure and for end-use equipment have been quantified for all regions in the Alternative Scenario. Carbon capture and storage and advanced nuclear technologies are discussed at the end of this chapter, but we have not taken these or any other breakthrough technologies into account.

Methodology

The Alternative Scenario considers those policies and measures that countries are currently considering or might reasonably be expected to adopt taking account of technical and cost factors, the political context and market barriers. The aim is to present a consistent picture of how global energy markets might evolve if governments decided to strengthen their environmental and energy-security policies. The projected energy savings and reductions in CO_2 emissions do not fully reflect the ultimate technical or economic potential. Even bigger reductions are possible, but they would require policy efforts that go beyond what governments are currently considering. The policy measures analysed have not been selected strictly according to their economic cost-effectiveness, but rather to reflect the current energy-policy debate. The main policies are outlined in the regional section of this chapter.[2]

For OECD regions, the Alternative Scenario analyses the impact of policies and measures that governments are currently considering and that could be adopted some time during the projection period. An example is given in Box 11.1. These policies "in the pipeline" are in addition to those that had already been implemented as of mid-2004, and which are included in the Reference Scenario.

---

1. *WEO-2000* contained an Alternative Scenario for power generation and transport in OECD countries. *WEO-2002* contained an OECD Alternative Policy Scenario, which considered the impact of all the policies and measures that were then under discussion in OECD countries. It did not take into account any initiatives being considered in non-OECD countries. Yet the developing countries are expected to contribute most over the projection period to rising energy production, demand and emissions of greenhouse gases. In order to capture the global effects of additional government actions, this *Outlook*, for the first time, presents an Alternative Policy Scenario covering all world regions.
2. More information on the policies analysed can be found in www.worldenergyoutlook.org.

> **Box 11.1: Example of OECD Policies Included in the Alternative Policy Scenario: the EU Renewables Target**
>
> The development of renewable energy is a key element of the energy policies of all European Union countries. A 1997 EU White Paper on Renewable Energy Sources proposed a target of 12% for the share of renewables in total energy consumption in 2010, compared with 6% in 1995. In 2001, the European Union adopted a directive aimed at increasing the share of renewables in electricity generation from 13.9% in 1997 to 22.1% in 2010. Under current policies, we do not expect the target to be met.
>
> In the Reference Scenario, the renewables share of EU electricity generation will reach 18.3% in 2010. The Alternative Policy Scenario assumes that additional policies will be put in place to meet the target. There are no targets for renewables beyond 2010 at EU level, although some countries have set national ones. The intention, however, is to continue the shift to renewables beyond this period. It is assumed that continuing support for renewables would increase their share in electricity generation to 34% in 2030. This much bigger share will be achieved not only through additional policies to promote renewables, but also by policies to reduce electricity consumption.

Many of the policies considered here push for faster deployment of more efficient and less polluting technologies. As these technologies are deployed in OECD countries, their unit costs fall, and they eventually become affordable for *all* countries.

As with OECD countries, the developing country policies assessed in the Alternative Scenario include those currently under discussion at the national level. In general, however, there are fewer such policies than in OECD countries, because environmental issues and energy-security concerns are lower on the agenda than in the OECD. But it is likely that many of these countries *will* devise new policies in the future to tackle problems in these areas. In most cases, the environmental policies would tackle local or regional pollution, though some countries could take climate-change effects into consideration in devising their policies. More efficient and less polluting technologies are assumed to become more widely and rapidly available to these countries, thanks to their faster development and deployment in OECD countries. As a result, global energy intensity falls more rapidly in this scenario than in the Reference Scenario.

The rates of efficiency gains vary with local conditions, including past efforts to encourage more efficient energy use and to reduce environmental damage. On average, the improvement in energy efficiency is assumed to be higher in the developing world than in OECD countries. This reflects a far larger potential for efficiency improvements, as well as a faster rate of technology transfer from the OECD. Depending on the region, the measures taken into account include a strengthening of existing policies, a wider coverage of existing policies and introduction of new policies. An example is given in Box 11.2.

> *Box 11.2:* **Example of Non-OECD Policies Included in the Alternative Policy Scenario: Vehicle Efficiency in China**
>
> China's Tenth Five-Year Plan for Energy Conservation and Resources (2001-2005) gives priority to improving energy efficiency. In line with the plan, the Chinese government has developed new motor-vehicle fuel-efficiency standards. When the rules take effect in 2005, the requirements for new cars will be, on average, 8% higher than in 2000. When the standards are fully implemented in 2008, fuel-efficiency requirements will be 7% to 10% higher, according to the vehicle's weight. By then, new passenger vehicles in China will be as efficient as those in Japan and *more* efficient than those in the United States. As these standards have already been enacted, they are taken into account in our Reference Scenario projections.
>
> In the Alternative Scenario, an additional 10% improvement is assumed in the efficiency of Chinese cars between 2008 and 2030. By about the 2020s, the efficiency of new cars sold in China will surpass the high standards currently in place in Europe. This scenario also assumes that similar fuel-efficiency standards will be applied to trucks and buses, which make up one-fifth of China's four-wheel vehicle fleet. The Alternative Scenario includes several other assumptions regarding China's transport sector, including an increased adoption of alternative-fuel vehicles and a modal shift from road to high-speed and intra-city rail traffic.

Many of the policies considered have effects at a very micro-level in the economy. The effects of mandatory efficiency standards, for example, cannot be estimated from past patterns of energy use, since these standards impose new technical constraints on the energy system. To analyse such measures, we have incorporated detailed "bottom-up" sub-models of the energy system into the IEA's World Energy Model.[3]

A key aspect of the model is the explicit representation of energy efficiency, of the different types of activity that drive energy demand and of the physical

---

3. See Annex C for a detailed description of the structure and main characteristics of the World Energy Model.

stock of capital. Capital-stock turnover is a key issue. The very long life of power plants, buildings and even cars limits the rate at which more efficient technology can be deployed. The detailed capital stock turnover sub-models within the World Energy Model factor in this very important consideration.

Key Results

Energy Demand[4]

In the Alternative Scenario, global primary energy demand in 2030 reaches 14 654 Mtoe – 1 671 Mtoe less than in the Reference Scenario, a difference of about 10% (Figure 11.1). The amount of energy "saved" in that year is roughly equal to total primary energy demand in the European Union today. Energy demand is projected to grow by 1.3% per year, 0.4 percentage points less than in the Reference Scenario. The reduction in demand for fossil fuels is even bigger, thanks to the use of more efficient technology and switching to carbon-free fuels. Demand for fossil fuels is 1 895 Mtoe, or 14%, lower in the Alternative Scenario, but fossil fuels still account for 78% of energy demand in 2030. On the other hand, the use of non-hydro renewables, excluding biomass, increases strongly. In 2030, their use is 30% higher than in the Reference Scenario. Biomass and nuclear energy also grow. None of those fuels emits any carbon dioxide. The impact of energy-saving policies on energy demand grows throughout the projection period, as the stock of energy capital is gradually replaced and new measures are introduced. Global energy savings achieved by 2010 are only 2%.

Coal use rises in both the Reference and Alternative Scenarios. But, at 2 744 Mtoe in 2030, it is 857 Mtoe, or almost a quarter, less in the Alternative Scenario than in the Reference Scenario. The saving is bigger than for any other fuel, both in absolute and in percentage terms. The average annual growth rate of coal demand is 0.5%, a full percentage point lower than in the Reference Scenario. Almost all the growth occurs during the first decade, with demand levelling off in the second half of the projection period. About 90% of the shortfall in primary coal demand comes from power generation. Coal use in that sector is driven down by lower electricity demand, by the increased thermal efficiency of coal-fired power plants – especially in developing countries – and by switching to other fuels. Savings in coal consumption are also significant in industry, especially iron and steel, and in coal transformation.

Primary oil demand rises to just under 5 000 Mtoe in 2030 in the Alternative Scenario, 610 Mtoe, or 11%, lower than in the Reference Scenario. The

---

4. Tables showing energy demand and CO_2 emissions by region in the Alternative Scenario can be found in the appendix to this chapter.

transport sector accounts for almost two-thirds of the savings. Increased fuel efficiency and faster penetration of alternative-fuel vehicles – powered by compressed natural gas, biofuels, fuel cells, or gasoline-powered hybrids – are the main factors behind the dip in transport oil demand. Industry, the residential and commercial sectors, and power generation each account for around 10% of oil savings.

Natural gas demand is 10% lower in 2030 than in the Reference Scenario. Again, the power sector accounts for most of the savings, almost three-quarters by 2030. The residential and commercial sectors account for an additional 20%, and industry for the rest.

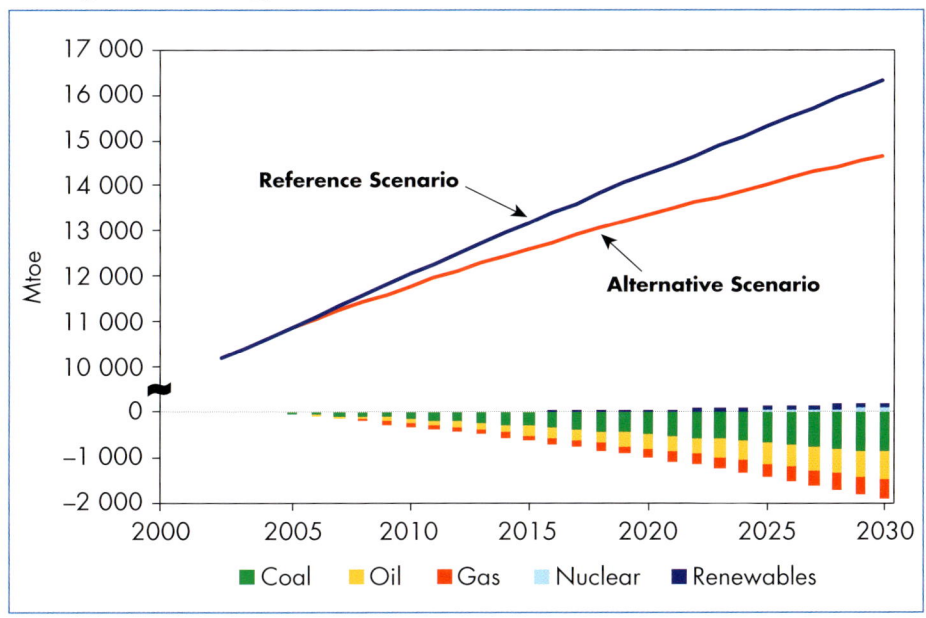

Figure 11.1: **Energy Demand in the Reference and Alternative Scenarios**

Renewables displace part of the relative reduction in fossil-fuel consumption. By 2030, global consumption of biomass is 43 Mtoe higher in the Alternative Scenario than in the Reference Scenario. This occurs despite reduced final consumption of biomass in developing countries, because of its more efficient use in industrial processes and in household cook-stoves. On the other hand, government incentives foster increased use of biomass in the power sector and in transportation, mainly in OECD countries. Consumption of other renewables increases even more, adding 75 Mtoe in 2030 – a 30% increase compared with the Reference Scenario. Power generation drives most of this increase, but solar water heaters and geothermal energy also contribute.

At final consumption level, electricity and heat demand are markedly lower in the Alternative Scenario. In 2030, electricity demand is 3 100 TWh, or 12%, lower than in the Reference Scenario. Energy-efficiency measures for industrial processes, appliances and lighting are the main causes of these savings in all regions. The residential sector accounts for 40% of the drop in electricity demand. Heat demand is 18 Mtoe, or 6%, lower in the Alternative Scenario. Reduction in heat demand is greatest in the transition economies, where district heating now is widespread and very inefficient.

Global energy intensity falls by 1.8% per year in the Alternative Scenario, compared with 1.4% in the Reference Scenario. As a result, intensity is 10% lower in 2030 in the Alternative Scenario. This decline is more pronounced than the 1.5% rate observed over the period 1990-2002. The difference in projected intensity between the two scenarios is more pronounced in developing countries and the transition economies (Figure 11.2). This is because of the larger potential in these regions for energy-efficiency improvements in end-use sectors and in power generation. In the OECD, energy intensity would fall by 1.5% per year, compared with 1.2% in the Reference Scenario.

Figure 11.2: **Change in Energy Intensity in the Reference and Alternative Scenarios, 2002-2030**

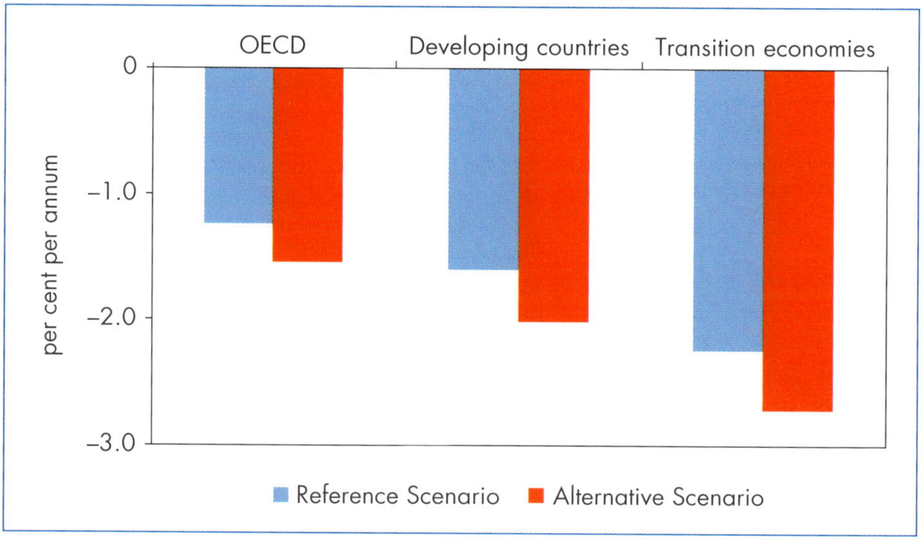

Implications for Energy Supply

The reduction in primary energy demand in the Alternative Scenario leads to major changes in fossil-fuel supply patterns and inter-regional trade. Global oil demand in 2030 is reduced by 12.8 mb/d[5] in the Alternative Scenario, an amount equal to the current combined production of Saudi Arabia, the United Arab Emirates and Nigeria. We estimate that this drop in demand would reduce the call on OPEC production by around 7 mb/d in 2030 – or 10% – compared with the Reference Scenario. In the Alternative Scenario, world oil prices would average 15% less than in the Reference Scenario, as a result of reduced pressure on supply. Net imports of oil in OECD countries would reach some 38 mb/d in 2030, a decrease of some 3 mb/d compared to the Reference Scenario. Chinese oil imports will be reduced significantly, by 12% compared to the Reference Scenario. The transport sector contributes the most to global oil-demand reduction, some 8 mb/d in 2030. Savings of oil in industry, and residential and commercial sectors are also significant, especially in developing countries (Figure 11.3).

Figure 11.3: **Reduction in Oil Demand by Sector in the Alternative Scenario*, 2030**

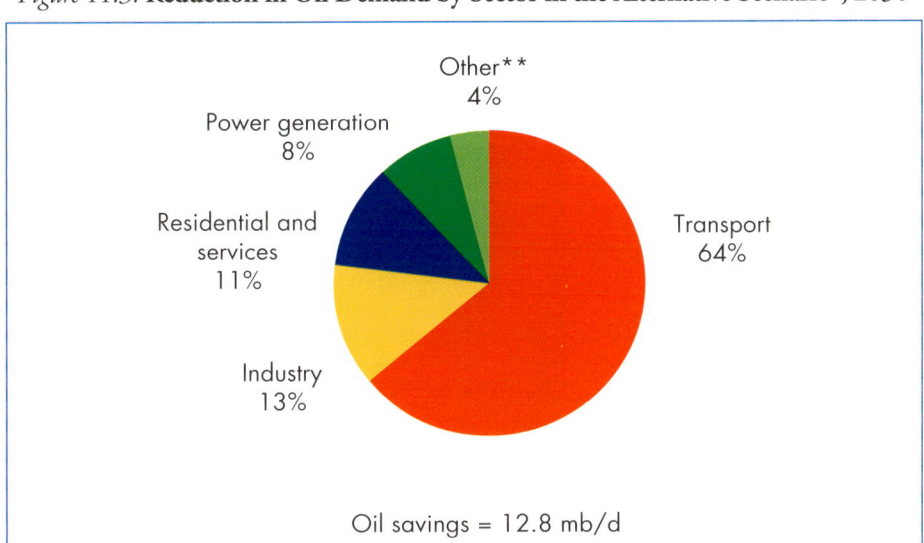

* Compared with the Reference Scenario.
** Includes non-energy use and other transformation.

Global demand for natural gas in 2030 would be some 500 billion cubic metres lower than in the Reference Scenario. This is roughly equal to the combined current production of Africa, the Middle East and Latin America. Lower demand, together with lower oil prices, would result in markedly lower

5. Oil demand savings do not include international marine bunkers.

gas prices – especially in North America. By 2030, gas imports to OECD North America would be some 80 bcm lower – equivalent to the output of eight large LNG regasification terminals (Figure 11.4).

The savings in gas consumption in OECD Europe would also be large, amounting to around 70 bcm in 2030. This would greatly reduce Europe's need to build LNG terminals. Imports of Russian gas would not be seriously affected. China, by contrast, would see a large *increase* in gas consumption compared with the Reference Scenario. This is the result of China's switch from coal to gas in both power generation and end-use sectors in an effort to cut local pollution.[6] China's need to import gas also increases substantially and its gas-import dependence is more than twice as high in the Alternative Scenario as in the Reference Scenario.

Figure 11.4: **Net Natural Gas Imports in Selected Regions in the Reference and Alternative Scenarios, 2030**

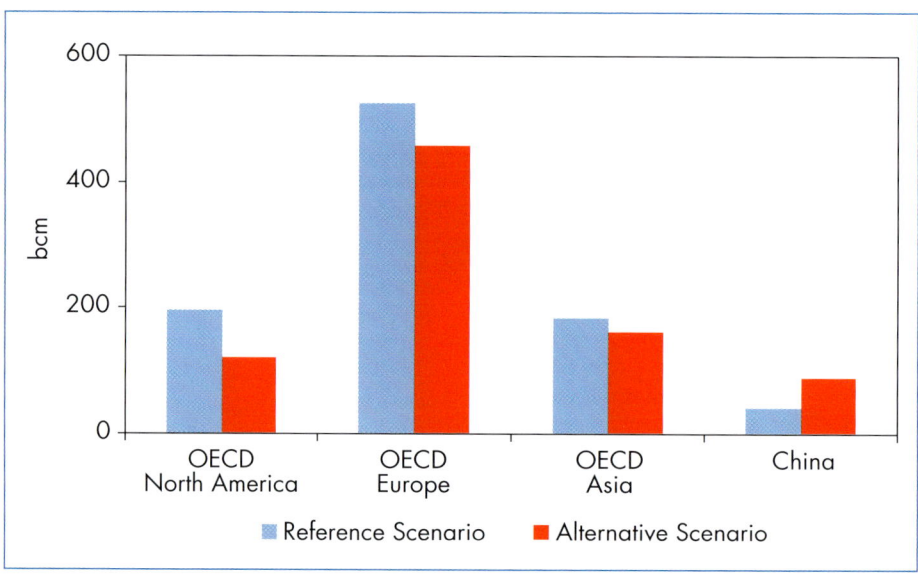

In 2030, the world's three largest coal producers – China, the United States and India – will bear two-thirds of the total reduction in coal output in the Alternative Scenario, but these countries will still account for over 60% of global coal production. China increases its coal production by over 530 Mt in

---

6. China aims to increase natural gas consumption to 50 bcm by 2005, and to achieve a 10% to 15% share of gas in the primary energy mix by 2020. In the Alternative Scenario, gas demand grows by 6.5% per year. However, gas still only meets about 7% of the country's energy needs in 2020. In the Reference Scenario, the share is only 5%.

the Alternative Scenario, but this is much lower than the 1 100 Mt increase projected in the Reference Scenario. In the Alternative Scenario, coal trade levels off after 2020, reaching around 800 Mt in 2030 – a quarter less than in the Reference Scenario. The main importing regions, notably OECD Europe and OECD Asia, see a big reduction in their domestic demand and imports.

Carbon-Dioxide Emissions

In the Alternative Scenario, energy-related CO_2 emissions are 31 686 million tonnes in 2030, 37% higher than current emissions. This is about 6 Gt, or 16%, lower than in the Reference Scenario (Figure 11.5). The reduction is comparable to the current combined emissions of the United States and Canada. The annual growth rate of emissions over the projection period falls from 1.8% in the Reference Scenario to 1.1%. The gap widens in the last decade of the projection period, during which the annual emissions growth rate is halved, from 1.4% to 0.7%. An increase in the share of carbon-free fuels in the fuel mix makes an important contribution to the reduction in emissions. By 2030, carbon-free fuels account for 22% of global primary energy demand in the Alternative Scenario, four percentage points more than in the Reference Scenario. Among the fossil fuels, coal sees the biggest drop in market share. On average, emissions of CO_2 per unit of energy consumed are 5% lower in 2030 than in 2002 and 6% lower than in the Reference Scenario in 2030.

Figure 11.5: **Global Energy-Related CO_2 Emissions in the Reference and Alternative Scenarios**

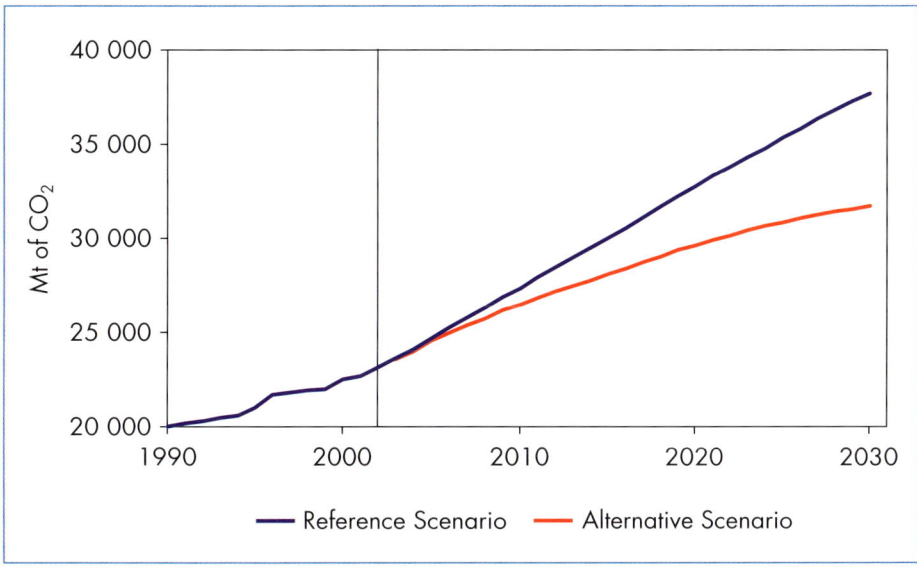

In the Alternative Scenario, CO_2 emissions in OECD countries peak at around 13 750 Mt in 2020 and then start to decline. In the transition economies, emissions growth slows dramatically in the 2020s, to only 0.2% per year. They nearly stabilise by 2030. In developing countries, emissions continue to rise throughout the *Outlook* period, even though their CO_2-emission savings are larger both in share and in volume than in the OECD and in the transition economies. OECD countries accounted for more than half of total emissions in 2002. In 2030, their share will drop to 42%. Developing countries account for almost half of the cumulative reduction in CO_2 emissions over the projection period compared with the Reference Scenario, OECD countries account for 42% and the transition economies for the rest. China alone accounts for 21%, a slightly bigger saving than that of OECD North America (Figure 11.6).

Figure 11.6: **Cumulative Reduction in Energy-Related CO_2 Emissions by Region in the Alternative Scenario*, 2002-2030**

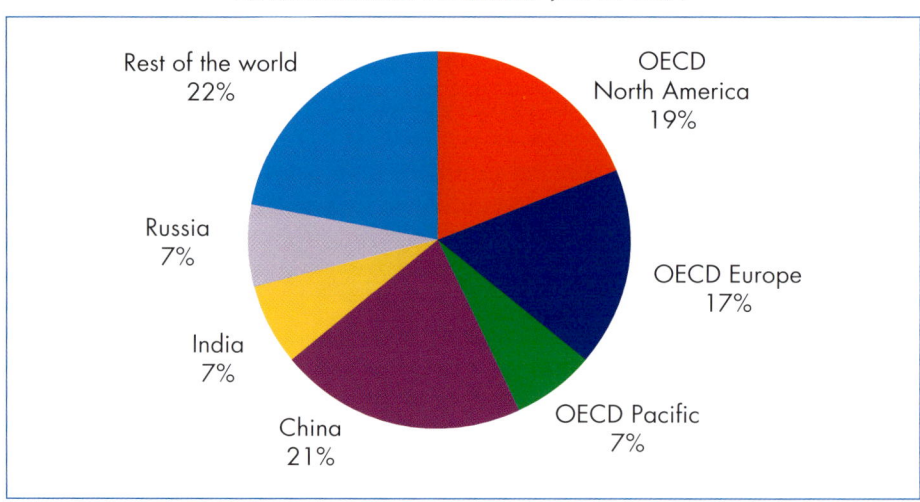

* Compared with the Reference Scenario.

The difference between growth rates of CO_2 emissions in the two scenarios is summarised in Figure 11.7. Measures to improve end-use efficiency explain almost 60% of the difference worldwide. These measures include more efficient vehicles, industrial processes and appliances, as well as stricter building standards. In the transition economies and in the developing countries, the role played by energy-efficiency measures is particularly large, reflecting the enormous potential for efficiency improvements there. The other big contributor to lower emissions is the increased share of renewables in power generation, accounting for 20% of the global reduction. The increased role of nuclear power accounts for an additional 10%. Fuel switching in end-uses and switching from coal to natural gas in power generation explain the rest.

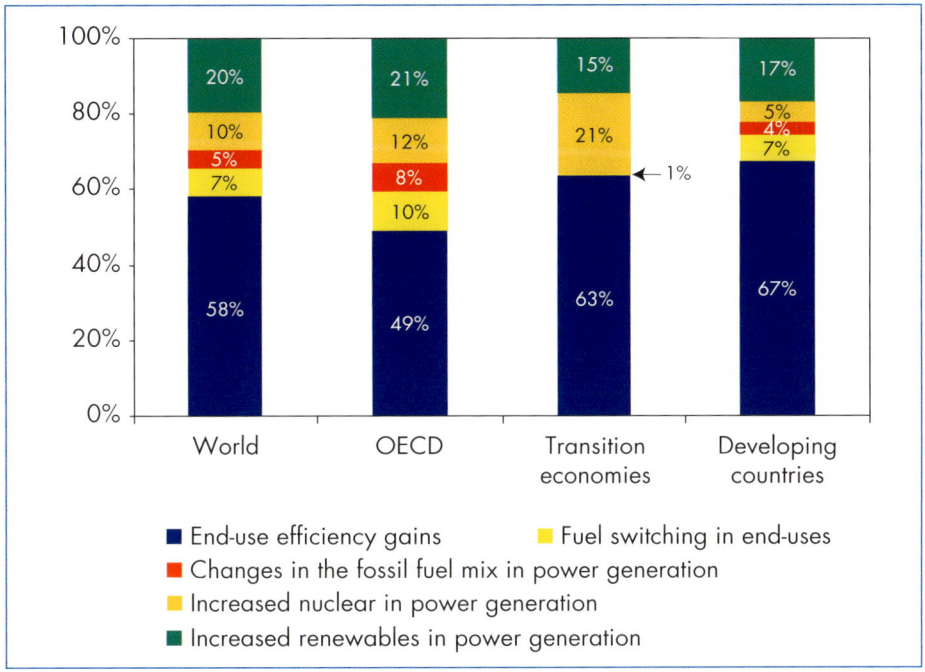

Figure 11.7: **Reduction in Energy-Related CO_2 Emissions in the Alternative Scenario* by Contributory Factor, 2002-2030**

* Compared with the Reference Scenario.

The electricity sector contributes the biggest reduction in CO_2 emissions in all regions (Table 11.1). In 2030, it accounts for two-thirds of the difference between the two scenarios. Of the 3 900 million tonnes reduction in electricity-sector emissions, nearly 40% comes from reduced electricity

Table 11.1: **Changes in Energy-Related CO_2 Emissions by Sector and by Region in the Alternative Scenario*, 2030** (Mt CO_2)

	OECD	Transition economies	Developing countries	World
Power generation	-1 627	- 340	-1 938	**-3 905**
Industry	- 134	- 78	- 371	**- 583**
Transport	- 557	- 59	- 381	**- 997**
Other	- 193	- 84	- 251	**- 528**
Total	**-2 511**	**- 561**	**-2 941**	**-6 013**

* Compared with the Reference Scenario.

demand. Higher thermal efficiency of power plants, changes in the fuel mix and smaller transmission losses account for the remaining 60%. Transport accounts for 17% of savings. The sectoral breakdown of emissions reductions does not vary much among regions. The main exception is industry in OECD countries, where efficiency improvements and emissions reductions are more modest than elsewhere.

Investment Outlook

The demand and supply trends in the Alternative Scenario would entail a dramatic change in the pattern of energy investment compared with the Reference Scenario.[7] The amounts of capital required for the entire energy chain – from energy production to end-uses – do not differ much between the two scenarios. In aggregate, larger capital needs on the demand side are entirely offset by lower needs on the supply side. But there are important differences at the regional and sectoral levels. Several major trends can be highlighted: much more investment will be needed in end-use equipment; the capital intensity of new power generation will be greater; and research and development on all energy technologies will have to increase.[8] Our projections have taken these factors into account. It is uncertain, however, whether all the investment invoked in the Alternative Scenario could actually be financed, especially in developing countries. This is mainly because end-users, who would have to invest more, are likely to find it harder to secure financing than would suppliers, who would need to invest less.

Investment by final consumers in energy equipment in the transport, industrial, residential and commercial sectors is more than $2 trillion *higher* in the Alternative Scenario. The capital costs of more efficient and cleaner end-use technology are generally higher, especially in the transport sector. But the result of such investment is to drive down energy demand, thereby reducing investment requirements for energy-supply infrastructure. These effects vary considerably among regions (Figure 11.8). OECD countries will see a net *increase* in their overall investment needs in the Alternative Scenario. The costs that developed countries will economise on the supply side will not compensate them for the expensive new investment they will have to make in end-use efficiency. In non-OECD regions, the reductions in supply-side investment more than outweigh increased capital spending on end-use equipment. The introduction of more efficient end-use technologies is less costly in non-OECD regions. Even so, the higher outlays demanded from final consumers may be an

---

7. The quantification of economic welfare in the two scenarios was beyond the scope of the analysis presented here. Thus, environmental costs, variable costs, losses in oil and gas revenues in exporting countries and other macroeconomic effects are not included in this analysis.
8. In recent years, R&D in end-use technologies, electricity transmission, renewables and alternative fuels has declined in many countries (IEA, 2003).

obstacle to the deployment of those technologies, especially in very poor countries. In both the Reference and Alternative Scenarios, capital requirements in the energy chain are very high outside the OECD. A sizable increase in foreign direct investment would be needed to finance them.

Figure 11.8: **Difference in Cumulative Energy Investment between the Reference and Alternative Scenarios by Region, 2003-2030**

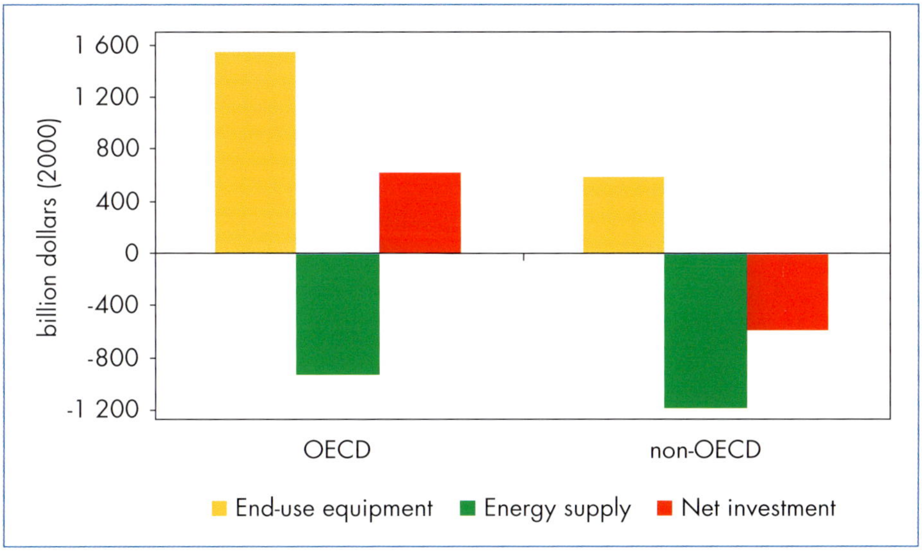

Demand-Side Investment

In the Alternative Scenario, consumers need to invest $2.1 trillion more in end-use technology than in the Reference Scenario.[9] More than two-thirds of this additional investment, or $1.6 trillion, is needed in OECD countries, where the capital cost of more efficient and cleaner technologies is highest. Transportation is by far the most capital-intensive end-use sector. Investment in transport increases by $1.1 trillion in the Alternative Scenario, more than half of the total additional demand-side investments foreseen. Investment in the residential, commercial and agriculture sectors is more than $600 billion higher than in the Reference Scenario, while industry has to invest an added $440 billion.

---

9. Most of our estimates of the capital cost of end-use technology are based on a co-operative effort between the Argonne Laboratory in the United States and the IEA (Hanson and Laitner, 2004). A number of independent sources were used for consistency-checking purposes. For certain fuel and end-uses, not covered by the Argonne study, the costs are based on IEA estimates. Given the vast regional and sectoral coverage of this scenario, there are many uncertainties surrounding these estimates.

Table 11.2: **Additional Demand-Side Investment in the Alternative Scenario*, 2003-2030** ($ billion)

	OECD	Non-OECD	World
Industry	255	186	442
of which electrical equipment	*143*	*114*	*257*
Transport	813	279	1 092
Other sectors	484	128	612
of which electrical equipment	*373*	*61*	*433*
Total	**1 552**	**594**	**2 145**

* Compared with the Reference Scenario.

The share of non-OECD countries in additional demand-side investment ranges from 20% to 40%, according to the sector. The capital cost of end-use technologies in developing and transition countries is always much lower than in OECD countries, because of lower labour costs. In the United States, for example, an air-conditioning unit might cost 40% less than one that is 25% more efficient. But the savings in running costs may not be big enough to compensate for the higher purchase price of the unit. In the Alternative Scenario, the capital costs of more efficient equipment are lower than in the Reference Scenario because of the faster deployment induced by aggressive new government policies. The air-conditioner will, therefore, be cheaper and more attractive to consumers.

Supply-Side Investment

In the Alternative Scenario, the worldwide investment requirement for energy-supply infrastructure over the period 2003-2030 is $13.8 trillion – $2.1 trillion, or 13%, less than in the Reference Scenario (Figure 11.9). The reduction in cumulative supply-side investment in developing countries amounts to about $900 billion, which is 12% lower than the Reference Scenario. The absolute amount is similar to that in OECD countries.

Savings in electricity-supply investment account for more than two-thirds of the overall reduction in investment in the Alternative Scenario. The capital needed for transmission and distribution networks is almost $1.2 trillion lower, thanks mainly to lower demand but also to the wider use of distributed generation. The fall in cumulative investment in power generation, at $300 billion, is proportionately much smaller. This is because the capital intensity of renewables, nuclear power and distributed generation is higher than that of fossil fuels. The capital cost per kWh produced is 14% higher on average than in the Reference Scenario. Although less new capacity is needed, the average cost of that capacity is higher (Figure 11.10).

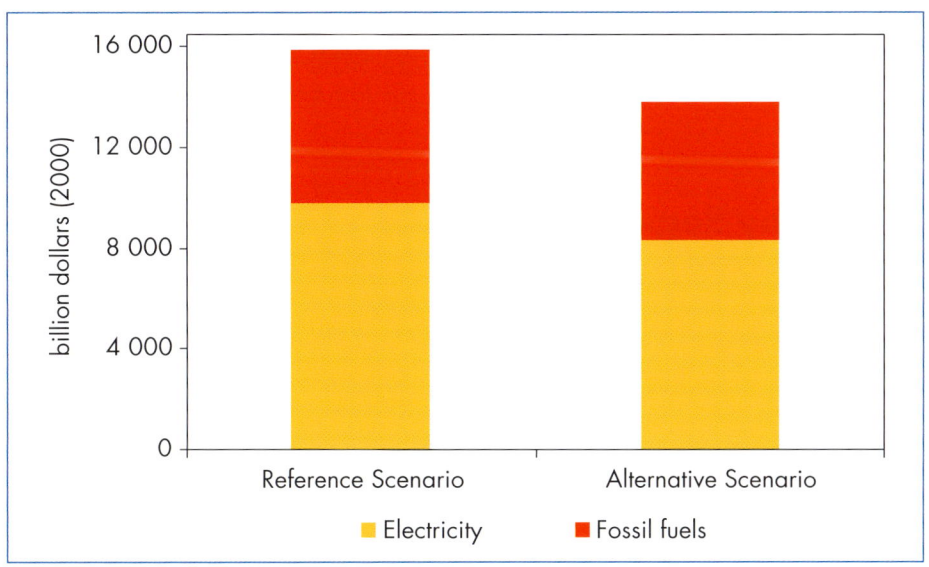

Figure 11.9: **Energy-Supply Investment in the Reference and Alternative Scenarios, 2003-2030**

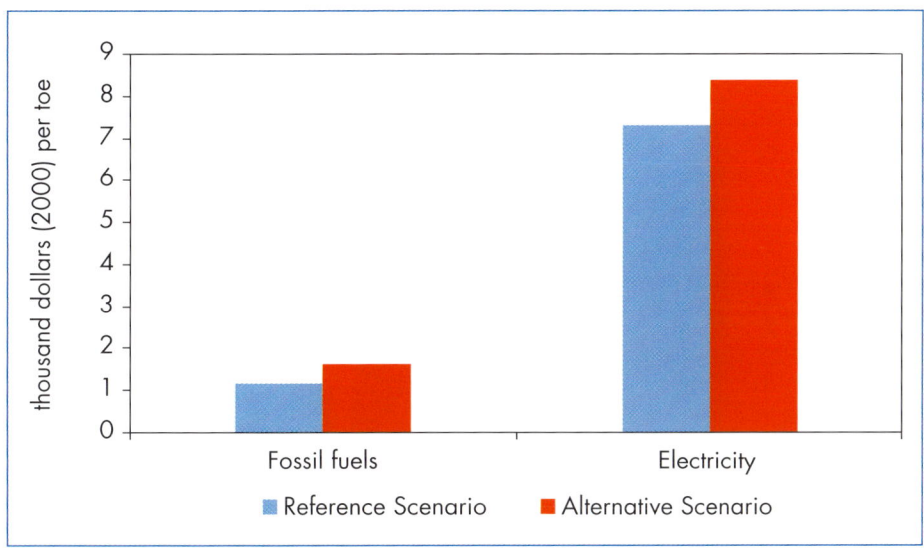

Figure 11.10: **Capital Intensity of Energy Supply in the Reference and Alternative Scenarios, 2003-2030**

Total investment in oil and gas is $550 billion, or 10%, lower in the Reference Scenario than it would otherwise be, mainly because there is less need to expand production. The impact will be greatest on high-cost marginal fields, such as deep-water offshore and non-conventional oilfields. Reduced requirements for transportation infrastructure contribute significantly to the $240 billion reduction in needed gas investment. Investment needs in the coal industry are reduced by 22%, from almost $400 billion in the Reference Scenario to around $310 billion. China accounts for about a third of that difference.

Results by Region

Developing countries account for more than half of the 1 671 Mtoe of energy saved in 2030 in the Alternative Scenario. Their share of energy savings varies slightly according to fuel (Figure 11.11). The reduction in coal use is biggest in developing countries, reflecting the large potential for improving the efficiency of their coal-fired power plants. India and China account for about 50% of the worldwide reduction in coal demand. Developing countries contribute almost half the reduction in oil demand, because of improvement in the fuel-efficiency of road vehicles and the increased efficiency of industrial processes. The OECD and the transition economies account for most of the projected gas savings, since their gas consumption is higher than that of developing countries.

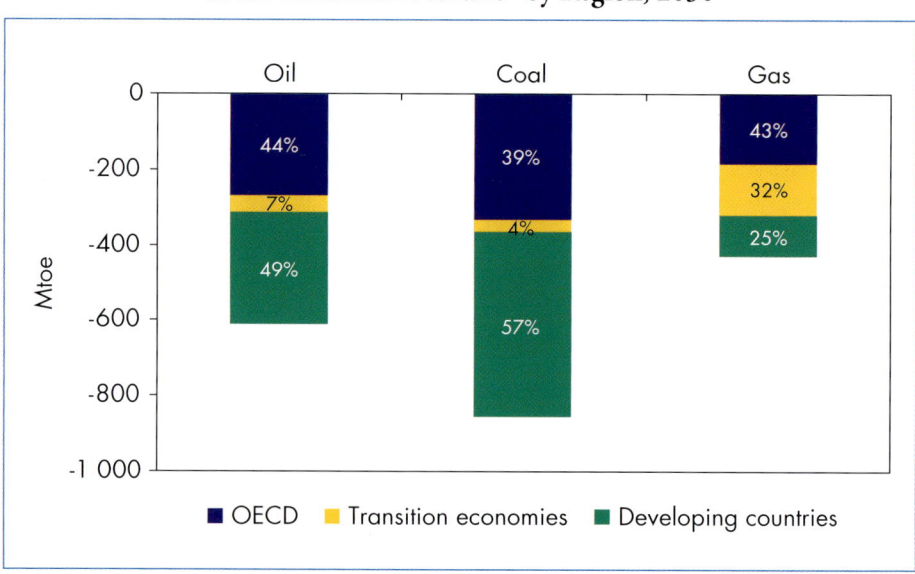

Figure 11.11: **Reduction in Demand for Fossil Fuels in the Alternative Scenario* by Region, 2030**

* Compared with the Reference Scenario.

The consumption of non-hydro renewables, excluding biomass, increases substantially in every region compared with the Reference Scenario. The increases are biggest in percentage terms in the developing countries, because they start from a very low base. Those countries account for almost 60% of the 75 Mtoe increase in world renewables consumption in 2030. Biomass consumption increases in OECD countries and in the transition economies, spurred by policies to boost its use in power generation and, to a lesser extent, in transport. Developing countries move away from traditional biomass more quickly than they do in the Reference Scenario, a trend that more than offsets the increase in biomass use in the power sector. In 2030, total biomass consumption in developing countries is 72 Mtoe, or 6%, lower in the Alternative Scenario than in the Reference Scenario. Global output of nuclear power increases by 400 TWh, or 14%. OECD countries and Russia together account for more than 80% of this increase.

OECD Regions and the EU

The adoption of policies and measures now under consideration in OECD countries and their application over the projection period would entail a decrease of 8% in energy demand by 2030 compared to the Reference Scenario. In the Alternative Scenario, CO_2 emissions peak around 2020, at around 10% higher than in 2002, and start to decline in the 2020s. By 2030, the total savings of CO_2, in comparison to the Reference Scenario, are nearly the same as today's emissions from France, Germany, Italy, Spain and the United Kingdom combined.

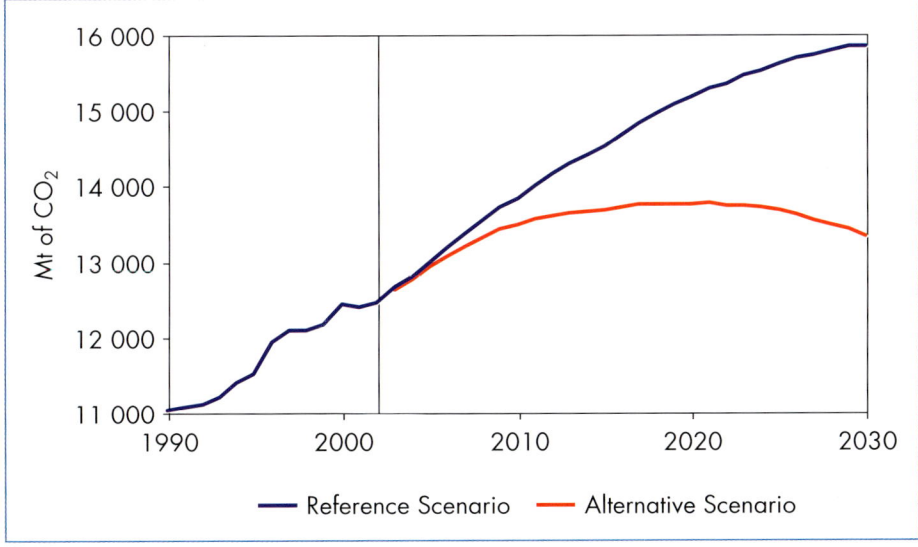

Figure 11.12: **OECD Energy-Related CO_2 Emissions in the Reference and Alternative Scenarios**

OECD North America

In the Alternative Scenario, North America accounts for almost one-fifth of the savings in the world's primary energy use and CO_2 emissions in 2030, and more than half of those in the OECD. Primary energy demand reaches 3 335 Mtoe in 2030, around 300 Mtoe, or 8%, lower than in the Reference Scenario. Coal demand increases in the first decade and then falls. On average, coal demand drops by 0.5% per year, reaching 500 Mtoe in 2030 – 14% less than in 2002. Coal use in power generation falls, almost entirely because of lower demand for electricity and an increase of renewables and nuclear power[10] in the fuel mix. Demand for natural gas is 870 Mtoe in 2030. This is 75 Mtoe, or 8%, lower than in the Reference Scenario. Power generation, industry and the residential and commercial sectors contribute evenly to the reduction in gas consumption. Transport policies, especially more stringent fuel-efficiency standards, account for almost all of the 125-Mtoe reduction in the region's oil demand in 2030. The consumption of biomass and other renewables is 70 Mtoe, or 30%, higher than in the Reference Scenario. Two-thirds of the increase comes from power generation. Solar water heaters in the residential sector and biofuels account for the rest.

North American carbon-dioxide emissions are 1 200 Mt, or 14%, lower than in the Reference Scenario in 2030. On average, emissions grow by 0.5% per year over the projection period, compared with 1% in the Reference Scenario. They level off in the 2020s.

European Union

In the Alternative Scenario, primary energy demand in the European Union reaches 1 870 Mtoe in 2030 – about 180 Mtoe, or 9%, lower than in the Reference Scenario. By 2030, the fuel mix in the Alternative Scenario looks very different from that in the Reference Scenario. Fossil fuels in aggregate account for 74% of primary energy demand, compared with 81% in the Reference Scenario. Coal consumption falls most. Renewables are 53 Mtoe higher, or 26%. In the Alternative Scenario, the share of Europe's oil savings is bigger than in other OECD regions. This reflects policies to improve vehicle fuel efficiency, to promote biofuels and to encourage mass transit. By 2030, demand for oil is cut by more than 100 Mtoe, or 14%, compared with the Reference Scenario. These results grow out of a combination of policies, most of which have been aimed at reaching the EU's Kyoto commitment.[11]

---

10. In the Alternative Scenario, 1 GW more nuclear power capacity in the United States is assumed.
11. To take effect, the Kyoto Protocol must be ratified by at least 55 nations, and the Annex I countries ratifying the Protocol must represent at least 55% of that group's total emissions in 1990.

Table 11.3: **Main Policies Considered in the Alternative Scenario in OECD North America**

Sector	Programme/measure	Impact
Power and heat	Credits, renewables portfolio standards and R&D for renewables-based electricity production	Renewables-based power generation increases
	Policies to promote combined heat and power generation	Increased share of electricity generation from CHP plants
	Support for faster deployment of more efficient and cleaner technology	Faster deployment of IGCC, fuel cells and renewables
Transport	Tighter vehicles fuel efficiency standards	New car and light-truck efficiency improves
	Increased R&D and tax breaks for alternative fuels and vehicles	Use of CNG, LPG, fuel-cell, hybrid powered vehicles and biofuels increases
Industry	Standards and certification for new motor systems	Improved efficiency of new motor systems
	Voluntary programmes and voluntary agreements to reduce industrial energy intensity	Improved efficiency of new technologies and accelerated deployment. Improved efficiency of energy use in factory buildings
	Tax incentives and low-interest loans for investment in efficient technologies	Accelerated deployment of new boilers, machine drives, and process-heat equipment
	Increased funding to R&D and demonstration programmes	Improved efficiency of new equipment entering the market after 2010
Residential and commercial	New equipment efficiency standards	Improved efficiency of new equipment
	Extended Energy Star buildings programme	Higher efficiency of lighting, air-conditioning and hot water in buildings, more efficient building fabrics and controls
	Energy efficiency programmes for utility companies	More efficient heat pumps and equipment
	More "whole-building" R&D	More efficient buildings
	Tighter commercial building codes from 2010	More efficient commercial buildings
	Credits for installation of solar water heater	Accelerated deployment of solar heaters

Table 11.4: **Main Policies Considered in the Alternative Scenario in the European Union**

Sector	Programme/measure	Impact
Power and heat	Renewable energy directive and extension	Renewables-based generation increases
	Policies to promote combined heat and power	Increased share of electricity generation from CHP plants
	Support for faster technology deployment	Faster deployment of renewables and fuel cells
	Extension of the life of nuclear plants in France and Sweden	More nuclear power production
Transport	Extended voluntary agreements with car manufacturers	New car and light-truck efficiency improves
	Increased support for alternative fuels	Increased use of biofuels
	White Paper on package of transport policies	Slower growth in passenger and freight transport and modal shift from road and aviation to rail and bus
Industry	Standards for new motor systems	Improved efficiency of new motor systems
	New voluntary programmes covering: – Information on and assistance in retrofitting, replacing and operating process equipment – Energy auditing, target setting and monitoring.	Improved efficiency of new technologies and accelerated deployment. Improved efficiency of energy use in factory buildings
	Tax incentives and low-interest loans for investment in new efficient technologies	Accelerated deployment of new boilers, machine drives, and process-heat equipment
	Increased funding to R&D and demonstration programmes	Improved efficiency of new equipment entering the market after 2015
Residential and commercial	Efficiency standards for lighting ballasts	More efficient lighting
	Voluntary agreements on equipment and measures to reduce the standby power of appliances	More efficient equipment and appliances
	Updated energy labels for washing machines and dishwashers	More efficient equipment
	More "whole-building" R&D	More efficient buildings
	Full implementation of the Energy Performance in Buildings Directive	More efficient buildings

The European Commission is promoting the use of renewables in power generation, in the transport sector (biofuels) and in buildings (solar heaters). A number of new measures in these areas have been taken into account in the Alternative Scenario (Table 11.4).[12] In addition, the emissions-trading scheme, which is to be introduced in early 2005, has been included in this scenario. The scheme will allow European companies to buy or sell emission allowances. It will cover almost half of Europe's emissions of CO_2, encompassing 12 000 installations.[13] The European Commission is still reviewing national allocation plans.

Carbon-dioxide emissions in Europe will peak at around 3 900 Mt in 2020, and then start to fall. In 2030, they will be 850 Mt, or 19%, lower than in the Reference Scenario. Emissions in 2010 would be some 450 Mt, or 13%, above the Kyoto target.[14] In the Reference Scenario, they are 18% higher.

OECD Pacific

Primary energy savings in OECD Pacific account for 16% of total OECD savings in the Alternative Scenario. In 2030, demand for fossil fuels is around 120 Mtoe, or 13%, lower than in the Reference Scenario. The change in the fuel mix is not as marked as in Europe. In 2030, the share of nuclear power, at 18%, is three percentage points higher in the Alternative Scenario. The share of renewables increases by two points to 8%. As in Europe, the region's carbon-dioxide emissions peak by 2020, at around 2 200 Mt, and then decline. Emissions are 380 Mt, or 16%, lower in 2030 than in the Reference Scenario.

Non-OECD Countries

Four countries outside the OECD – Russia, India, China and Brazil – account for almost 40% of all the energy and CO_2 emissions saved worldwide in 2030. The results for these countries are illustrated in Figure 11.13.

Russia

Primary energy demand in Russia will be 774 Mtoe in 2030, 13% lower than in the Reference Scenario. Less use of gas in power generation, resulting from lower electricity demand as well as decreasing losses in transmission and distribution, will account for most of this difference. In 2030, losses of electricity are cut by a quarter and heat losses are halved. The implementation

---

12. The proposed EU directive for Eco-Design Requirements in Energy-Using Products has not been included in the Alternative Scenario, since its implementation has not begun and the baselines for targets have not yet been specified.
13. These include power plants, oil refineries, coke ovens, iron and steel plants, and factories making cement, glass, lime, brick, ceramics, pulp and paper.
14. The target date for the 8% emissions-reductions under the Kyoto Protocol is the period 2008-2012.

Table 11.5: **Main Policies Considered in the Alternative Scenario in OECD Pacific**

Sector	Programme/measure	Impact
Power and heat	Tax incentives, green certificates and R&D for renewables-based generation	Renewables-based generation increases
	Policies to promote combined heat and power	Share of electricity generation from CHP plants increases
	Increased government support for nuclear power in Japan and Korea	Nuclear power production increases
Transport	Extended Top Runner programme and similar	New car and light-truck fuel efficiency improves
	Increased R&D and tax credits for alternative fuels vehicles	Increases use of CNG, LPG, fuel-cell and hybrid powered vehicles and biofuels
	Improved efficiency of city logistics, urban-road pricing (Japan) and expansion of high-speed rail (Japan and Korea)	Slower growth in passenger and freight road transport, modal shift to mass transport
Industry	Tighter standards and certification for new motor systems	Improved efficiency of new motor systems
	Extended voluntary-agreements programmes in Korea, Australia and New Zealand	Faster deployment of more efficient technologies
	Tax incentives and low-interest loans for investment in new efficient technologies	Accelerated deployment of new boilers, machine drives and process-heat equipment
	Increased funding to R&D and demonstration programmes	Improved efficiency of new equipment entering the market after 2010
Residential and commercial	Building codes for new commercial buildings	More efficient new commercial buildings
	Subsidies for heat-pump water heaters and efficient gas water heaters	More efficient water heaters
	Top Runner efficiency standards for equipment	More efficient equipment
	Promotion of business and home energy management systems (Japan)	Improved control of energy services and lower energy use
	Government financing of energy service companies (Japan)	Lower energy use in existing commercial buildings

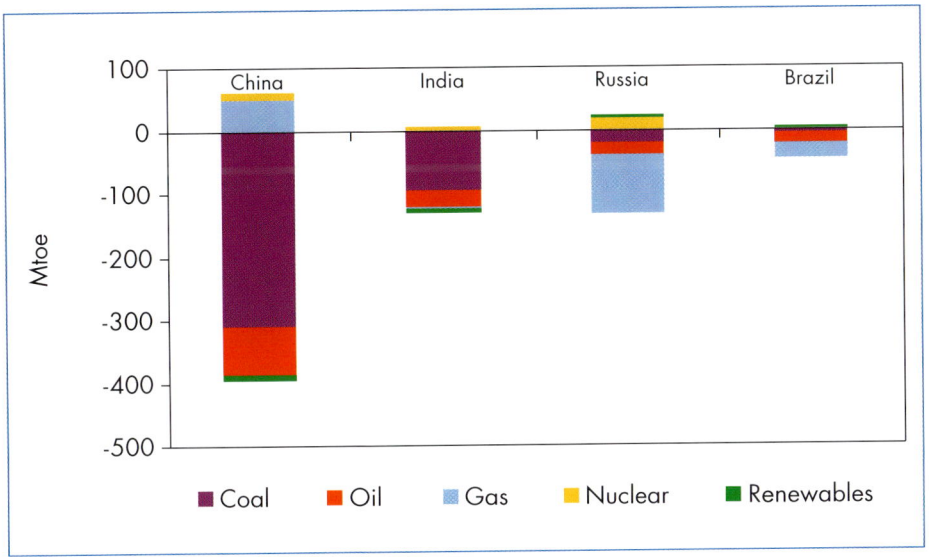

Figure 11.13: **Change in Energy Demand in the Alternative Scenario in the Largest Non-OECD Countries*, 2030**

* Compared with the Reference Scenario.

of stricter building standards and the use of more efficient industrial equipment help to drive down electricity consumption. In total, gas used in power and heat generation is reduced by 28%. Overall primary gas demand is 94 Mtoe, or 19%, lower than in the Reference Scenario. The fall in domestic demand frees up Russian gas for export, and exports are indeed slightly higher, thanks to a large increase in demand from China. Switching away from coal in the power sector increases China's gas demand substantially. Russian gas exports to OECD Europe are more or less unchanged.

Russia's energy-related carbon-dioxide emissions will rise, but less rapidly than in the Reference Scenario. Emissions reach 1 714 Mt in 2030 – which is 350 Mt, or 17%, less than in the Reference Scenario and well below Russia's 1990 level of just under 2 330 Mt.

China

Primary energy demand in China will be 2 205 Mtoe in 2030 – 330 Mtoe, or 13%, lower than in the Reference Scenario. China alone will account for one-fifth of global energy savings in 2030. Its share in the global reduction in coal use is even bigger, at one-third. More than 80% of the reduction in China's coal demand will come from the power sector. This is the result of lower final electricity demand, switching from coal to natural gas and increased use of nuclear power. Increases in renewables-based electricity production and

Table 11.6: **Main Policies Considered in the Alternative Scenario in Russia**

Sector	Programme/measure*	Impact
Power and heat	Implementation and extension of the federal energy-efficiency programme	Increase in thermal efficiency of power plants
	Policies to reduce transmission and distribution losses	Lower electricity transmission, distribution and district heating losses
	Policies to promote combined heat and power	Increased share of electricity and heat from CHP plants
	Increased government support for nuclear power	Increased nuclear power production
	Promotion of renewables-based generation	Increased production from wind and bioenergy
Transport	Accelerated introduction of more efficient vehicles	Increased fuel efficiency of new vehicles
Industry	Better enforcement of standards, tax incentives, low-interest loans	Improved efficiency of manufacturing
	Mandatory energy auditing of industrial facilities	Improved efficiency of industrial processes and of factory buildings
Residential and commercial	Full implementation and extension of existing mandatory efficiency standards for new appliances and equipment	Improved efficiency of new appliances and equipment
	Full implementation of federal and regional energy codes for new buildings and mandatory monitoring of energy use in existing buildings	More efficient buildings, leading to lower lighting, heating and cooling loads

* Policy measures primarily derived from the Federal Comprehensive Programme "Energy-Efficient Economy", the "Development Strategy of Russian Power Sector" and the Russian *Energy Strategy to 2020*.

in the efficiency of coal-fired plants will also drive down demand for coal. Primary gas demand is 50 Mtoe higher in 2030 in the Alternative Scenario.

The primary fuel mix in 2030 is substantially different from that in the Reference Scenario. The share of coal is six percentage points lower, at just under half. The share of gas is up by three points, to 9%. The fall in energy demand and the change in the fuel mix towards less carbon-intensive fuels result in a 1 300 Mt, or 18%, drop in CO_2 emissions in 2030.

Table 11.7: **Main Policies Considered in the Alternative Scenario in China**

Sector	Programme/measure	Impact
Power and heat	Refurbishment of existing coal-fired plants	Increased thermal efficiency of old coal-fired plants
	Expanded support for more efficient and cleaner coal-fired plants	Increased thermal efficiency of new coal-fired plants
	Expanded government support for gas-fired plants	Increased gas-fired generation
	Extended support for renewables-based generation	Increased renewables-based generation
	Policies to promote combined heat and power	Increased share of electricity generation from CHP plants
	More government support for nuclear power	Increased nuclear power production
Transport	Tighter vehicle-fuel efficiency standards	Improved vehicle-fuel efficiency
	Increased R&D and tax credits for clean vehicles	Increased use of CNG, LPG, fuel-cell and hybrid powered vehicles, and biofuels
	Expansion of intra- and inter-city railway networks	Slower growth in passenger vehicle transport and modal shift to mass transport
Industry	Energy-efficiency standards for industrial equipment	Improved energy efficiency of boilers, furnaces, electric motors, fans, pumps and transformers
	Voluntary agreements including energy auditing, target setting and monitoring	Improved efficiency of new technologies and accelerated deployment
	Tax incentives and low-interest loans for investment in new efficient technologies	Accelerated deployment of new boilers, machine drives, and process-heat equipment
	Environmental restrictions on coal use	Switching from coal to gas
	Further restructuring of state-owned and small producers	Investment in larger-scale and more efficient processes
Residential and commercial	Tighter efficiency standards for appliances and equipment	More efficient appliances and equipment
	New mandatory energy labelling for domestic appliances, broadening and updating voluntary energy labelling	More efficient refrigerators, air-conditioners and other appliances and equipment
	China Green Lights Programme	Improved efficiency of new lighting equipment
	Building codes for residential and commercial buildings	More efficient new buildings, leading to lower lighting, heating and cooling loads

India

India's primary energy demand in 2030 is 125 Mtoe lower than in the Reference Scenario, a difference of 12%. The amount of energy saved is slightly higher than in Russia. As in China, most of these savings are in coal. Coal demand is 100 Mtoe lower in 2030. The power sector accounts for almost all of this reduction. Reduced losses in transmission, lower electricity demand, the increased efficiency of coal-fired plants and more reliance on renewables and nuclear power explain the reduction.[15]

Other factors contribute significantly to changes in energy demand in India. They include the introduction of improved biomass cook-stoves and more use of biofuels in the transport sector. In the Alternative Scenario, biomass consumption is slightly lower in 2030 than in the Reference Scenario. A decline in the use of traditional fuels by households more than offsets the increased use of biomass in electricity generation and of biofuels for transport.

In 2030, the primary fuel mix in India looks very different from the one depicted in the Reference Scenario. The share of coal falls, while those of gas, nuclear power and other renewables increase. Total CO_2 emissions are 436 Mt, or 19%, lower.

Brazil

Primary energy demand in Brazil will reach 330 Mtoe by 2030 – 44 Mtoe, or 12%, lower than in the Reference Scenario. Most of the saving is in the form of natural gas, demand for which is 23 Mtoe lower. The power sector accounts for 90% of this reduction. Lower electricity demand and increased use of renewables in power generation explain this drop. In 2030, oil demand is 17 Mtoe, or 10%, lower than in the Reference Scenario. The popularity of "flex-fuel" vehicles (which can run on gasoline or ethanol or a mixture of both), higher efficiency of conventional vehicles and increased use of biodiesel result in a 10-Mtoe reduction in oil demand for transport.[16] Lower demand in power generation and industry also contributes to the remaining drop in oil demand.

Total CO_2 emissions are 118 Mt, or 18%, lower than in the Reference Scenario. In 2030, emissions of CO_2 per unit of energy consumed are, nonetheless, higher than in 2002. Renewables today account for 38% of primary energy use in Brazil. Their share falls by three percentage points even in the Alternative Scenario.

---

15. Almost 600 million people in India lack access to electricity. Reducing inefficiencies in electricity supply and end-use would make available more electricity to those who are currently deprived. Chapter 10 provides a detailed analysis of the link between energy and development.
16. By 2030, the share of flex-fuel cars in new-car sales increases from 18% now to 35% in the Alternative Scenario, compared with 26% in the Reference Scenario.

Table 11.8: **Main Policies Considered in the Alternative Scenario in India**

Sector	Programme/measure	Impact
Power	Refurbishment of existing coal-fired plants	More efficient old coal-fired plants
	Support for more efficient and cleaner new coal-fired plants	More efficient new coal-fired plants
	Incentives to promote renewables-based generation	Increased renewables-based generation
	Policies to reduce transmission and distribution losses	Fewer transmission and distribution losses
	More government support for nuclear power	Increased nuclear power production
Transport	Measures to accelerate the introduction of less polluting vehicles and fuels	More efficient new vehicles; faster deployment of CNG, LPG, biofuels Faster replacement of old, polluting vehicles
Industry *	Standards and certification for new motor systems	More efficient motor systems
	Voluntary agreements covering energy auditing, target setting and monitoring	Faster deployment of more efficient technologies
	Tax incentives and low-interest loans for efficient technologies	Accelerated deployment of new boilers, machine drives, and process-heat equipment
	Restructuring of state-owned industries	More investment in larger-scale, more efficient processes
Residential and commercial *	Efficiency standards and new mandatory energy labelling for new appliances and equipment	More efficient appliances and equipment
	Measures to improve the efficiency of lighting equipment	More efficient lighting
	Building codes for commercial and large residential buildings	More efficient buildings, leading to lower lighting and cooling loads
	Financing schemes and promotional campaigns for solar water heaters and improved cook-stoves	More solar water heating, more use of LPG and more efficient biomass cook-stoves

* Policy measures primarily derived from the Energy Conservation Act.

Table 11.9: **Main Policies Considered in the Alternative Scenario in Brazil**

Sector	Programme/measure	Impact
Power	More support for renewables-based generation	Increased renewables-based generation
	Programmes to cut transmission and distribution losses	Fewer transmission and distribution losses
Transport	Extended biodiesel programme	Increased use of biofuels
	Bigger tax incentives and more R&D for alternative fuel vehicles	Faster deployment of flex-fuel, CNG and LPG
Industry	Energy-efficiency standards, labelling, and certification	More efficient motor systems
	Tax incentives and low-interest loans for efficient technologies	More efficient industrial equipment
Residential and commercial	Voluntary programmes covering energy auditing, target setting and monitoring	Faster deployment of more efficient technologies
	Measures to improve the efficiency of lighting equipment	More efficient lighting
	Building codes for new and renovated commercial and larger residential buildings	More efficient buildings, leading to lower lighting, heating and cooling loads

Results by Sector

Power Generation

Policy Assumptions and Effects

There are several options for reducing fossil-fuel consumption and greenhouse-gas emissions in the power-generation sector. The most important policies and measures considered in the Alternative Scenario are:

- Incentives and regulations to boost the use of renewables.
- Programmes to improve the performance of existing power stations and networks.
- Programmes to improve the efficiency and reduce the cost of advanced technologies in power generation.
- Policies to boost the production of nuclear power.

- Incentives to promote the use of combined heat and power generation (CHP).

Many governments currently favour using renewables as a way to reduce CO_2 emissions. Most OECD countries have national targets for increasing the use of renewables. In the Alternative Scenario, it is assumed that policies are put in place to ensure that these targets are met. Several developing countries also have programmes to promote renewables. China recently announced its intention to increase output from small hydro, biomass, wind and solar power plants to 60 GW in 2010 and to 161 GW in 2020. In Brazil, the ProInfa federal programme provides incentives for the development of alternative sources of energy. Many developing countries have begun to focus on renewables in rural electrification. The rate of deployment of renewables in developing countries is higher in this scenario. This is because more rigorous policies in OECD countries lead to faster technology development and deployment, and lower costs worldwide.

Several countries, particularly in the OECD, are assumed to increase incentives for using combined heat and power. Most new CHP capacity is likely to be used for on-site generation in industry. The share of electricity produced from CHP plants is in general from one to three percentage points higher in the Alternative Scenario than in the Reference Scenario.

The Alternative Scenario assumes that advanced power-generation technologies will become available earlier than in the Reference Scenario. Gas turbines and combined-cycle gas-turbine plants are two to three percentage points more efficient in 2030 in this scenario. The average efficiency of coal-fired plants reaches 55% by 2030, compared with 52% in the Reference Scenario. Coal-gasification technologies grow more competitive. Fuel cells become economic in some cases by 2015, rather than by 2020, as in the Reference Scenario.

The efficiency of fossil-fired electricity generation in the developing regions is currently much lower than in the OECD. Some developing countries have programmes to rehabilitate their power stations and improve their performance. The Alternative Scenario assumes that the efficiency of existing coal-fired power stations in India and China improves by two percentage points thanks to expanded programmes of this sort.

The Alternative Scenario assumes that measures will be adopted to accelerate the construction of nuclear plants, *only* in those countries that already have nuclear reactors in the Reference Scenario. A number of countries plan to expand the use of nuclear power. Japan, Korea, Russia, China and India have specific development targets. Extensions to the lives of existing reactors from 40 to 60 years are assumed in France and Sweden in this scenario.

Summary of Results

In the Alternative Scenario, world electricity generation in 2030 is 13% lower than in the Reference Scenario. The reduction comes from end-use efficiency improvements, from reduced losses in transmission and distribution and from greater use of distributed generation. The difference between the two scenarios is roughly equal to the current electricity output of the United States.

The power-generation fuel mix is considerably different. In the Reference Scenario, fossil fuels account for 70% of electricity generation in 2030. In the Alternative Scenario, the share of fossil fuels falls to 61%, while the shares of carbon-free fuels rise substantially (Figure 11.14).

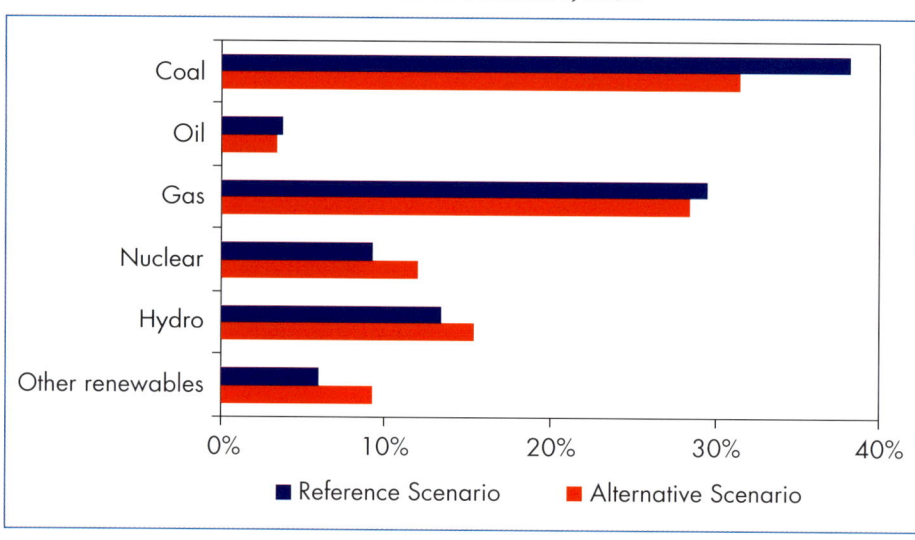

Figure 11.14: **Fuel Shares in Electricity Generation in the Reference and Alternative Scenarios, 2030**

In the Reference Scenario, coal's share in electricity generation remains almost unchanged up to 2030, at a little less than 40%. In the Alternative Scenario, coal gradually loses market share, dropping to less than a third of total generation by 2030. At 8 700 TWh, coal-fired generation is 28% lower than in the Reference Scenario (Table 11.10). The decline is sharpest in the OECD, where the share of coal drops to 25% in 2030, compared with 33% in the Reference Scenario. Coal-based electricity generation is 15% less than in 2002, because many coal-fired plants are retired and replaced with plants using other fuels. China and India also see their coal-fired generation reduced by more than a quarter in 2030 compared to the Reference Scenario. Nevertheless, these two countries still account for 45% of the world's coal-fired generation in 2030.

Table 11.10: **Changes in Electricity Generation by Fuel in the Alternative Scenario*, (TWh)**

	2010	2020	2030
Coal	-352	-1 787	-3 392
Oil	-71	-163	-243
Gas	-239	-638	-1 481
Nuclear	15	154	400
Hydro	0	10	19
Other renewables	109	301	692
Total	**-538**	**-2 122**	**-4 004**

* Compared with the Reference Scenario.

Gas-fired electricity generation, excluding hydrogen, is 1 666 TWh, or 19%, lower in 2030 than in the Reference Scenario, although the share of gas in total generation drops only slightly. Within the OECD, the largest reductions in gas-fired generation occur in Europe and Japan, where renewables and nuclear energy play a large role. In Russia, gas-fired power plants produce a quarter less electricity in 2030 than in the Reference Scenario. In the Reference Scenario, Russian gas-fired generation nearly doubles between 2002 and 2030 and its share increases from 43% to 53%. In the Alternative Scenario, it increases at a much slower pace and its share increases slightly because electricity demand is much lower and because nuclear power substitutes for gas. Global electricity generation from fuel cells using hydrogen from reformed natural gas is 530 TWh, twice as high as in the Reference Scenario in 2030.

Nuclear power capacity expands to 428 GW in 2030, about 50 GW more than in the Reference Scenario. Nuclear power production is 14% higher. The largest increases in output occur outside the OECD, notably in Russia, where nuclear production is 40% higher in 2030 compared with the Reference Scenario. Nuclear production rises by 16% in China and by 21% in India. All three countries have ambitious nuclear programmes and plans for nuclear plant construction.

In the Alternative Scenario, hydroelectric generation in 2030 is slightly higher than in the Reference Scenario. In the Reference Scenario, hydropower's share in world generation drops from 16% in 2002 to 13% in 2030. In the Alternative Scenario, its share falls only by one percentage point, to 15%. The shares of non-hydro renewables increase much more, from an aggregate 6% in 2030 in the Reference Scenario to 9% in the Alternative Scenario. The strongest increase is in OECD Europe, driven by the European Union's strong support for renewables (Figure 11.15). Electricity generation using non-hydro renewables is almost ten times higher in 2030 in the Alternative Scenario than in 2002, and more than a third higher than in the Reference Scenario.

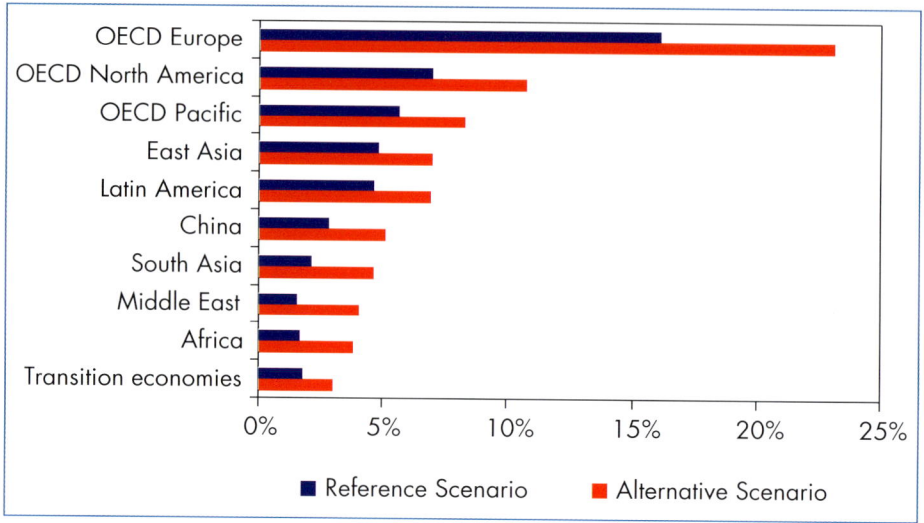

Figure 11.15: **Share of Non-Hydro Renewables in Electricity Generation in the Reference and Alternative Scenarios by Region, 2030**

In the OECD, power-sector CO_2 emissions in 2030 are 26% lower than in the Reference Scenario and 5% lower than in 2002. About 40% of the emissions savings comes from reduced demand and the rest from changes in the fuel mix. In 2030, power plants emit 20% less carbon dioxide per kWh produced in the Alternative Scenario (Figure 11.16).

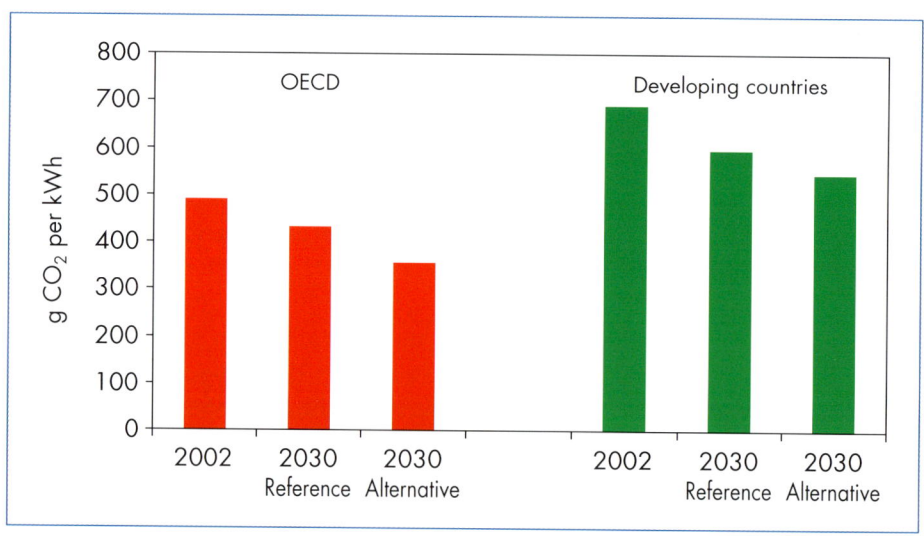

Figure 11.16: **CO_2 Emissions per kWh of Electricity Generated in the Reference and Alternative Scenarios**

Electricity prices are higher in the Alternative Scenario than those in the Reference Scenario. The higher cost of renewables and of combined heat and power plants will outweigh economies gained from the use of more efficient technologies and lower gas prices. For example, renewables add $8 to $9 per MWh in the Alternative Scenario in Europe. As a result, wholesale electricity prices will be 12% higher in Europe in 2030. These additional costs do not include the cost of intermittence and grid connections, which will also be passed on to the consumer.

Transport

Policy Assumptions and Effects

There are three policy areas in the transport sector considered in the Alternative Scenario:

- Improved vehicle-fuel efficiency.
- Increased sales of alternative-fuel vehicles and fuels.
- Demand-side measures to reduce demand for mobility and encourage a switch to less energy-intensive modes of transport.

The fuel efficiency of new vehicles varies greatly among countries (Figure 11.17). In general, vehicles are less fuel-efficient in developing countries. Vehicles manufactured in these countries do not usually incorporate state-of-the-art efficiency technologies. The Alternative Scenario assumes that OECD countries will do more to increase their own vehicle-fuel efficiency. New measures considered here include the European Union's voluntary agreement with car manufacturers and the Japanese Top-Runner programme. Fuel-efficiency in the United States and Canada, for example, is nearly 20% better in 2030 than in the Reference Scenario, and this reduction does not assume any major tightening of current Corporate Average Fuel Economy (CAFE) standards. Many developing countries and transition economies are considering introducing standards and other policies to increase efficiency. Average vehicle-fuel efficiency in those countries is assumed in the Alternative Scenario to improve by an additional 10% to 15% by 2030, as a result of both new standards and technology spillover from OECD countries. Most vehicle designs in non-OECD countries are imported from OECD countries. Several factors, including the slow rate of replacement of vehicle fleets, the gap between test and on-road efficiency values, and the "rebound" effect,[17] reduce the impact of improved fuel efficiency on actual energy demand and CO_2 emissions.

---

17. Increased fuel efficiency and lower fuel costs can lead to more kilometres driven and, therefore, higher fuel consumption. This "rebound" effect is taken into account in our World Energy Model.

Figure 11.17: **Average Vehicle Fuel Efficiency for New Light Duty Vehicles in Selected Regions, 2002**

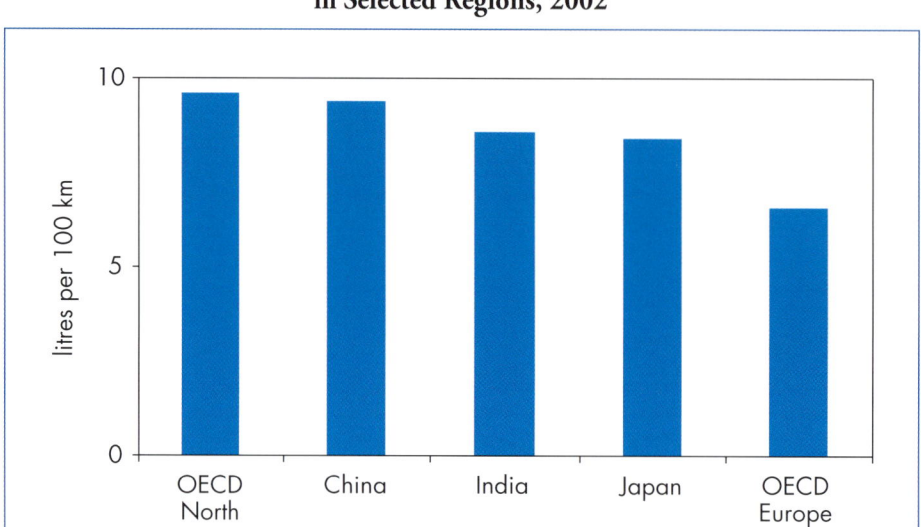

The use of alternative fuels for road transport – essentially natural gas and biofuels – increases more quickly in the Alternative Scenario, mainly owing to tax incentives and regulatory measures. After 2010, hybrid and alternative-fuel vehicles contribute to improved fuel efficiency and, in the case of hybrids, also help reduce carbon emissions below the Reference Scenario projections. In the United States and Canada, for example, hybrid and fuel-cell powered vehicles are expected to make up 15% of the stock of cars and light trucks in 2030.

Measures to slow traffic growth, to shift away from road travel to less intensive transport modes are taken into account in the European Union and in OECD Pacific. In the European Union, road passenger travel is reduced by around 5% and road freight by 8%. In Japan, road passenger travel is cut by 6% and freight by 10%, half of that amount shifting to rail.

Summary of Results

In 2030, global demand for oil in transport is 390 Mtoe (8 mb/d), or 12%, lower than in the Reference Scenario (Table 11.11). The expected oil savings in 2030 are comparable to the current consumption of oil for transport in OECD Europe. CO_2 emissions are cut by around 1 000 Mt, or 11%. The reduction in CO_2 emissions is almost as big as India's emissions today.

Table 11.11: **Changes in Transport Energy Consumption and CO_2 Emissions in the Alternative Scenario***

	2010	2020	2030
Total energy	-2.0%	-5.9%	-9.6%
Of which oil	-2.6%	-7.5%	-12.4%
other fuels	10.7%	25.6%	43.5%
CO_2 emissions	-2.2%	-6.8%	-11.4%

* Compared with the Reference Scenario.

The largest decline in transport energy consumption in percentage terms occur in the developing countries and the transition economies (Table 11.12). The car markets in OECD regions will further mature over the projection period and the rate of growth in the number of new vehicles entering the fleet will slow. Rapid motorisation will continue in developing countries, especially in China and India. By 2030, the number of vehicles in use in non-OECD regions will approach that in the OECD.[18] In the Alternative Scenario, new, more efficient vehicles quickly raise the average fuel efficiency of the vehicle fleet in these countries. Vehicles are generally older and less efficient in developing countries, so the impact of replacing them with more efficient new cars will be bigger there.

Table 11.12: **Changes in Transport Energy Consumption and CO_2 Emissions in the Alternative Scenario* by Region**

	2010	2020	2030	2010	2020	2030
	Energy consumption (%)			CO_2 emissions (%)		
OECD	-1.6	-5.7	-9.0	-2.0	-6.9	-11.5
Developing countries	-2.6	-6.3	-10.4	-2.8	-6.8	-11.4
Transition economies	-2.2	-5.8	-10.4	-2.3	-6.2	-11.2
World	**-2.0**	**-5.9**	**-9.6**	**-2.2**	**-6.8**	**-11.4**
	Energy consumption (Mtoe)			CO_2 emissions (Mt)		
OECD	-23	-94	-162	-77	-308	-557
Developing countries	-17	-59	-130	-49	-170	-381
Transition economies	-3	-11	-23	-8	-27	-59
World	**-44**	**-163**	**-315**	**-133**	**-505**	**-997**

* Compared with the Reference Scenario.

18. See Chapter 3 for a discussion about car ownership.

Figure 11.18.: **Oil Demand for Transport in the Reference and Alternative Scenarios by Region**

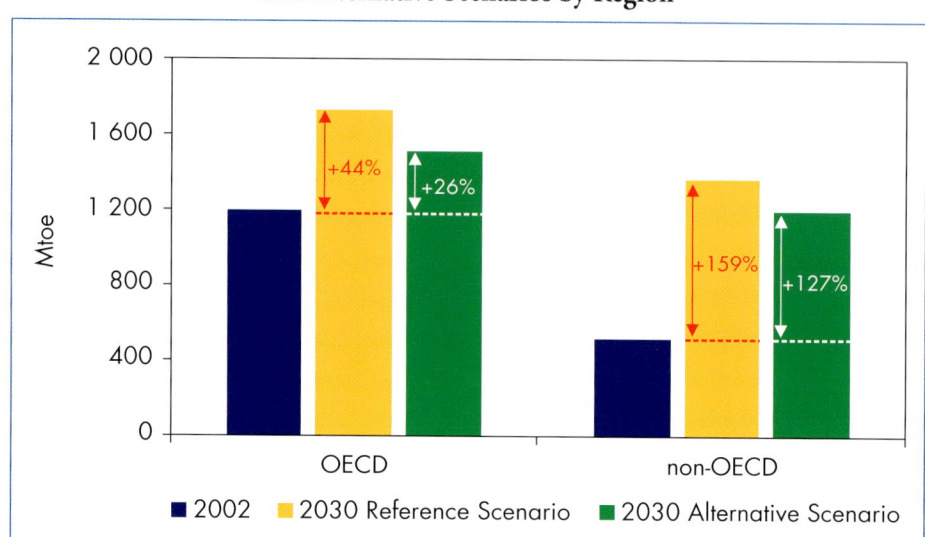

Even so, the largest *absolute* savings occur in the OECD (Figure 11.18). The fall in transport demand in the Alternative Scenario is largely the result of the faster introduction of more efficient vehicles into the car fleet than is the case in the Reference Scenario. The efficiency gains are driven by tighter vehicle fuel-efficiency standards and other government measures aimed at improving efficiency.

There is considerable scope for biofuels such as ethanol and biodiesel to replace oil in the transport sector (IEA, 2004). The Alternative Scenario assumes that government policies boost the share of biofuels in worldwide road transport-fuel consumption to around 4% in 2030, more than doubling the level of consumption in the Reference Scenario. The rate of increase varies greatly among countries, reflecting different degrees of interest in biofuels. In OECD countries, biofuel consumption reaches 55 Mtoe in 2030, four time more than in the Reference Scenario.[19] The increase in OECD Europe is the biggest anywhere in the OECD. In the developing countries, biofuels remain important in Brazil and start to play a significant role in India. The supply of suitable crops could limit the development of biofuels in some regions, as the appropriate crops typically require large areas of land. The market for pure biofuel vehicles is likely to be small in most countries. In most cases, biofuels will continue to be blended with conventional gasoline or diesel. Current technology allows cars to run on gasoline blended with up to 10% biofuel, without reconfiguring the cars' engines.

19. Biofuel consumption in the OECD was 2.2 Mtoe in 2002.

Shifts from road to rail transport will increase the energy used by railways in OECD countries, but that increase will be very small compared with the savings the same trend will achieve in road oil consumption.

In all regions, the *percentage* reduction in transport sector's CO_2 emissions projected in the Alternative Scenario is bigger than percentage energy savings, because of widespread switching to less carbon-intensive fuels – natural gas and biofuels in road transport and electricity in rail transport. The reduction in carbon emissions is largest in OECD countries, where fuel-switching is more extensive than elsewhere. OECD emissions reduction will also be the largest among regions in absolute terms. Their share of the global reduction in transport emissions will nonetheless decline from 58% in 2010 to 56% in 2030, as oil savings grow in developing countries.

Industry

Policy Assumptions and Effects

Estimating the impact of industrial policies on energy use and CO_2 emissions is difficult because of data limitations and the heterogeneity of the processes and technologies in use. The analytic approach used in this section differs somewhat between the OECD and non-OECD regions.

In the OECD, the Alternative Scenario analyses the impact of new policies to improve energy efficiency in process heat, in steam generation, in motive power and in buildings. Policies affecting steam generation and process heat can reduce industrial energy consumption significantly. Policies on motive power can produce significant savings of electricity. The main policy types considered in this section of the scenario are standards and certification for new motor systems; voluntary programmes to improve the efficiency of new technologies and to accelerate the deployment of new boilers, machine drives and process-heat equipment; and research and development to improve the efficiency of new equipment entering the market after 2015.

In the non-OECD regions, the analysis of efficiency improvements focuses on iron and steel manufacturing, ammonia, ethylene and propylene, aromatics, cement and pulp and paper. For each process, it is assumed that the efficiency of new capital stock will approach that of the current stock in OECD countries. Changes in the process mix are based on the assumption that state-owned firms will be restructured and privatised more quickly than the Reference Scenario foresees, stimulating investments in larger-scale and more efficient processes. These policies are of particular importance in China and India. A switch from coal to more efficient gas-based processes is assumed in China only.[20]

---

20. Policies are already in place in major cities such as Beijing and Shanghai to replace coal with gas in order to reduce local air pollution. In the Alternative Scenario, these policy efforts are assumed to be strengthened.

In the energy-intensive sectors in both OECD and non-OECD countries, energy use per tonne of output is calculated for different processes. Typically the energy efficiency of each of these processes differs. Regional differences in the potential for improving energy efficiency have been identified by disaggregating energy use by process. The improvements in efficiency in the Alternative Scenario are derived from changes in the energy efficiency of each process and changes in the mix of processes used.

Summary of Results

In the Alternative Scenario, global industrial energy demand is 9%, or almost 300 Mtoe, lower than in the Reference Scenario by 2030 (Table 11.13). The difference between the two scenarios corresponds to the current energy consumption of the industry sector in China. Reduced consumption of electricity accounts for 30% of total savings, or 90 Mtoe, while oil accounts for 76 Mtoe in savings (26%), coal for 65 Mtoe (22%) and gas for 47 Mtoe (16%). Improved efficiency in developing countries contributes more than half of global savings. OECD countries account for a third and transition economies for the rest. Improvements in efficiency average 7% in the OECD, 9.5% in developing regions and 11% in transition economies.

Table 11.13: **Change in Industrial Energy Consumption in the Alternative Scenario*, 2030**

	OECD	Transition economies	Developing countries	World
Change in industrial energy consumption (%)				
Coal	−9.1	−12.4	−16.9	−14.9
Oil	−3.1	−13.5	−13.8	−8.9
Gas	−8.1	−12.4	0.0	−5.7
Electricity	−9.2	−8.9	−11.5	−10.3
Heat	−6.0	−9.8	12.3	−1.4
Renewables	−4.5	0.0	−6.8	−5.7
Total	**−6.7**	**−11.3**	**−9.9**	**−8.7**
Contribution to total change by fuel (Mtoe)				
Coal	−9	−4	−52	−65
Oil	−12	−7	−57	−76
Gas	−32	−16	0	−47
Electricity	−36	−7	−47	−90
Heat	−1	−5	5	−1
Renewables	−6	0	−9	−15
Total	**−95**	**−39**	**−161**	**−295**

* Compared with the Reference Scenario.

A large part of the reduction of coal use by industry in developing countries results from the substitution of natural gas for coal in China. Increased use of heat from combined heat and power plants in China boosts overall heat consumption in developing countries by around 5 Mtoe in 2030, offsetting some of the savings in other fuels. In the Reference Scenario, the share of gas in industrial energy use remains high in the transition economies throughout the *Outlook* period. Efficiency improvements in industrial processes in the Alternative Scenario yield large *savings* in gas use in this region, amounting to 15.5 Mtoe in 2030 and representing 40% of the total energy saved by the region's industry.

In the OECD, electricity contributes 38% of total savings, primarily as a result of policies aimed at improving the efficiency of motor systems. Gas accounts for about a third and oil for most of the rest. Oil savings are largest in the OECD Europe region, driven by improvements in process-heat and boiler efficiencies. In OECD North America, oil savings are modest because a large share of the oil in industry is used as a feedstock for chemicals. No policies are considered in the Alternative Scenario that would reduce feedstock use. Despite the importance of feedstock in the chemical industry, that industry still contributes significantly to total industrial savings in all regions, because of its large share in total industrial energy use (Figure 11.19). In the OECD, the iron and steel industry sees incremental intensity gains of between 9% and 11% by 2030 compared with the Reference Scenario. In absolute terms, the "other industries" category contributes as much as half the total savings of industrial energy in the OECD regions.

Figure 11.19: **Reduction in Industrial Energy Demand by Sector in the Alternative Scenario*, 2030**

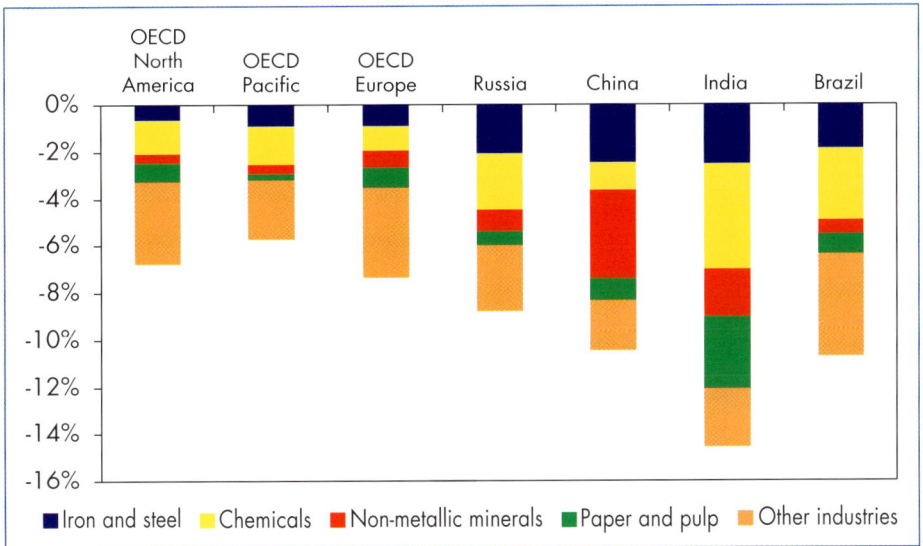

* Compared with the Reference Scenario.

Efficiency gains in the iron and steel industry in Russia, China and Brazil are roughly of the same magnitude as in the OECD regions. In India, the efficiency improvement is very substantial. In 2030, 25% less energy will be required to produce one tonne of steel in India that is projected in the Reference Scenario. This results from industry consolidation. In China and India, the efficiency of the production of non-metallic minerals will increase considerably, providing more than a third of China's total savings of industrial energy use by 2030.

Residential and Services Sectors

Policy Assumptions and Effects

The Alternative Scenario evaluates the impact on energy use of new policy measures in residential and commercial buildings for a variety of end-uses. In the residential sector, these include lighting, electric appliances, space heating, water heating, cooking and air-conditioning. In the services sector, lighting, space heating, air-conditioning and ventilation are assessed, as well as miscellaneous electric equipment.

In the OECD, equipment standards, building codes and voluntary agreements are analysed. In some cases, mandatory labelling schemes are also considered. The voluntary agreements include financing schemes for efficiency investments, endorsement labelling and "whole-building" programmes. Financing schemes include direct consumer rebates, low-interest loans and energy-saving performance contracting. Accelerated research and development efforts by governments are also taken into account.

In recent years, many non-OECD countries have adopted policies aimed at improving the energy efficiency of new equipment and buildings, but few have yet attained the energy efficiency prevalent in the OECD. The Alternative Scenario assumes that policies currently "in the pipeline" are adopted and are complemented by new measures. As a result, these countries are assumed to achieve efficiencies that approach those of the OECD. Our scenario does not assume a faster transition from traditional to modern commercial energy sources than that projected in the Reference Scenario. The rate of electrification and access to gas networks is the same in both scenarios. But measures aimed at promoting the use of commercial energy in equipment and buildings are assumed to be stepped up. As in the OECD region, the most important measures are energy labelling and mandatory minimum energy-efficiency standards. For buildings, stricter mandatory codes, building certification and energy-rating schemes are assumed.

Many non-OECD countries have already established energy labelling and minimum efficiency standards. Other countries are planning to implement such programmes. In the Alternative Scenario, it is assumed that existing programmes are broadened to cover more equipment types. Standards are also raised to levels closer to those found in the OECD today for new equipment sold between 2010 and 2030. However, efficiency standards and labels are *not* assumed to reach life-cycle least-cost efficiency levels, which would bring even greater efficiency gains. Where there is a large spread in the efficiency levels of a specific product type among OECD countries, we have assumed that the lower levels are attained in non-OECD countries.

Very few non-OECD countries have adopted measures to improve the energy performance of buildings. In the Alternative Scenario, it is assumed that building codes are adopted for new commercial and residential buildings. It is also assumed that certain policy measures are implemented to encourage higher efficiency in existing commercial buildings. These include energy-performance certification and energy-rating schemes for buildings. Solar water heating in houses is also assumed to expand more quickly than it does in the Reference Scenario.

Summary of Results

In the Alternative Scenario, global energy use in the residential and services sectors combined is 451 Mtoe, or 11%, lower in 2030 than in the Reference Scenario. This is comparable to current residential and service consumption in OECD Europe. The residential sector accounts for two-thirds of those savings. More electricity is saved than any other energy sources. It accounts for more than a third of the total. Biomass savings are also impressive. Biomass use is 136 Mtoe, or 14%, lower than in the Reference Scenario, thanks to faster switching away from inefficient and polluting cook-stoves in poor developing regions.

In both the residential and services sectors, energy savings are higher in developing countries than in OECD. These differences reflect for the most part the greater potential for efficiency improvements in developing and transitional economies than in OECD countries, where more efficiency policies have already been implemented and where there is easier access to capital for making energy-efficiency investments.

In the *residential* sector, global coal use is reduced the most in percentage terms, by 24% compared to the Reference Scenario. But because of the small share of coal in residential energy demand the contribution to total savings is modest, at 12 Mtoe, or 3%, of total savings. The fall in the use of biomass contributes 39% of total savings, followed by electricity (31%), and gas (14%). In the *services* sector, savings of electricity contribute more than half of the 136 Mtoe of overall savings by 2030. Most of the remaining savings are equally split between oil and gas.

Table 11.14: **Change in Residential and Services Sector Energy Consumption in the Alternative Scenario*, 2030**

	OECD	Transition economies	Developing countries	World
Change in residential and services energy consumption (%)				
Coal	-4.6	-16.2	-28.7	-19.6
Oil	-11.3	-15.6	-10.8	-11.2
Gas	-8.3	-13.1	-9.7	-9.4
Electricity	-12.4	-16.7	-15.9	-14.1
Heat	-0.5	-9.4	-24.2	-8.9
Renewables	10.1	0.9	-12.0	-9.8
Total	**-9.1**	**-12.4**	**-12.9**	**-11.4**
Contribution to total change by fuel (Mtoe)				
Coal	-0.2	-2	-9	-12
Oil	-21	-5	-41	-67
Gas	-41	-18	-18	-77
Electricity	-83	-10	-86	-179
Heat	-0.2	-11	-5	-16
Renewables	9	0.1	-110	-100
Total	**-136**	**-46**	**-269**	**-451**

* Compared with the Reference Scenario.

Sizeable cuts in gas consumption in OECD Europe and OECD North America and of oil use in the OECD Pacific region result from an assumed strengthening of building codes. Oil and gas are the main heating fuels in these regions. Tougher building codes in Russia will result in 12% savings of energy demand for residential and commercial space heating by 2030.

In most developing countries, residential and commercial energy use is primarily for water heating and cooking as their generally warmer climate obviates heating. The most important commercial fuel is liquefied petroleum gas (LPG). A few developing countries also have natural gas in urban areas. LPG consumption is slightly higher in the Alternative Scenario because of policies promoting more efficient LPG cook-stoves as an alternative to biomass.

Roughly half of the savings in global electricity demand in the services and residential sectors comes from developing countries. These countries are poised for a boom in the sale of electrical appliances and equipment, and offer a significant opportunity for saving electricity through energy labelling and minimum efficiency standards. In the Alternative Scenario, standards for household and office appliances improve the appliances' efficiency by 10% to

30% compared with the Reference Scenario. Standards and labelling result in an 8% saving in electricity use in residential and commercial buildings in India and 7% in Russia (Figure 11.20). Several OECD and developing countries have adopted policies to encourage solar energy – mainly solar water heaters – but further government action will be necessary to boost solar markets. In the Alternative Scenario, solar energy use reaches 69 Mtoe in 2030, twice as much as in the Reference Scenario.[21]

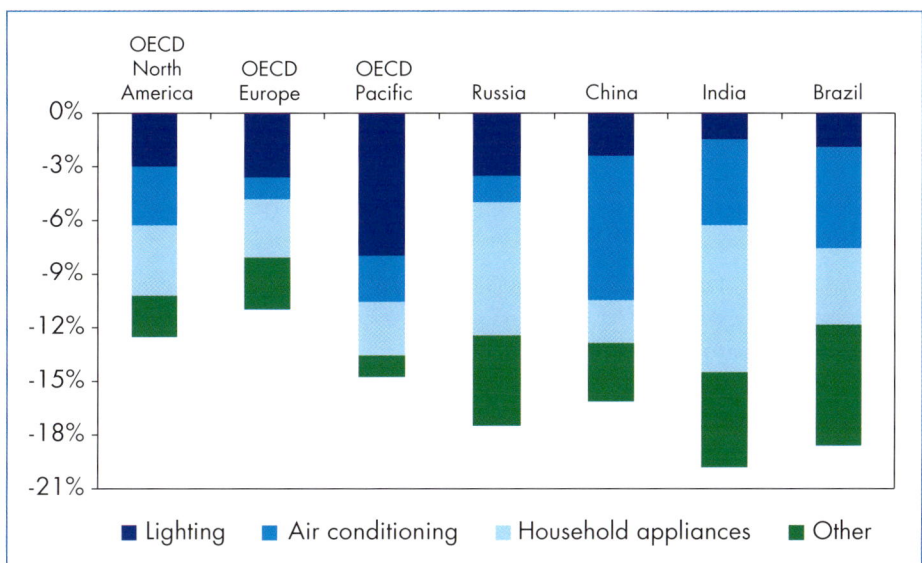

Figure 11.20: **Reduction in Electricity Demand in the Residential and Services Sectors in the Alternative Scenario*, 2030**

* Compared with the Reference Scenario.

Beyond the Alternative Policy Scenario

The Alternative Scenario analyses the effect on energy demand of faster deployment of many different types of supply and end-use technologies. They range from hybrid vehicles to power-generation fuel cells, from solar water heaters to IGCC. But the impact of *breakthrough* technologies is not analysed here. Very advanced energy technologies under development today could, however, radically modify the trends described in this chapter, inducing far bigger reductions in CO_2 emissions. Among these technologies, carbon sequestration and advanced nuclear reactors appear most likely to change the long-term energy outlook.

21. Chapter 7 provides a more detailed analysis of solar water heaters.

CO_2 capture and storage (CCS) involves the separation of the gas emitted when fossil fuels are burned, then its transport and storage in the earth or the ocean. For carbon capture to be widely used, these technologies need to be further developed and demonstrated. CCS technologies are not assumed to be deployed on a large scale before 2030. If they were, the energy-market trends described in the Reference and Alternative Scenarios might be very different. In particular, low-cost CCS would boost the prospects for coal in power generation.

The term CCS covers a wide range of technologies and storage options. It can be applied to all three fossil fuels – coal, oil and natural gas – as well as to biomass. The current cost of capturing carbon through chemical absorption and storing it in aquifers is estimated at over $50 per tonne of CO_2.[22] For power plants, using CCS would increase the cost of the electricity produced by two to three cents per kWh. But big reductions in the cost of the technology are envisaged. New, more energy-efficient power plants with better integrated capture systems could reduce the efficiency losses associated with carbon capture and lower costs. The use of CO_2 in enhanced oil- and gas-recovery techniques could in certain situations offset part of the capture cost. Some optimistic estimates put the net cost of CCS at only one cent per kWh.

Electricity generation is the main sector in which CCS could play a significant role. But CCS could also be used in manufacturing industry and in the production of transportation fuels. Reducing emissions from such large-scale sources is usually cheaper and easier than reducing them from such small-scale sources as cars and residential heating equipment.

Some stored CO_2 could eventually leak into the atmosphere. Determining the potential for such leakage from storage sites will depend on careful analysis of underground geological structures, cap-rock integrity and well-capping methods. Monitoring of leakage would be required and would add to the overall cost of CCS. In recent years, significant progress has been achieved in monitoring underground CO_2 storage at pilot projects in Norway and Canada. The results suggest that leakage is small, increasing the credibility of underground storage as a viable strategy. On the other hand, storage in deep oceans raises larger environmental concerns and remains highly controversial. Gaining public acceptance for CO_2 storage will be a key prerequisite for the success of CCS technology. The financing and construction of CO_2 storage demonstration projects, including enhanced oil recovery, with CO_2 storage to monitor long-term storage and possible leaks should be given highest priority.

A simple calculation illustrates the potential impact of CCS. In the Alternative Scenario, about 136 GW of new coal-fired power-generation capacity and 38 GW of new CCGT capacity are expected to be built in OECD

---

22. A forthcoming IEA publication discusses the framework conditions for a significant market penetration of CCS through 2050.

countries between 2015 and 2030. New capacity additions in the transition economies and the developing countries will be even larger. If all new capacity built in OECD countries after 2015 were equipped with CO_2 capture technology, and if this were to be matched by a similar amount in non-OECD countries, CCS would cover 5% of total world power generating capacity in 2030. The additional investment cost would be between $200 billion and $220 billion.[23] By 2030, the reduction in CO_2 emissions would be between 1.5 and 2 gigatonnes, depending on the utilisation rate of the power plants and the energy consumed in capturing and pressurising the CO_2. Assuming a reduction of 1.75 Gt, the total emissions reduction compared with the Reference Scenario would be 21% compared with 16% in the Alternative Scenario (Figure 11.21).[24]

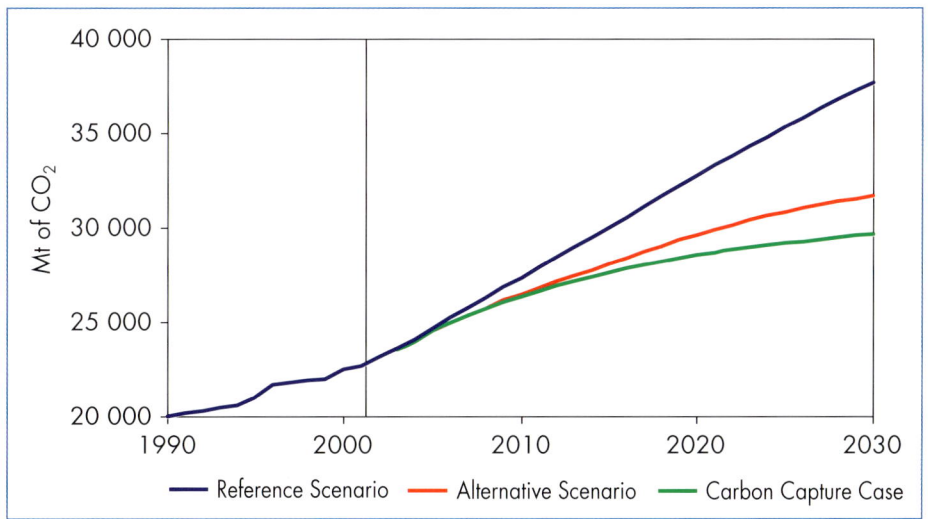

Figure 11.21: **Global Energy-Related CO_2 Emissions in the Reference and Alternative Scenarios and the CCS Case**

There is widespread interest worldwide in a new generation of nuclear reactors. But public opposition is likely to persist in many countries, because of continuing concerns about nuclear waste and the proliferation of nuclear weapons. An international task force[25] has agreed on six nuclear reactor

23. The cost of CCS is assumed to be $675 per kW for coal-fired power plants and $450 per kW for gas-fired power plants.
24. This analysis does not take into account the cost-effectiveness of CCS compared with other options for reducing emissions, such as energy efficiency and renewables.
25. The United States, Argentina, Brazil, Canada, France, Japan, South Korea, South Africa, Switzerland and the United Kingdom constitute the Generation IV International Forum.

technologies that should be developed, with the aim of deploying them commercially by 2030. In addition, the International Atomic Energy Agency is leading an international project, called Innovative Nuclear Reactors and Fuel Cycles (INPRO). Spending on the development of new designs and technology improvements for all the major reactor types is estimated to exceed $1.5 billion per year. If any of the technologies being developed becomes economically competitive during the projection period and if proliferation, waste and safety issues are adequately addressed, nuclear power could play a much bigger role in power generation than it is assigned in either of our scenarios. In this case, global CO_2 emissions could be much lower.

APPENDIX TO CHAPTER 11: TABLES FOR WORLD ALTERNATIVE POLICY SCENARIO PROJECTIONS

General Note to the Tables

The following tables show the World Alternative Policy Scenario projections of energy demand and CO_2 emissions for the following regions:

- World
- OECD
- OECD North America
- OECD Pacific
- OECD Europe
- European Union
- Transition economies
- Russia
- Developing countries
- China
- India
- Brazil

The definitions for regions and fuels can be found in Annex E.

The tables show the Alternative Scenario projections for total primary energy supply (TPES) by fuel for the years 2020 and 2030. In addition, fuel-by-fuel projections for power generation and heat plants, as well as for overall total final consumption (TFC), are given. However, a full disaggregation of the TPES data is, unlike in the Reference Scenario, not presented. Consequently, the sum of the data for power generation and heat plants and TFC is not the same as the TPES number.

Alternative Scenario CO_2 emission projections are presented with the same level of disaggregation as energy demand, i.e. overall emissions, emissions related to power generation and heat plants, and emissions related to TFC.

In addition to the level of energy demand and CO_2 emissions, the tables show the fuel mix shares for each category for the years 2002, 2020 and 2030. Average annual growth between 2002 and 2030 is also presented.

Both in the text of this book and in the tables, rounding may cause some differences between the total and the sum of the individual components.

Alternative Scenario: World

	Energy Demand (Mtoe)			Shares (%)			Growth (% p.a.)
	2002	2020	2030	2002	2020	2030	2002-2030
Total Primary Energy Supply*	10 200	13 345	14 654	100	100	100	1.3
Coal	2 389	2 726	2 744	23	20	19	0.5
Oil*	3 530	4 600	4 995	35	34	34	1.2
Gas	2 190	3 254	3 701	21	24	25	1.9
Nuclear	692	816	868	7	6	6	0.8
Hydro	224	322	367	2	2	3	1.8
Biomass and Waste	1 199	1 433	1 648	11	11	11	1.4
Other Renewables	55	195	330	0.5	1	2	6.6
Power Generation and Heat Plants	3 764	5 183	5 809	100	100	100	1.6
Coal	1 641	2 004	2 066	44	39	36	0.8
Oil	288	273	231	8	5	4	-0.8
Gas	796	1 379	1 622	21	27	28	2.6
Nuclear	692	816	868	18	16	15	0.8
Hydro	224	322	367	6	6	6	1.8
Biomass and Waste	76	241	401	2	5	7	6.1
Other Renewables	47	148	254	1	3	4	6.2
Total Final Consumption	7 075	9 181	10 110	100	100	100	1.3
Coal	502	480	449	7	5	4	-0.4
Oil	3 041	4 050	4 472	43	44	44	1.4
of which transport	*1 737*	*2 423*	*2 724*	*25*	*26*	*27*	*1.6*
Gas	1 150	1 507	1 652	16	16	16	1.3
Electricity	1 139	1 694	1 995	16	18	20	2.0
Heat	237	264	276	3	3	3	0.6
Biomass and Waste	999	1 139	1 190	14	12	12	0.6
Other Renewables	8	46	77	0.1	0.5	0.8	8.7

	CO_2 Emissions (Mt)			Shares (%)			Growth (% p.a.)
	2002	2020	2030	2002	2020	2030	2002-2030
Total CO_2 Emissions	23 116	29 583	31 686	100	100	100	1.1
Coal	9 023	10 394	10 468	39	35	33	0.5
Oil	9 174	11 856	12 885	40	40	41	1.2
Gas	4 919	7 333	8 333	21	25	26	1.9
Power Generation and Heat Plants	9 417	12 202	12 865	100	100	100	1.1
Coal	6 636	8 086	8 308	70	66	65	0.8
Oil	926	884	748	10	7	6	-0.8
Gas	1 856	3 232	3 809	20	26	30	2.6
Total Final Consumption	12 479	15 779	17 038	100	100	100	1.1
Coal	2 246	2 171	2 031	18	14	12	-0.4
Oil	7 630	10 207	11 304	61	65	66	1.4
of which transport	*4 762*	*6 628*	*7 454*	*38*	*42*	*44*	*1.6*
Gas	2 603	3 401	3 703	21	22	22	1.3

* International marine bunkers are not included.

Alternative Scenario: OECD

	Energy Demand (Mtoe)			Shares (%)			Growth (% p.a.)
	2002	2020	2030	2002	2020	2030	2002-2030
Total Primary Energy Supply	5 346	6 231	6 372	100	100	100	0.6
Coal	1 095	1 012	859	20	16	13	-0.9
Oil	2 167	2 437	2 454	41	39	39	0.4
Gas	1 171	1 561	1 645	22	25	26	1.2
Nuclear	593	621	625	11	10	10	0.2
Hydro	106	125	132	2	2	2	0.8
Biomass and Waste	181	355	467	3	6	7	3.4
Other Renewables	33	121	190	0.6	2	3	6.5
Power Generation and Heat Plants	2 139	2 545	2 557	100	100	100	0.6
Coal	891	842	706	42	33	28	-0.8
Oil	117	83	50	5	3	2	-3.0
Gas	347	613	668	16	24	26	2.4
Nuclear	593	621	625	28	24	24	0.2
Hydro	106	125	132	5	5	5	0.8
Biomass and Waste	60	166	222	3	7	9	4.8
Other Renewables	26	96	155	1	4	6	6.6
Total Final Consumption	3 691	4 361	4 552	100	100	100	0.8
Coal	119	101	91	3	2	2	-1.0
Oil	1 945	2 243	2 292	53	51	50	0.6
of which transport	*1 207*	*1 466*	*1 518*	*33*	*34*	*33*	*0.8*
Gas	728	840	863	20	19	19	0.6
Electricity	726	906	962	20	21	21	1.0
Heat	47	60	65	1	1	1	1.2
Biomass and Waste	119	187	243	3	4	5	2.6
Other Renewables	7	24	35	0.2	0.6	0.8	6.1

	CO_2 Emissions (Mt)			Shares (%)			Growth (% p.a.)
	2002	2020	2030	2002	2020	2030	2002-2030
Total CO_2 Emissions	12 446	13 737	13 322	100	100	100	0.2
Coal	4 221	3 940	3 325	34	29	25	-0.8
Oil	5 550	6 202	6 210	45	45	47	0.4
Gas	2 676	3 595	3 787	21	26	28	1.2
Power Generation and Heat Plants	4 793	5 110	4 563	100	100	100	-0.2
Coal	3 609	3 414	2 851	75	67	62	-0.8
Oil	384	279	172	8	5	4	-2.8
Gas	801	1 418	1 540	17	28	34	2.4
Total Final Consumption	7 018	7 930	8 043	100	100	100	0.5
Coal	550	473	426	8	6	5	-0.9
Oil	4 800	5 518	5 625	68	70	70	0.6
of which transport	*3 332*	*4 040*	*4 188*	*47*	*51*	*52*	*0.8*
Gas	1 668	1 939	1 993	24	24	25	0.6

Alternative Scenario: OECD North America

	Energy Demand (Mtoe)			Shares (%)			Growth (% p.a.)
	2002	2020	2030	2002	2020	2030	2002-2030
Total Primary Energy Supply	2 698	3 216	3 335	100	100	100	0.8
Coal	578	553	500	21	17	15	-0.5
Oil	1 079	1 305	1 354	40	41	41	0.8
Gas	651	832	871	24	26	26	1.0
Nuclear	232	251	235	9	8	7	0.1
Hydro	52	60	62	2	2	2	0.6
Biomass and Waste	90	161	224	3	5	7	3.3
Other Renewables	16	53	89	0.6	2	3	6.4
Power Generation and Heat Plants	1 065	1 278	1 295	100	100	100	0.7
Coal	524	506	456	49	40	35	-0.5
Oil	45	46	34	4	4	3	-1.0
Gas	171	305	333	16	24	26	2.4
Nuclear	232	251	235	22	20	18	0.1
Hydro	52	60	62	5	5	5	0.6
Biomass and Waste	28	67	101	3	5	8	4.6
Other Renewables	13	43	73	1	3	6	6.3
Total Final Consumption	1 841	2 224	2 353	100	100	100	0.9
Coal	31	31	29	2	1	1	-0.3
Oil	976	1 188	1 248	53	53	53	0.9
of which transport	*690*	*858*	*905*	*37*	*39*	*38*	*1.0*
Gas	407	445	450	22	20	19	0.4
Electricity	358	450	481	19	20	20	1.1
Heat	7	8	8	0	0	0	0.8
Biomass and Waste	59	92	121	3	4	5	2.6
Other Renewables	2	10	15	0.1	0.5	0.7	7.2

	CO_2 Emissions (Mt)			Shares (%)			Growth (% p.a.)
	2002	2020	2030	2002	2020	2030	2002-2030
Total CO_2 Emissions	6 480	7 402	7 403	100	100	100	0.5
Coal	2 219	2 143	1 940	34	29	26	-0.5
Oil	2 774	3 353	3 471	43	45	47	0.8
Gas	1 487	1 906	1 992	23	26	27	1.0
Power Generation and Heat Plants	2 613	2 856	2 682	100	100	100	0.1
Coal	2 069	1 996	1 801	79	70	67	-0.5
Oil	152	162	122	6	6	5	-0.8
Gas	392	699	759	15	24	28	2.4
Total Final Consumption	3 506	4 099	4 250	100	100	100	0.7
Coal	147	144	137	4	4	3	-0.3
Oil	2 426	2 934	3 082	69	72	73	0.9
of which transport	*1 946*	*2 415*	*2 552*	*56*	*59*	*60*	*1.0*
Gas	933	1 021	1 031	27	25	24	0.4

Alternative Scenario: OECD Pacific

	Energy Demand (Mtoe)			Shares (%)			Growth (% p.a.)
	2002	2020	2030	2002	2020	2030	2002-2030
Total Primary Energy Supply	852	1 015	1 041	100	100	100	0.7
Coal	196	196	168	23	19	16	-0.5
Oil	398	427	419	47	42	40	0.2
Gas	113	173	186	13	17	18	1.8
Nuclear	108	159	189	13	16	18	2.0
Hydro	11	13	13	1	1	1	0.7
Biomass and Waste	20	32	40	2	3	4	2.5
Other Renewables	6	16	26	0.7	2	3	5.3
Power Generation and Heat Plants	362	453	468	100	100	100	0.9
Coal	132	131	104	36	29	22	-0.8
Oil	33	19	9	9	4	2	-4.6
Gas	64	103	110	18	23	23	1.9
Nuclear	108	159	189	30	35	40	2.0
Hydro	11	13	13	3	3	3	0.7
Biomass and Waste	10	18	24	3	4	5	3.0
Other Renewables	5	11	18	1	2	4	5.1
Total Final Consumption	583	689	706	100	100	100	0.7
Coal	34	34	33	6	5	5	-0.1
Oil	354	399	401	61	58	57	0.4
of which transport	*159*	*193*	*196*	*27*	*28*	*28*	*0.8*
Gas	51	69	74	9	10	11	1.3
Electricity	129	161	168	22	23	24	1.0
Heat	5	6	7	1	1	1	1.3
Biomass and Waste	9	14	16	2	2	2	1.9
Other Renewables	1	5	7	0.2	0.8	1	6.1

	CO_2 Emissions (Mt)			Shares (%)			Growth (% p.a.)
	2002	2020	2030	2002	2020	2030	2002-2030
Total CO_2 Emissions	2 022	2 198	2 070	100	100	100	0.1
Coal	761	750	627	38	34	30	-0.7
Oil	993	1 049	1 016	49	48	49	0.1
Gas	267	400	427	13	18	21	1.7
Power Generation and Heat Plants	823	862	729	100	100	100	-0.4
Coal	571	561	446	69	65	61	-0.9
Oil	104	60	27	13	7	4	-4.7
Gas	149	241	257	18	28	35	2.0
Total Final Consumption	1 111	1 255	1 263	100	100	100	0.5
Coal	158	157	151	14	13	12	-0.2
Oil	841	947	950	76	75	75	0.4
of which transport	*437*	*531*	*542*	*39*	*42*	*43*	*0.8*
Gas	111	151	162	10	12	13	1.4

Alternative Scenario: OECD Europe

	Energy Demand (Mtoe)			Shares (%)			Growth (% p.a.)
	2002	2020	2030	2002	2020	2030	2002-2030
Total Primary Energy Supply	1 795	1 999	1 996	100	100	100	0.4
Coal	321	262	191	18	13	10	-1.8
Oil	689	705	681	38	35	34	-0.0
Gas	407	556	589	23	28	29	1.3
Nuclear	253	211	200	14	11	10	-0.8
Hydro	43	52	56	2	3	3	1.0
Biomass and Waste	71	162	204	4	8	10	3.8
Other Renewables	11	51	75	0.6	3	4	7.1
Power Generation and Heat Plants	712	814	794	100	100	100	0.4
Coal	235	206	146	33	25	18	-1.7
Oil	40	18	7	6	2	1	-5.8
Gas	112	205	225	16	25	28	2.5
Nuclear	253	211	200	36	26	25	-0.8
Hydro	43	52	56	6	6	7	1.0
Biomass and Waste	21	80	97	3	10	12	5.6
Other Renewables	8	42	63	1	5	8	7.5
Total Final Consumption	1 267	1 448	1 492	100	100	100	0.6
Coal	54	36	29	4	2	2	-2.2
Oil	615	656	644	49	45	43	0.2
of which transport	*357*	*416*	*416*	*28*	*29*	*28*	*0.5*
Gas	270	325	338	21	22	23	0.8
Electricity	238	294	312	19	20	21	1.0
Heat	36	46	50	3	3	3	1.2
Biomass and Waste	50	82	106	4	6	7	2.7
Other Renewables	3	9	12	0.2	0.6	0.8	5.1

	CO_2 Emissions (Mt)			Shares (%)			Growth (% p.a.)
	2002	2020	2030	2002	2020	2030	2002-2030
Total CO_2 Emissions	3 945	4 137	3 850	100	100	100	-0.1
Coal	1 241	1 047	758	31	25	20	-1.7
Oil	1 782	1 801	1 723	45	44	45	-0.1
Gas	922	1 289	1 368	23	31	36	1.4
Power Generation and Heat Plants	1 357	1 392	1 152	100	100	100	- 0.6
Coal	969	857	604	71	62	52	-1.7
Oil	127	57	24	9	4	2	-5.8
Gas	260	478	524	19	34	45	2.5
Total Final Consumption	2 402	2 576	2 530	100	100	100	0.2
Coal	244	172	139	10	7	5	-2.0
Oil	1 534	1 637	1 592	64	64	63	0.1
of which transport	*949*	*1 094*	*1 094*	*40*	*42*	*43*	*0.5*
Gas	624	767	799	26	30	32	0.9

Alternative Scenario: European Union

	Energy Demand (Mtoe)			Shares (%)			Growth (% p.a.)
	2002	2020	2030	2002	2020	2030	2002-2030
Total Primary Energy Supply	1 690	1 881	1 872	100	100	100	0.4
Coal	303	242	177	18	13	9	-1.9
Oil	648	663	636	38	35	34	-0.1
Gas	389	541	574	23	29	31	1.4
Nuclear	251	205	195	15	11	10	-0.9
Hydro	26	31	33	2	2	2	0.9
Biomass and Waste	65	152	189	4	8	10	3.9
Other Renewables	8	46	68	0.5	2	4	8.0
Power Generation and Heat Plants	676	771	749	100	100	100	0.4
Coal	228	197	142	34	26	19	-1.7
Oil	40	18	7	6	2	1	-6.0
Gas	105	203	224	15	26	30	2.8
Nuclear	251	205	195	37	27	26	-0.9
Hydro	26	31	33	4	4	4	0.9
Biomass and Waste	20	76	89	3	10	12	5.5
Other Renewables	7	41	59	1	5	8	8.0
Total Final Consumption	1 186	1 360	1 401	100	100	100	0.6
Coal	45	26	20	4	2	1	-2.9
Oil	574	616	602	48	45	43	0.2
of which transport	*339*	*396*	*394*	*29*	*29*	*28*	*0.5*
Gas	266	319	332	22	23	24	0.8
Electricity	218	270	287	18	20	20	1.0
Heat	37	48	52	3	4	4	1.2
Biomass and Waste	45	76	100	4	6	7	2.9
Other Renewables	1	6	9	0.1	0.4	0.6	7.5

	CO_2 Emissions (Mt)			Shares (%)			Growth (% p.a.)
	2002	2020	2030	2002	2020	2030	2002-2030
Total CO_2 Emissions	3 731	3 906	3 639	100	100	100	-0.1
Coal	1 170	959	698	31	25	19	-1.8
Oil	1 677	1 694	1 607	45	43	44	-0.2
Gas	884	1 253	1 334	24	32	37	1.5
Power Generation and Heat Plants	1 308	1 338	1 127	100	100	100	-0.5
Coal	937	810	583	72	61	52	-1.7
Oil	128	57	23	10	4	2	-6.0
Gas	243	471	521	19	35	46	2.8
Total Final Consumption	2 249	2 419	2 371	100	100	100	0.2
Coal	207	132	101	9	5	4	-2.5
Oil	1 432	1 537	1 489	64	64	63	0.1
of which transport	*901*	*1 039*	*1 035*	*40*	*43*	*44*	*0.5*
Gas	611	750	781	27	31	33	0.9

Alternative Scenario: Transition Economies

	Energy Demand (Mtoe)			Shares (%)			Growth (% p.a.)
	2002	2020	2030	2002	2020	2030	2002-2030
Total Primary Energy Supply	1 030	1 252	1 312	100	100	100	0.9
Coal	194	202	184	19	16	14	-0.2
Oil	222	295	321	22	24	24	1.3
Gas	504	613	644	49	49	49	0.9
Nuclear	69	86	91	7	7	7	1.0
Hydro	24	31	32	2	2	2	1.0
Biomass and Waste	16	20	32	2	2	2	2.5
Other Renewables	0	4	10	0.0	0.4	0.7	14.2
Power Generation and Heat Plants	527	592	603	100	100	100	0.5
Coal	131	134	116	25	23	19	-0.5
Oil	33	32	26	6	5	4	-0.8
Gas	265	300	314	50	51	52	0.6
Nuclear	69	86	91	13	15	15	1.0
Hydro	24	31	32	5	5	5	1.0
Biomass and Waste	5	5	15	1	1	2	4.3
Other Renewables	0	4	9	0.0	0.7	2	14.4
Total Final Consumption	669	833	888	100	100	100	1.0
Coal	39	43	44	6	5	5	0.4
Oil	160	219	247	24	26	28	1.6
of which transport	*86*	*123*	*144*	*13*	*15*	*16*	*1.9*
Gas	210	277	291	31	33	33	1.2
Electricity	90	123	137	13	15	15	1.5
Heat	160	155	152	24	19	17	-0.2
Biomass and Waste	11	15	17	2	2	2	1.5
Other Renewables	0	0	1	0.0	0.0	0.1	11.9

	CO_2 Emissions (Mt)			Shares (%)			Growth (% p.a.)
	2002	2020	2030	2002	2020	2030	2002-2030
Total CO_2 Emissions	2 444	2 891	2 940	100	100	100	0.7
Coal	756	790	718	31	27	24	-0.2
Oil	556	732	789	23	25	27	1.3
Gas	1 132	1 369	1 433	46	47	49	0.8
Power Generation and Heat Plants	1 270	1 360	1 299	100	100	100	0.1
Coal	544	554	478	43	41	37	-0.5
Oil	106	103	85	8	8	7	-0.8
Gas	620	704	736	49	52	57	0.6
Total Final Consumption	1 080	1 401	1 502	100	100	100	1.2
Coal	208	233	237	19	17	16	0.5
Oil	398	546	614	37	39	41	1.6
of which transport	*214*	*309*	*360*	*20*	*22*	*24*	*1.9*
Gas	474	622	651	44	44	43	1.1

Alternative Scenario: Russia

	Energy Demand (Mtoe)			Shares (%)			Growth (% p.a.)
	2002	2020	2030	2002	2020	2030	2002-2030
Total Primary Energy Supply	**619**	**723**	**774**	**100**	**100**	**100**	**0.8**
Coal	107	105	96	17	15	12	-0.4
Oil	128	160	178	21	22	23	1.2
Gas	326	379	396	53	52	51	0.7
Nuclear	37	53	67	6	7	9	2.1
Hydro	14	17	17	2	2	2	0.7
Biomass and Waste	7	6	12	1	1	2	2.1
Other Renewables	0	3	7	0.0	0.5	0.9	15.0
Power Generation and Heat Plants	**348**	**368**	**381**	**100**	**100**	**100**	**0.3**
Coal	79	74	65	23	20	17	-0.7
Oil	21	21	18	6	6	5	-0.5
Gas	193	197	198	55	53	52	0.1
Nuclear	37	53	67	11	14	18	2.1
Hydro	14	17	17	4	5	5	0.7
Biomass and Waste	4	3	8	1	1	2	2.7
Other Renewables	0	3	7	0.0	0.9	2	15.0
Total Final Consumption	**412**	**495**	**533**	**100**	**100**	**100**	**0.9**
Coal	20	22	23	5	4	4	0.4
Oil	90	114	132	22	23	25	1.4
of which transport	*50*	*69*	*84*	*12*	*14*	*16*	*1.8*
Gas	116	160	174	28	32	33	1.5
Electricity	53	69	77	13	14	14	1.3
Heat	131	126	123	32	26	23	-0.2
Biomass and Waste	3	3	4	1	1	1	1.1
Other Renewables	0	0	0	0.0	0.0	0.0	-

	CO$_2$ Emissions (Mt)			Shares (%)			Growth (% p.a.)
	2002	2020	2030	2002	2020	2030	2002-2030
Total CO$_2$ Emissions	**1 488**	**1 673**	**1 714**	**100**	**100**	**100**	**0.5**
Coal	439	433	398	29	26	23	-0.3
Oil	319	397	436	21	24	25	1.1
Gas	731	843	879	49	50	51	0.7
Power Generation and Heat Plants	**849**	**841**	**796**	**100**	**100**	**100**	**-0.2**
Coal	331	312	274	39	37	34	-0.7
Oil	68	68	59	8	8	7	-0.5
Gas	450	460	464	53	55	58	0.1
Total Final Consumption	**581**	**750**	**825**	**100**	**100**	**100**	**1.3**
Coal	106	119	123	18	16	15	0.5
Oil	216	273	314	37	36	38	1.3
of which transport	*119*	*162*	*197*	*20*	*22*	*24*	*1.8*
Gas	259	358	388	45	48	47	1.5

Alternative Scenario: Developing Countries

	Energy Demand (Mtoe)			Shares (%)			Growth (% p.a.)
	2002	2020	2030	2002	2020	2030	2002-2030
Total Primary Energy Supply	3 824	5 863	6 970	100	100	100	2.2
Coal	1 099	1 512	1 701	29	26	24	1.6
Oil	1 142	1 868	2 221	30	32	32	2.4
Gas	515	1 080	1 412	13	18	20	3.7
Nuclear	30	109	153	1	2	2	6.0
Hydro	94	167	203	2	3	3	2.8
Biomass and Waste	922	1 058	1 149	24	18	16	0.8
Other Renewables	21	69	131	0.6	1	2	6.7
Power Generation and Heat Plants	1 097	2 046	2 649	100	100	100	3.2
Coal	619	1 028	1 244	56	50	47	2.5
Oil	137	158	155	13	8	6	0.4
Gas	184	465	640	17	23	24	4.5
Nuclear	30	109	153	3	5	6	6.0
Hydro	94	167	203	9	8	8	2.8
Biomass and Waste	11	70	165	1	3	6	10.0
Other Renewables	21	48	90	2	2	3	5.4
Total Final Consumption	2 714	3 988	4 670	100	100	100	2.0
Coal	343	336	314	13	8	7	-0.3
Oil	936	1 588	1 933	34	40	41	2.6
of which transport	*445*	*833*	*1 062*	*16*	*21*	*23*	*3.2*
Gas	212	390	498	8	10	11	3.1
Electricity	323	665	896	12	17	19	3.7
Heat	30	49	59	1	1	1	2.5
Biomass and Waste	869	937	929	32	24	20	0.2
Other Renewables	1	21	41	0.0	0.5	0.9	15.2

	CO_2 Emissions (Mt)			Shares (%)			Growth (% p.a.)
	2002	2020	2030	2002	2020	2030	2002-2030
Total CO_2 Emissions	8 226	12 955	15 424	100	100	100	2.3
Coal	4 047	5 665	6 425	49	44	42	1.7
Oil	3 068	4 921	5 887	37	38	38	2.4
Gas	1 111	2 369	3 112	14	18	20	3.7
Power Generation and Heat Plants	3 354	5 732	7 003	100	100	100	2.7
Coal	2 483	4 119	4 979	74	72	71	2.5
Oil	436	503	491	13	9	7	0.4
Gas	435	1 110	1 532	13	19	22	4.6
Total Final Consumption	4 381	6 448	7 493	100	100	100	1.9
Coal	1 488	1 465	1 369	34	23	18	-0.3
Oil	2 432	4 143	5 065	56	64	68	2.7
of which transport	*1 215*	*2 279*	*2 906*	*28*	*35*	*39*	*3.2*
Gas	460	840	1 059	11	13	14	3.0

Alternative Scenario: China

	Energy Demand (Mtoe)			Shares (%)			Growth (% p.a.)
	2002	2020	2030	2002	2020	2030	2002-2030
Total Primary Energy Supply	1 242	1 877	2 205	100	100	100	2.1
Coal	713	962	1 045	57	51	47	1.4
Oil	247	464	559	20	25	25	3.0
Gas	36	128	207	3	7	9	6.5
Nuclear	7	53	84	1	3	4	9.6
Hydro	25	50	63	2	3	3	3.4
Biomass and Waste	216	206	219	17	11	10	0.0
Other Renewables	0	14	28	0.0	0.7	1	-
Power Generation and Heat Plants	428	845	1 082	100	100	100	3.4
Coal	374	632	738	87	75	68	2.5
Oil	17	18	15	4	2	1	-0.4
Gas	5	47	85	1	6	8	10.8
Nuclear	7	53	84	2	6	8	9.6
Hydro	25	50	63	6	6	6	3.4
Biomass and Waste	1	39	80	0	5	7	15.7
Other Renewables	0	6	17	0.0	0.7	2	-
Total Final Consumption	823	1 157	1 344	100	100	100	1.8
Coal	240	230	211	29	20	16	-0.5
Oil	204	395	498	25	34	37	3.2
of which transport	*80*	*196*	*275*	*10*	*17*	*20*	*4.5*
Gas	22	62	96	3	5	7	5.4
Electricity	112	246	329	14	21	24	3.9
Heat	30	49	59	4	4	4	2.5
Biomass and Waste	215	167	139	26	14	10	-1.5
Other Renewables	0	8	11	0.0	0.7	0.8	-

	CO_2 Emissions (Mt)			Shares (%)			Growth (% p.a.)
	2002	2020	2030	2002	2020	2030	2002-2030
Total CO_2 Emissions	3 307	5 053	5 856	100	100	100	2.1
Coal	2 621	3 632	3 976	79	72	68	1.5
Oil	618	1 142	1 423	19	23	24	3.0
Gas	69	279	457	2	6	8	7.0
Power Generation and Heat Plants	1 576	2 722	3 231	100	100	100	2.6
Coal	1 507	2 546	2 973	96	94	92	2.5
Oil	57	62	52	4	2	2	-0.4
Gas	12	114	206	1	4	6	10.8
Total Final Consumption	1 591	2 131	2 397	100	100	100	1.5
Coal	1 045	1 015	935	66	48	39	-0.4
Oil	507	1 007	1 293	32	47	54	3.4
of which transport	*223*	*549*	*769*	*14*	*26*	*32*	*4.5*
Gas	40	110	169	2	5	7	5.3

Alternative Scenario: India

	Energy Demand (Mtoe)			Shares (%)			Growth (% p.a.)
	2002	2020	2030	2002	2020	2030	2002-2030
Total Primary Energy Supply	538	762	902	100	100	100	1.9
Coal	178	226	267	33	30	30	1.5
Oil	119	201	241	22	26	27	2.6
Gas	23	62	87	4	8	10	5.0
Nuclear	5	25	35	1	3	4	7.2
Hydro	5	16	18	1	2	2	4.4
Biomass and Waste	208	227	244	39	30	27	0.6
Other Renewables	0	5	9	0.0	0.7	1	14.6
Power Generation and Heat Plants	160	277	372	100	100	100	3.1
Coal	130	181	227	81	65	61	2.0
Oil	7	8	7	5	3	2	0.0
Gas	11	38	56	7	14	15	6.0
Nuclear	5	25	35	3	9	9	7.2
Hydro	5	16	18	3	6	5	4.4
Biomass and Waste	1	7	25	1	3	7	11.9
Other Renewables	0	2	4	0.1	0.7	1	11.1
Total Final Consumption	382	531	611	100	100	100	1.7
Coal	36	33	29	9	6	5	-0.7
Oil	97	178	220	25	34	36	3.0
of which transport	*33*	*51*	*63*	*9*	*10*	*10*	*2.3*
Gas	9	21	28	2	4	5	4.0
Electricity	33	75	109	9	14	18	4.4
Heat	0	0	0	0	0	0	-
Biomass and Waste	207	220	219	54	41	36	0.2
Other Renewables	0	3	5	0.0	0.6	0.9	-

	CO_2 Emissions (Mt)			Shares (%)			Growth (% p.a.)
	2002	2020	2030	2002	2020	2030	2002-2030
Total CO_2 Emissions	1 016	1 501	1 818	100	100	100	2.1
Coal	671	855	1 014	66	57	56	1.5
Oil	293	504	608	29	34	33	2.6
Gas	52	142	196	5	9	11	4.8
Power Generation and Heat Plants	553	819	1 037	100	100	100	2.3
Coal	505	704	883	91	86	85	2.0
Oil	23	25	23	4	3	2	0.0
Gas	26	90	131	5	11	13	6.0
Total Final Consumption	432	648	746	100	100	100	2.0
Coal	165	148	128	38	23	17	-0.9
Oil	246	454	561	57	70	75	3.0
of which transport	*94*	*144*	*176*	*22*	*22*	*24*	*2.3*
Gas	21	46	57	5	7	8	3.6

Alternative Scenario: Brazil

	Energy Demand (Mtoe)			Shares (%)			Growth (% p.a.)
	2002	2020	2030	2002	2020	2030	2002-2030
Total Primary Energy Supply	188	272	328	100	100	100	2.0
Coal	13	13	16	7	5	5	0.7
Oil	88	129	154	47	48	47	2.0
Gas	12	25	35	6	9	11	3.9
Nuclear	4	6	6	2	2	2	2.0
Hydro	25	39	46	13	14	14	2.3
Biomass and Waste	46	58	67	25	21	20	1.3
Other Renewables	0	2	4	0.0	0.6	1.1	46.1
Power Generation and Heat Plants	39	58	73	100	100	100	2.3
Coal	2	0	1	6	0	2	-2.1
Oil	3	1	1	8	1	1	-4.0
Gas	3	6	11	7	10	14	5.2
Nuclear	4	6	6	9	11	8	2.0
Hydro	25	39	46	63	68	62	2.3
Biomass and Waste	2	5	7	6	9	9	3.8
Other Renewables	0	1	2	0.0	0.9	3	42.9
Total Final Consumption	160	236	284	100	100	100	2.1
Coal	6	8	10	4	3	3	1.7
Oil	80	122	147	50	52	52	2.2
of which transport	*43*	*72*	*88*	*27*	*31*	*31*	*2.6*
Gas	7	16	21	5	7	7	3.9
Electricity	27	40	50	17	17	18	2.3
Heat	0	0	0	0	0	0	-
Biomass and Waste	40	48	55	25	21	19	1.1
Other Renewables	0	1	2	0.0	0.4	0.6	-

	CO_2 Emissions (Mt)			Shares (%)			Growth (% p.a.)
	2002	2020	2030	2002	2020	2030	2002-2030
Total CO2 Emissions	303	443	547	100	100	100	2.1
Coal	41	39	50	14	9	9	0.7
Oil	238	352	422	79	79	77	2.1
Gas	23	52	75	8	12	14	4.3
Power Generation and Heat Plants	24	16	32	100	100	100	1.0
Coal	11	0	5	45	0	15	-3.0
Oil	9	2	3	39	12	9	-4.2
Gas	4	14	25	16	88	76	6.9
Total Final Consumption	257	403	488	100	100	100	2.3
Coal	26	36	42	10	9	9	1.7
Oil	217	336	403	84	83	83	2.2
of which transport	*127*	*212*	*259*	*49*	*53*	*53*	*2.6*
Gas	14	32	43	6	8	9	4.0

ANNEX A

TABLES FOR REFERENCE SCENARIO PROJECTIONS

General Note to the Tables

For OECD countries and for most non-OECD countries, the analysis of energy demand is based on data up to 2002, published in mid-2004 in *Energy Balances of OECD Countries* and in *Energy Balances of Non-OECD Countries*.

The tables show projections of energy demand, electricity generation and capacity, and CO_2 emissions for the following regions:
– World
– OECD
– OECD North America
– United States and Canada
– Mexico
– OECD Pacific
– OECD Asia
– OECD Oceania
– OECD Europe
– European Union
– Transition economies
– Russia
– Developing countries
– China
– East Asia
– Indonesia
– South Asia
– India
– Latin America
– Brazil
– Middle East
– Africa

The definitions for regions, fuels and sectors are in Annex E.

Both in the text of this book and in the tables, rounding may cause some differences between the total and the sum of the individual components.

Reference Scenario: World

	Energy Demand (Mtoe)					Shares (%)					Growth Rates (% per annum)			
	1971	2002	2010	2020	2030	1971	2002	2010	2020	2030	1971-2002	2002-2010	2002-2020	2002-2030
Total Primary Energy Supply	5 536	10 345	12 194	14 404	16 487	100	100	100	100	100	2.0	2.1	1.9	1.7
Coal	1 407	2 389	2 763	3 193	3 601	25	23	23	22	22	1.7	1.8	1.6	1.5
Oil	2 413	3 676	4 308	5 074	5 766	44	36	35	35	35	1.4	2.0	1.8	1.6
of which international marine bunkers	*106*	*146*	*148*	*152*	*162*	*2*	*1*	*1*	*1*	*1*	*1.0*	*0.2*	*0.2*	*0.4*
Gas	892	2 190	2 703	3 451	4 130	16	21	22	24	25	2.9	2.7	2.6	2.3
Nuclear	29	692	778	776	764	1	7	6	5	5	10.8	1.5	0.6	0.4
Hydro	104	224	276	321	365	2	2	2	2	2	2.5	2.6	2.0	1.8
Biomass and Waste	687	1 119	1 264	1 428	1 605	12	11	10	10	10	1.6	1.5	1.4	1.3
Other Renewables	4	55	101	162	256	0	1	1	1	2	8.5	8.0	6.2	5.7
Power Generation and Heat Plants	1 211	3 764	4 613	5 629	6 630	100	100	100	100	100	3.7	2.6	2.3	2.0
Coal	594	1 641	1 990	2 410	2 815	49	44	43	43	42	3.3	2.4	2.2	1.9
Oil	270	288	290	305	281	22	8	6	5	4	0.2	0.1	0.3	-0.1
Gas	207	796	1 076	1 512	1 932	17	21	23	27	29	4.4	3.8	3.6	3.2
Nuclear	29	692	778	776	764	2	18	17	14	12	10.8	1.5	0.6	0.4
Hydro	104	224	276	321	365	9	6	6	6	6	2.5	2.6	2.0	1.8
Biomass and Waste	4	76	115	165	258	0	2	2	3	4	9.9	5.4	4.4	4.5
Other Renewables	4	47	87	139	215	0	1	2	2	3	8.3	8.1	6.2	5.6
Other Transformation, Own Use and Losses	526	1 025	1 204	1 407	1 584						2.2	2.0	1.8	1.6
of which electricity	*59*	*244*	*302*	*379*	*460*						*4.7*	*2.7*	*2.5*	*2.3*
Total Final Consumption	4 200	7 075	8 267	9 754	11 176	100	100	100	100	100	1.7	2.0	1.8	1.6
Coal	617	502	516	522	526	15	7	6	5	5	-0.7	0.4	0.2	0.2
Oil	1 893	3 041	3 610	4 326	5 005	45	43	44	44	45	1.5	2.2	2.0	1.8
Gas	604	1 150	1 336	1 564	1 758	14	16	16	16	16	2.1	1.9	1.7	1.5
Electricity	377	1 139	1 436	1 835	2 263	9	16	17	19	20	3.6	2.9	2.7	2.5
Heat	68	237	254	275	294	2	3	3	3	3	4.1	0.9	0.8	0.8
Biomass and Waste	641	999	1 101	1 210	1 290	15	14	13	12	12	1.4	1.2	1.1	0.9
Other Renewables	0	8	13	23	41	0	0.1	0.2	0.2	0.4	9.9	7.4	6.3	6.2

Industry	1 505	2 236	2 578	2 999	3 374	100	100	100	100	100	1.3	1.8	1.6	1.5
Coal	348	380	407	425	437	23	17	16	14	13	0.3	0.9	0.6	0.5
Oil	499	604	682	780	861	33	27	26	26	26	0.6	1.5	1.4	1.3
Gas	333	522	615	731	828	22	23	24	24	25	1.5	2.1	1.9	1.7
Electricity	196	471	583	730	877	13	21	23	24	26	2.9	2.7	2.5	2.2
Heat	46	98	102	106	109	3	4	4	4	3	2.4	0.5	0.4	0.4
Renewables	83	162	190	227	262	6	7	7	8	8	2.2	2.0	1.9	1.7
Transport	856	1 827	2 230	2 755	3 273	100	100	100	100	100	2.5	2.5	2.3	2.1
Oil	797	1 737	2 120	2 621	3 110	93	95	95	95	95	2.5	2.5	2.3	2.1
Other Fuels	60	89	109	135	163	7	5	5	5	5	1.3	2.6	2.3	2.2
Other Sectors	1 605	2 770	3 169	3 675	4 175	100	100	100	100	100	1.8	1.7	1.6	1.5
Coal	223	98	88	78	72	14	4	3	2	2	-2.6	-1.4	-1.2	-1.1
Oil	455	502	565	650	733	28	18	18	18	18	0.3	1.5	1.4	1.4
Gas	229	562	643	740	825	14	20	20	20	20	2.9	1.7	1.5	1.4
Electricity	151	637	816	1 060	1 332	9	23	26	29	32	4.8	3.1	2.9	2.7
Heat	12	138	151	167	183	1	5	5	5	4	8.2	1.2	1.1	1.0
Renewables	535	833	906	980	1 030	33	30	29	27	25	1.4	1.1	0.9	0.8
Non-Energy Use	233	243	291	326	354						0.1	2.3	1.7	1.4
Electricity Generation (TWh)	5 217	16 074	20 185	25 752	31 657	100	100	100	100	100	3.7	2.9	2.7	2.5
Coal	2 095	6 241	7 692	9 766	12 091	40	39	38	38	38	3.6	2.6	2.5	2.4
Oil	1 096	1 181	1 187	1 274	1 182	21	7	6	5	4	0.2	0.1	0.4	0.0
Gas	696	3 070	4 427	6 827	9 329	13	19	22	27	29	4.9	4.7	4.5	4.0
Nuclear	111	2 654	2 985	2 975	2 929	2	17	15	12	9	10.8	1.5	0.6	0.4
Hydro	1 206	2 610	3 212	3 738	4 248	23	16	16	15	13	2.5	2.6	2.0	1.8
Biomass and Waste	9	207	326	438	627	0	1	2	2	2	10.8	5.8	4.3	4.0
Other Renewables	5	111	356	733	1 250	0	1	2	3	4	10.5	15.7	11.1	9.0
Traditional Biomass (included above)	492	765	827	883	907	9	7	7	6	6	1.4	1.0	0.8	0.6
Total Primary Energy Supply (excluding traditional biomass)	5 044	9 580	11 367	13 521	15 580						2.1	2.2	1.9	1.8

Reference Scenario: World

	Capacity (GW)				Shares (%)				Growth Rates (% per annum)		
	2002	2010	2020	2030	2002	2010	2020	2030	2002-2010	2002-2020	2002-2030
Total Capacity	**3 719**	**4 539**	**5 822**	**7 303**	**100**	**100**	**100**	**100**	**2.5**	**2.5**	**2.4**
Coal	1 135	1 337	1 691	2 156	31	29	29	30	2.1	2.2	2.3
Oil	454	499	524	453	12	11	9	6	1.2	0.8	0.0
Gas	893	1 211	1 824	2 564	24	27	31	35	3.9	4.0	3.8
Nuclear	359	385	382	376	10	8	6	5	0.9	0.4	0.2
Hydro	801	934	1 076	1 216	22	21	18	17	1.9	1.7	1.5
of which pumped storage	*89*	*92*	*96*	*99*	*2*	*2*	*2*	*1*	*0.5*	*0.4*	*0.4*
Renewables (excluding hydro)	77	172	326	539	2	4	6	7	10.7	8.4	7.2

Renewables Breakdown

	Capacity (GW)				Shares (%)				Growth Rates (% per annum)		
	2002	2010	2020	2030	2002	2010	2020	2030	2002-2010	2002-2020	2002-2030
Renewables (excluding hydro)	**77**	**172**	**326**	**539**	**100**	**100**	**100**	**100**	**10.7**	**8.4**	**7.2**
Biomass and Waste	34	53	71	101	44	31	22	19	5.6	4.2	4.0
Wind	32	97	206	328	42	57	63	61	15.0	10.9	8.7
Geothermal	9	13	18	25	12	8	5	5	4.3	3.5	3.6
Solar	1	8	29	76	1	5	9	14	28.7	20.3	16.6
Tide/Wave	0	1	3	9	0	1	1	2	17.9	13.2	12.4

	Electricity Generation (TWh)				Shares (%)				Growth Rates (% per annum)		
	2002	2010	2020	2030	2002	2010	2020	2030	2002-2010	2002-2020	2002-2030
Renewables (excluding hydro)	**317**	**682**	**1 171**	**1 877**	**100**	**100**	**100**	**100**	**7.9**	**6.7**	**6.1**
Biomass and Waste	207	326	438	627	65	48	37	33	4.6	3.8	3.8
Wind	52	251	559	929	16	37	48	50	17.0	12.6	10.1
Geothermal	57	89	119	167	18	13	10	9	4.6	3.8	3.7
Solar	1	12	43	119	0	2	4	6	27.9	20.6	17.2
Tide/Wave	1	5	12	35	0	1	1	2	19.0	14.3	13.2

Reference Scenario: World

	CO$_2$ Emissions (Mt)					Shares (%)					Growth Rates (% per annum)				
	1971	2002	2010	2020	2030	1971	2002	2010	2020	2030	1971-2002	2002-2010	2002-2020	2002-2030
Total CO$_2$ Emissions	13 958	23 579	27 817	33 226	38 214	100	100	100	100	100	1.7	2.1	1.9	1.7
change since 1990 (%)		*17.7*	*38.9*	*65.9*	*90.8*									
Coal	5 207	9 023	10 517	12 244	13 866	37	38	38	37	36	1.8	1.9	1.7	1.5
Oil	6 691	9 637	11 209	13 202	15 035	48	41	40	40	39	1.2	1.9	1.8	1.6
of which international marine bunkers	*342*	*463*	*471*	*483*	*515*	*2*	*2*	*2*	*1*	*1*	*1.0*	*0.2*	*0.2*	*0.4*
Gas	2 060	4 919	6 091	7 780	9 313	15	21	22	23	24	2.8	2.7	2.6	2.3
Power Generation and Heat Plants	3 731	9 417	11 494	14 258	16 771	100	100	100	100	100	3.0	2.5	2.3	2.1
Coal	2 385	6 636	8 040	9 731	11 334	64	70	70	68	68	3.4	2.4	2.1	1.9
Oil	862	926	937	987	908	23	10	8	7	5	0.2	0.1	0.4	-0.1
Gas	483	1 856	2 517	3 540	4 529	13	20	22	25	27	4.4	3.9	3.7	3.2
Transformation, Own Use and Losses	759	1 220	1 405	1 648	1 865						1.5	1.8	1.7	1.5
Total Final Consumption	9 125	12 479	14 447	16 837	19 063	100	100	100	100	100	1.0	1.8	1.7	1.5
Coal	2 687	2 246	2 331	2 365	2 386	29	18	16	14	13	-0.6	0.5	0.3	0.2
Oil	5 085	7 630	9 089	10 941	12 725	56	61	63	65	67	1.3	2.2	2.0	1.8
Gas	1 353	2 603	3 028	3 530	3 952	15	21	21	21	21	2.1	1.9	1.7	1.5
Industry	3 672	4 076	4 579	5 122	5 567	100	100	100	100	100	0.3	1.5	1.3	1.1
Coal	1 580	1 758	1 894	1 975	2 032	43	43	41	39	36	0.3	0.9	0.6	0.5
Oil	1 332	1 175	1 342	1 553	1 734	36	29	29	30	31	-0.4	1.7	1.6	1.4
Gas	760	1 143	1 344	1 594	1 801	21	28	29	31	32	1.3	2.0	1.9	1.6
Transport	2 351	4 914	5 977	7 375	8 739	100	100	100	100	100	2.4	2.5	2.3	2.1
Oil	2 166	4 762	5 797	7 161	8 495	92	97	97	97	97	2.6	2.5	2.3	2.1
Other Fuels	185	152	180	214	244	8	3	3	3	3	-0.6	2.1	1.9	1.7
Other Sectors	2 943	3 248	3 603	4 022	4 417	100	100	100	100	100	0.3	1.3	1.2	1.1
Coal	943	418	372	331	300	32	13	10	8	7	-2.6	-1.4	-1.3	-1.2
Oil	1 449	1 501	1 706	1 948	2 189	49	46	47	48	50	0.1	1.6	1.5	1.4
Gas	551	1 329	1 525	1 743	1 929	19	41	42	43	44	2.9	1.7	1.5	1.3
Non-Energy Use	160	241	288	318	340	100	100	100	100	100	1.3	2.2	1.5	1.2

Reference Scenario: OECD

	Energy Demand (Mtoe)					Shares (%)					Growth Rates (% per annum)			
	1971	2002	2010	2020	2030	1971	2002	2010	2020	2030	1971-2002	2002-2010	2002-2020	2002-2030
Total Primary Energy Supply	**3 387**	**5 346**	**5 970**	**6 550**	**6 953**	**100**	**100**	**100**	**100**	**100**	**1.5**	**1.4**	**1.1**	**0.9**
Coal	812	1 095	1 170	1 213	1 192	24	20	20	19	17	1.0	0.8	0.6	0.3
Oil	1 730	2 167	2 372	2 594	2 725	51	41	40	40	39	0.7	1.1	1.0	0.8
Gas	655	1 171	1 379	1 635	1 830	19	22	23	25	26	1.9	2.1	1.9	1.6
Nuclear	27	593	642	599	557	1	11	11	9	8	10.5	1.0	0.1	-0.2
Hydro	75	106	121	125	131	2	2	2	2	2	1.1	1.6	0.9	0.8
Biomass and Waste	83	181	223	282	359	2	3	4	4	5	2.5	2.6	2.5	2.5
Other Renewables	4	33	63	103	159	0	1	1	2	2	6.8	8.4	6.5	5.8
Power Generation and Heat Plants	**823**	**2 139**	**2 448**	**2 698**	**2 845**	**100**	**100**	**100**	**100**	**100**	**3.1**	**1.7**	**1.3**	**1.0**
Coal	410	891	973	1 032	1 022	50	42	40	38	36	2.5	1.1	0.8	0.5
Oil	187	117	102	97	67	23	5	4	4	2	-1.5	-1.8	-1.0	-2.0
Gas	117	347	476	650	787	14	16	19	24	28	3.6	4.1	3.6	3.0
Nuclear	27	593	642	599	557	3	28	26	22	20	10.5	1.0	0.1	-0.2
Hydro	75	106	121	125	131	9	5	5	5	5	1.1	1.6	0.9	0.8
Biomass and Waste	3	60	81	107	147	0	3	3	4	5	10.5	3.8	3.3	3.3
Other Renewables	4	26	53	88	134	0	1	2	3	5	6.3	9.4	7.0	6.0
Other Transformation, Own Use and Losses	**335**	**410**	**432**	**449**	**464**						**0.7**	**0.6**	**0.5**	**0.4**
of which electricity	*52*	*114*	*127*	*139*	*147*						*2.6*	*1.4*	*1.1*	*0.9*
Total Final Consumption	**2 580**	**3 691**	**4 125**	**4 587**	**4 945**	**100**	**100**	**100**	**100**	**100**	**1.2**	**1.4**	**1.2**	**1.0**
Coal	300	119	117	106	100	12	3	3	2	2	-2.9	-0.3	-0.6	-0.6
Oil	1 434	1 945	2 160	2 385	2 545	56	53	52	52	51	1.0	1.3	1.1	1.0
Gas	470	728	800	873	922	18	20	19	19	19	1.4	1.2	1.0	0.8
Electricity	277	726	846	974	1 078	11	20	20	21	22	3.2	1.9	1.6	1.4
Heat	18	47	54	61	67	1	1	1	1	1	3.2	1.6	1.4	1.2
Biomass and Waste	80	119	140	173	210	3	3	3	4	4	1.3	2.1	2.1	2.0
Other Renewables	0	7	9	15	25	0.0	0.2	0.2	0.3	0.5	9.5	4.2	4.4	4.8

Industry	971	1 105	1 211	1 324	1 412	100	100	100	100	100	0.4	1.1	1.0	0.9
Coal	182	102	105	98	94	19	9	9	7	7	-1.8	0.3	-0.2	-0.3
Oil	376	345	364	385	396	39	31	30	29	28	-0.3	0.7	0.6	0.5
Gas	228	307	337	367	391	23	28	28	28	28	1.0	1.2	1.0	0.9
Electricity	137	273	310	354	388	14	25	26	27	27	2.2	1.6	1.5	1.3
Heat	10	16	17	19	20	1	1	1	1	1	1.6	1.1	1.0	0.9
Renewables	38	63	78	100	124	4	6	6	8	9	1.6	2.6	2.6	2.4
Transport	638	1 240	1 429	1 637	1 797	100	100	100	100	100	2.2	1.8	1.6	1.3
Oil	611	1 207	1 388	1 588	1 736	96	97	97	97	97	2.2	1.8	1.5	1.3
Other Fuels	26	34	40	50	60	4	3	3	3	3	0.8	2.2	2.2	2.1
Other Sectors	876	1 216	1 335	1 463	1 565	100	100	100	100	100	1.1	1.2	1.0	0.9
Coal	111	16	11	7	5	13	1	1	1	0	-6.0	-4.3	-4.3	-4.1
Oil	355	265	258	251	243	40	22	19	17	16	-0.9	-0.4	-0.3	-0.3
Gas	225	399	438	476	499	26	33	33	33	32	1.9	1.2	1.0	0.8
Electricity	135	444	525	608	677	15	37	39	42	43	3.9	2.1	1.8	1.5
Heat	8	31	36	41	46	1	3	3	3	3	4.3	1.8	1.6	1.4
Renewables	42	60	67	79	95	5	5	5	5	6	1.2	1.4	1.5	1.6
Non-Energy Use	95	130	151	163	171						1.0	1.9	1.3	1.0
Electricity Generation (TWh)	3 821	9 757	11 302	12 941	14 243	100	100	100	100	100	3.1	1.9	1.6	1.4
Coal	1 514	3 733	4 079	4 528	4 736	40	38	36	35	33	3.0	1.1	1.1	0.9
Oil	821	558	472	451	309	21	6	4	3	2	-1.2	-2.1	-1.2	-2.1
Gas	497	1 709	2 374	3 315	4 145	13	18	21	26	29	4.1	4.2	3.8	3.2
Nuclear	104	2 276	2 462	2 300	2 137	3	23	22	18	15	10.5	1.0	0.1	-0.2
Hydro	874	1 230	1 402	1 453	1 529	23	13	12	11	11	1.1	1.6	0.9	0.8
Biomass and Waste	6	168	237	307	405	0	2	2	2	3	11.1	4.3	3.4	3.2
Other Renewables	5	82	276	587	983	0	1	2	5	7	9.4	16.3	11.5	9.3
Traditional Biomass (included above)	0	0	0	0	0	0	0	0	0	0				
Total Primary Energy Supply (excluding traditional biomass)	3 387	5 346	5 970	6 550	6 953						1.5	1.4	1.1	0.9

Annex A - Tables for Reference Scenario Projections

Reference Scenario: OECD

	Capacity (GW)				Shares (%)				Growth Rates (% per annum)		
	2002	2010	2020	2030	2002	2010	2020	2030	2002-2010	2002-2020	2002-2030
Total Capacity	2 177	2 464	2 875	3 305	100	100	100	100	**1.5**	**1.5**	**1.5**
Coal	621	621	695	763	29	25	24	23	0.0	0.6	0.7
Oil	237	247	228	134	11	10	8	4	0.5	-0.2	-2.0
Gas	535	695	930	1 225	25	28	32	37	3.3	3.1	3.0
Nuclear	302	314	291	271	14	13	10	8	0.5	-0.2	-0.4
Hydro	420	451	467	491	19	18	16	15	0.9	0.6	0.6
of which pumped storage	86	89	92	95	4	3	3	3	0.5	0.4	0.4
Renewables (excluding hydro)	61	136	264	422	3	6	9	13	10.5	8.5	7.1

Renewables Breakdown

	Capacity (GW)				Shares (%)				Growth Rates (% per annum)		
	2002	2010	2020	2030	2002	2010	2020	2030	2002-2010	2002-2020	2002-2030
Renewables (excluding hydro)	61	136	264	422	100	100	100	100	**10.5**	**8.5**	**7.1**
Biomass and Waste	26	37	49	65	42	28	19	15	4.7	3.6	3.3
Wind	28	81	173	266	46	60	65	63	14.2	10.6	8.3
Geothermal	6	8	10	14	9	6	4	3	4.4	3.4	3.2
Solar	1	8	28	69	2	6	11	16	29.6	20.7	16.5
Tide/Wave	0	1	3	9	1	1	1	2	17.9	13.2	12.4

	Electricity Generation (TWh)				Shares (%)				Growth Rates (% per annum)		
	2002	2010	2020	2030	2002	2010	2020	2030	2002-2010	2002-2020	2002-2030
Renewables (excluding hydro)	251	513	895	1 387	100	100	100	100	**7.4**	**6.6**	**5.9**
Biomass and Waste	168	237	307	405	67	46	34	29	3.5	3.1	3.0
Wind	48	206	463	749	19	40	52	54	15.8	12.0	9.6
Geothermal	33	54	70	92	13	10	8	7	5.1	3.8	3.5
Solar	1	12	43	107	0	2	5	8	28.9	21.1	17.1
Tide/Wave	1	5	12	34	0	1	1	2	19.0	14.3	13.1

Reference Scenario: OECD

	CO_2 Emissions (Mt)					Shares (%)					Growth Rates (% per annum)			
	1971	2002	2010	2020	2030	1971	2002	2010	2020	2030	1971-2002	2002-2010	2002-2020	2002-2030
Total CO_2 Emissions	9 378	12 446	13 813	15 151	15 833	100	100	100	100	100	0.9	1.3	1.1	0.9
change since 1990 (%)		12.9	25.3	37.5	43.7									
Coal	3 149	4 221	4 560	4 750	4 668	34	34	33	31	29	0.9	1.0	0.7	0.4
Oil	4 747	5 550	6 080	6 639	6 955	51	45	44	44	44	0.5	1.1	1.0	0.8
Gas	1 482	2 676	3 173	3 762	4 210	16	21	23	25	27	1.9	2.2	1.9	1.6
Power Generation and Heat Plants	2 523	4 793	5 389	6 023	6 191	100	100	100	100	100	2.1	1.5	1.3	0.9
Coal	1 652	3 609	3 949	4 194	4 147	65	75	73	70	67	2.6	1.1	0.8	0.5
Oil	596	384	338	326	229	24	8	6	5	4	-1.4	-1.6	-0.9	-1.8
Gas	274	801	1 103	1 502	1 815	11	17	20	25	29	3.5	4.1	3.6	3.0
Transformation, Own Use and Losses	559	635	689	713	743	100	100	100	100	100	0.4	1.0	0.6	0.6
Total Final Consumption	6 296	7 018	7 735	8 416	8 900	100	100	100	100	100	0.4	1.2	1.0	0.9
Coal	1 403	550	551	500	468	22	8	7	6	5	-3.0	0.0	-0.5	-0.6
Oil	3 842	4 800	5 339	5 904	6 308	61	68	69	70	71	0.7	1.3	1.2	1.0
Gas	1 050	1 668	1 844	2 011	2 124	17	24	24	24	24	1.5	1.3	1.0	0.9
Industry	2 354	1 723	1 844	1 907	1 949	100	100	100	100	100	-1.0	0.9	0.6	0.4
Coal	881	478	501	467	444	37	28	27	24	23	-2.0	0.6	-0.1	-0.3
Oil	951	557	590	619	633	40	32	32	32	32	-1.7	0.7	0.6	0.5
Gas	522	687	752	820	872	22	40	41	43	45	0.9	1.1	1.0	0.9
Transport	1 803	3 384	3 880	4 438	4 856	100	100	100	100	100	2.1	1.7	1.5	1.3
Oil	1 727	3 332	3 822	4 371	4 781	96	98	99	98	98	2.1	1.7	1.5	1.3
Other Fuels	76	52	58	67	75	4	2	1	2	2	-1.2	1.4	1.4	1.3
Other Sectors	2 066	1 801	1 883	1 933	1 950	100	100	100	100	100	-0.4	0.6	0.4	0.3
Coal	472	68	47	30	20	23	4	2	2	1	-6.1	-4.5	-4.5	-4.2
Oil	1 106	804	802	778	751	54	45	43	40	39	-1.0	0.0	-0.2	-0.2
Gas	487	930	1 034	1 125	1 178	24	52	55	58	60	2.1	1.3	1.1	0.8
Non-Energy Use	72	110	128	138	145	100	100	100	100	100	1.4	1.9	1.3	1.0

Annex A - Tables for Reference Scenario Projections

Reference Scenario: OECD North America

	Energy Demand (Mtoe)					Shares (%)					Growth Rates (% per annum)			
	1971	2002	2010	2020	2030	1971	2002	2010	2020	2030	1971-2002	2002-2010	2002-2020	2002-2030
Total Primary Energy Supply	1 779	2 698	3 035	3 371	3 634	100	100	100	100	100	1.4	1.5	1.2	1.1
Coal	297	578	633	653	672	17	21	21	19	18	2.2	1.1	0.7	0.5
Oil	826	1 079	1 218	1 369	1 478	46	40	40	41	41	0.9	1.5	1.3	1.1
Gas	557	651	743	861	946	31	24	24	26	26	0.5	1.7	1.6	1.3
Nuclear	12	232	245	250	233	1	9	8	7	6	10.1	0.7	0.4	0.0
Hydro	38	52	58	60	62	2	2	2	2	2	1.1	1.4	0.8	0.6
Biomass and Waste	49	90	108	130	165	3	3	4	4	5	2.0	2.3	2.0	2.2
Other Renewables	1	16	30	47	77	0	1	1	1	2	11.7	8.5	6.3	5.9
Power Generation and Heat Plants	400	1 065	1 214	1 355	1 450	100	100	100	100	100	3.2	1.6	1.3	1.1
Coal	192	524	576	602	623	48	49	47	44	43	3.3	1.2	0.8	0.6
Oil	61	45	45	52	41	15	4	4	4	3	-1.0	0.0	0.8	-0.3
Gas	97	171	231	308	363	24	16	19	23	25	1.8	3.8	3.3	2.7
Nuclear	12	232	245	250	233	3	22	20	18	16	10.1	0.7	0.4	0.0
Hydro	38	52	58	60	62	9	5	5	4	4	1.1	1.4	0.8	0.6
Biomass and Waste	0	28	33	41	62	0	3	3	3	4	16.5	1.9	2.1	2.8
Other Renewables	1	13	26	41	66	0	1	2	3	5	11.1	8.9	6.4	5.9
Other Transformation, Own Use and Losses	159	214	236	249	261						1.0	1.2	0.8	0.7
of which electricity	*26*	*55*	*62*	*68*	*73*						*2.5*	*1.5*	*1.2*	*1.0*
Total Final Consumption	1 388	1 841	2 075	2 330	2 549	100	100	100	100	100	0.9	1.5	1.3	1.2
Coal	84	31	36	32	32	6	2	2	1	1	-3.1	1.8	0.2	0.0
Oil	728	976	1 103	1 246	1 365	52	53	53	53	54	1.0	1.5	1.4	1.2
Gas	385	407	434	467	489	28	22	21	20	19	0.2	0.8	0.8	0.7
Electricity	143	358	418	484	542	10	19	20	21	21	3.0	1.9	1.7	1.5
Heat	0	7	7	8	9	0	0	0	0	0	-	1.4	1.2	1.1
Biomass and Waste	48	59	73	86	101	3	3	3	4	4	0.7	2.5	2.1	1.9
Other Renewables	0	2	3	6	11	0.0	0.1	0.2	0.3	0.4	-	5.6	5.7	5.9

						Shares (%)					Growth rates (% p.a.)			
Industry	457	490	545	604	656	100	100	100	100	100	0.2	1.3	1.2	1.0
Coal	67	29	34	31	31	15	6	6	5	5	-2.7	2.1	0.5	0.3
Oil	117	129	141	155	164	26	26	26	26	25	0.3	1.2	1.0	0.9
Gas	182	174	184	198	212	40	35	34	33	32	-0.1	0.7	0.7	0.7
Electricity	59	114	131	154	175	13	23	24	26	27	2.2	1.8	1.7	1.5
Heat	0	5	6	7	7	0	1	1	1	1	-	1.3	1.2	1.0
Renewables	32	39	49	58	67	7	8	9	10	10	0.6	2.8	2.2	1.9
Transport	420	713	818	946	1 057	100	100	100	100	100	1.7	1.7	1.6	1.4
Oil	402	690	791	913	1 016	96	97	97	96	96	1.8	1.7	1.6	1.4
Other Fuels	18	23	27	33	41	4	3	3	4	4	0.8	2.0	2.0	2.1
Other Sectors	468	565	621	682	733	100	100	100	100	100	0.6	1.2	1.1	0.9
Coal	18	3	2	1	0	4	0	0	0	0	-6.1	-3.5	-5.8	-5.6
Oil	165	84	80	81	81	35	15	13	12	11	-2.2	-0.6	-0.2	-0.1
Gas	186	212	228	243	249	40	38	37	36	34	0.4	0.9	0.7	0.6
Electricity	83	244	286	329	366	18	43	46	48	50	3.5	2.0	1.7	1.5
Heat	0	0	0	1	2	0	0	0	0	0	-	1.7	1.5	1.2
Renewables	16	21	24	28	34	3	4	4	4	5	0.8	1.6	1.6	1.8
Non-Energy Use	44	73	90	98	103						1.7	2.6	1.6	1.2
Electricity Generation (TWh)	1 956	4 809	5 580	6 417	7 154	100	100	100	100	100	2.9	1.9	1.6	1.4
Coal	805	2 191	2 437	2 650	2 896	41	46	44	41	40	3.3	1.3	1.1	1.0
Oil	253	193	192	230	183	13	4	3	4	3	-0.9	0.0	1.0	-0.2
Gas	413	816	1 162	1 600	1 955	21	17	21	25	27	2.2	4.5	3.8	3.2
Nuclear	45	890	940	960	895	2	19	17	15	13	10.1	0.7	0.4	0.0
Hydro	440	609	679	702	726	23	13	12	11	10	1.1	1.4	0.8	0.6
Biomass and Waste	0	79	91	106	148	0	2	2	2	2	20.3	1.7	1.6	2.3
Other Renewables	1	32	79	170	351	0	1	3	3	5	13.8	11.9	9.7	8.9
Traditional Biomass (included above)	0	0	0	0	0	0	0	0	0	0	-	-	-	-
Total Primary Energy Supply (excluding traditional biomass)	1 779	2 698	3 035	3 371	3 634						1.4	1.5	1.2	1.1

Reference Scenario: OECD North America

	Capacity (GW)				Shares (%)				Growth Rates (% per annum)		
	2002	2010	2020	2030	2002	2010	2020	2030	2002-2010	2002-2020	2002-2030
Total Capacity	1 042	1 154	1 338	1 551	100	100	100	100	1.3	1.4	1.4
Coal	336	333	385	447	32	29	29	29	-0.1	0.7	1.0
Oil	83	85	88	54	8	7	7	4	0.4	0.3	-1.5
Gas	317	403	492	601	30	35	37	39	3.1	2.5	2.3
Nuclear	110	116	119	111	11	10	9	7	0.7	0.4	0.0
Hydro	175	181	186	192	17	16	14	12	0.4	0.3	0.3
of which pumped storage	*19*	*19*	*19*	*19*	*2*	*2*	*1*	*1*	*0.0*	*0.0*	*0.0*
Renewables (excluding hydro)	20	35	69	146	2	3	5	9	7.1	7.1	7.4

Renewables Breakdown

	Capacity (GW)				Shares (%)				Growth Rates (% per annum)		
	2002	2010	2020	2030	2002	2010	2020	2030	2002-2010	2002-2020	2002-2030
Renewables (excluding hydro)	20	35	69	146	100	100	100	100	7.1	7.1	7.4
Biomass and Waste	11	13	15	22	55	36	22	15	1.8	1.8	2.5
Wind	5	13	36	81	23	38	52	56	13.7	12.0	10.7
Geothermal	4	6	8	11	19	17	12	7	5.4	4.1	3.7
Solar	0	3	10	32	2	8	14	22	26.8	18.9	16.6
Tide/Wave	0	0	0	0	0	0	0	0	-16.4	-7.7	1.3

	Electricity Generation (TWh)				Shares (%)				Growth Rates (% per annum)		
	2002	2010	2020	2030	2002	2010	2020	2030	2002-2010	2002-2020	2002-2030
Renewables (excluding hydro)	111	169	276	499	100	100	100	100	4.3	4.6	5.1
Biomass and Waste	79	91	106	148	71	53	38	30	1.4	1.5	2.1
Wind	11	35	100	227	10	21	36	46	12.4	11.7	10.7
Geothermal	20	39	53	71	18	23	19	14	6.6	4.9	4.2
Solar	1	5	17	53	1	3	6	11	22.9	17.8	15.9
Tide/Wave	0	0	0	0	0	0	0	0	-12.7	-6.6	1.5

Reference Scenario: OECD North America

	CO_2 Emissions (Mt)					Shares (%)					Growth Rates (% per annum)			
	1971	2002	2010	2020	2030	1971	2002	2010	2020	2030	1971-2002	2002-2010	2002-2020	2002-2030
Total CO_2 Emissions	4 735	6 480	7 283	8 042	8 596	100	100	100	100	100	**1.0**	**1.5**	**1.2**	**1.0**
change since 1990 (%)		*16.4*	*30.9*	*44.5*	*54.5*									
Coal	1 146	2 219	2 446	2 534	2 614	24	34	34	32	30	2.2	1.2	0.7	0.6
Oil	2 311	2 774	3 134	3 536	3 819	49	43	43	44	44	0.6	1.5	1.4	1.1
Gas	1 278	1 487	1 703	1 972	2 163	27	23	23	25	25	0.5	1.7	1.6	1.3
Power Generation and Heat Plants	1 167	2 613	2 960	3 267	3 435	100	100	100	100	100	**2.6**	**1.6**	**1.2**	**1.0**
Coal	744	2 069	2 274	2 379	2 461	64	79	77	73	72	3.4	1.2	0.8	0.6
Oil	196	152	156	183	147	17	6	5	6	4	-0.8	0.3	1.0	-0.1
Gas	227	392	530	706	827	19	15	18	22	24	1.8	3.9	3.3	2.7
Transformation, Own Use and Losses	349	361	427	456	488	100	100	100	100	100	**0.1**	**2.1**	**1.3**	**1.1**
Total Final Consumption	3 219	3 506	3 896	4 319	4 674	100	100	100	100	100	**0.3**	**1.3**	**1.2**	**1.0**
Coal	383	147	169	153	150	12	4	4	4	3	-3.0	1.8	0.2	0.1
Oil	1 936	2 426	2 731	3 096	3 403	60	69	70	72	73	0.7	1.5	1.4	1.2
Gas	899	933	996	1 070	1 121	28	27	26	25	24	0.1	0.8	0.8	0.7
Industry	1 003	696	761	799	840	100	100	100	100	100	**-1.2**	**1.1**	**0.8**	**0.7**
Coal	314	136	161	149	147	31	20	21	19	17	-2.7	2.1	0.5	0.3
Oil	264	170	189	207	219	26	24	25	26	26	-1.4	1.3	1.1	0.9
Gas	425	389	411	444	474	42	56	54	56	56	-0.3	0.7	0.7	0.7
Transport	1 191	1 994	2 283	2 630	2 925	100	100	100	100	100	**1.7**	**1.7**	**1.6**	**1.4**
Oil	1 150	1 946	2 231	2 570	2 859	97	98	98	98	98	1.7	1.7	1.6	1.4
Other Fuels	41	48	53	60	66	3	2	2	2	2	0.5	1.2	1.3	1.2
Other Sectors	1 002	754	775	807	821	100	100	100	100	100	**-0.9**	**0.4**	**0.4**	**0.3**
Coal	69	10	7	3	2	7	1	1	0	0	-6.1	-3.5	-5.8	-5.6
Oil	499	247	236	237	238	50	33	30	29	29	-2.2	-0.6	-0.2	-0.1
Gas	434	496	532	567	581	43	66	69	70	71	0.4	0.9	0.7	0.6
Non-Energy Use	22	62	76	83	87	100	100	100	100	100	**3.4**	**2.5**	**1.6**	**1.2**

Annex A - Tables for Reference Scenario Projections

Reference Scenario: United States and Canada

	Energy Demand (Mtoe)						Shares (%)					Growth Rates (% per annum)			
	1971	2002	2010	2020	2030		1971	2002	2010	2020	2030	1971-2002	2002-2010	2002-2020	2002-2030
Total Primary Energy Supply	1 735	2 540	2 840	3 120	3 316		100	100	100	100	100	1.2	1.4	1.1	1.0
Coal	295	570	625	645	662		17	22	22	21	20	2.1	1.2	0.7	0.5
Oil	800	985	1 106	1 231	1 315		46	39	39	39	40	0.7	1.5	1.2	1.0
Gas	548	612	692	787	845		32	24	24	25	25	0.4	1.5	1.4	1.2
Nuclear	12	229	242	248	231		1	9	9	8	7	10.1	0.7	0.4	0.0
Hydro	37	50	55	56	58		2	2	2	2	2	1.0	1.1	0.6	0.5
Biomass and Waste	43	82	99	116	143		2	3	3	4	4	2.1	2.3	2.0	2.0
Other Renewables	1	11	22	37	62		0	0	1	1	2	10.4	9.3	7.1	6.4
Power Generation and Heat Plants	394	1 014	1 146	1 261	1 323		100	100	100	100	100	3.1	1.5	1.2	1.0
Coal	192	518	570	596	615		49	51	50	47	47	3.3	1.2	0.8	0.6
Oil	58	25	22	24	12		15	2	2	2	1	-2.7	-1.4	-0.3	-2.7
Gas	95	155	207	271	308		24	15	18	22	23	1.6	3.7	3.2	2.5
Nuclear	12	229	242	248	231		3	23	21	20	17	10.1	0.7	0.4	0.0
Hydro	37	50	55	56	58		9	5	5	4	4	1.0	1.1	0.6	0.5
Biomass and Waste	0	28	31	35	47		0	3	3	3	4	16.4	1.5	1.3	1.9
Other Renewables	1	9	19	31	51		0	1	2	2	4	9.6	10.0	7.4	6.6
Other Transformation, Own Use and Losses	153	183	200	205	210							0.6	1.1	0.6	0.5
of which electricity	*25*	*51*	*56*	*60*	*63*							*2.3*	*1.3*	*0.9*	*0.8*
Total Final Consumption	1 353	1 748	1 959	2 181	2 362		100	100	100	100	100	0.8	1.4	1.2	1.1
Coal	83	30	35	32	31		6	2	2	1	1	-3.2	1.9	0.2	0.0
Oil	708	915	1 028	1 151	1 247		52	52	52	53	53	0.8	1.5	1.3	1.1
Gas	379	397	422	450	466		28	23	22	21	20	0.2	0.7	0.7	0.6
Electricity	140	344	398	456	504		10	20	20	21	21	2.9	1.8	1.6	1.4
Heat	0	7	7	8	9		0	0	0	0	0	-	1.4	1.2	1.1
Biomass and Waste	42	52	65	79	94		3	3	3	4	4	0.7	2.8	2.3	2.1
Other Renewables	0	2	3	6	11		0.0	0.1	0.2	0.3	0.4	-	5.7	5.7	5.9

Industry	444	461	508	557	598	100	100	100	100	100	0.1	1.2	1.1	0.9
Coal	66	28	33	31	30	15	6	7	5	5	-2.7	2.3	0.5	0.3
Oil	114	119	129	142	149	26	26	25	25	25	0.1	1.0	1.0	0.8
Gas	176	165	173	183	192	40	36	34	33	32	-0.2	0.5	0.6	0.5
Electricity	58	105	119	137	153	13	23	23	25	26	2.0	1.6	1.5	1.3
Heat	0	5	6	7	7	0	1	1	1	1	-	-	1.2	1.0
Renewables	31	38	48	57	66	7	8	9	10	11	0.6	3.0	2.3	2.0
Transport	409	675	769	880	970	100	100	100	100	100	1.6	1.6	1.5	1.3
Oil	391	652	742	847	929	96	97	97	96	96	1.7	1.6	1.5	1.3
Other Fuels	18	23	27	33	40	4	3	3	4	4	0.8	2.0	2.0	2.0
Other Sectors	458	541	594	649	693	100	100	100	100	100	0.5	1.2	1.0	0.9
Coal	18	3	2	1	0	4	0	0	0	0	-6.1	-3.5	-5.8	-5.6
Oil	161	73	68	67	67	35	13	12	10	10	-2.5	-0.8	-0.4	-0.3
Gas	186	212	226	241	246	41	39	38	37	35	0.4	0.8	0.7	0.5
Electricity	82	238	279	318	351	18	44	47	49	51	3.5	2.0	1.6	1.4
Heat	0	1	1	1	2	0	0	0	0	0	-	1.7	1.5	1.2
Renewables	11	15	17	21	27	3	3	3	3	4	0.8	2.0	2.1	2.3
Non-Energy Use	43	72	88	96	101						1.7	2.6	1.6	1.3
Electricity Generation (TWh)	1 925	4 594	5 286	5 999	6 599	100	100	100	100	100	2.8	1.8	1.5	1.3
Coal	805	2 165	2 410	2 621	2 861	42	47	46	44	43	3.2	1.3	1.1	1.0
Oil	242	113	97	107	53	13	2	2	2	1	-2.4	-1.9	-0.3	-2.7
Gas	407	747	1 056	1 413	1 664	21	16	20	24	25	2.0	4.4	3.6	2.9
Nuclear	45	880	930	950	885	2	19	18	16	13	10.1	0.7	0.4	0.0
Hydro	426	584	636	655	679	22	13	12	11	10	1.0	1.1	0.6	0.5
Biomass and Waste	0	79	89	100	133	0	2	2	2	2	20.3	1.5	1.3	1.9
Other Renewables	1	27	68	154	325	0	1	1	3	5	13.1	12.4	10.2	9.3
Traditional Biomass (included above)	0	0	0	0	0	0	0	0	0	0	-	-	-	-
Total Primary Energy Supply (excluding traditional biomass)	1 735	2 540	2 840	3 120	3 316						1.2	1.4	1.1	1.0

Annex A - Tables for Reference Scenario Projections

Reference Scenario: United States and Canada

		Capacity (GW)				Shares (%)			Growth Rates (% per annum)		
	2002	2010	2020	2030	2002	2010	2020	2030	2002-2010	2002-2020	2002-2030
Total Capacity	995	1 091	1 249	1 432	100	100	100	100	**1.1**	**1.2**	**1.3**
Coal	333	329	380	441	33	30	30	31	-0.1	0.7	1.0
Oil	64	64	62	28	6	6	5	2	0.1	-0.1	-2.9
Gas	305	382	452	538	31	35	36	38	2.8	2.2	2.0
Nuclear	109	115	117	109	11	11	9	8	0.7	0.4	0.0
Hydro	166	169	173	179	17	15	14	12	0.2	0.2	0.3
of which pumped storage	*19*	*19*	*19*	*19*	*2*	*2*	*2*	*1*	*0.0*	*0.0*	*0.0*
Renewables (excluding hydro)	19	32	64	137	2	3	5	10	6.9	7.0	7.3

Renewables Breakdown

		Capacity (GW)				Shares (%)			Growth Rates (% per annum)		
	2002	2010	2020	2030	2002	2010	2020	2030	2002-2010	2002-2020	2002-2030
Renewables (excluding hydro)	19	32	64	137	100	100	100	100	**6.9**	**7.0**	**7.3**
Biomass and Waste	11	12	14	19	56	38	22	14	1.8	1.6	2.1
Wind	5	12	34	79	25	38	53	58	12.9	11.7	10.6
Geothermal	3	5	6	8	16	14	10	6	5.4	4.3	3.6
Solar	0	3	9	30	2	9	15	22	27.1	18.8	16.5
Tide/Wave	0	0	0	0	0	0	0	0	-16.4	-7.7	1.3

		Electricity Generation (TWh)				Shares (%)			Growth Rates (% per annum)		
	2002	2010	2020	2030	2002	2010	2020	2030	2002-2010	2002-2020	2002-2030
Renewables (excluding hydro)	105	156	254	458	100	100	100	100	**4.0**	**4.5**	**5.0**
Biomass and Waste	79	89	100	133	75	57	39	29	1.2	1.2	1.8
Wind	11	33	96	219	10	21	38	48	11.7	11.5	10.5
Geothermal	15	30	43	55	14	19	17	12	7.2	5.4	4.4
Solar	1	5	16	50	1	3	6	11	23.3	17.8	15.9
Tide/Wave	0	0	0	0	0	0	0	0	-12.7	-6.6	1.5

Reference Scenario: United States and Canada

	CO_2 Emissions (Mt)					Shares (%)					Growth Rates (% per annum)			
	1971	2002	2010	2020	2030	1971	2002	2010	2020	2030	1971-2002	2002-2010	2002-2020	2002-2030
Total CO_2 Emissions	4 637	6 121	6 839	7 472	7 894	100	100	100	100	100	0.9	1.4	1.1	0.9
change since 1990 (%)		16.1	29.7	41.7	49.7									
Coal	1 140	2 188	2 416	2 501	2 576	25	36	35	33	33	2.1	1.2	0.7	0.6
Oil	2 239	2 523	2 828	3 151	3 361	48	41	41	42	43	0.4	1.4	1.2	1.0
Gas	1 257	1 410	1 596	1 820	1 956	27	23	23	24	25	0.4	1.6	1.4	1.2
Power Generation and Heat Plants	1 154	2 489	2 807	3 063	3 188	100	100	100	100	100	2.5	1.5	1.2	0.9
Coal	744	2 044	2 249	2 351	2 429	64	82	80	77	76	3.3	1.2	0.8	0.6
Oil	187	82	74	78	38	16	3	3	3	1	-2.6	-1.4	-0.3	-2.7
Gas	223	363	484	634	721	19	15	17	21	23	1.6	3.7	3.2	2.5
Transformation, Own Use and Losses	338	312	365	379	395	100	100	100	100	100	-0.3	2.0	1.1	0.8
Total Final Consumption	3 145	3 320	3 667	4 029	4 311	100	100	100	100	100	0.2	1.3	1.1	0.9
Coal	379	141	164	147	144	12	4	4	4	3	-3.1	1.9	0.2	0.1
Oil	1 881	2 266	2 534	2 848	3 095	60	68	69	71	72	0.6	1.4	1.3	1.1
Gas	885	913	969	1 034	1 072	28	28	26	26	25	0.1	0.7	0.7	0.6
Industry	974	649	703	730	758	100	100	100	100	100	-1.3	1.0	0.7	0.6
Coal	309	130	156	143	141	32	20	22	20	19	-2.7	2.3	0.5	0.3
Oil	254	147	160	175	185	26	23	23	24	24	-1.7	1.0	1.0	0.8
Gas	411	371	388	411	432	42	57	55	56	57	-0.3	0.5	0.6	0.5
Transport	1 161	1 892	2 153	2 456	2 696	100	100	100	100	100	1.6	1.6	1.5	1.3
Oil	1 120	1 844	2 100	2 396	2 630	96	97	98	98	98	1.6	1.6	1.5	1.3
Other Fuels	41	48	53	60	66	4	3	2	2	2	0.5	1.2	1.3	1.2
Other Sectors	990	721	740	766	776	100	100	100	100	100	-1.0	0.3	0.3	0.3
Coal	69	10	7	3	2	7	1	1	0	0	-6.1	-3.5	-5.8	-5.6
Oil	488	217	203	201	199	49	30	27	26	26	-2.6	-0.8	-0.4	-0.3
Gas	433	495	529	562	575	44	69	72	73	74	0.4	0.8	0.7	0.5
Non-Energy Use	19	58	71	77	82	100	100	100	100	100	3.6	2.6	1.6	1.2

Annex A - Tables for Reference Scenario Projections

Reference Scenario: Mexico

	Energy Demand (Mtoe)					Shares (%)					Growth Rates (% per annum)			
	1971	2002	2010	2020	2030	1971	2002	2010	2020	2030	1971-2002	2002-2010	2002-2020	2002-2030
Total Primary Energy Supply	43	157	195	250	318	100	100	100	100	100	**4.2**	**2.7**	**2.6**	**2.5**
Coal	2	8	8	8	10	3	5	4	3	3	5.3	0.5	0.6	0.9
Oil	26	94	112	138	163	60	60	57	55	51	4.2	2.3	2.2	2.0
Gas	9	38	52	74	101	20	24	27	30	32	4.9	3.8	3.7	3.5
Nuclear	0	3	3	3	3	0	2	1	1	1	-	0.2	0.1	0.0
Hydro	1	2	4	4	4	3	1	2	2	1	1.8	6.9	3.6	2.3
Biomass and Waste	6	8	9	13	22	14	5	5	5	7	1.0	1.9	2.8	3.7
Other Renewables	0	5	8	10	15	0	3	4	4	5	-	6.5	4.1	4.1
Power Generation and Heat Plants	6	52	68	94	128	100	100	100	100	100	**7.4**	**3.5**	**3.4**	**3.3**
Coal	0	6	6	7	8	1	12	9	7	6	17.7	0.9	0.8	1.0
Oil	3	20	22	28	29	48	38	33	30	23	6.5	1.5	2.0	1.4
Gas	2	16	24	37	55	29	31	35	39	43	7.5	5.1	4.9	4.5
Nuclear	0	3	3	3	3	0	5	4	3	2	-	0.2	0.1	0.0
Hydro	1	2	4	4	4	22	4	5	4	3	1.8	6.9	3.6	2.3
Biomass and Waste	0	1	2	6	15	0	2	3	6	12	-	11.8	11.5	10.9
Other Renewables	0	5	8	10	14	0	9	11	10	11	-	6.5	4.1	4.1
Other Transformation, Own Use and Losses	6	31	37	43	51						**5.7**	**2.1**	**1.9**	**1.8**
of which electricity	*0*	*4*	*6*	*8*	*10*						*7.4*	*3.9*	*3.6*	*3.3*
Total Final Consumption	35	93	116	148	187	100	100	100	100	100	**3.2**	**2.8**	**2.6**	**2.5**
Coal	1	1	1	1	1	3	1	1	1	0	-0.3	-2.1	-0.7	-0.3
Oil	19	61	75	95	118	56	66	65	64	63	3.8	2.7	2.5	2.4
Gas	6	9	12	17	23	18	10	11	12	12	1.4	3.5	3.5	3.2
Electricity	2	14	20	28	37	6	15	17	19	20	6.2	4.0	3.8	3.5
Heat	0	0	0	0	0	0	0	0	0	0	-	-	-	-
Biomass and Waste	6	7	7	7	7	17	8	6	5	4	0.6	0.1	0.0	-0.1
Other Renewables	0	0	0	0	1	0.0	0.1	0.1	0.2	0.3	-	4.5	5.2	5.8

Industry	13	29	37	47	58	100	100	100	100	100	100	2.6	3.0	2.7	2.4
Coal	1	1	1	1	1	8	3	2	2	2	2	-0.3	-2.1	-0.7	-0.3
Oil	4	10	12	14	15	28	34	33	29	29	25	3.2	2.9	1.8	1.4
Gas	6	9	11	15	19	44	29	29	32	32	34	1.2	3.2	3.2	3.0
Electricity	1	9	12	17	22	10	30	32	36	36	38	6.4	4.0	3.7	3.3
Heat	0	0	0	0	0	0	0	0	0	0	0	-	-	-	-
Renewables	1	1	1	1	1	9	4	3	2	2	1	-0.1	-2.5	-2.3	-2.4
Transport	11	39	49	66	87	100	100	100	100	100	100	4.2	3.1	3.0	2.9
Oil	11	39	49	66	87	100	100	100	100	100	100	4.2	3.1	3.0	2.9
Other Fuels	0.0	0.1	0.2	0.3	0.3	0	0	0	0	0	0	5.6	7.0	5.3	4.1
Other Sectors	10	24	28	33	40	100	100	100	100	100	100	2.9	1.9	1.9	1.9
Coal	0	0	0	0	0	0	0	0	0	0	0	-	-	-	-
Oil	4	11	12	13	14	41	47	44	40	40	36	3.4	0.9	1.0	0.9
Gas	0	1	1	2	3	3	4	5	6	6	8	3.3	6.1	5.3	4.8
Electricity	1	6	8	11	15	7	23	27	33	33	38	6.8	4.0	4.0	3.7
Heat	0	0	0	0	0	0	0	0	0	0	0	-	-	-	-
Renewables	5	6	7	7	7	49	26	24	20	20	18	0.9	0.7	0.5	0.5
Non-Energy Use	1	1	2	2	2							0.8	1.5	1.0	0.8
Electricity Generation (TWh)	31	215	294	418	556	100	100	100	100	100	100	6.4	4.0	3.8	3.4
Coal	0	26	27	30	35	1	12	9	7	7	6	17.9	0.5	0.7	1.1
Oil	11	79	95	123	130	36	37	32	29	29	23	6.6	2.3	2.5	1.8
Gas	5	69	106	187	292	18	32	36	45	45	53	8.5	5.5	5.7	5.3
Nuclear	0	10	10	10	10	0	5	3	2	2	2	-	0.2	0.1	0.0
Hydro	14	25	43	47	47	46	12	14	11	11	9	1.8	6.9	3.6	2.3
Biomass and Waste	0	0	2	6	15	0	0	1	1	1	3	-	19.7	15.0	13.1
Other Renewables	0	5	11	16	26	0	3	4	4	4	5	-	9.2	6.1	5.8
Traditional Biomass (included above)	0	0	0	0	0	0	0	0	0	0	0	-	-	-	-
Total Primary Energy Supply (excluding traditional biomass)	43	157	195	250	318							4.2	2.7	2.6	2.5

Annex A - Tables for Reference Scenario Projections

Reference Scenario: Mexico

	Capacity (GW)				Shares (%)				Growth Rates (% per annum)		
	2002	2010	2020	2030	2002	2010	2020	2030	2002-2010	2002-2020	2002-2030
Total Capacity	47	63	90	119	100	100	100	100	3.7	3.7	3.4
Coal	4	4	5	6	8	7	5	5	0.9	1.2	1.4
Oil	19	21	25	26	40	34	28	22	1.5	1.7	1.2
Gas	12	21	40	63	26	34	45	53	7.4	6.9	6.0
Nuclear	1	1	1	1	3	2	2	1	0.0	0.0	0.0
Hydro	10	12	14	14	21	20	15	12	3.2	2.0	1.3
of which pumped storage	0	0	0	0	0	0	0	0	-	-	-
Renewables (excluding hydro)	1	2	4	9	3	4	5	7	9.3	7.6	7.5

Renewables Breakdown

	Capacity (GW)				Shares (%)				Growth Rates (% per annum)		
	2002	2010	2020	2030	2002	2010	2020	2030	2002-2010	2002-2020	2002-2030
Renewables (excluding hydro)	1	2	4	9	100	100	100	100	9.3	7.6	7.5
Biomass and Waste	0	0	1	2	27	14	21	26	1.0	6.2	7.4
Wind	0	1	2	3	0	31	35	29	98.9	41.5	27.3
Geothermal	1	1	2	3	72	53	35	28	5.2	3.4	4.0
Solar	0	0	0	2	1	2	9	17	16.0	20.3	18.2
Tide/Wave	0	0	0	0	0	0	0	0	-	-	-

	Electricity Generation (TWh)				Shares (%)				Growth Rates (% per annum)		
	2002	2010	2020	2030	2002	2010	2020	2030	2002-2010	2002-2020	2002-2030
Renewables (excluding hydro)	6	13	22	41	100	100	100	100	8.2	6.7	6.6
Biomass and Waste	0	2	6	15	8	15	27	36	15.5	13.4	12.1
Wind	0	2	4	8	0	16	21	19	63.1	32.6	22.8
Geothermal	5	9	11	15	91	67	49	38	4.9	3.4	3.6
Solar	0	0	1	3	1	1	4	7	14.6	17.3	16.3
Tide/Wave	0	0	0	0	0	0	0	0	-	-	-

Reference Scenario: Mexico

	CO$_2$ Emissions (Mt)					Shares (%)					Growth Rates (% per annum)			
	1971	2002	2010	2020	2030	1971	2002	2010	2020	2030	1971-2002	2002-2010	2002-2020	2002-2030
Total CO$_2$ Emissions	97	359	444	570	702	100	100	100	100	100	4.3	2.7	2.6	2.4
change since 1990 (%)		23.0	52.0	95.4	140.5									
Coal	5	31	31	33	38	5	9	7	6	5	5.9	0.0	0.4	0.7
Oil	72	251	306	385	458	74	70	69	68	65	4.1	2.5	2.4	2.2
Gas	20	77	107	152	207	21	21	24	27	29	4.4	4.2	3.8	3.6
Power Generation and Heat Plants	13	124	154	204	247	100	100	100	100	100	7.6	2.7	2.8	2.5
Coal	0	25	26	28	32	1	20	17	13	13	19.0	0.5	0.6	0.9
Oil	9	70	82	104	109	69	57	54	51	44	6.9	2.0	2.2	1.6
Gas	4	29	46	72	106	30	23	30	35	43	6.7	5.8	5.2	4.7
Transformation, Own Use and Losses	11	49	62	77	93	100	100	100	100	100	5.0	2.9	2.5	2.3
Total Final Consumption	74	186	228	290	362	100	100	100	100	100	3.0	2.6	2.5	2.4
Coal	5	6	5	5	6	7	3	2	2	2	0.6	-2.2	-0.6	-0.2
Oil	55	160	197	248	309	74	86	86	85	85	3.5	2.7	2.5	2.4
Gas	14	20	26	37	48	19	11	12	13	13	1.2	3.4	3.4	3.2
Industry	29	47	58	70	82	100	100	100	100	100	1.6	2.5	2.2	2.0
Coal	5	6	5	5	6	17	12	9	8	7	0.6	-2.2	-0.6	-0.2
Oil	11	23	29	32	34	37	49	50	46	42	2.6	2.9	1.8	1.4
Gas	13	18	24	32	42	46	39	41	47	51	1.1	3.2	3.2	3.0
Transport	30	102	130	174	229	100	100	100	100	100	4.0	3.1	3.0	2.9
Oil	30	102	130	174	229	100	100	100	100	100	4.0	3.1	3.0	2.9
Other Fuels	0	0	0	0	0	0	0	0	0	0	-	-	-	-
Other Sectors	12	32	35	41	46	100	100	100	100	100	3.2	1.3	1.3	1.3
Coal	0	0	0	0	0	0	0	0	0	0	-	-	-	-
Oil	11	30	33	36	39	94	95	92	89	86	3.2	0.9	1.0	0.9
Gas	1	2	3	4	6	6	5	8	11	14	2.9	6.1	5.3	4.8
Non-Energy Use	3	5	5	5	6	100	100	100	100	100	1.5	1.5	1.0	0.8

Reference Scenario: OECD Pacific

	Energy Demand (Mtoe)					Shares (%)					Growth Rates (% per annum)			
	1971	2002	2010	2020	2030	1971	2002	2010	2020	2030	1971-2002	2002-2010	2002-2020	2002-2030
Total Primary Energy Supply	346	852	971	1 074	1 132	100	100	100	100	100	3.0	1.6	1.3	1.0
Coal	84	196	210	226	221	24	23	22	21	20	2.8	0.9	0.8	0.4
Oil	240	398	426	449	453	69	47	44	42	40	1.6	0.8	0.7	0.5
Gas	5	113	151	189	216	2	13	16	18	19	10.4	3.7	2.9	2.3
Nuclear	2	108	139	153	172	1	13	14	14	15	13.6	3.2	2.0	1.7
Hydro	9	11	12	13	13	3	1	1	1	1	0.4	1.2	0.9	0.7
Biomass and Waste	4	20	24	30	37	1	2	2	3	3	5.7	2.5	2.5	2.3
Other Renewables	1	6	9	13	20	0	1	1	1	2	5.7	4.6	4.3	4.3
Power Generation and Heat Plants	92	362	432	488	519	100	100	100	100	100	4.5	2.2	1.7	1.3
Coal	27	132	143	157	151	29	36	33	32	29	5.3	1.1	1.0	0.5
Oil	51	33	28	22	12	56	9	6	5	2	-1.4	-2.2	-2.2	-3.4
Gas	2	64	90	115	133	2	18	21	24	26	12.6	4.3	3.3	2.7
Nuclear	2	108	139	153	172	2	30	32	31	33	13.6	3.2	2.0	1.7
Hydro	9	11	12	13	13	10	3	3	3	3	0.4	1.2	0.9	0.7
Biomass and Waste	0	10	13	18	22	0	3	3	4	4	20.1	3.1	3.0	2.8
Other Renewables	1	5	7	10	15	1	1	2	2	3	4.7	4.8	4.3	4.3
Other Transformation, Own Use and Losses	39	57	58	63	67						1.2	0.3	0.6	0.6
of which electricity	*5*	*15*	*17*	*19*	*19*						*3.9*	*1.7*	*1.2*	*0.9*
Total Final Consumption	254	583	658	725	763	100	100	100	100	100	2.7	1.5	1.2	1.0
Coal	30	34	35	36	36	12	6	5	5	5	0.4	0.5	0.4	0.2
Oil	178	354	389	418	430	70	61	59	58	56	2.2	1.2	0.9	0.7
Gas	8	51	62	74	81	3	9	9	10	11	6.4	2.5	2.1	1.7
Electricity	35	129	153	175	189	14	22	23	24	25	4.3	2.1	1.7	1.4
Heat	0	5	5	6	7	0	1	1	1	1	-	2.0	1.7	1.4
Biomass and Waste	4	9	11	13	15	1	2	2	2	2	3.2	1.9	1.8	1.6
Other Renewables	0	1	2	3	5	0.0	0.2	0.3	0.4	0.7	-	4.3	4.5	4.7

	1980	2000	2010	2020	2030	1980	2000	2010	2020	2030	1980-2000	2000-2010	2000-2020	2000-2030
Industry	136	227	248	269	279	100	100	100	100	100	1.7	1.1	1.0	0.7
Coal	22	33	34	35	35	16	14	14	13	13	1.3	0.6	0.4	0.3
Oil	86	104	110	116	117	63	46	44	43	42	0.6	0.7	0.6	0.4
Gas	3	23	28	32	35	2	10	11	12	13	6.7	2.2	1.8	1.5
Electricity	23	58	66	75	80	17	26	27	28	29	3.0	1.7	1.4	1.1
Heat	0	2	2	3	3	0	1	1	1	1	-	2.4	2.0	1.6
Renewables	2	6	7	8	8	1	3	3	3	3	4.7	1.4	1.2	1.1
Transport	53	161	186	207	219	100	100	100	100	100	3.7	1.8	1.4	1.1
Oil	51	159	183	203	214	97	98	98	98	98	3.7	1.8	1.4	1.1
Other Fuels	2	3	3	4	5	3	2	2	2	2	1.3	2.8	2.6	2.2
Other Sectors	49	175	203	227	242	100	100	100	100	100	4.2	1.8	1.4	1.2
Coal	8	1	1	1	1	15	0	0	0	0	-7.1	-1.2	-1.0	-0.9
Oil	25	72	76	77	77	50	41	37	34	32	3.5	0.7	0.4	0.3
Gas	4	27	34	41	45	9	16	17	18	18	6.1	2.7	2.2	1.8
Electricity	11	69	84	98	106	21	39	42	43	44	6.2	2.5	2.0	1.6
Heat	0	2	2	3	3	0	1	1	1	1	-	2.4	2.1	1.8
Renewables	2	5	6	8	10	4	3	3	3	4	2.7	2.9	2.9	3.0
Non-Energy Use	16	20	21	22	23						0.7	0.7	0.5	0.4
Electricity Generation (TWh)	462	1 677	1 978	2 252	2 415	100	100	100	100	100	4.2	2.1	1.7	1.3
Coal	85	597	642	718	702	18	36	32	32	29	6.5	0.9	1.0	0.6
Oil	250	180	141	112	61	54	11	7	5	3	-1.1	-3.0	-2.6	-3.8
Gas	7	323	464	598	700	2	19	23	27	29	13.0	4.6	3.5	2.8
Nuclear	8	414	534	588	662	2	25	27	26	27	13.6	3.2	2.0	1.7
Hydro	110	126	138	147	154	24	8	7	7	6	0.4	1.2	0.9	0.7
Biomass and Waste	0	29	40	48	59	0	2	2	2	2	16.4	4.0	2.7	2.5
Other Renewables	1	7	19	41	78	0	0	1	2	3	5.8	12.8	10.1	8.9
Traditional Biomass (included above)	0	0	0	0	0						-	-	-	-
Total Primary Energy Supply (excluding traditional biomass)	346	852	971	1 074	1 132						3.0	1.6	1.3	1.0

Reference Scenario: OECD Pacific

	Capacity (GW)				Shares (%)				Growth Rates (% per annum)	
	2002	2010	2020	2030	2002	2010	2020	2030	2002-2010	2002-2020	2002-2030
Total Capacity	392	460	529	595	100	100	100	100	1.9	1.6	1.5
Coal	92	101	116	119	23	22	22	20	1.2	1.3	0.9
Oil	81	84	72	45	21	18	14	8	0.6	-0.6	-2.0
Gas	91	122	165	225	23	27	31	38	3.7	3.3	3.3
Nuclear	59	70	77	87	15	15	15	15	2.2	1.5	1.4
Hydro	63	69	73	77	16	15	14	13	1.0	0.8	0.7
of which pumped storage	28	31	33	35	7	7	6	6	0.9	0.8	0.7
Renewables (excluding hydro)	6	13	26	42	2	5	5	7	9.8	8.2	7.1

Renewables Breakdown

	Capacity (GW)				Shares (%)				Growth Rates (% per annum)	
	2002	2010	2020	2030	2002	2010	2020	2030	2002-2010	2002-2020	2002-2030
Renewables (excluding hydro)	6	13	26	42	100	100	100	100	9.8	8.2	7.1
Biomass and Waste	5	6	7	9	75	48	29	22	3.8	2.7	2.5
Wind	0	2	5	10	7	15	18	24	20.9	14.0	12.0
Geothermal	1	1	1	2	15	9	5	4	2.0	1.9	2.3
Solar	0	4	12	20	2	28	48	49	49.4	27.8	19.3
Tide/Wave	0	0	0	0	0	0	0	0	-	-	-

	Electricity Generation (TWh)				Shares (%)				Growth Rates (% per annum)	
	2002	2010	2020	2030	2002	2010	2020	2030	2002-2010	2002-2020	2002-2030
Renewables (excluding hydro)	36	59	89	137	100	100	100	100	4.9	4.6	4.5
Biomass and Waste	29	40	48	59	80	68	54	43	3.2	2.5	2.4
Wind	1	6	14	33	3	10	16	24	19.6	14.4	12.6
Geothermal	6	8	9	12	17	13	10	9	2.1	1.9	2.3
Solar	0	6	18	32	0	10	20	24	86.6	44.8	30.5
Tide/Wave	0	0	0	0	0	0	0	0	-	-	-

Reference Scenario: OECD Pacific

	CO$_2$ Emissions (Mt)					Shares (%)					Growth Rates (% per annum)			
	1971	2002	2010	2020	2030	1971	2002	2010	2020	2030	1971-2002	2002-2010	2002-2020	2002-2030
Total CO$_2$ Emissions	**951**	**2 022**	**2 228**	**2 426**	**2 451**	**100**	**100**	**100**	**100**	**100**	**2.5**	**1.2**	**1.0**	**0.7**
change since 1990 (%)		*32.8*	*46.4*	*59.4*	*61.0*									
Coal	293	761	813	878	846	31	38	36	36	35	3.1	0.8	0.8	0.4
Oil	646	993	1 062	1 110	1 107	68	49	48	46	45	1.4	0.8	0.6	0.4
Gas	13	267	353	439	498	1	13	16	18	20	10.3	3.6	2.8	2.3
Power Generation and Heat Plants	**290**	**823**	**916**	**1 018**	**999**	**100**	**100**	**100**	**100**	**100**	**3.4**	**1.3**	**1.2**	**0.7**
Coal	123	571	619	679	649	43	69	68	67	65	5.1	1.0	1.0	0.5
Oil	163	104	87	70	38	56	13	10	7	4	-1.4	-2.2	-2.2	-3.5
Gas	4	149	210	269	312	1	18	23	26	31	12.5	4.4	3.4	2.7
Transformation, Own Use and Losses	**43**	**88**	**86**	**84**	**82**	**100**	**100**	**100**	**100**	**100**	**2.3**	**-0.2**	**-0.3**	**-0.2**
Total Final Consumption	**618**	**1 111**	**1 226**	**1 324**	**1 371**	**100**	**100**	**100**	**100**	**100**	**1.9**	**1.2**	**1.0**	**0.8**
Coal	154	158	161	165	164	25	14	13	12	12	0.1	0.2	0.2	0.1
Oil	455	841	929	998	1 029	74	76	76	75	75	2.0	1.3	1.0	0.7
Gas	8	111	136	162	177	1	10	11	12	13	8.8	2.5	2.1	1.7
Industry	**327**	**368**	**394**	**415**	**422**	**100**	**100**	**100**	**100**	**100**	**0.4**	**0.8**	**0.7**	**0.5**
Coal	114	152	159	163	162	35	41	40	39	38	0.9	0.5	0.4	0.2
Oil	207	167	175	183	184	63	45	45	44	44	-0.7	0.6	0.5	0.3
Gas	6	50	59	69	76	2	13	15	17	18	7.2	2.3	1.9	1.5
Transport	**147**	**438**	**505**	**561**	**592**	**100**	**100**	**100**	**100**	**100**	**3.6**	**1.8**	**1.4**	**1.1**
Oil	145	437	504	560	591	98	100	100	100	100	3.6	1.8	1.4	1.1
Other Fuels	3	1	1	1	1	2	0	0	0	0	-2.8	-2.5	0.0	0.5
Other Sectors	**136**	**292**	**314**	**334**	**343**	**100**	**100**	**100**	**100**	**100**	**2.5**	**0.9**	**0.8**	**0.6**
Coal	37	6	2	2	2	27	2	1	1	1	-5.6	-12.4	-6.1	-4.1
Oil	96	225	237	241	241	71	77	75	72	70	2.8	0.6	0.4	0.2
Gas	3	61	76	91	100	2	21	24	27	29	10.8	2.8	2.3	1.8
Non-Energy Use	**8**	**12**	**13**	**14**	**14**	**100**	**100**	**100**	**100**	**100**	**1.3**	**0.7**	**0.6**	**0.5**

Reference Scenario: OECD Asia

	Energy Demand (Mtoe)					Shares (%)					Growth Rates (% per annum)			
	1971	2002	2010	2020	2030	1971	2002	2010	2020	2030	1971-2002	2002-2010	2002-2020	2002-2030
Total Primary Energy Supply	**286**	**721**	**824**	**909**	**957**	**100**	**100**	**100**	**100**	**100**	**3.0**	**1.7**	**1.3**	**1.0**
Coal	62	146	158	172	168	22	20	19	19	18	2.8	1.0	0.9	0.5
Oil	211	357	378	396	395	74	50	46	44	41	1.7	0.7	0.6	0.4
Gas	3	88	120	151	174	1	12	15	17	18	11.1	4.0	3.1	2.5
Nuclear	2	108	139	153	172	1	15	17	17	18	13.6	3.2	2.0	1.7
Hydro	7	7	8	9	9	3	1	1	1	1	0.0	1.5	1.1	0.9
Biomass and Waste	0	11	15	19	24	0	2	2	2	3	-	3.7	3.2	2.8
Other Renewables	0	4	6	9	14	0	1	1	1	1	-	5.8	5.0	4.6
Power Generation and Heat Plants	**76**	**300**	**364**	**412**	**440**	**100**	**100**	**100**	**100**	**100**	**4.6**	**2.4**	**1.8**	**1.4**
Coal	14	87	96	107	102	19	29	26	26	23	6.0	1.2	1.2	0.6
Oil	51	32	27	21	11	67	11	7	5	3	-1.5	-2.4	-2.3	-3.7
Gas	1	56	79	100	117	1	18	22	24	27	13.5	4.5	3.3	2.7
Nuclear	2	108	139	153	172	3	36	38	37	39	13.6	3.2	2.0	1.7
Hydro	7	7	8	9	9	10	2	2	2	2	0.0	1.5	1.1	0.9
Biomass and Waste	0	8	11	14	18	0	3	3	4	4	-	4.4	3.7	3.2
Other Renewables	0	3	5	7	10	0	1	1	2	2	-	6.0	4.9	4.3
Other Transformation, Own Use and Losses	**32**	**51**	**51**	**55**	**59**						**1.5**	**0.1**	**0.5**	**0.5**
of which electricity	*4*	*12*	*14*	*15*	*15*						*3.7*	*1.7*	*1.2*	*0.8*
Total Final Consumption	**212**	**498**	**559**	**612**	**640**	**100**	**100**	**100**	**100**	**100**	**2.8**	**1.5**	**1.2**	**0.9**
Coal	24	30	32	33	33	11	6	6	5	5	0.7	0.7	0.5	0.3
Oil	152	312	341	364	372	71	63	61	59	58	2.3	1.1	0.9	0.6
Gas	6	37	46	54	59	3	7	8	9	9	5.9	2.6	2.1	1.7
Electricity	30	110	130	148	158	14	22	23	24	25	4.3	2.1	1.7	1.3
Heat	0	5	5	6	7	0	1	1	1	1	-	2.0	1.7	1.4
Biomass and Waste	0	3	4	5	6	0	1	1	1	1	-	2.1	2.0	1.8
Other Renewables	0	1	1	2	4	0.0	0.2	0.3	0.4	0.7	-	5.0	5.1	5.3

Industry	119	197	214	230	238	100	100	100	100	100	**1.6**	**1.1**	**0.9**	**0.7**
Coal	17	29	31	32	32	14	15	15	14	14	1.8	0.7	0.5	0.3
Oil	79	100	106	111	112	66	51	49	48	47	0.8	0.7	0.6	0.4
Gas	2	14	17	19	21	2	7	8	8	9	6.2	2.3	1.9	1.5
Electricity	21	50	56	63	67	18	25	26	27	28	2.7	1.6	1.3	1.1
Heat	0	2	2	3	3	0	1	1	1	1	-	2.4	2.0	1.6
Renewables	0	2	2	2	2	0	1	1	1	1	-	0.3	0.3	0.2
Transport	39	128	147	163	171	100	100	100	100	100	**3.9**	**1.8**	**1.4**	**1.0**
Oil	37	126	145	160	168	96	99	98	98	98	4.0	1.8	1.4	1.0
Other Fuels	2	2	2	3	3	4	1	2	2	2	0.5	2.7	2.5	2.1
Other Sectors	41	155	179	200	212	100	100	100	100	100	**4.4**	**1.8**	**1.4**	**1.1**
Coal	7	1	1	0	0	17	0	0	0	0	-7.9	-0.8	-0.7	-0.6
Oil	22	69	73	74	74	54	44	41	37	35	3.7	0.7	0.4	0.3
Gas	4	23	29	35	38	10	15	16	17	18	5.8	2.8	2.3	1.8
Electricity	8	58	72	83	89	19	38	40	41	42	6.8	2.6	1.9	1.5
Heat	0	2	2	3	3	0	1	1	1	1	-	2.4	2.1	1.8
Renewables	0	2	3	5	7	0	1	2	2	3	-	4.3	4.2	4.2
Non-Energy Use	13	18	18	19	19						**1.0**	**0.5**	**0.4**	**0.3**
Electricity Generation (TWh)	393	1 415	1 665	1 886	2 011	100	100	100	100	100	**4.2**	**2.1**	**1.6**	**1.3**
Coal	46	422	454	511	491	12	30	27	27	24	7.4	0.9	1.1	0.5
Oil	248	177	136	107	56	63	12	8	6	3	-1.1	-3.2	-2.7	-4.0
Gas	6	286	398	507	597	1	20	24	27	30	13.6	4.2	3.2	2.7
Nuclear	8	414	534	588	662	2	29	32	31	33	13.6	3.2	2.0	1.7
Hydro	86	86	96	104	110	22	6	6	5	5	0.0	1.5	1.1	0.9
Biomass and Waste	0	26	35	42	51	0	2	2	2	3	-	3.7	2.6	2.4
Other Renewables	0	4	12	27	45	0	0	1	1	2	-	14.8	11.5	9.2
Traditional Biomass (included above)	0	1	1	2	2	0	0	0	0	0		3.8	3.3	2.9
Total Primary Energy Supply (excluding traditional biomass)	286	720	823	908	954						**3.0**	**1.7**	**1.3**	**1.0**

Annex A - Tables for Reference Scenario Projections

Reference Scenario: OECD Asia

	Capacity (GW)				Shares (%)				Growth Rates (% per annum)		
	2002	2010	2020	2030	2002	2010	2020	2030	2002-2010	2002-2020	2002-2030
Total Capacity	336	393	450	503	100	100	100	100	1.9	1.6	1.4
Coal	63	73	85	88	19	18	19	18	1.8	1.6	1.2
Oil	78	81	69	42	23	21	15	8	0.5	-0.7	-2.2
Gas	82	105	141	196	24	27	31	39	3.2	3.1	3.2
Nuclear	59	70	77	87	18	18	17	17	2.2	1.5	1.4
Hydro	50	55	59	63	15	14	13	12	1.1	0.9	0.8
of which pumped storage	*27*	*29*	*31*	*33*	*8*	*7*	*7*	*7*	*1.0*	*0.8*	*0.7*
Renewables (excluding hydro)	4	9	19	28	1	2	4	6	10.5	8.7	6.9

Renewables Breakdown

	Capacity (GW)				Shares (%)				Growth Rates (% per annum)		
	2002	2010	2020	2030	2002	2010	2020	2030	2002-2010	2002-2020	2002-2030
Renewables (excluding hydro)	4	9	19	28	100	100	100	100	10.5	8.7	6.9
Biomass and Waste	3	5	6	8	81	52	32	27	4.6	3.2	2.8
Wind	0	1	3	6	7	11	13	21	17.3	12.9	11.3
Geothermal	1	1	1	1	12	7	5	4	3.7	3.0	2.6
Solar	0	3	10	13	0	30	50	48	120.7	52.2	32.5
Tide/Wave	0	0	0	0	0	0	0	0	-	-	-

	Electricity Generation (TWh)				Shares (%)				Growth Rates (% per annum)		
	2002	2010	2020	2030	2002	2010	2020	2030	2002-2010	2002-2020	2002-2030
Renewables (excluding hydro)	30	47	69	95	100	100	100	100	4.5	4.2	3.9
Biomass and Waste	26	35	42	51	87	75	61	53	2.9	2.3	2.2
Wind	0	3	8	19	1	7	12	20	21.6	15.9	13.5
Geothermal	3	5	6	7	11	10	9	8	3.6	3.0	2.6
Solar	0	4	13	17	0	8	18	18	90.1	46.6	30.4
Tide/Wave	0	0	0	0	0	0	0	0	-	-	-

Reference Scenario: OECD Asia

	CO$_2$ Emissions (Mt)						Shares (%)					Growth Rates (% per annum)			
	1971	2002	2010	2020	2030	1971	2002	2010	2020	2030	1971-2002	2002-2010	2002-2020	2002-2030
Total CO$_2$ Emissions	794	1 647	1 814	1 971	1 984	100	100	100	100	100	2.4	1.2	1.0	0.7
change since 1990 (%)		32.7	46.1	58.7	59.8									
Coal	215	563	606	660	636	27	34	33	33	32	3.1	0.9	0.9	0.4
Oil	570	873	923	956	943	72	53	51	49	48	1.4	0.7	0.5	0.3
Gas	9	212	285	354	405	1	13	16	18	20	10.9	3.8	2.9	2.3
Power Generation and Heat Plants	238	619	694	776	759	100	100	100	100	100	3.1	1.4	1.3	0.7
Coal	74	388	426	475	451	31	63	61	61	59	5.5	1.2	1.1	0.5
Oil	161	102	84	67	35	68	16	12	9	5	-1.5	-2.4	-2.4	-3.8
Gas	3	129	184	234	273	1	21	26	30	36	13.5	4.5	3.3	2.7
Transformation, Own Use and Losses	41	71	66	63	60						1.8	-0.9	-0.6	-0.6
Total Final Consumption	515	957	1 054	1 132	1 164	100	100	100	100	100	2.0	1.2	0.9	0.7
Coal	126	145	149	153	153	24	15	14	14	13	0.4	0.3	0.3	0.2
Oil	383	731	805	859	880	74	76	76	76	76	2.1	1.2	0.9	0.7
Gas	6	81	100	120	131	1	9	10	11	11	8.9	2.7	2.2	1.7
Industry	279	326	347	364	368	100	100	100	100	100	0.5	0.8	0.6	0.4
Coal	88	139	147	151	151	32	43	42	41	41	1.5	0.6	0.5	0.3
Oil	186	157	164	171	171	67	48	47	47	47	-0.6	0.6	0.5	0.3
Gas	4	30	36	42	46	2	9	10	12	13	6.6	2.4	1.9	1.5
Transport	109	348	402	445	466	100	100	100	100	100	3.8	1.8	1.4	1.0
Oil	107	348	402	445	466	98	100	100	100	100	3.9	1.8	1.4	1.1
Other Fuels	2	0	0	0	0	2	0	0	0	0	-	-	-	-
Other Sectors	122	273	294	311	319	100	100	100	100	100	2.6	0.9	0.7	0.6
Coal	35	5	2	2	1	29	2	1	1	1	-5.9	-11.4	-5.6	-3.8
Oil	85	216	227	232	232	70	79	77	75	73	3.1	0.6	0.4	0.3
Gas	2	51	64	78	85	1	19	22	25	27	11.7	2.8	2.3	1.8
Non-Energy Use	5	10	11	11	11	100	100	100	100	100	2.3	0.4	0.4	0.3

Reference Scenario: OECD Oceania

	Energy Demand (Mtoe)					Shares (%)					Growth Rates (% per annum)			
	1971	2002	2010	2020	2030	1971	2002	2010	2020	2030	1971-2002	2002-2010	2002-2020	2002-2030
Total Primary Energy Supply	**59**	**131**	**147**	**165**	**176**	**100**	**100**	**100**	**100**	**100**	**2.6**	**1.4**	**1.3**	**1.1**
Coal	22	50	52	55	53	37	38	36	33	30	2.7	0.5	0.5	0.2
Oil	29	41	48	53	57	48	31	32	32	33	1.2	1.9	1.5	1.2
Gas	2	26	31	38	42	3	20	21	23	24	8.8	2.5	2.3	1.8
Nuclear	0	0	0	0	0	0	0	0	0	0	-	-	-	-
Hydro	2	3	4	4	4	4	3	2	2	2	1.6	0.5	0.4	0.3
Biomass and Waste	4	9	9	11	13	6	7	6	7	7	2.9	0.9	1.4	1.4
Other Renewables	1	2	3	3	6	2	2	2	2	4	2.3	2.3	2.6	3.9
Power Generation and Heat Plants	**17**	**62**	**68**	**76**	**79**	**100**	**100**	**100**	**100**	**100**	**4.3**	**1.1**	**1.2**	**0.9**
Coal	13	45	48	50	49	75	73	70	66	61	4.2	0.7	0.6	0.3
Oil	0	1	1	1	1	3	1	1	1	1	0.8	6.1	2.6	1.9
Gas	1	8	11	15	17	3	14	16	20	21	9.3	3.6	3.4	2.5
Nuclear	0	0	0	0	0	0	0	0	0	0	-	-	-	-
Hydro	2	3	4	4	4	13	6	5	5	5	1.6	0.5	0.4	0.3
Biomass and Waste	0	3	2	3	4	0	4	4	4	5	15.1	-1.3	0.7	1.3
Other Renewables	1	2	2	3	5	6	3	3	3	7	1.3	2.4	2.7	4.3
Other Transformation, Own Use and Losses	**7**	**6**	**7**	**8**	**8**						**-0.3**	**2.1**	**1.3**	**1.0**
of which electricity	*1*	*3*	*4*	*4*	*4*						*4.3*	*1.6*	*1.3*	*1.0*
Total Final Consumption	**42**	**86**	**99**	**113**	**123**	**100**	**100**	**100**	**100**	**100**	**2.4**	**1.8**	**1.5**	**1.3**
Coal	6	4	3	3	3	14	4	4	3	3	-1.5	-0.6	-0.5	-0.4
Oil	26	42	48	54	58	62	49	49	48	47	1.6	1.6	1.3	1.1
Gas	1	14	17	20	22	3	16	17	17	18	7.8	2.3	1.9	1.6
Electricity	5	19	23	27	31	12	23	23	24	25	4.4	2.3	2.0	1.6
Heat	0	0	0	0	0	0	0	0	0	0	-	-	-	-
Biomass and Waste	4	6	7	8	9	8	7	7	7	7	1.7	1.8	1.7	1.5
Other Renewables	0	0	1	1	1	0.0	0.5	0.5	0.6	0.8	-	2.6	2.8	3.0

Industry	17	29	34	38	42	100	100	100	100	100	1.9	1.7	1.5	1.2
Coal	5	3	3	3	3	31	12	10	8	7	-1.4	-0.5	-0.5	-0.4
Oil	7	4	4	5	5	41	13	13	12	12	-1.8	1.2	1.0	0.8
Gas	1	10	11	13	15	6	33	34	35	35	7.7	2.1	1.8	1.5
Electricity	2	9	10	12	13	12	29	30	31	31	4.8	2.2	1.8	1.5
Heat	0	0	0	0	0	0	0	0	0	0	-	-	-	-
Renewables	2	4	5	5	6	9	13	13	14	14	3.1	1.9	1.7	1.5
Transport	14	34	39	44	48	100	100	100	100	100	2.9	1.7	1.5	1.2
Oil	14	33	38	43	46	99	98	98	97	97	2.9	1.7	1.4	1.2
Other Fuels	0	1	1	1	1	1	2	2	3	3	5.2	3.2	2.8	2.5
Other Sectors	8	20	24	27	31	100	100	100	100	100	3.0	2.0	1.7	1.5
Coal	0	0	0	0	0	6	1	1	1	0	-2.5	-2.1	-1.7	-1.5
Oil	2	3	3	3	3	29	15	13	11	10	0.8	0.4	0.1	0.0
Gas	0	4	5	6	7	5	20	21	22	22	7.8	2.5	2.1	1.8
Electricity	3	10	13	15	17	36	52	54	55	56	4.2	2.4	2.1	1.8
Heat	0	0	0	0	0	0	0	0	0	0	-	-	-	-
Renewables	2	2	3	3	4	25	12	11	12	12	0.6	1.5	1.5	1.5
Non-Energy Use	3	2	3	3	4						-0.8	1.9	1.5	1.2
Electricity Generation (TWh)	69	262	313	366	404	100	100	100	100	100	4.4	2.2	1.9	1.6
Coal	38	176	188	207	211	56	67	60	56	52	5.0	0.9	0.9	0.7
Oil	2	4	5	5	5	3	1	2	1	1	1.9	3.3	1.5	1.2
Gas	2	36	65	91	103	3	14	21	25	25	10.2	7.5	5.2	3.8
Nuclear	0	0	0	0	0	0	0	0	0	0	-	-	-	-
Hydro	25	40	42	43	44	36	15	13	12	11	1.6	0.5	0.4	0.3
Biomass and Waste	0	3	5	6	8	0	1	2	2	2	8.0	6.9	3.9	3.7
Other Renewables	1	3	7	14	33	2	1	2	4	8	3.3	10.2	8.3	8.5
Traditional Biomass (included above)	0	0	0	0	0	0	0	0	0	0	-	-	-	-
Total Primary Energy Supply (excluding traditional biomass)	59	131	147	165	176						2.6	1.4	1.3	1.1

Annex A - Tables for Reference Scenario Projections

Reference Scenario: OECD Oceania

	2002	Capacity (GW) 2010	2020	2030	Shares (%) 2002	2010	2020	2030	Growth Rates (% per annum) 2002-2010	2002-2020	2002-2030
Total Capacity	56	67	79	93	100	100	100	100	2.2	1.9	1.8
Coal	29	29	31	31	51	43	39	33	0.0	0.5	0.3
Oil	3	3	4	4	5	5	5	4	3.3	1.9	1.2
Gas	10	18	24	30	18	26	30	32	7.6	5.1	4.0
Nuclear	0	0	0	0	0	0	0	0	-	-	-
Hydro	13	14	14	14	23	20	18	15	0.6	0.4	0.4
of which pumped storage	1	2	2	2	3	2	2	2	0.5	0.5	0.5
Renewables (excluding hydro)	2	3	6	14	3	5	8	15	8.0	7.1	7.5

Renewables Breakdown

	2002	Capacity (GW) 2010	2020	2030	Shares (%) 2002	2010	2020	2030	Growth Rates (% per annum) 2002-2010	2002-2020	2002-2030
Renewables (excluding hydro)	2	3	6	14	100	100	100	100	8.0	7.1	7.5
Biomass and Waste	1	1	1	2	62	37	22	13	1.4	1.2	1.5
Wind	0	1	2	4	8	27	31	32	26.4	15.8	13.1
Geothermal	0	0	0	1	23	12	7	5	0.2	1.9
Solar	0	1	3	7	8	24	40	50	24.9	17.4	15.0
Tide/Wave	0	0	0	0	0	0	0	0	-	-	-

	2002	Electricity Generation (TWh) 2010	2020	2030	Shares (%) 2002	2010	2020	2030	Growth Rates (% per annum) 2002-2010	2002-2020	2002-2030
Renewables (excluding hydro)	6	12	20	41	100	100	100	100	7.0	6.0	6.5
Biomass and Waste	3	5	6	8	46	40	29	19	5.4	3.5	3.4
Wind	1	3	6	14	8	21	29	33	17.6	12.8	11.5
Geothermal	3	3	3	5	45	23	15	12	0.1	0.4	2.0
Solar	0	2	5	15	0	16	27	36	81.4	41.8	30.5
Tide/Wave	0	0	0	0	0	0	0	0	-	-	-

Reference Scenario: OECD Oceania

	CO$_2$ Emissions (Mt)					Shares (%)					Growth Rates (% per annum)			
	1971	2002	2010	2020	2030	1971	2002	2010	2020	2030	1971-2002	2002-2010	2002-2020	2002-2030
Total CO$_2$ Emissions	157	374	414	456	468	100	100	100	100	100	**2.8**	**1.3**	**1.1**	**0.8**
change since 1990 (%)		33.4	47.4	62.3	66.5									
Coal	77	199	207	218	211	49	53	50	48	45	3.1	0.5	0.5	0.2
Oil	75	121	139	154	164	48	32	33	34	35	1.5	1.8	1.4	1.1
Gas	4	55	68	84	93	3	15	16	19	20	8.5	2.6	2.4	1.9
Power Generation and Heat Plants	52	204	222	242	240	100	100	100	100	100	**4.5**	**1.1**	**1.0**	**0.6**
Coal	49	183	193	204	197	95	89	87	84	82	4.3	0.7	0.6	0.3
Oil	2	2	3	3	3	3	1	1	1	1	0.8	6.1	2.6	1.9
Gas	1	19	26	35	39	2	10	12	15	16	9.3	3.6	3.4	2.5
Transformation, Own Use and Losses	2	17	20	21	21	100	100	100	100	100	**6.6**	**2.0**	**1.1**	**0.8**
Total Final Consumption	102	154	172	193	207	100	100	100	100	100	**1.3**	**1.4**	**1.3**	**1.1**
Coal	28	14	12	12	12	27	9	7	6	6	-2.2	-1.4	-0.9	-0.6
Oil	72	110	124	139	149	70	72	72	72	72	1.4	1.5	1.3	1.1
Gas	2	30	36	42	46	2	19	21	22	22	8.5	2.3	1.9	1.6
Industry	48	42	46	51	54	100	100	100	100	100	**-0.4**	**1.1**	**1.0**	**0.8**
Coal	26	13	12	12	12	54	30	26	23	21	-2.2	-0.5	-0.5	-0.4
Oil	21	10	11	12	13	43	24	24	24	23	-2.3	1.2	1.0	0.8
Gas	2	20	23	27	30	3	46	50	53	55	8.6	2.1	1.8	1.5
Transport	38	90	102	116	126	100	100	100	100	100	**2.8**	**1.6**	**1.4**	**1.2**
Oil	38	89	101	115	124	99	99	99	99	99	2.8	1.7	1.4	1.2
Other Fuels	0	1	1	1	1	1	1	1	1	1	3.6	-0.6	0.9	1.1
Other Sectors	13	19	21	23	24	100	100	100	100	100	**1.2**	**1.0**	**1.0**	**0.8**
Coal	2	1	0	0	0	15	5	1	0	0	-2.6	-21.3	-10.8	-7.5
Oil	11	9	9	9	9	79	46	44	40	36	-0.6	0.4	0.1	0.0
Gas	1	9	12	14	15	7	49	55	60	63	7.9	2.5	2.1	1.8
Non-Energy Use	3	2	2	3	3	100	100	100	100	100	**-1.4**	**1.9**	**1.5**	**1.2**

Reference Scenario: OECD Europe

	Energy Demand (Mtoe)					Shares (%)					Growth Rates (% per annum)			
	1971	2002	2010	2020	2030	1971	2002	2010	2020	2030	1971-2002	2002-2010	2002-2020	2002-2030
Total Primary Energy Supply	1 263	1 795	1 964	2 105	2 187	100	100	100	100	100	1.1	1.1	0.9	0.7
Coal	431	321	327	333	299	34	18	17	16	14	-0.9	0.2	0.2	-0.3
Oil	664	689	729	776	794	53	38	37	37	36	0.1	0.7	0.7	0.5
Gas	93	407	485	584	668	7	23	25	28	31	4.9	2.2	2.0	1.8
Nuclear	13	253	257	196	151	1	14	13	9	7	10.0	0.2	-1.4	-1.8
Hydro	28	43	50	52	56	2	2	3	2	3	1.4	2.1	1.1	1.0
Biomass and Waste	31	71	91	122	157	2	4	5	6	7	2.7	3.1	3.0	2.8
Other Renewables	3	11	24	43	62	0	1	1	2	3	4.6	10.2	7.7	6.3
Power Generation and Heat Plants	330	712	802	855	876	100	100	100	100	100	2.5	1.5	1.0	0.7
Coal	191	235	254	272	248	58	33	32	32	28	0.7	1.0	0.8	0.2
Oil	74	40	30	23	14	22	6	4	3	2	-2.0	-3.6	-3.0	-3.7
Gas	19	112	156	226	291	6	16	19	26	33	5.9	4.2	4.0	3.5
Nuclear	13	253	257	196	151	4	36	32	23	17	10.0	0.2	-1.4	-1.8
Hydro	28	43	50	52	56	8	6	6	6	6	1.4	2.1	1.1	1.0
Biomass and Waste	2	21	34	48	63	1	3	4	6	7	7.2	6.3	4.7	4.0
Other Renewables	2	8	21	37	54	1	1	3	4	6	4.1	12.2	8.8	6.9
Other Transformation, Own Use and Losses	137	140	137	137	136						0.1	-0.2	-0.1	-0.1
of which electricity	*21*	*43*	*48*	*52*	*54*						*2.3*	*1.2*	*1.0*	*0.8*
Total Final Consumption	938	1 267	1 392	1 532	1 634	100	100	100	100	100	1.0	1.2	1.1	0.9
Coal	185	54	46	38	32	20	4	3	2	2	-3.9	-2.1	-2.0	-1.9
Oil	528	615	668	722	750	56	49	48	47	46	0.5	1.0	0.9	0.7
Gas	78	270	303	332	351	8	21	22	22	22	4.1	1.5	1.1	0.9
Electricity	100	238	274	316	348	11	19	20	21	21	2.8	1.8	1.6	1.4
Heat	18	36	41	46	51	2	3	3	3	3	2.3	1.5	1.4	1.2
Biomass and Waste	28	50	57	73	94	3	4	4	5	6	1.9	1.5	2.1	2.2
Other Renewables	0	3	4	5	8	0.0	0.2	0.3	0.4	0.5	6.8	2.9	3.3	3.7

Industry	378	389	417	451	477	100	100	100	100	100	0.1	0.9	0.8	0.7
Coal	93	41	36	32	28	25	10	9	7	6	-2.7	-1.4	-1.4	-1.4
Oil	173	112	113	114	114	46	29	27	25	24	-1.4	0.1	0.1	0.1
Gas	43	110	125	137	143	11	28	30	30	30	3.1	1.6	1.2	0.9
Electricity	55	100	113	125	133	15	26	27	28	28	2.0	1.5	1.2	1.0
Heat	10	8	9	9	10	3	2	2	2	2	-0.5	0.6	0.6	0.6
Renewables	5	18	22	34	49	1	5	5	8	10	4.4	2.5	3.7	3.7
Transport	165	365	424	484	521	100	100	100	100	100	2.6	1.9	1.6	1.3
Oil	158	357	414	472	506	96	98	97	97	97	2.7	1.9	1.6	1.3
Other Fuels	7	8	10	12	15	4	2	3	3	3	0.6	2.8	2.4	2.2
Other Sectors	359	476	511	554	590	100	100	100	100	100	0.9	0.9	0.8	0.8
Coal	85	13	9	6	4	24	3	2	1	1	-5.9	-4.7	-4.4	-4.1
Oil	165	110	102	93	84	46	23	20	17	14	-1.3	-0.9	-0.9	-0.9
Gas	35	159	176	193	205	10	33	35	35	35	5.0	1.3	1.1	0.9
Electricity	42	132	154	182	205	12	28	30	33	35	3.8	2.0	1.8	1.6
Heat	8	28	32	37	41	2	6	6	7	7	3.9	1.8	1.6	1.4
Renewables	24	35	38	43	51	7	7	7	8	9	1.2	1.0	1.2	1.3
Non-Energy Use	35	37	39	43	45						0.1	0.9	0.9	0.8
Electricity Generation (TWh)	1 403	3 271	3 743	4 272	4 674	100	100	100	100	100	2.8	1.7	1.5	1.3
Coal	624	945	1 000	1 160	1 138	45	29	27	27	24	1.3	0.7	1.1	0.7
Oil	318	185	139	109	65	23	6	4	3	1	-1.7	-3.6	-2.9	-3.7
Gas	77	570	748	1 117	1 490	5	17	20	26	32	6.7	3.5	3.8	3.5
Nuclear	51	972	987	752	580	4	30	26	18	12	10.0	0.2	-1.4	-1.8
Hydro	324	496	585	604	649	23	15	14	14	14	1.4	2.1	1.1	1.0
Biomass and Waste	6	60	106	154	198	0	2	3	4	4	7.8	7.4	5.4	4.4
Other Renewables	3	43	179	377	554	0	1	5	9	12	8.8	19.5	12.8	9.6
Traditional Biomass (included above)	0	0	0	0	0	0	0	0	0	0	-	-	-	-
Total Primary Energy Supply (excluding traditional biomass)	1 263	1 795	1 964	2 105	2 187						1.1	1.1	0.9	0.7

Reference Scenario: OECD Europe

	Capacity (GW)				Shares (%)				Growth Rates (% per annum)		
	2002	2010	2020	2030	2002	2010	2020	2030	2002-2010	2002-2020	2002-2030
Total Capacity	**743**	**850**	**1 007**	**1 159**	**100**	**100**	**100**	**100**	**1.6**	**1.6**	**1.5**
Coal	193	186	194	197	26	22	19	17	-0.4	0.0	0.1
Oil	74	78	68	34	10	9	7	3	0.6	-0.5	-2.7
Gas	126	170	273	399	17	20	27	34	3.8	4.4	4.2
Nuclear	133	127	95	73	18	15	9	6	-0.6	-1.9	-2.1
Hydro	182	202	208	222	25	24	21	19	1.3	0.7	0.7
of which pumped storage	*38*	*39*	*40*	*41*	*5*	*5*	*4*	*4*	*0.3*	*0.3*	*0.3*
Renewables (excluding hydro)	35	88	169	234	5	10	17	20	12.3	9.2	7.0

Renewables Breakdown

	Capacity (GW)				Shares (%)				Growth Rates (% per annum)		
	2002	2010	2020	2030	2002	2010	2020	2030	2002-2010	2002-2020	2002-2030
Renewables (excluding hydro)	**35**	**88**	**169**	**234**	**100**	**100**	**100**	**100**	**12.3**	**9.2**	**7.0**
Biomass and Waste	10	18	27	34	30	21	16	15	7.5	5.4	4.4
Wind	23	66	132	174	66	75	78	74	14.1	10.2	7.5
Geothermal	1	1	1	1	3	1	1	1	2.4	1.1	1.3
Solar	0	1	7	17	1	1	4	7	14.4	17.1	14.5
Tide/Wave	0	1	3	8	1	1	2	4	22.1	15.1	13.5

	Electricity Generation (TWh)				Shares (%)				Growth Rates (% per annum)		
	2002	2010	2020	2030	2002	2010	2020	2030	2002-2010	2002-2020	2002-2030
Renewables (excluding hydro)	**103**	**285**	**531**	**752**	**100**	**100**	**100**	**100**	**10.7**	**8.5**	**6.9**
Biomass and Waste	60	106	154	198	58	37	29	26	5.8	4.8	4.1
Wind	36	165	349	490	35	58	66	65	16.5	12.1	9.1
Geothermal	6	7	7	9	6	3	1	1	1.7	0.9	1.1
Solar	0	1	8	22	0	0	2	3	15.7	18.2	15.5
Tide/Wave	1	5	12	33	1	2	2	4	24.4	16.9	14.8

Reference Scenario: OECD Europe

	CO$_2$ Emissions (Mt)					Shares (%)					Growth Rates (% per annum)			
	1971	2002	2010	2020	2030	1971	2002	2010	2020	2030	1971-2002	2002-2010	2002-2020	2002-2030
Total CO$_2$ Emissions	**3 692**	**3 945**	**4 302**	**4 683**	**4 785**	**100**	**100**	**100**	**100**	**100**	**0.2**	**1.1**	**1.0**	**0.7**
change since 1990 (%)		*0.3*	*9.4*	*19.1*	*21.7*									
Coal	1 711	1 241	1 300	1 339	1 208	46	31	30	29	25	-1.0	0.6	0.4	-0.1
Oil	1 790	1 782	1 885	1 993	2 028	48	45	44	43	42	0.0	0.7	0.6	0.5
Gas	191	922	1 117	1 351	1 549	5	23	26	29	32	5.2	2.4	2.1	1.9
Power Generation and Heat Plants	**1 066**	**1 357**	**1 513**	**1 737**	**1 757**	**100**	**100**	**100**	**100**	**100**	**0.8**	**1.4**	**1.4**	**0.9**
Coal	784	969	1 055	1 136	1 037	74	71	70	65	59	0.7	1.1	0.9	0.2
Oil	237	127	95	74	44	22	9	6	4	2	-2.0	-3.6	-3.0	-3.7
Gas	44	260	363	527	676	4	19	24	30	38	5.9	4.2	4.0	3.5
Transformation, Own Use and Losses	**167**	**187**	**176**	**173**	**173**	**100**	**100**	**100**	**100**	**100**	**0.4**	**-0.7**	**-0.4**	**-0.3**
Total Final Consumption	**2 460**	**2 402**	**2 613**	**2 772**	**2 855**	**100**	**100**	**100**	**100**	**100**	**-0.1**	**1.1**	**0.8**	**0.6**
Coal	866	244	221	182	154	35	10	8	7	5	-4.0	-1.2	-1.6	-1.6
Oil	1 451	1 534	1 680	1 811	1 875	59	64	64	65	66	0.2	1.1	0.9	0.7
Gas	143	624	713	780	826	6	26	27	28	29	4.9	1.7	1.2	1.0
Industry	**1 025**	**658**	**689**	**692**	**687**	**100**	**100**	**100**	**100**	**100**	**-1.4**	**0.6**	**0.3**	**0.2**
Coal	453	190	181	156	135	44	29	26	22	20	-2.8	-0.6	-1.1	-1.2
Oil	480	220	226	230	230	47	33	33	33	33	-2.5	0.3	0.2	0.2
Gas	91	248	282	307	322	9	38	41	44	47	3.3	1.6	1.2	0.9
Transport	**465**	**952**	**1 092**	**1 247**	**1 339**	**100**	**100**	**100**	**100**	**100**	**2.3**	**1.7**	**1.5**	**1.2**
Oil	432	949	1 088	1 241	1 331	93	100	100	100	99	2.6	1.7	1.5	1.2
Other Fuels	32	3	4	6	8	7	0	0	0	1	-7.4	4.7	3.8	3.4
Other Sectors	**928**	**756**	**793**	**792**	**785**	**100**	**100**	**100**	**100**	**100**	**-0.7**	**0.6**	**0.3**	**0.1**
Coal	366	52	37	24	17	39	7	5	3	2	-6.1	-4.0	-4.1	-4.0
Oil	511	331	329	300	272	55	44	42	38	35	-1.4	-0.1	-0.5	-0.7
Gas	51	373	427	467	497	5	49	54	59	63	6.6	1.7	1.3	1.0
Non-Energy Use	**42**	**35**	**39**	**42**	**44**	**100**	**100**	**100**	**100**	**100**	**-0.6**	**1.2**	**0.9**	**0.8**

Reference Scenario: European Union

	Energy Demand (Mtoe)					Shares (%)					Growth Rates (% per annum)			
	1971	2002	2010	2020	2030	1971	2002	2010	2020	2030	1971-2002	2002-2010	2002-2020	2002-2030
Total Primary Energy Supply	1 211	1 690	1 848	1 976	2 048	100	100	100	100	100	1.1	1.1	0.9	0.7
Coal	426	303	307	307	274	35	18	17	16	13	-1.1	0.2	0.1	-0.4
Oil	633	648	687	729	743	52	38	37	37	36	0.1	0.7	0.7	0.5
Gas	93	389	468	565	649	8	23	25	29	32	4.7	2.3	2.1	1.8
Nuclear	13	251	251	190	146	1	15	14	10	7	10.0	0.0	-1.5	-1.9
Hydro	20	26	30	31	33	2	2	2	2	2	0.9	1.7	1.1	0.8
Biomass and Waste	25	65	84	115	147	2	4	5	6	7	3.1	3.3	3.2	3.0
Other Renewables	2	8	21	39	57	0	0	1	2	3	3.9	12.6	9.2	7.2
Power Generation and Heat Plants	319	676	763	807	823	100	100	100	100	100	2.5	1.5	1.0	0.7
Coal	190	228	246	258	234	60	34	32	32	28	0.6	1.0	0.7	0.1
Oil	72	40	32	23	13	23	6	4	3	2	-1.9	-2.9	-3.0	-4.0
Gas	19	105	152	221	286	6	15	20	27	35	5.7	4.8	4.2	3.7
Nuclear	13	251	251	190	146	4	37	33	24	18	10.0	0.0	-1.5	-1.9
Hydro	20	26	30	31	33	6	4	4	4	4	0.9	1.7	1.1	0.8
Biomass and Waste	2	20	33	48	60	1	3	4	6	7	7.0	6.6	5.0	4.0
Other Renewables	2	7	19	36	51	1	1	2	4	6	3.5	13.5	9.6	7.4
Other Transformation, Own Use and Losses	134	129	126	124	120						-0.1	-0.3	-0.3	-0.3
of which electricity	*20*	*40*	*44*	*48*	*50*						*2.2*	*1.1*	*1.0*	*0.8*
Total Final Consumption	892	1 186	1 303	1 435	1 533	100	100	100	100	100	0.9	1.2	1.1	0.9
Coal	182	45	35	27	22	20	4	3	2	1	-4.4	-3.0	-2.7	-2.5
Oil	500	574	625	677	702	56	48	48	47	46	0.4	1.1	0.9	0.7
Gas	77	266	297	324	344	9	22	23	23	22	4.1	1.4	1.1	0.9
Electricity	92	218	251	288	318	10	18	19	20	21	2.8	1.7	1.5	1.4
Heat	18	37	43	48	53	2	3	3	3	3	2.4	1.6	1.4	1.3
Biomass and Waste	22	45	51	67	87	3	4	4	5	6	2.3	1.6	2.3	2.4
Other Renewables	0	1	2	3	6	0.0	0.1	0.1	0.2	0.4	8.4	5.4	5.6	6.1

Industry	366	360	385	416	443	100	100	100	100	100	0.0	0.8	0.8	0.7
Coal	92	33	27	22	18	25	9	7	5	4	-3.3	-2.3	-2.2	-2.0
Oil	166	103	103	104	104	45	29	27	25	23	-1.5	0.0	0.1	0.0
Gas	43	109	124	135	142	12	30	32	33	32	3.1	1.6	1.2	1.0
Electricity	51	91	101	113	121	14	25	26	27	27	1.9	1.4	1.2	1.0
Heat	10	8	9	9	10	3	2	2	2	2	-0.5	0.6	0.6	0.6
Renewables	5	17	21	33	48	1	5	5	8	11	4.3	2.6	3.8	3.8
Transport	155	346	402	460	494	100	100	100	100	100	2.6	1.9	1.6	1.3
Oil	149	339	393	449	481	96	98	98	98	97	2.7	1.9	1.6	1.3
Other Fuels	6	8	9	11	13	4	2	2	2	3	0.8	2.3	2.2	2.1
Other Sectors	338	446	480	519	554	100	100	100	100	100	0.9	0.9	0.9	0.8
Coal	84	12	8	5	3	25	3	2	1	1	-6.1	-5.0	-4.8	-4.7
Oil	155	99	92	85	77	46	22	19	16	14	-1.4	-0.9	-0.9	-0.9
Gas	34	156	172	187	199	10	35	36	36	36	5.0	1.2	1.0	0.9
Electricity	38	121	142	167	188	11	27	30	32	34	3.8	2.0	1.8	1.6
Heat	8	29	34	39	44	2	7	7	8	8	4.1	1.9	1.6	1.4
Renewables	18	29	31	37	44	5	6	7	7	8	1.5	1.1	1.4	1.5
Non-Energy Use	34	34	37	40	42						0.0	0.9	0.9	0.7
Electricity Generation (TWh)	1 296	2 986	3 417	3 894	4 272	100	100	100	100	100	2.7	1.7	1.5	1.3
Coal	621	920	969	1 099	1 076	48	31	28	28	25	1.3	0.7	1.0	0.6
Oil	311	182	143	107	59	24	6	4	3	1	-1.7	-3.0	-2.9	-3.9
Gas	77	521	715	1 071	1 458	6	17	21	28	34	6.4	4.0	4.1	3.7
Nuclear	49	961	964	728	560	4	32	28	19	13	10.0	0.0	-1.5	-1.9
Hydro	229	302	347	366	382	18	10	10	9	9	0.9	1.7	1.1	0.8
Biomass and Waste	6	58	104	153	194	0	2	3	4	5	7.8	7.5	5.5	4.4
Other Renewables	3	41	175	371	542	0	1	5	10	13	8.7	19.7	12.9	9.6
Traditional Biomass (included above)	0	0	0	0	0	0	0	0	0	0				
Total Primary Energy Supply (excluding traditional biomass)	1 211	1 690	1 848	1 976	2 048						1.1	1.1	0.9	0.7

Reference Scenario: European Union

	2002	Capacity (GW) 2010	Capacity (GW) 2020	Capacity (GW) 2030	Shares (%) 2002	Shares (%) 2010	Shares (%) 2020	Shares (%) 2030	Growth Rates (% per annum) 2002-2010	Growth Rates (% per annum) 2002-2020	Growth Rates (% per annum) 2002-2030
Total Capacity	**681**	**771**	**918**	**1 062**	**100**	**100**	**100**	**100**	**1.5**	**1.6**	**1.5**
Coal	187	180	181	183	27	23	20	17	-0.5	-0.2	-0.1
Oil	78	81	67	32	11	10	7	3	0.5	-0.8	-3.1
Gas	119	165	269	399	17	21	29	38	4.2	4.6	4.4
Nuclear	133	124	93	71	20	16	10	7	-0.8	-2.0	-2.2
Hydro	130	134	140	146	19	17	15	14	0.4	0.4	0.4
of which pumped storage	*36*	*37*	*38*	*39*	*5*	*5*	*4*	*4*	*0.3*	*0.3*	*0.3*
Renewables (excluding hydro)	34	87	168	230	5	11	18	22	12.3	9.2	7.0

Renewables Breakdown

	2002	Capacity (GW) 2010	Capacity (GW) 2020	Capacity (GW) 2030	Shares (%) 2002	Shares (%) 2010	Shares (%) 2020	Shares (%) 2030	Growth Rates (% per annum) 2002-2010	Growth Rates (% per annum) 2002-2020	Growth Rates (% per annum) 2002-2030
Renewables (excluding hydro)	**34**	**87**	**168**	**230**	**100**	**100**	**100**	**100**	**12.3**	**9.2**	**7.0**
Biomass and Waste	10	18	26	34	29	21	16	15	7.7	5.6	4.4
Wind	23	66	131	170	67	76	78	74	14.0	10.1	7.4
Geothermal	1	1	1	1	2	1	1	0	2.5	1.1	1.1
Solar	0	1	7	17	1	1	4	7	15.1	17.6	14.7
Tide/Wave	0	1	3	8	1	1	2	4	22.1	15.1	13.5

	2002	Electricity Generation (TWh) 2010	Electricity Generation (TWh) 2020	Electricity Generation (TWh) 2030	Shares (%) 2002	Shares (%) 2010	Shares (%) 2020	Shares (%) 2030	Growth Rates (% per annum) 2002-2010	Growth Rates (% per annum) 2002-2020	Growth Rates (% per annum) 2002-2030
Renewables (excluding hydro)	**100**	**279**	**523**	**736**	**100**	**100**	**100**	**100**	**10.8**	**8.6**	**6.9**
Biomass and Waste	58	104	153	194	58	37	29	26	6.0	4.9	4.1
Wind	36	163	344	480	36	59	66	65	16.4	12.0	9.0
Geothermal	5	6	6	6	5	2	1	1	1.9	0.9	0.9
Solar	0	1	8	22	0	0	2	3	16.1	18.5	15.7
Tide/Wave	1	5	12	33	1	2	2	5	24.4	16.9	14.8

Reference Scenario: European Union

	CO$_2$ Emissions (Mt)					Shares (%)					Growth Rates (% per annum)			
	1971	2002	2010	2020	2030	1971	2002	2010	2020	2030	1971-2002	2002-2010	2002-2020	2002-2030
Total CO$_2$ Emissions	3 570	3 731	4 071	4 399	4 488	100	100	100	100	100	0.1	1.1	0.9	0.7
change since 1990 (%)		-0.1	9.1	17.8	20.2									
Coal	1 687	1 170	1 213	1 220	1 091	47	31	30	28	24	-1.2	0.5	0.2	-0.2
Oil	1 692	1 677	1 778	1 872	1 894	47	45	44	43	42	0.0	0.7	0.6	0.4
Gas	191	884	1 081	1 307	1 503	5	24	27	30	33	5.1	2.5	2.2	1.9
Power Generation and Heat Plants	1 054	1 308	1 466	1 650	1 669	100	100	100	100	100	0.7	1.4	1.3	0.9
Coal	779	937	1 012	1 063	963	74	72	69	64	58	0.6	1.0	0.7	0.1
Oil	232	128	101	74	40	22	10	7	4	2	-1.9	-2.9	-3.0	-4.0
Gas	44	243	353	514	666	4	19	24	31	40	5.7	4.8	4.2	3.7
Transformation, Own Use and Losses	163	173	159	151	144	100	100	100	100	100	0.2	-1.1	-0.7	-0.6
Total Final Consumption	2 352	2 249	2 447	2 597	2 675	100	100	100	100	100	-0.1	1.1	0.8	0.6
Coal	848	207	179	139	113	36	9	7	5	4	-4.5	-1.8	-2.2	-2.1
Oil	1 362	1 432	1 572	1 698	1 757	58	64	64	65	66	0.2	1.2	1.0	0.7
Gas	143	611	696	760	805	6	27	28	29	30	4.8	1.6	1.2	1.0
Industry	993	602	626	626	622	100	100	100	100	100	-1.6	0.5	0.2	0.1
Coal	445	157	143	116	97	45	26	23	19	16	-3.3	-1.2	-1.7	-1.7
Oil	457	200	204	206	206	46	33	33	33	33	-2.6	0.3	0.2	0.1
Gas	91	245	278	304	319	9	41	44	49	51	3.2	1.6	1.2	1.0
Transport	433	903	1 036	1 184	1 270	100	100	100	100	100	2.4	1.7	1.5	1.2
Oil	403	901	1 032	1 178	1 262	93	100	100	100	99	2.6	1.7	1.5	1.2
Other Fuels	30	3	4	6	7	7	0	0	0	1	-7.4	4.8	3.8	3.5
Other Sectors	885	709	747	746	740	100	100	100	100	100	-0.7	0.6	0.3	0.2
Coal	360	47	33	21	13	41	7	4	3	2	-6.4	-4.2	-4.4	-4.5
Oil	474	299	300	275	249	54	42	40	37	34	-1.5	0.1	-0.5	-0.7
Gas	51	363	413	450	478	6	51	55	60	65	6.5	1.6	1.2	1.0
Non-Energy Use	41	34	38	41	42	100	100	100	100	100	-0.6	1.2	1.0	0.8

Reference Scenario: Transition Economies

	Energy Demand (Mtoe)					Shares (%)					Growth Rates (% per annum)			
	1971	2002	2010	2020	2030	1971	2002	2010	2020	2030	1971-2002	2002-2010	2002-2020	2002-2030
Total Primary Energy Supply	814	1 030	1 186	1 358	1 499	100	100	100	100	100	0.8	1.8	1.6	1.3
Coal	302	194	219	227	217	37	19	18	17	15	-1.4	1.5	0.9	0.4
Oil	278	222	265	312	362	34	22	22	23	24	-0.7	2.2	1.9	1.8
Gas	197	504	578	685	782	24	49	49	50	52	3.1	1.7	1.7	1.6
Nuclear	2	69	77	80	71	0	7	6	6	5	12.9	1.2	0.8	0.1
Hydro	13	24	29	31	32	2	2	2	2	2	2.1	2.3	1.3	1.0
Biomass and Waste	24	16	17	19	25	3	2	1	1	2	-1.3	0.8	1.1	1.7
Other Renewables	0	0	3	4	8	0	0	0	0	1	-	36.7	17.3	13.6
Power Generation and Heat Plants	263	527	591	656	702	100	100	100	100	100	2.3	1.4	1.2	1.0
Coal	120	131	150	153	140	46	25	25	23	20	0.3	1.7	0.9	0.2
Oil	47	33	35	31	27	18	6	6	5	4	-1.1	0.7	-0.4	-0.7
Gas	81	265	293	353	415	31	50	50	54	59	3.9	1.3	1.6	1.6
Nuclear	2	69	77	80	71	1	13	13	12	10	12.9	1.2	0.8	0.1
Hydro	13	24	29	31	32	5	5	5	5	5	2.1	2.3	1.3	1.0
Biomass and Waste	0	5	4	4	8	0	1	1	1	1	-	-1.3	-0.2	2.2
Other Renewables	0	0	3	4	8	0	0	0	1	1	-	37.9	17.8	13.9
Other Transformation, Own Use and Losses	69	147	168	187	202						2.5	1.7	1.3	1.1
of which electricity	*0*	*37*	*44*	*51*	*58*							*1.9*	*1.7*	*1.5*
Total Final Consumption	594	669	766	887	997	100	100	100	100	100	0.4	1.7	1.6	1.4
Coal	114	39	43	47	51	19	6	6	5	5	-3.4	1.0	1.0	0.9
Oil	230	160	190	235	283	39	24	25	26	28	-1.2	2.2	2.2	2.1
Gas	115	210	251	294	324	19	31	33	33	33	2.0	2.3	1.9	1.6
Electricity	61	90	108	132	154	10	13	14	15	15	1.3	2.4	2.2	2.0
Heat	50	160	161	164	167	8	24	21	19	17	3.8	0.1	0.2	0.2
Biomass and Waste	24	11	13	15	17	4	2	2	2	2	-2.4	1.5	1.5	1.5
Other Renewables	0	0	0	0	0	0	0	0	0	0	-	21.7	11.1	8.3

						Shares (%)					Growth rates			
Industry	273	235	269	310	341	100	100	100	100	100	-0.5	1.7	1.6	1.3
Coal	43	26	30	33	36	16	11	11	11	10	-1.6	1.5	1.3	1.1
Oil	59	30	36	44	52	22	13	13	14	15	-2.2	2.2	2.2	2.0
Gas	91	76	94	112	125	33	32	35	36	37	-0.6	2.8	2.2	1.8
Electricity	43	42	52	64	76	16	18	19	21	22	-0.1	2.6	2.4	2.2
Heat	37	60	57	53	51	13	25	21	17	15	1.6	-0.6	-0.6	-0.6
Renewables	0	1	1	2	2	0	1	1	1	1	-	1.9	1.9	1.8
Transport	99	124	150	186	225	100	100	100	100	100	0.7	2.4	2.3	2.2
Oil	85	86	105	134	168	86	70	70	72	75	0.0	2.5	2.5	2.4
Other Fuels	14	38	45	52	57	14	30	30	28	25	3.2	2.2	1.8	1.5
Other Sectors	120	285	320	362	400	100	100	100	100	100	2.8	1.4	1.3	1.2
Coal	57	11	12	13	14	48	4	4	4	4	-5.1	0.2	0.6	0.7
Oil	51	31	35	40	45	42	11	11	11	11	-1.6	1.5	1.4	1.3
Gas	1	93	110	128	141	1	33	34	35	35	16.3	2.0	1.8	1.5
Electricity	5	40	48	59	69	4	14	15	16	17	6.7	2.3	2.2	2.0
Heat	4	100	105	111	117	3	35	33	31	29	11.2	0.6	0.6	0.6
Renewables	2	10	11	13	14	2	4	4	4	4	5.3	1.4	1.3	1.3
Non-Energy Use	101	26	27	29	31						-4.4	0.7	0.7	0.7
Electricity Generation (TWh)	870	1 485	1 745	2 134	2 469	100	100	100	100	100	1.7	2.0	2.0	1.8
Coal	387	324	381	409	394	45	22	22	19	16	-0.6	2.0	1.3	0.7
Oil	152	57	69	65	61	17	4	4	3	2	-3.1	2.5	0.8	0.3
Gas	175	556	652	976	1 324	20	37	37	46	54	3.8	2.0	3.2	3.1
Nuclear	6	264	292	305	272	1	18	17	14	11	12.9	1.2	0.8	0.1
Hydro	149	281	338	355	373	17	19	19	17	15	2.1	2.3	1.3	1.0
Biomass and Waste	0	3	4	5	11	0	0	0	0	0	-	2.6	2.6	4.7
Other Renewables	0	0	10	19	33	0	0	1	1	1	-	50.0	24.2	17.3
Traditional Biomass (included above)	0	0	0	0	0	0	0	0	0	0	-	-	-	-
Total Primary Energy Supply (excluding traditional biomass)	814	1 030	1 186	1 358	1 499						0.8	1.8	1.6	1.3

Reference Scenario: Transition Economies

	2002	Capacity (GW) 2010	2020	2030	Shares (%) 2002	2010	2020	2030	Growth Rates (% per annum) 2002-2010	2002-2020	2002-2030
Total Capacity	411	445	525	617	100	100	100	100	**1.0**	**1.4**	**1.5**
Coal	111	108	93	87	27	24	18	14	-0.3	-1.0	-0.9
Oil	38	39	38	31	9	9	7	5	0.4	0.0	-0.7
Gas	130	149	236	332	32	33	45	54	1.7	3.4	3.4
Nuclear	40	41	43	38	10	9	8	6	0.5	0.5	-0.1
Hydro	91	104	109	114	22	23	21	19	1.7	1.0	0.8
of which pumped storage	3	3	3	4	1	1	1	1	0.5	0.5	0.5
Renewables (excluding hydro)	1	4	7	13	0	1	1	2	15.5	10.4	8.9

Renewables Breakdown

	2002	Capacity (GW) 2010	2020	2030	Shares (%) 2002	2010	2020	2030	Growth Rates (% per annum) 2002-2010	2002-2020	2002-2030
Renewables (excluding hydro)	1	4	7	13	100	100	100	100	**15.5**	**10.4**	**8.9**
Biomass and Waste	1	1	1	2	67	23	15	16	1.0	1.5	3.4
Wind	0	3	5	9	26	68	77	72	30.3	17.4	12.9
Geothermal	0	0	1	1	7	9	7	11	20.2	10.9	10.9
Solar	0	0	0	1	0	0	0	1	-	-	-
Tide/Wave	0	0	0	0	0	0	0	0	-	-	-

	2002	Electricity Generation (TWh) 2010	2020	2030	Shares (%) 2002	2010	2020	2030	Growth Rates (% per annum) 2002-2010	2002-2020	2002-2030
Renewables (excluding hydro)	3	13	23	44	100	100	100	100	**14.8**	**10.2**	**8.9**
Biomass and Waste	3	4	5	11	89	27	20	25	2.0	2.3	4.4
Wind	0	7	16	26	5	54	67	59	46.0	25.6	18.4
Geothermal	0	2	3	7	6	18	13	15	27.6	14.3	12.2
Solar	0	0	0	0	0	0	0	0	-	-	-
Tide/Wave	0	0	0	0	0	0	0	0	-	-	-

Reference Scenario: Transition Economies

	CO_2 Emissions (Mt)					Shares (%)					Growth Rates (% per annum)			
	1971	2002	2010	2020	2030	1971	2002	2010	2020	2030	1971-2002	2002-2010	2002-2020	2002-2030
Total CO_2 Emissions	2 242	2 444	2 808	3 200	3 501	100	100	100	100	100	0.3	1.7	1.5	1.3
change since 1990 (%)		-34.5	-24.7	-14.2	-6.2									
Coal	968	756	852	890	854	43	31	30	28	24	-0.8	1.5	0.9	0.4
Oil	788	556	662	775	895	35	23	24	24	26	-1.1	2.2	1.9	1.7
Gas	487	1 132	1 294	1 534	1 752	22	46	46	48	50	2.8	1.7	1.7	1.6
Power Generation and Heat Plants	825	1 270	1 421	1 560	1 639	100	100	100	100	100	1.4	1.4	1.1	0.9
Coal	483	544	622	634	579	59	43	44	41	35	0.4	1.7	0.8	0.2
Oil	151	106	112	99	87	18	8	8	6	5	-1.1	0.7	-0.4	-0.7
Gas	190	620	687	827	973	23	49	48	53	59	3.9	1.3	1.6	1.6
Transformation, Own Use and Losses	95	94	120	138	154	100	100	100	100	100	0.0	3.0	2.1	1.8
Total Final Consumption	1 323	1 080	1 267	1 502	1 709	100	100	100	100	100	-0.7	2.0	1.9	1.7
Coal	448	208	227	253	271	34	19	18	17	16	-2.4	1.1	1.1	1.0
Oil	612	398	474	587	707	46	37	37	39	41	-1.4	2.2	2.2	2.1
Gas	262	474	566	662	730	20	44	45	44	43	1.9	2.2	1.9	1.6
Industry	563	400	472	558	618	100	100	100	100	100	-1.1	2.1	1.9	1.6
Coal	176	157	175	198	212	31	39	37	36	34	-0.4	1.4	1.3	1.1
Oil	181	81	96	120	140	32	20	20	22	23	-2.6	2.2	2.2	2.0
Gas	206	162	201	240	266	37	41	42	43	43	-0.8	2.7	2.2	1.8
Transport	220	285	348	436	531	100	100	100	100	100	0.8	2.5	2.4	2.2
Oil	182	214	263	336	419	83	75	75	77	79	0.5	2.6	2.5	2.4
Other Fuels	38	70	86	100	111	17	25	25	23	21	2.0	2.4	2.0	1.6
Other Sectors	472	378	428	488	538	100	100	100	100	100	-0.7	1.6	1.4	1.3
Coal	229	46	47	51	56	48	12	11	10	10	-5.1	0.2	0.6	0.7
Oil	188	91	102	116	130	40	24	24	24	24	-2.3	1.4	1.4	1.3
Gas	55	242	280	322	352	12	64	65	66	65	4.9	1.8	1.6	1.4
Non-Energy Use	69	17	19	20	22	100	100	100	100	100	-4.4	1.0	0.9	0.9

Reference Scenario: Russia

	Energy Demand (Mtoe)					Shares (%)					Growth Rates (% per annum)			
	1971	2002	2010	2020	2030	1971	2002	2010	2020	2030	1971-2002	2002-2010	2002-2020	2002-2030
Total Primary Energy Supply	n.a.	619	708	802	885	n.a.	100	100	100	100	n.a.	**1.7**	**1.5**	**1.3**
Coal	n.a.	107	118	125	117	n.a.	17	17	16	13	n.a.	1.3	0.9	0.3
Oil	n.a.	128	149	171	199	n.a.	21	21	21	22	n.a.	1.9	1.6	1.6
Gas	n.a.	326	371	433	489	n.a.	53	52	54	55	n.a.	1.7	1.6	1.5
Nuclear	n.a.	37	45	47	48	n.a.	6	6	6	5	n.a.	2.3	1.3	0.9
Hydro	n.a.	14	16	17	17	n.a.	2	2	2	2	n.a.	1.7	1.0	0.7
Biomass and Waste	n.a.	7	6	6	9	n.a.	1	1	1	1	n.a.	-0.9	-0.3	0.8
Other Renewables	n.a.	0	2	3	6	n.a.	0	0	0	1	n.a.	42.7	19.2	14.6
Power Generation and Heat Plants	n.a.	348	383	419	449	n.a.	100	100	100	100	n.a.	**1.2**	**1.0**	**0.9**
Coal	n.a.	79	87	91	82	n.a.	23	23	22	18	n.a.	1.2	0.8	0.1
Oil	n.a.	21	22	19	16	n.a.	6	6	5	4	n.a.	0.3	-0.5	-0.9
Gas	n.a.	193	209	239	275	n.a.	55	54	57	61	n.a.	1.0	1.2	1.3
Nuclear	n.a.	37	45	47	48	n.a.	11	12	11	11	n.a.	2.3	1.3	0.9
Hydro	n.a.	14	16	17	17	n.a.	4	4	4	4	n.a.	1.7	1.0	0.7
Biomass and Waste	n.a.	4	3	3	5	n.a.	1	1	1	1	n.a.	-2.4	-1.6	0.6
Other Renewables	n.a.	0	2	3	6	n.a.	0	1	1	1	n.a.	42.7	19.2	14.6
Other Transformation, Own Use and Losses	n.a.	85	96	106	115						n.a.	**1.6**	**1.2**	**1.1**
of which electricity	*n.a.*	*22*	*25*	*28*	*31*						*n.a.*	*1.5*	*1.4*	*1.2*
Total Final Consumption	n.a.	412	466	530	589	n.a.	100	100	100	100	n.a.	**1.5**	**1.4**	**1.3**
Coal	n.a.	20	22	24	25	n.a.	5	5	5	4	n.a.	1.1	1.0	0.9
Oil	n.a.	90	104	125	150	n.a.	22	22	24	26	n.a.	1.9	1.8	1.8
Gas	n.a.	116	143	171	190	n.a.	28	31	32	32	n.a.	2.7	2.2	1.8
Electricity	n.a.	53	62	74	85	n.a.	13	13	14	14	n.a.	2.0	1.8	1.7
Heat	n.a.	131	131	133	135	n.a.	32	28	25	23	n.a.	0.1	0.1	0.1
Biomass and Waste	n.a.	3	3	3	4	n.a.	1	1	1	1	n.a.	1.0	1.0	1.1
Other Renewables	n.a.	0	0	0	0	n.a.	0	0	0	0	n.a.	-	-	-

Industry	n.a.	146	167	188	202	n.a.	100	100	100	100	1.7	1.4	1.2
Coal	n.a.	11	13	14	15	n.a.	7	8	8	7	2.0	1.5	1.1
Oil	n.a.	17	19	22	24	n.a.	11	12	12	12	1.9	1.6	1.3
Gas	n.a.	41	54	67	74	n.a.	28	33	36	37	3.5	2.7	2.1
Electricity	n.a.	27	33	41	47	n.a.	19	20	22	23	2.5	2.2	2.0
Heat	n.a.	49	47	44	41	n.a.	34	28	23	20	-0.7	-0.7	-0.6
Renewables	n.a.	0	0	1	1	n.a.	0	0	0	0	2.1	2.2	2.3
Transport	n.a.	84	101	124	149	n.a.	100	100	100	100	2.3	2.1	2.1
Oil	n.a.	50	60	76	96	n.a.	60	59	61	65	2.2	2.3	2.4
Other Fuels	n.a.	34	41	48	53	n.a.	40	41	39	35	2.3	1.9	1.6
Other Sectors	n.a.	174	189	209	228	n.a.	100	100	100	100	1.1	1.0	1.0
Coal	n.a.	8	8	9	10	n.a.	4	4	4	4	0.1	0.6	0.9
Oil	n.a.	16	18	19	21	n.a.	9	9	9	9	1.1	0.9	0.9
Gas	n.a.	46	54	63	70	n.a.	27	28	30	31	1.9	1.7	1.5
Electricity	n.a.	20	23	27	31	n.a.	11	12	13	13	1.7	1.6	1.6
Heat	n.a.	81	84	89	94	n.a.	47	45	43	41	0.5	0.5	0.5
Renewables	n.a.	2	2	2	2	n.a.	1	1	1	1	0.4	0.4	0.4
Non-Energy Use	n.a.	8	8	9	10						0.5	0.5	0.8
Electricity Generation (TWh)	n.a.	889	1 028	1 200	1 361	n.a.	100	100	100	100	1.8	1.7	1.5
Coal	n.a.	170	201	222	206	n.a.	19	20	18	15	2.1	1.5	0.7
Oil	n.a.	27	35	33	28	n.a.	3	3	3	2	3.1	1.0	0.0
Gas	n.a.	385	427	560	720	n.a.	43	42	47	53	1.3	2.1	2.3
Nuclear	n.a.	142	170	179	184	n.a.	16	17	15	14	2.3	1.3	0.9
Hydro	n.a.	162	186	194	200	n.a.	18	18	16	15	1.7	1.0	0.7
Biomass and Waste	n.a.	3	3	3	6	n.a.	0	0	0	0	1.0	0.4	2.6
Other Renewables	n.a.	0	6	10	18	n.a.	0	1	1	1	56.5	26.0	18.2
Traditional Biomass (included above)	n.a.	0	0	0	0	n.a.	0	0	0	0	n.a.	-	-
Total Primary Energy Supply (excluding traditional biomass)	n.a.	619	708	802	885	n.a.					1.7	1.5	1.3

Annex A - Tables for Reference Scenario Projections 475

Reference Scenario: Russia

	2002	Capacity (GW) 2010	Capacity (GW) 2020	Capacity (GW) 2030	Shares (%) 2002	Shares (%) 2010	Shares (%) 2020	Shares (%) 2030	Growth Rates (% per annum) 2002-2010	Growth Rates (% per annum) 2002-2020	Growth Rates (% per annum) 2002-2030
Total Capacity	223	237	268	306	100	100	100	100	0.8	1.0	1.1
Coal	52	51	45	44	23	21	17	14	-0.3	-0.8	-0.6
Oil	15	15	15	13	7	6	6	4	0.5	0.1	-0.4
Gas	89	95	128	163	40	40	48	53	0.8	2.0	2.2
Nuclear	21	24	25	26	9	10	9	8	1.6	1.0	0.7
Hydro	45	50	52	53	20	21	19	17	1.3	0.8	0.6
of which pumped storage	1	1	1	1	1	1	0	0	0.5	0.5	0.5
Renewables (excluding hydro)	1	2	4	6	0	1	1	2	11.2	7.8	6.9

Renewables Breakdown

	2002	Capacity (GW) 2010	Capacity (GW) 2020	Capacity (GW) 2030	Shares (%) 2002	Shares (%) 2010	Shares (%) 2020	Shares (%) 2030	Growth Rates (% per annum) 2002-2010	Growth Rates (% per annum) 2002-2020	Growth Rates (% per annum) 2002-2030
Renewables (excluding hydro)	1	2	4	6	100	100	100	100	11.2	7.8	6.9
Biomass and Waste	1	1	1	1	77	33	20	18	0.0	0.0	1.6
Wind	0	1	3	4	16	52	67	62	29.1	16.8	12.3
Geothermal	0	0	1	1	7	15	13	20	21.7	11.6	11.0
Solar	0	0	0	0	0	0	0	0	-	-	-
Tide/Wave	0	0	0	0	0	0	0	0	-	-	-

	2002	Electricity Generation (TWh) 2010	Electricity Generation (TWh) 2020	Electricity Generation (TWh) 2030	Shares (%) 2002	Shares (%) 2010	Shares (%) 2020	Shares (%) 2030	Growth Rates (% per annum) 2002-2010	Growth Rates (% per annum) 2002-2020	Growth Rates (% per annum) 2002-2030
Renewables (excluding hydro)	3	9	13	23	100	100	100	100	11.6	7.8	7.1
Biomass and Waste	3	3	3	6	95	34	23	25	0.8	0.4	2.5
Wind	0	4	7	12	0	40	55	49	89.1	42.7	28.7
Geothermal	0	2	3	6	5	26	22	26	31.1	15.9	13.0
Solar	0	0	0	0	0	0	0	0	-	-	-
Tide/Wave	0	0	0	0	0	0	0	0	-	-	-

Reference Scenario: Russia

	CO$_2$ Emissions (Mt)					Shares (%)					Growth Rates (% per annum)			
	1971	2002	2010	2020	2030	1971	2002	2010	2020	2030	1971-2002	2002-2010	2002-2020	2002-2030
Total CO$_2$ Emissions	n.a.	1 488	1 688	1 905	2 062	n.a.	100	100	100	100	n.a.	1.6	1.4	1.2
change since 1990 (%)		-32.7	-23.7	-13.9	-6.8									
Coal	n.a.	439	485	516	483	n.a.	29	29	27	23	n.a.	1.3	0.9	0.3
Oil	n.a.	319	372	423	484	n.a.	21	22	22	23	n.a.	2.0	1.6	1.5
Gas	n.a.	731	830	967	1 094	n.a.	49	49	51	53	n.a.	1.6	1.6	1.5
Power Generation and Heat Plants	n.a.	849	923	1 006	1 039	n.a.	100	100	100	100	n.a.	1.0	0.9	0.7
Coal	n.a.	331	366	385	345	n.a.	39	40	38	33	n.a.	1.2	0.8	0.1
Oil	n.a.	68	69	62	53	n.a.	8	8	6	5	n.a.	0.3	-0.5	-0.9
Gas	n.a.	450	488	559	642	n.a.	53	53	56	62	n.a.	1.0	1.2	1.3
Transformation, Own Use and Losses	n.a.	58	76	88	103	n.a.	100	100	100	100	n.a.	3.6	2.4	2.1
Total Final Consumption	n.a.	581	688	811	920	n.a.	100	100	100	100	n.a.	2.1	1.9	1.7
Coal	n.a.	106	118	130	137	n.a.	18	17	16	15	n.a.	1.3	1.1	0.9
Oil	n.a.	216	251	300	360	n.a.	37	36	37	39	n.a.	1.9	1.8	1.8
Gas	n.a.	259	319	381	424	n.a.	45	46	47	46	n.a.	2.6	2.2	1.8
Industry	n.a.	199	245	287	310	n.a.	100	100	100	100	n.a.	2.7	2.1	1.6
Coal	n.a.	70	82	91	93	n.a.	35	33	32	30	n.a.	2.0	1.5	1.1
Oil	n.a.	44	51	58	63	n.a.	22	21	20	20	n.a.	1.9	1.6	1.3
Gas	n.a.	85	113	139	153	n.a.	43	46	48	49	n.a.	3.5	2.7	2.1
Transport	n.a.	185	223	275	334	n.a.	100	100	100	100	n.a.	2.3	2.2	2.1
Oil	n.a.	119	141	179	228	n.a.	64	64	65	68	n.a.	2.2	2.3	2.4
Other Fuels	n.a.	66	81	96	107	n.a.	36	36	35	32	n.a.	2.6	2.1	1.7
Other Sectors	n.a.	188	210	239	265	n.a.	100	100	100	100	n.a.	1.4	1.3	1.2
Coal	n.a.	31	31	34	39	n.a.	16	15	14	15	n.a.	0.1	0.6	0.9
Oil	n.a.	49	54	58	63	n.a.	26	26	24	24	n.a.	1.1	0.9	0.9
Gas	n.a.	108	126	147	164	n.a.	57	60	61	62	n.a.	1.9	1.7	1.5
Non-Energy Use	n.a.	10	10	10	10	n.a.	100	100	100	100	n.a.	-0.2	-0.1	0.1

Reference Scenario: Developing Countries

	Energy Demand (Mtoe)					Shares (%)					Growth Rates (% per annum)			
	1971	2002	2010	2020	2030	1971	2002	2010	2020	2030	1971-2002	2002-2010	2002-2020	2002-2030
Total Primary Energy Supply	1 228	3 824	4 890	6 344	7 873	100	100	100	100	100	3.7	3.1	2.9	2.6
Coal	293	1 099	1 374	1 754	2 192	24	29	28	28	28	4.4	2.8	2.6	2.5
Oil	299	1 142	1 523	2 016	2 517	24	30	31	32	32	4.4	3.7	3.2	2.9
Gas	40	515	746	1 131	1 518	3	13	15	18	19	8.6	4.7	4.5	3.9
Nuclear	0	30	60	96	135	0	1	1	2	2	15.6	9.2	6.7	5.6
Hydro	16	94	127	166	202	1	2	3	3	3	6.0	3.7	3.2	2.7
Biomass and Waste	580	922	1 024	1 127	1 221	47	24	21	18	16	1.5	1.3	1.1	1.0
Other Renewables	0	21	35	54	89	0	1	1	1	1	-	6.4	5.3	5.2
Power Generation and Heat Plants	126	1 097	1 574	2 274	3 082	100	100	100	100	100	7.2	4.6	4.1	3.8
Coal	64	619	866	1 225	1 652	51	56	55	54	54	7.6	4.3	3.9	3.6
Oil	36	137	153	177	187	29	13	10	8	6	4.4	1.4	1.4	1.1
Gas	8	184	306	509	730	6	17	19	22	24	10.7	6.6	5.8	5.0
Nuclear	0	30	60	96	135	0	3	4	4	4	15.6	9.2	6.7	5.6
Hydro	16	94	127	166	202	13	9	8	7	7	6.0	3.7	3.2	2.7
Biomass and Waste	1	11	30	54	103	1	1	2	2	3	7.1	13.0	9.0	8.2
Other Renewables	0	21	31	47	73	0	2	2	2	2	-	5.3	4.6	4.6
Other Transformation, Own Use and Losses	123	468	604	771	918						4.4	3.3	2.8	2.4
of which electricity	7	93	132	189	256						8.5	4.5	4.1	3.7
Total Final Consumption	1 025	2 714	3 376	4 281	5 233	100	100	100	100	100	3.2	2.8	2.6	2.4
Coal	203	343	357	369	376	20	13	11	9	7	1.7	0.5	0.4	0.3
Oil	229	936	1 261	1 705	2 177	22	34	37	40	42	4.7	3.8	3.4	3.1
Gas	18	212	285	397	512	2	8	8	9	10	8.3	3.8	3.5	3.2
Electricity	38	323	482	729	1 030	4	12	14	17	20	7.1	5.1	4.6	4.2
Heat	0	30	39	50	60	0	1	1	1	1	-	3.5	2.9	2.5
Biomass and Waste	537	869	948	1 022	1 063	52	32	28	24	20	1.6	1.1	0.9	0.7
Other Renewables	0	1	4	8	16	0.0	0.0	0.1	0.2	0.3	-	22.3	13.6	11.3

Industry	260	896	1 098	1 365	1 621	100	100	100	100	100	4.1	2.6	2.4	2.1
Coal	122	251	272	293	308	47	28	25	21	19	2.3	1.0	0.9	0.7
Oil	63	230	281	350	414	24	26	26	26	26	4.2	2.6	2.4	2.1
Gas	14	139	184	251	313	6	15	17	18	19	7.6	3.6	3.4	2.9
Electricity	16	157	221	312	413	6	17	20	23	25	7.7	4.4	3.9	3.5
Heat	0	22	28	33	38	0	2	3	2	2	-	2.9	2.3	1.9
Renewables	45	97	111	125	135	17	11	10	9	8	2.6	1.7	1.4	1.2
Transport	119	463	651	932	1 251	100	100	100	100	100	4.5	4.4	4.0	3.6
Oil	100	445	627	899	1 206	84	96	96	96	96	4.9	4.4	4.0	3.6
Other Fuels	19	18	24	33	46	16	4	4	4	4	-0.2	4.0	3.6	3.4
Other Sectors	609	1 269	1 514	1 849	2 210	100	100	100	100	100	2.4	2.2	2.1	2.0
Coal	55	70	64	58	52	9	6	4	3	2	0.8	-1.0	-1.0	-1.0
Oil	49	206	273	359	446	8	16	18	19	20	4.7	3.6	3.1	2.8
Gas	4	70	95	136	185	1	6	6	7	8	10.1	4.0	3.8	3.5
Electricity	11	153	244	393	586	2	12	16	21	26	9.0	6.0	5.4	4.9
Heat	0	7	11	16	21	0	1	1	1	1	-	5.4	4.6	4.0
Renewables	490	763	828	888	920	81	60	55	48	42	1.4	1.0	0.8	0.7
Non-Energy Use	37	87	113	134	151						2.8	3.3	2.4	2.0
Electricity Generation (TWh)	526	4 832	7 139	10 677	14 945	100	100	100	100	100	7.4	5.0	4.5	4.1
Coal	194	2 183	3 232	4 829	6 961	37	45	45	45	47	8.1	5.0	4.5	4.2
Oil	122	566	646	759	812	23	12	9	7	5	5.1	1.7	1.6	1.3
Gas	24	806	1 401	2 536	3 860	5	17	20	24	26	12.0	7.2	6.6	5.8
Nuclear	1	114	231	370	520	0	2	3	3	3	15.6	9.2	6.7	5.6
Hydro	183	1 099	1 472	1 931	2 346	35	23	21	18	16	6.0	3.7	3.2	2.7
Biomass and Waste	2	35	85	126	211	0	1	1	1	1	9.3	11.7	7.3	6.6
Other Renewables	0	28	70	127	235	0	1	1	1	2	-	12.2	8.7	7.9
Traditional Biomass (included above)	492	765	827	883	907	40	20	17	14	12	1.4	1.0	0.8	0.6
Total Primary Energy Supply (excluding traditional biomass)	736	3 059	4 063	5 461	6 966						4.7	3.6	3.3	3.0

Reference Scenario: Developing Countries

	\multicolumn{4}{c	}{Capacity (GW)}	\multicolumn{4}{c	}{Shares (%)}	\multicolumn{3}{c}{Growth Rates (% per annum)}
	2002	2010	2020	2030	2002	2010	2020	2030	2002-2010	2002-2020	2002-2030
Total Capacity	1 132	1 630	2 422	3 381	100	100	100	100	**4.7**	**4.3**	**4.0**
Coal	403	608	904	1 305	36	37	37	39	5.3	4.6	4.3
Oil	179	213	258	287	16	13	11	8	2.2	2.1	1.7
Gas	229	367	658	1 007	20	23	27	30	6.1	6.0	5.4
Nuclear	17	30	48	67	1	2	2	2	7.7	6.0	5.1
Hydro	290	379	500	611	26	23	21	18	3.4	3.1	2.7
of which pumped storage	0	0	0	0	0	0	0	0	-	-	-
Renewables (excluding hydro)	14	33	55	104	1	2	2	3	10.9	7.8	7.4

Renewables Breakdown

	\multicolumn{4}{c	}{Capacity (GW)}	\multicolumn{4}{c	}{Shares (%)}	\multicolumn{3}{c}{Growth Rates (% per annum)}
	2002	2010	2020	2030	2002	2010	2020	2030	2002-2010	2002-2020	2002-2030
Renewables (excluding hydro)	14	33	55	104	100	100	100	100	**10.9**	**7.8**	**7.4**
Biomass and Waste	7	14	21	34	50	44	37	33	9.1	6.1	5.7
Wind	3	13	28	53	24	41	50	51	18.7	12.4	10.4
Geothermal	4	5	7	10	25	14	12	9	3.4	3.5	3.6
Solar	0	0	0	7	0	0	0	6	0.0	0.0	17.9
Tide/Wave	0	0	0	0	0	0	0	0	-	-	-

	\multicolumn{4}{c	}{Electricity Generation (TWh)}	\multicolumn{4}{c	}{Shares (%)}	\multicolumn{3}{c}{Growth Rates (% per annum)}
	2002	2010	2020	2030	2002	2010	2020	2030	2002-2010	2002-2020	2002-2030
Renewables (excluding hydro)	63	156	253	447	100	100	100	100	**9.4**	**7.2**	**6.7**
Biomass and Waste	35	85	126	211	56	55	50	47	9.2	6.6	6.1
Wind	4	38	80	154	7	24	32	35	24.1	15.7	12.6
Geothermal	24	33	46	68	37	21	18	15	3.3	3.4	3.6
Solar	0	0	0	12	0	0	0	3	3.7	1.8	17.9
Tide/Wave	0	0	0	0	0	0	0	0	-	-	-

Reference Scenario: Developing Countries

	CO$_2$ Emissions (Mt)					Shares (%)					Growth Rates (% per annum)			
	1971	2002	2010	2020	2030	1971	2002	2010	2020	2030	1971-2002	2002-2010	2002-2020	2002-2030
Total CO$_2$ Emissions	**1 995**	**8 226**	**10 726**	**14 392**	**18 365**	**100**	**100**	**100**	**100**	**100**	**4.7**	**3.4**	**3.2**	**2.9**
change since 1990 (%)		*56.0*	*103.4*	*172.9*	*248.2*									
Coal	1 090	4 047	5 105	6 603	8 344	55	49	48	46	45	4.3	2.9	2.8	2.6
Oil	814	3 068	3 996	5 305	6 670	41	37	37	37	36	4.4	3.4	3.1	2.8
Gas	91	1 111	1 625	2 484	3 351	5	14	15	17	18	8.4	4.9	4.6	4.0
Power Generation and Heat Plants	**383**	**3 354**	**4 684**	**6 676**	**8 941**	**100**	**100**	**100**	**100**	**100**	**7.2**	**4.3**	**3.9**	**3.6**
Coal	250	2 483	3 470	4 903	6 608	65	74	74	73	74	7.7	4.3	3.9	3.6
Oil	115	436	487	562	592	30	13	10	8	7	4.4	1.4	1.4	1.1
Gas	18	435	728	1 211	1 742	5	13	16	18	19	10.8	6.6	5.9	5.1
Transformation, Own Use and Losses	**105**	**491**	**597**	**798**	**969**	**100**	**100**	**100**	**100**	**100**	**5.1**	**2.5**	**2.7**	**2.5**
Total Final Consumption	**1 507**	**4 381**	**5 445**	**6 919**	**8 455**	**100**	**100**	**100**	**100**	**100**	**3.5**	**2.8**	**2.6**	**2.4**
Coal	835	1 488	1 553	1 612	1 647	55	34	29	23	19	1.9	0.5	0.4	0.4
Oil	631	2 432	3 275	4 450	5 710	42	56	60	64	68	4.4	3.8	3.4	3.1
Gas	40	460	618	857	1 098	3	11	11	12	13	8.2	3.7	3.5	3.2
Industry	**755**	**1 954**	**2 263**	**2 657**	**3 000**	**100**	**100**	**100**	**100**	**100**	**3.1**	**1.9**	**1.7**	**1.5**
Coal	522	1 123	1 217	1 310	1 376	69	57	54	49	46	2.5	1.0	0.9	0.7
Oil	200	536	655	813	962	26	27	29	31	32	3.2	2.5	2.3	2.1
Gas	32	294	391	534	663	4	15	17	20	22	7.4	3.6	3.4	2.9
Transport	**328**	**1 245**	**1 748**	**2 501**	**3 353**	**100**	**100**	**100**	**100**	**100**	**4.4**	**4.3**	**4.0**	**3.6**
Oil	258	1 215	1 712	2 454	3 295	79	98	98	98	98	5.1	4.4	4.0	3.6
Other Fuels	70	30	36	48	58	21	2	2	2	2	-2.7	2.4	2.6	2.4
Other Sectors	**405**	**1 068**	**1 292**	**1 601**	**1 930**	**100**	**100**	**100**	**100**	**100**	**3.2**	**2.4**	**2.3**	**2.1**
Coal	242	305	279	250	223	60	29	22	16	12	0.7	-1.1	-1.1	-1.1
Oil	154	607	802	1 054	1 308	38	57	62	66	68	4.5	3.5	3.1	2.8
Gas	8	157	211	297	398	2	15	16	19	21	10.0	3.8	3.6	3.4
Non-Energy Use	**19**	**114**	**141**	**159**	**173**	**100**	**100**	**100**	**100**	**100**	**5.9**	**2.7**	**1.9**	**1.5**

Reference Scenario: China

	Energy Demand (Mtoe)					Shares (%)					Growth Rates (% per annum)			
	1971	2002	2010	2020	2030	1971	2002	2010	2020	2030	1971-2002	2002-2010	2002-2020	2002-2030
Total Primary Energy Supply	405	1 242	1 622	2 072	2 539	100	100	100	100	100	3.7	3.4	2.9	2.6
Coal	192	713	904	1 119	1 354	47	57	56	54	53	4.3	3.0	2.5	2.3
Oil	43	247	375	503	636	11	20	23	24	25	5.8	5.4	4.0	3.4
Gas	3	36	59	107	158	1	3	4	5	6	8.2	6.4	6.3	5.4
Nuclear	0	7	21	47	73	0	1	1	2	3	-	15.9	11.6	9.0
Hydro	3	25	33	50	63	1	2	2	2	2	7.6	3.6	3.9	3.4
Biomass and Waste	164	216	227	236	236	41	17	14	11	9	0.9	0.6	0.5	0.3
Other Renewables	0	0	5	10	20	0	0	0	0	1	-	-	-	-
Power Generation and Heat Plants	37	428	653	951	1 267	100	100	100	100	100	8.2	5.4	4.5	4.0
Coal	30	374	551	757	991	82	87	84	80	78	8.4	5.0	4.0	3.5
Oil	4	17	18	18	15	11	4	3	2	1	4.5	1.0	0.6	-0.3
Gas	0	5	14	40	63	0	1	2	4	5	-	14.3	12.5	9.6
Nuclear	0	7	21	47	73	0	2	3	5	6	-	15.9	11.6	9.0
Hydro	3	25	33	50	63	7	6	5	5	5	7.6	3.6	3.9	3.4
Biomass and Waste	0	1	14	33	51	0	0	2	3	4	-	33.7	19.5	13.9
Other Renewables	0	0	2	5	11	0	0	0	1	1	-	-	-	-
Other Transformation, Own Use and Losses	28	175	232	268	297						6.1	3.6	2.4	1.9
of which electricity	2	32	49	72	97						*9.2*	*5.5*	*4.6*	*4.0*
Total Final Consumption	352	823	1 016	1 262	1 528	100	100	100	100	100	2.8	2.7	2.4	2.2
Coal	139	240	248	253	254	39	29	24	20	17	1.8	0.4	0.3	0.2
Oil	38	204	301	426	563	11	25	30	34	37	5.6	5.0	4.2	3.7
Gas	1	22	33	52	77	0	3	3	4	5	9.2	5.2	4.9	4.6
Electricity	10	112	179	273	382	3	14	18	22	25	8.0	6.0	5.1	4.5
Heat	0	30	39	50	60	0	4	4	4	4	-	3.5	2.9	2.5
Biomass and Waste	164	215	213	203	184	47	26	21	16	12	0.9	-0.1	-0.3	-0.5
Other Renewables	0	0	2	4	9	0.0	0.0	0.2	0.3	0.6	-	-	-	-

Industry	104	328	399	478	550	100	100	100	100	100	**3.8**	**2.5**	**2.1**	**1.9**
Coal	84	164	178	190	197	81	50	45	40	36	2.2	1.0	0.8	0.7
Oil	18	58	72	86	98	18	18	18	18	18	3.8	2.6	2.2	1.9
Gas	1	13	18	25	31	1	4	5	5	6	7.7	3.9	3.4	3.0
Electricity	0	70	103	144	186	0	21	26	30	34	20.7	4.9	4.1	3.6
Heat	0	22	28	33	38	0	7	7	7	7	-	2.9	2.3	1.9
Renewables	0	0	0	0	0	0	0	0	0	0	-	-	-	-
Transport	14	87	136	212	306	100	100	100	100	100	**6.0**	**5.8**	**5.1**	**4.6**
Oil	8	80	129	204	296	55	92	95	96	97	7.7	6.2	5.4	4.8
Other Fuels	6	7	7	8	10	45	8	5	4	3	0.3	0.5	0.9	1.2
Other Sectors	218	372	432	514	608	100	100	100	100	100	**1.7**	**1.9**	**1.8**	**1.8**
Coal	45	58	54	49	43	21	16	12	9	7	0.8	-1.0	-1.0	-1.0
Oil	9	50	74	103	132	4	13	17	20	22	5.7	5.0	4.1	3.5
Gas	0	8	15	27	45	0	2	3	5	7	16.9	7.0	6.6	6.2
Electricity	0	34	64	113	175	0	9	15	22	29	17.2	8.4	6.9	6.0
Heat	0	7	11	16	21	0	2	2	3	3	-	5.4	4.6	4.0
Renewables	164	215	215	207	192	75	58	50	40	31	0.9	0.0	-0.2	-0.4
Non-Energy Use	16	36	49	59	64						**2.7**	**3.8**	**2.7**	**2.1**
Electricity Generation (TWh)	144	1 675	2 653	4 018	5 573	100	100	100	100	100	**8.2**	**5.9**	**5.0**	**4.4**
Coal	98	1 293	2 030	2 910	4 035	68	77	77	72	72	8.7	5.8	4.6	4.1
Oil	16	50	59	65	53	11	3	2	2	1	3.6	2.3	1.5	0.3
Gas	0	17	55	196	315	0	1	2	5	6	-	15.9	14.6	11.0
Nuclear	0	25	82	180	280	0	2	3	4	5	-	15.9	11.6	9.0
Hydro	30	288	383	578	734	21	17	14	14	13	7.6	3.6	3.9	3.4
Biomass and Waste	0	2	31	58	84	0	0	1	1	1	-	37.2	19.2	13.5
Other Renewables	0	0	13	31	72	0	0	0	1	1	-	-	-	-

Traditional Biomass (included above)	164	215	213	203	183	41	17	13	10	7	0.9	-0.1	-0.3	-0.6
Total Primary Energy Supply (excluding traditional biomass)	241	1 028	1 409	1 869	2 356						**4.8**	**4.0**	**3.4**	**3.0**

Annex A - Tables for Reference Scenario Projections

Reference Scenario: China

	2002	Capacity (GW) 2010	2020	2030	Shares (%) 2002	2010	2020	2030	Growth Rates (% per annum) 2002-2010	2002-2020	2002-2030
Total Capacity	360	565	855	1 187	100	100	100	100	**5.8**	**4.9**	**4.4**
Coal	247	394	560	776	69	70	65	65	6.0	4.7	4.2
Oil	17	20	21	17	5	3	2	1	2.1	1.3	0.1
Gas	8	23	67	111	2	4	8	9	13.1	12.2	9.6
Nuclear	4	10	22	35	1	2	3	3	11.0	9.5	7.7
Hydro	82	109	165	210	23	19	19	18	3.6	4.0	3.4
of which pumped storage	0	0	0	0	0	0	0	0	-	-	-
Renewables (excluding hydro)	2	10	20	38	0	2	2	3	23.4	14.2	11.5

Renewables Breakdown

	2002	Capacity (GW) 2010	2020	2030	Shares (%) 2002	2010	2020	2030	Growth Rates (% per annum) 2002-2010	2002-2020	2002-2030
Renewables (excluding hydro)	2	10	20	38	100	100	100	100	**23.4**	**14.2**	**11.5**
Biomass and Waste	1	6	10	14	54	57	49	36	24.3	13.6	9.9
Wind	1	4	10	22	43	40	48	58	22.3	15.0	12.7
Geothermal	0	0	0	1	3	2	3	3	21.2	13.6	11.1
Solar	0	0	0	1	0	0	0	3	0.0	0.0	39.1
Tide/Wave	0	0	0	0	0	0	0	0	-	-	-

	2002	Electricity Generation (TWh) 2010	2020	2030	Shares (%) 2002	2010	2020	2030	Growth Rates (% per annum) 2002-2010	2002-2020	2002-2030
Renewables (excluding hydro)	2	43	89	155	100	100	100	100	**33.3**	**19.7**	**14.9**
Biomass and Waste	2	31	58	84	100	71	65	54	28.8	17.1	12.5
Wind	0	11	28	63	0	25	31	41	-	-	-
Geothermal	0	2	3	7	0	4	4	4	-	-	-
Solar	0	0	0	2	0	0	0	1	-	-	-
Tide/Wave	0	0	0	0	0	0	0	0	-	-	-

Reference Scenario: China

	CO$_2$ Emissions (Mt)					Shares (%)					Growth Rates (% per annum)			
	1971	2002	2010	2020	2030	1971	2002	2010	2020	2030	1971-2002	2002-2010	2002-2020	2002-2030
Total CO$_2$ Emissions	**809**	**3 307**	**4 386**	**5 708**	**7 144**	**100**	**100**	**100**	**100**	**100**	**4.6**	**3.6**	**3.1**	**2.8**
change since 1990 (%)		*44.5*	*91.6*	*149.3*	*212.1*									
Coal	678	2 621	3 381	4 243	5 194	84	79	77	74	73	4.5	3.2	2.7	2.5
Oil	124	618	883	1 233	1 610	15	19	20	22	23	5.3	4.6	3.9	3.5
Gas	7	69	122	231	340	1	2	3	4	5	7.5	7.4	6.9	5.9
Power Generation and Heat Plants	**132**	**1 576**	**2 316**	**3 210**	**4 195**	**100**	**100**	**100**	**100**	**100**	**8.3**	**4.9**	**4.0**	**3.6**
Coal	118	1 507	2 220	3 049	3 990	90	96	96	95	95	8.6	5.0	4.0	3.5
Oil	13	57	62	64	53	10	4	3	2	1	4.8	1.0	0.6	-0.3
Gas	0	12	34	98	152	0	1	1	3	4	-	14.3	12.5	9.6
Transformation, Own Use and Losses	**6**	**140**	**170**	**204**	**232**	**100**	**100**	**100**	**100**	**100**	**10.7**	**2.4**	**2.1**	**1.8**
Total Final Consumption	**671**	**1 591**	**1 900**	**2 294**	**2 717**	**100**	**100**	**100**	**100**	**100**	**2.8**	**2.2**	**2.1**	**1.9**
Coal	559	1 045	1 086	1 116	1 125	83	66	57	49	41	2.0	0.5	0.4	0.3
Oil	109	507	755	1 086	1 458	16	32	40	47	54	5.1	5.1	4.3	3.8
Gas	3	40	59	91	134	0	2	3	4	5	8.3	5.0	4.8	4.5
Industry	**406**	**885**	**984**	**1 080**	**1 149**	**100**	**100**	**100**	**100**	**100**	**2.5**	**1.3**	**1.1**	**0.9**
Coal	346	741	804	858	892	85	84	82	79	78	2.5	1.0	0.8	0.7
Oil	57	119	146	175	199	14	13	15	16	17	2.4	2.6	2.2	1.9
Gas	3	25	34	47	58	1	3	3	4	5	6.9	3.9	3.4	3.0
Transport	**47**	**244**	**383**	**592**	**852**	**100**	**100**	**100**	**100**	**100**	**5.5**	**5.8**	**5.0**	**4.6**
Oil	22	223	361	570	828	47	91	94	96	97	7.7	6.2	5.4	4.8
Other Fuels	25	21	22	23	24	53	9	6	4	3	-0.5	0.2	0.3	0.4
Other Sectors	**216**	**405**	**466**	**552**	**647**	**100**	**100**	**100**	**100**	**100**	**2.1**	**1.8**	**1.7**	**1.7**
Coal	189	245	228	208	187	87	61	49	38	29	0.8	-0.9	-0.9	-1.0
Oil	27	146	215	301	387	12	36	46	55	60	5.6	5.0	4.1	3.5
Gas	0	14	23	43	73	0	3	5	8	11	16.9	7.0	6.6	6.2
Non-Energy Use	**3**	**57**	**66**	**70**	**70**	**100**	**100**	**100**	**100**	**100**	**10.3**	**1.8**	**1.1**	**0.7**

Annex A - Tables for Reference Scenario Projections

Reference Scenario: East Asia

	Energy Demand (Mtoe)					Shares (%)					Growth Rates (% per annum)			
	1971	2002	2010	2020	2030	1971	2002	2010	2020	2030	1971-2002	2002-2010	2002-2020	2002-2030
Total Primary Energy Supply	159	533	712	955	1 188	100	100	100	100	100	**4.0**	**3.7**	**3.3**	**2.9**
Coal	21	85	122	192	272	14	16	17	20	23	4.5	4.7	4.7	4.3
Oil	47	215	285	382	467	30	40	40	40	39	5.0	3.6	3.3	2.8
Gas	2	93	139	197	241	1	17	20	21	20	14.1	5.2	4.3	3.5
Nuclear	0	9	14	16	18	0	2	2	2	2	-	5.7	3.1	2.4
Hydro	2	7	11	13	17	1	1	2	1	1	4.6	5.3	3.5	3.1
Biomass and Waste	87	107	118	123	132	55	20	17	13	11	0.7	1.2	0.8	0.8
Other Renewables	0	17	22	32	41	0	3	3	3	3	-	3.5	3.5	3.2
Power Generation and Heat Plants	13	153	224	334	447	100	100	100	100	100	**8.4**	**4.9**	**4.4**	**3.9**
Coal	3	44	74	138	214	21	29	33	41	48	9.5	6.6	6.5	5.8
Oil	8	28	32	35	31	62	18	14	10	7	4.1	1.9	1.3	0.5
Gas	0	45	67	96	113	3	30	30	29	25	17.1	5.0	4.3	3.3
Nuclear	0	9	14	16	18	0	6	6	5	4	-	5.7	3.1	2.4
Hydro	2	7	11	13	17	14	5	5	4	4	4.6	5.3	3.5	3.1
Biomass and Waste	0	2	4	4	13	0	1	2	1	3	-	7.8	3.9	6.8
Other Renewables	0	17	22	32	41	0	11	10	9	9	-	3.4	3.4	3.1
Other Transformation, Own Use and Losses	31	68	96	125	150						**2.5**	**4.4**	**3.4**	**2.9**
of which electricity	*1*	*9*	*13*	*19*	*26*						*8.3*	*4.8*	*4.4*	*3.8*
Total Final Consumption	120	369	476	625	766	100	100	100	100	100	**3.7**	**3.3**	**3.0**	**2.6**
Coal	17	37	40	44	47	15	10	8	7	6	2.4	1.1	1.0	0.9
Oil	34	169	233	323	409	29	46	49	52	53	5.3	4.1	3.7	3.2
Gas	1	21	31	44	57	1	6	6	7	7	10.4	5.0	4.2	3.7
Electricity	4	47	71	109	149	4	13	15	17	19	8.0	5.3	4.7	4.2
Heat	0	0	0	0	0	0	0	0	0	0	-	-	-	-
Biomass and Waste	63	95	101	104	103	52	26	21	17	13	1.4	0.8	0.5	0.3
Other Renewables	0	0	0	0	1	0.0	0.0	0.0	0.0	0.1	-	-	-	-

Industry	**27**	**127**	**161**	**206**	**251**	**100**	**100**	**100**	**100**	**100**	**5.2**	**3.0**	**2.7**	**2.5**
Coal	13	33	36	40	44	49	26	23	20	17	3.0	1.4	1.2	1.1
Oil	8	45	55	67	78	30	35	34	32	31	5.7	2.6	2.2	2.0
Gas	1	18	26	38	49	3	14	16	18	19	10.7	5.1	4.3	3.7
Electricity	2	22	31	46	62	8	17	19	22	25	7.8	4.5	4.2	3.8
Heat	0	0	0	0	0	0	0	0	0	0	-	-	-	-
Renewables	3	10	12	15	18	10	8	8	7	7	4.3	2.5	2.3	2.1
Transport	**16**	**91**	**137**	**205**	**269**	**100**	**100**	**100**	**100**	**100**	**5.8**	**5.2**	**4.6**	**3.9**
Oil	16	91	136	204	267	99	100	100	100	99	5.8	5.2	4.6	3.9
Other Fuels	0	0	0	1	2	1	0	0	0	1	1.4	11.1	9.3	7.9
Other Sectors	**65**	**136**	**162**	**196**	**227**	**100**	**100**	**100**	**100**	**100**	**2.4**	**2.3**	**2.1**	**1.9**
Coal	0	1	1	1	1	0	1	1	0	0	6.4	1.3	1.3	1.2
Oil	6	27	34	43	52	9	20	21	22	23	5.1	2.7	2.6	2.4
Gas	0	3	5	6	8	0	2	3	3	4	14.1	4.9	4.0	3.5
Electricity	1	23	38	61	85	1	17	23	31	37	11.3	6.3	5.5	4.7
Heat	0	0	0	0	0	0	0	0	0	0	-	-	-	-
Renewables	58	81	85	85	81	89	60	53	44	36	1.1	0.6	0.3	0.0
Non-Energy Use	**12**	**15**	**16**	**17**	**19**						**0.7**	**0.9**	**0.9**	**0.8**
Electricity Generation (TWh)	**59**	**653**	**980**	**1 490**	**2 032**	**100**	**100**	**100**	**100**	**100**	**8.1**	**5.2**	**4.7**	**4.1**
Coal	9	183	313	610	1 002	16	28	32	41	49	10.1	7.0	6.9	6.3
Oil	29	117	138	153	139	49	18	14	10	7	4.6	2.1	1.5	0.6
Gas	0	206	306	452	532	0	32	31	30	26	23.7	5.0	4.5	3.4
Nuclear	0	35	55	62	69	0	5	6	4	3	-	5.7	3.1	2.4
Hydro	21	83	126	155	193	35	13	13	10	10	4.6	5.3	3.5	3.1
Biomass and Waste	0	8	11	11	32	0	1	1	1	2	-	4.1	1.9	5.0
Other Renewables	0	20	31	47	66	0	3	3	3	3	-	5.4	4.8	4.3
Traditional Biomass (included above)	60	85	88	88	83	38	16	12	9	7	1.1	0.5	0.2	-0.1
Total Primary Energy Supply (excluding traditional biomass)	**99**	**448**	**623**	**867**	**1 105**						**5.0**	**4.2**	**3.7**	**3.3**

Annex A - Tables for Reference Scenario Projections

Reference Scenario: East Asia

	Capacity (GW)				Shares (%)				Growth Rates (% per annum)		
	2002	2010	2020	2030	2002	2010	2020	2030	2002-2010	2002-2020	2002-2030
Total Capacity	184	267	399	546	100	100	100	100	**4.7**	**4.4**	**4.0**
Coal	38	69	134	221	20	26	34	41	8.0	7.3	6.5
Oil	40	46	52	50	22	17	13	9	1.8	1.4	0.7
Gas	64	90	137	175	35	34	34	32	4.2	4.3	3.6
Nuclear	5	8	9	9	3	3	2	2	5.7	3.1	2.4
Hydro	32	46	57	71	17	17	14	13	4.7	3.2	2.9
of which pumped storage	0	0	0	0	0	0	0	0	-	-	-
Renewables (excluding hydro)	5	7	11	19	3	3	3	3	5.4	4.6	4.9

Renewables Breakdown

	Capacity (GW)				Shares (%)				Growth Rates (% per annum)		
	2002	2010	2020	2030	2002	2010	2020	2030	2002-2010	2002-2020	2002-2030
Renewables (excluding hydro)	5	7	11	19	100	100	100	100	**5.4**	**4.6**	**4.9**
Biomass and Waste	2	2	2	5	34	24	17	27	1.2	0.6	4.0
Wind	0	2	4	7	4	26	37	35	31.4	17.8	13.0
Geothermal	3	4	5	7	62	50	47	35	2.6	3.0	2.8
Solar	0	0	0	1	0	0	0	3	-	-	-
Tide/Wave	0	0	0	0	0	0	0	0	-	-	-

	Electricity Generation (TWh)				Shares (%)				Growth Rates (% per annum)		
	2002	2010	2020	2030	2002	2010	2020	2030	2002-2010	2002-2020	2002-2030
Renewables (excluding hydro)	28	42	58	97	100	100	100	100	**4.0**	**3.7**	**4.2**
Biomass and Waste	8	11	11	32	28	26	19	33	3.3	1.7	4.7
Wind	0	6	12	19	2	13	20	20	28.5	17.7	13.3
Geothermal	20	26	35	45	70	61	61	47	2.5	2.9	2.8
Solar	0	0	0	1	0	0	0	1	-	-	-
Tide/Wave	0	0	0	0	0	0	0	0	-	-	-

Reference Scenario: East Asia

	CO$_2$ Emissions (Mt)					Shares (%)					Growth Rates (% per annum)			
	1971	2002	2010	2020	2030	1971	2002	2010	2020	2030	1971-2002	2002-2010	2002-2020	2002-2030
Total CO$_2$ Emissions	**207**	**1 055**	**1 459**	**2 092**	**2 701**	**100**	**100**	**100**	**100**	**100**	**5.4**	**4.1**	**3.9**	**3.4**
change since 1990 (%)		*79.0*	*147.5*	*254.9*	*358.3*									
Coal	82	323	454	724	1 035	40	31	31	35	38	4.5	4.4	4.6	4.2
Oil	121	538	718	963	1 173	59	51	49	46	43	4.9	3.7	3.3	2.8
Gas	3	194	286	405	493	1	18	20	19	18	14.3	5.0	4.2	3.4
Power Generation and Heat Plants	**36**	**368**	**549**	**880**	**1 208**	**100**	**100**	**100**	**100**	**100**	**7.8**	**5.1**	**5.0**	**4.3**
Coal	10	175	291	545	845	28	48	53	62	70	9.6	6.6	6.5	5.8
Oil	25	87	101	111	100	69	24	18	13	8	4.1	1.9	1.3	0.5
Gas	1	106	157	224	264	2	29	29	25	22	17.1	5.0	4.3	3.3
Transformation, Own Use and Losses	**7**	**72**	**98**	**128**	**152**						**7.8**	**4.0**	**3.2**	**2.7**
Total Final Consumption	**164**	**616**	**812**	**1 084**	**1 341**	**100**	**100**	**100**	**100**	**100**	**4.4**	**3.5**	**3.2**	**2.8**
Coal	72	147	161	176	187	44	24	20	16	14	2.3	1.1	1.0	0.9
Oil	90	426	588	819	1 038	55	69	72	76	77	5.2	4.1	3.7	3.2
Gas	2	42	62	89	116	1	7	8	8	9	10.2	5.0	4.3	3.7
Industry	**87**	**271**	**327**	**394**	**456**	**100**	**100**	**100**	**100**	**100**	**3.7**	**2.4**	**2.1**	**1.9**
Coal	57	130	145	162	175	65	48	44	41	38	2.7	1.4	1.2	1.1
Oil	29	106	130	158	185	33	39	40	40	41	4.3	2.6	2.2	2.0
Gas	2	34	51	74	96	2	13	16	19	21	10.5	5.1	4.4	3.8
Transport	**35**	**236**	**354**	**530**	**693**	**100**	**100**	**100**	**100**	**100**	**6.3**	**5.2**	**4.6**	**3.9**
Oil	35	236	354	530	693	99	100	100	100	100	6.4	5.2	4.6	3.9
Other Fuels	0	0	0	0	0	1	0	0	0	0	-4.4	2.8	2.2	1.8
Other Sectors	**39**	**107**	**128**	**157**	**187**	**100**	**100**	**100**	**100**	**100**	**3.3**	**2.3**	**2.2**	**2.0**
Coal	14	16	15	13	11	37	15	12	8	6	0.4	-1.2	-1.2	-1.3
Oil	24	83	102	129	157	62	78	80	82	84	4.1	2.7	2.5	2.3
Gas	1	8	11	15	19	1	7	9	10	10	9.0	4.7	3.9	3.4
Non-Energy Use	**2**	**2**	**3**	**4**	**4**	**100**	**100**	**100**	**100**	**100**	**0.5**	**2.6**	**2.4**	**2.1**

Reference Scenario: Indonesia

	Energy Demand (Mtoe)					Shares (%)					Growth Rates (% per annum)			
	1971	2002	2010	2020	2030	1971	2002	2010	2020	2030	1971-2002	2002-2010	2002-2020	2002-2030
Total Primary Energy Supply	35	156	204	269	333	100	100	100	100	100	5.0	3.4	3.1	2.7
Coal	0	18	25	41	63	0	12	12	15	19	17.2	4.0	4.7	4.6
Oil	8	56	77	103	126	24	36	38	38	38	6.3	4.0	3.4	2.9
Gas	0	33	49	70	86	1	21	24	26	26	17.2	5.0	4.2	3.5
Nuclear	0	0	0	0	0	0	0	0	0	0	-	-	-	-
Hydro	0	1	2	2	2	0	1	1	1	1	-	-	-	-
Biomass and Waste	26	41	45	44	43	73	26	22	16	13	6.5	7.5	4.2	3.5
Other Renewables	0	7	7	10	14	1	4	4	4	4	1.6	1.1	0.3	0.2
											9.3	0.8	2.3	2.4
Power Generation and Heat Plants	1	29	44	71	102	100	100	100	100	100	10.7	5.2	5.0	4.5
Coal	0	12	18	33	54	0	42	41	47	53	-	4.8	5.7	5.4
Oil	1	6	7	7	7	90	20	16	11	7	5.4	2.8	1.4	0.6
Gas	0	5	10	18	22	0	17	23	25	21	-	9.2	7.3	5.4
Nuclear	0	0	0	0	0	0	0	0	0	0	-	-	-	-
Hydro	0	1	2	2	2	10	3	3	3	2	6.5	7.5	4.2	3.5
Biomass and Waste	0	0	1	1	5	0	0	3	2	5	-	-	-	-
Other Renewables	0	5	6	9	12	0	18	13	12	12	-	0.9	2.8	2.9
Other Transformation, Own Use and Losses	3	22	31	40	48						6.7	4.2	3.4	2.8
of which electricity	*0*	*2*	*3*	*4*	*6*						*12.3*	*5.3*	*4.9*	*4.4*
Total Final Consumption	31	114	145	184	220	100	100	100	100	100	4.3	3.0	2.7	2.4
Coal	0	6	7	8	9	0	5	5	4	4	14.2	1.8	1.7	1.6
Oil	6	47	64	88	111	21	41	44	48	50	6.6	4.0	3.6	3.1
Gas	0	13	18	25	32	0	11	13	13	14	17.0	4.4	3.7	3.3
Electricity	0	7	12	21	31	0	7	9	11	14	13.4	6.6	5.9	5.2
Heat	0	0	0	0	0	0	0	0	0	0	-	-	-	-
Biomass and Waste	24	41	44	42	38	78	36	30	23	17	1.8	0.7	0.1	-0.3
Other Renewables	0	0	0	0	0	0.0	0.0	0.0	0.0	0.1	-	-	-	-

Industry	2	31	40	52	64	100	100	100	100	100	9.9	3.2	2.9	2.6
Coal	0	6	7	8	9	4	18	16	15	14	15.7	1.8	1.7	1.6
Oil	1	10	12	14	16	89	31	29	27	25	6.3	2.5	2.1	1.9
Gas	0	11	15	20	26	6	34	37	38	40	16.3	4.2	3.6	3.2
Electricity	0	3	5	8	11	2	10	12	15	17	16.7	5.4	5.2	4.6
Heat	0	0	0	0	0	0	0	0	0	0	-	-	-	-
Renewables	0	2	2	2	3	0	6	5	5	4	-	1.0	1.0	1.0
Transport	3	24	36	53	67	100	100	100	100	100	7.2	5.2	4.5	3.8
Oil	3	24	36	52	67	99	100	100	100	99	7.2	5.2	4.5	3.8
Other Fuels	0	0	0	0	0	1	0	0	0	1	0.0	14.6	11.1	9.1
Other Sectors	26	59	68	78	87	100	100	100	100	100	2.6	1.9	1.6	1.4
Coal	0	0	0	0	0	0	0	0	0	0	-	1.5	1.6	1.4
Oil	2	13	16	21	26	8	22	23	26	30	5.9	2.7	2.7	2.6
Gas	0	2	3	5	6	0	4	5	6	7	-	5.1	4.1	3.5
Electricity	0	4	8	13	20	0	7	11	17	23	12.1	7.5	6.3	5.5
Heat	0	0	0	0	0	0	0	0	0	0	-	-	-	-
Renewables	24	39	41	40	35	91	67	61	51	40	1.6	0.7	0.1	-0.4
Non-Energy Use	0	1	1	1	2						7.4	3.0	2.8	2.5
Electricity Generation (TWh)	2	108	177	294	428	100	100	100	100	100	13.1	6.4	5.7	5.0
Coal	0	43	67	135	237	0	40	38	46	55	-	5.8	6.6	6.3
Oil	1	25	33	34	31	40	23	18	12	7	11.1	3.3	1.7	0.8
Gas	0	24	48	88	106	0	22	27	30	25	-	9.2	7.5	5.5
Nuclear	0	0	0	0	0	0	0	0	0	0	-	-	-	-
Hydro	1	10	18	21	26	60	9	10	7	6	6.5	7.5	4.2	3.5
Biomass and Waste	0	0	3	3	9	0	0	2	1	2	-	-	-	-
Other Renewables	0	6	8	13	19	0	6	5	4	4	-	3.3	4.2	4.0
Traditional Biomass (included above)	24	39	41	40	35	69	25	20	15	10	1.6	0.7	0.1	-0.4
Total Primary Energy Supply (excluding traditional biomass)	11	117	163	229	298						8.0	4.2	3.8	3.4

Reference Scenario: Indonesia

	Capacity (GW)				Shares (%)				Growth Rates (% per annum)		
	2002	2010	2020	2030	2002	2010	2020	2030	2002-2010	2002-2020	2002-2030
Total Capacity	37	50	75	109	100	100	100	100	3.8	4.0	3.9
Coal	6	10	21	41	16	20	28	37	6.7	7.2	7.1
Oil	15	16	18	18	41	33	24	17	0.9	1.0	0.7
Gas	10	15	26	36	27	31	35	33	5.6	5.5	4.7
Nuclear	0	0	0	0	0	0	0	0	-	-	-
Hydro	4	6	7	9	12	12	9	8	4.1	2.7	2.5
of which pumped storage	0	0	0	0	0	0	0	0	-	-	-
Renewables (excluding hydro)	1	2	3	5	4	4	4	5	4.2	4.4	4.8

Renewables Breakdown

	Capacity (GW)				Shares (%)				Growth Rates (% per annum)		
	2002	2010	2020	2030	2002	2010	2020	2030	2002-2010	2002-2020	2002-2030
Renewables (excluding hydro)	1	2	3	5	100	100	100	100	4.2	4.4	4.8
Biomass and Waste	0	0	1	1	23	25	17	26	5.2	2.5	5.2
Wind	0	1	1	2	4	26	36	34	31.6	17.9	13.1
Geothermal	1	1	1	2	73	48	47	36	-0.9	1.8	2.2
Solar	0	0	0	0	0	0	0	4	-	-	-
Tide/Wave	0	0	0	0	0	0	0	0	-	-	-

	Electricity Generation (TWh)				Shares (%)				Growth Rates (% per annum)		
	2002	2010	2020	2030	2002	2010	2020	2030	2002-2010	2002-2020	2002-2030
Renewables (excluding hydro)	6	11	16	28	100	100	100	100	6.0	4.9	5.1
Biomass and Waste	0	3	3	9	0	27	19	32	-	-	-
Wind	0	2	3	5	0	13	20	19	-	-	-
Geothermal	6	7	10	13	100	59	61	48	0.5	2.3	2.5
Solar	0	0	0	0	0	0	0	1	-	-	-
Tide/Wave	0	0	0	0	0	0	0	0	-	-	-

Reference Scenario: Indonesia

	CO$_2$ Emissions (Mt)					Shares (%)					Growth Rates (% per annum)			
	1971	2002	2010	2020	2030	1971	2002	2010	2020	2030	1971-2002	2002-2010	2002-2020	2002-2030
Total CO$_2$ Emissions	25	303	417	599	783	100	100	100	100	100	**8.4**	**4.1**	**3.9**	**3.4**
change since 1990 (%)		*118.5*	*200.8*	*332.1*	*464.5*									
Coal	1	70	95	160	243	2	23	23	27	31	17.2	4.0	4.7	4.6
Oil	24	165	221	295	362	97	54	53	49	46	6.4	3.7	3.3	2.8
Gas	0	68	101	145	179	1	22	24	24	23	18.7	5.0	4.3	3.5
Power Generation and Heat Plants	4	78	116	194	281	100	100	100	100	100	**10.4**	**5.2**	**5.2**	**4.7**
Coal	0	48	70	130	209	0	62	60	67	74	-	4.8	5.7	5.4
Oil	4	18	23	23	22	100	23	19	12	8	5.3	2.8	1.4	0.6
Gas	0	12	24	41	50	0	15	20	21	18	-	9.2	7.3	5.4
Transformation, Own Use and Losses	2	41	53	69	81	100	100	100	100	100	**10.4**	**3.2**	**2.9**	**2.5**
Total Final Consumption	20	185	249	336	421	100	100	100	100	100	**7.5**	**3.8**	**3.4**	**3.0**
Coal	0	22	26	30	34	2	12	10	9	8	14.0	1.8	1.7	1.6
Oil	19	138	188	258	325	97	75	76	77	77	6.6	4.0	3.6	3.1
Gas	0	25	35	48	62	1	14	14	14	15	16.3	4.4	3.7	3.3
Industry	5	72	90	111	132	100	100	100	100	100	**9.0**	**2.8**	**2.4**	**2.2**
Coal	0	22	26	30	34	5	31	28	27	26	15.4	1.8	1.7	1.6
Oil	4	30	37	44	50	90	42	41	40	38	6.4	2.5	2.1	1.9
Gas	0	20	27	37	48	5	27	31	33	36	15.4	4.2	3.6	3.2
Transport	8	69	103	152	195	100	100	100	100	100	**7.2**	**5.2**	**4.5**	**3.8**
Oil	8	69	103	152	195	99	100	100	100	100	7.2	5.2	4.5	3.8
Other Fuels	0	0	0	0	0	1	0	0	0	0	-1.6	4.0	3.1	2.5
Other Sectors	6	43	55	72	92	100	100	100	100	100	**6.3**	**3.0**	**2.9**	**2.7**
Coal	0	0	0	0	0	0	0	0	0	0	-	1.5	1.6	1.4
Oil	6	38	47	61	79	100	88	86	85	85	5.9	2.7	2.7	2.6
Gas	0	5	8	11	14	0	12	14	15	15	-	5.1	4.1	3.5
Non-Energy Use	0	1	1	1	2	100	100	100	100	100	**3.7**	**3.0**	**2.8**	**2.5**

Reference Scenario: South Asia

	Energy Demand (Mtoe)					Shares (%)					Growth Rates (% per annum)			
	1971	2002	2010	2020	2030	1971	2002	2010	2020	2030	1971-2002	2002-2010	2002-2020	2002-2030
Total Primary Energy Supply	**211**	**644**	**797**	**1 024**	**1 283**	**100**	**100**	**100**	**100**	**100**	**3.7**	**2.7**	**2.6**	**2.5**
Coal	36	181	211	277	369	17	28	26	27	29	5.3	1.9	2.4	2.6
Oil	28	144	197	271	345	13	22	25	26	27	5.5	3.9	3.6	3.2
Gas	3	52	82	134	190	2	8	10	13	15	9.3	5.9	5.4	4.8
Nuclear	0	6	14	21	32	0	1	2	2	3	9.5	12.1	7.5	6.4
Hydro	3	8	15	21	24	1	1	2	2	2	3.5	7.6	5.2	4.0
Biomass and Waste	141	252	278	299	318	67	39	35	29	25	1.9	1.2	0.9	0.8
Other Renewables	0	0	1	2	4	0	0	0	0	0	-	21.5	13.0	10.5
Power Generation and Heat Plants	**14**	**180**	**247**	**361**	**513**	**100**	**100**	**100**	**100**	**100**	**8.5**	**4.0**	**3.9**	**3.8**
Coal	8	130	158	222	312	57	72	64	61	61	9.4	2.5	3.0	3.2
Oil	2	13	18	23	26	13	7	7	6	5	6.6	3.8	3.2	2.5
Gas	1	22	38	70	102	9	12	15	19	20	9.7	7.0	6.6	5.6
Nuclear	0	6	14	21	32	2	3	6	6	6	9.5	12.1	7.5	6.4
Hydro	3	8	15	21	24	20	5	6	6	5	3.5	7.6	5.2	4.0
Biomass and Waste	0	1	3	3	13	0	1	1	1	3	-	13.4	5.7	9.1
Other Renewables	0	0	1	2	3	0	0	0	1	1	-	20.4	12.3	9.7
Other Transformation, Own Use and Losses	**7**	**54**	**63**	**78**	**96**						**6.9**	**2.0**	**2.1**	**2.1**
of which electricity	*1*	*21*	*29*	*41*	*56*						*10.2*	*4.0*	*3.8*	*3.5*
Total Final Consumption	**196**	**470**	**578**	**726**	**880**	**100**	**100**	**100**	**100**	**100**	**2.9**	**2.6**	**2.4**	**2.3**
Coal	25	38	39	41	42	13	8	7	6	5	1.3	0.3	0.4	0.4
Oil	23	116	163	231	302	12	25	28	32	34	5.4	4.4	3.9	3.5
Gas	2	26	39	59	81	1	5	7	8	9	8.6	5.5	4.8	4.2
Electricity	5	39	61	99	150	3	8	11	14	17	6.7	5.7	5.3	4.9
Heat	0	0	0	0	0	0	0	0	0	0	-	-	-	-
Biomass and Waste	141	250	274	295	304	72	53	47	41	35	1.9	1.1	0.9	0.7
Other Renewables	0	0	0	0	1	0.0	0.0	0.0	0.0	0.1	-	-	-	-

Industry	37	125	161	218	275	100	100	100	100	100	4.0	3.3	3.2	2.9
Coal	12	29	31	34	36	32	24	19	16	13	3.1	0.8	0.8	0.8
Oil	5	32	46	73	98	14	26	29	34	36	6.2	4.6	4.7	4.1
Gas	2	20	30	43	58	5	16	19	20	21	7.9	5.2	4.4	3.9
Electricity	3	17	23	33	45	9	13	14	15	16	5.2	4.2	3.8	3.6
Heat	0	0	0	0	0	0	0	0	0	0	-	-	-	-
Renewables	15	26	30	34	37	40	21	19	16	14	1.9	1.7	1.5	1.3
Transport	17	47	64	91	127	100	100	100	100	100	3.4	3.8	3.8	3.6
Oil	9	46	62	89	122	53	98	98	98	97	5.5	3.8	3.7	3.5
Other Fuels	8	1	1	2	4	47	2	2	2	3	-7.4	6.9	5.8	6.7
Other Sectors	141	285	334	395	454	100	100	100	100	100	2.3	2.0	1.8	1.7
Coal	6	9	8	7	6	4	3	2	2	1	1.3	-1.7	-1.6	-1.5
Oil	6	26	38	51	62	5	9	12	13	14	4.6	5.1	3.8	3.2
Gas	0	6	9	15	21	0	2	3	4	5	13.7	6.5	5.7	4.9
Electricity	1	20	35	62	99	1	7	10	16	22	8.8	6.9	6.4	5.8
Heat	0	0	0	0	0	0	0	0	0	0	-	-	-	-
Renewables	126	224	244	261	266	90	79	73	66	59	1.9	1.1	0.8	0.6
Non-Energy Use	3	13	19	21	24						5.1	4.5	2.8	2.3
Electricity Generation (TWh)	71	703	1 047	1 633	2 394	100	100	100	100	100	7.7	5.1	4.8	4.5
Coal	26	425	546	839	1 302	37	60	52	51	54	9.4	3.2	3.8	4.1
Oil	6	51	73	98	109	9	7	7	6	5	7.0	4.7	3.7	2.8
Gas	4	106	184	353	525	6	15	18	22	22	11.0	7.2	6.9	5.9
Nuclear	1	22	54	80	123	2	3	5	5	5	9.5	12.1	7.5	6.4
Hydro	33	95	171	239	284	46	14	16	15	12	3.5	7.6	5.2	4.0
Biomass and Waste	0	2	10	10	24	0	1	1	1	1	-	21.2	8.9	9.1
Other Renewables	0	3	9	16	27	0	1	1	1	1	-	16.6	10.4	8.6
Traditional Biomass (included above)	126	224	244	260	265	60	35	31	25	21	1.9	1.1	0.8	0.6
Total Primary Energy Supply (excluding traditional biomass)	85	419	553	764	1 018						5.3	3.5	3.4	3.2

Reference Scenario: South Asia

	2002	Capacity (GW) 2010	2020	2030	2002	Shares (%) 2010	2020	2030	Growth Rates (% per annum) 2002-2010	2002-2020	2002-2030
Total Capacity	142	205	321	464	100	100	100	100	**4.7**	**4.6**	**4.3**
Coal	69	85	129	200	48	41	40	43	2.6	3.5	3.9
Oil	12	20	28	34	9	10	9	7	6.4	4.8	3.7
Gas	24	42	81	125	17	20	25	27	7.0	6.9	6.0
Nuclear	3	7	11	16	2	3	3	4	11.7	7.4	6.3
Hydro	32	46	65	77	22	23	20	17	4.9	4.1	3.2
of which pumped storage	0	0	0	0	0	0	0	0	-	-	-
Renewables (excluding hydro)	2	5	7	13	2	2	2	3	10.6	6.7	6.7

Renewables Breakdown

	2002	Capacity (GW) 2010	2020	2030	2002	Shares (%) 2010	2020	2030	Growth Rates (% per annum) 2002-2010	2002-2020	2002-2030
Renewables (excluding hydro)	2	5	7	13	100	100	100	100	**10.6**	**6.7**	**6.7**
Biomass and Waste	0	2	2	4	18	30	21	27	18.0	7.6	8.3
Wind	2	3	6	8	80	68	77	62	8.4	6.5	5.7
Geothermal	0	0	0	0	0	1	1	1	31.6	17.8	12.8
Solar	0	0	0	1	2	1	0	9	0.0	0.0	13.5
Tide/Wave	0	0	0	0	0	0	0	0	-	-	-

	2002	Electricity Generation (TWh) 2010	2020	2030	2002	Shares (%) 2010	2020	2030	Growth Rates (% per annum) 2002-2010	2002-2020	2002-2030
Renewables (excluding hydro)	5	19	25	51	100	100	100	100	**14.7**	**8.8**	**8.3**
Biomass and Waste	2	10	10	24	44	51	38	47	16.6	8.0	8.5
Wind	3	9	15	24	56	47	59	47	12.7	9.1	7.6
Geothermal	0	0	1	1	0	2	2	2	-	-	-
Solar	0	0	0	2	0	0	0	4	11.6	83.6	52.1
Tide/Wave	0	0	0	0	0	0	0	0	-	5.7	17.2

Reference Scenario: South Asia

	CO$_2$ Emissions (Mt)					Shares (%)					Growth Rates (% per annum)			
	1971	2002	2010	2020	2030	1971	2002	2010	2020	2030	1971-2002	2002-2010	2002-2020	2002-2030
Total CO$_2$ Emissions	**222**	**1 165**	**1 483**	**2 052**	**2 730**	**100**	**100**	**100**	**100**	**100**	**5.5**	**3.1**	**3.2**	**3.1**
change since 1990 (%)		73.3	120.7	205.4	306.3									
Coal	145	683	795	1 050	1 407	65	59	54	51	52	5.1	1.9	2.4	2.6
Oil	70	367	505	700	897	31	32	34	34	33	5.5	4.1	3.7	3.2
Gas	7	115	184	302	426	3	10	12	15	16	9.3	6.0	5.5	4.8
Power Generation and Heat Plants	**40**	**599**	**759**	**1 099**	**1 535**	**100**	**100**	**100**	**100**	**100**	**9.1**	**3.0**	**3.4**	**3.4**
Coal	32	506	614	863	1 214	78	85	81	79	79	9.4	2.5	3.0	3.2
Oil	6	41	55	72	82	14	7	7	7	5	6.5	3.9	3.2	2.5
Gas	3	52	89	163	239	7	9	12	15	16	9.7	7.0	6.6	5.6
Transformation, Own Use and Losses	**7**	**33**	**36**	**39**	**43**	**100**	**100**	**100**	**100**	**100**	**4.9**	**0.9**	**0.9**	**0.9**
Total Final Consumption	**174**	**533**	**689**	**915**	**1 153**	**100**	**100**	**100**	**100**	**100**	**3.7**	**3.3**	**3.0**	**2.8**
Coal	111	175	178	184	190	64	33	26	20	16	1.5	0.2	0.3	0.3
Oil	59	301	423	600	785	34	56	61	66	68	5.4	4.4	3.9	3.5
Gas	4	57	88	131	178	2	11	13	14	15	8.7	5.5	4.7	4.1
Industry	**74**	**259**	**324**	**428**	**529**	**100**	**100**	**100**	**100**	**100**	**4.1**	**2.8**	**2.8**	**2.6**
Coal	57	140	150	163	174	77	54	46	38	33	2.9	0.8	0.8	0.8
Oil	13	75	108	171	228	18	29	33	40	43	5.7	4.6	4.6	4.0
Gas	4	44	66	95	128	5	17	20	22	24	8.0	5.2	4.4	3.9
Transport	**53**	**130**	**174**	**250**	**342**	**100**	**100**	**100**	**100**	**100**	**3.0**	**3.7**	**3.7**	**3.5**
Oil	23	130	174	249	341	43	100	100	100	100	5.8	3.8	3.7	3.5
Other Fuels	30	0	0	0	0	57	0	0	0	0	-13.1	0.0	0.0	0.0
Other Sectors	**44**	**124**	**163**	**206**	**247**	**100**	**100**	**100**	**100**	**100**	**3.4**	**3.4**	**2.8**	**2.5**
Coal	24	35	28	21	16	54	28	17	10	7	1.2	-2.7	-2.7	-2.7
Oil	20	76	113	149	181	46	61	69	72	73	4.4	5.0	3.8	3.1
Gas	0	13	22	36	50	1	11	13	17	20	13.7	6.5	5.7	4.9
Non-Energy Use	**3**	**20**	**28**	**31**	**34**	**100**	**100**	**100**	**100**	**100**	**6.3**	**4.6**	**2.6**	**2.0**

Annex A - Tables for Reference Scenario Projections

Reference Scenario: India

	Energy Demand (Mtoe)					Shares (%)					Growth Rates (% per annum)			
	1971	2002	2010	2020	2030	1971	2002	2010	2020	2030	1971-2002	2002-2010	2002-2020	2002-2030
Total Primary Energy Supply	**182**	**538**	**656**	**829**	**1 026**	**100**	**100**	**100**	**100**	**100**	**3.6**	**2.5**	**2.4**	**2.3**
Coal	36	178	207	271	362	20	33	32	33	35	5.3	1.9	2.4	2.6
Oil	22	119	160	215	267	12	22	24	26	26	5.5	3.8	3.4	2.9
Gas	1	23	37	63	90	0	4	6	8	9	12.6	6.3	5.9	5.0
Nuclear	0	5	13	19	29	0	1	2	2	3	9.4	12.4	7.5	6.4
Hydro	2	5	11	16	18	1	1	2	2	2	2.7	9.3	6.0	4.3
Biomass and Waste	121	208	227	243	258	66	39	35	29	25	1.8	1.1	0.9	0.8
Other Renewables	0	0	1	2	3	0	0	0	0	0	-	21.3	12.8	10.2
Power Generation and Heat Plants	**13**	**160**	**213**	**309**	**438**	**100**	**100**	**100**	**100**	**100**	**8.5**	**3.7**	**3.7**	**3.7**
Coal	8	130	157	219	308	64	81	73	71	70	9.4	2.4	3.0	3.1
Oil	2	7	9	11	10	12	5	4	4	2	5.1	3.0	2.4	1.2
Gas	0	11	20	40	59	2	7	9	13	14	12.9	7.5	7.4	6.2
Nuclear	0	5	13	19	29	2	3	6	6	7	9.4	12.4	7.5	6.4
Hydro	2	5	11	16	18	19	3	5	5	4	2.7	9.3	6.0	4.3
Biomass and Waste	0	1	3	3	11	0	1	1	1	2	-	14.4	6.2	8.6
Other Renewables	0	0	1	2	3	0	0	0	1	1	-	20.4	12.2	9.6
Other Transformation, Own Use and Losses	**6**	**48**	**56**	**68**	**83**						**6.9**	**1.9**	**2.0**	**2.0**
of which electricity	1	19	25	35	48						10.7	3.8	3.6	3.5
Total Final Consumption	**169**	**382**	**462**	**569**	**678**	**100**	**100**	**100**	**100**	**100**	**2.7**	**2.4**	**2.2**	**2.1**
Coal	25	36	37	38	39	15	9	8	7	6	1.2	0.3	0.3	0.3
Oil	18	97	136	189	241	11	25	29	33	35	5.5	4.4	3.8	3.3
Gas	0	9	15	21	27	0	2	3	4	4	11.9	5.8	4.6	3.9
Electricity	5	33	51	82	124	3	9	11	14	18	6.5	5.5	5.2	4.9
Heat	0	0	0	0	0	0	0	0	0	0	-	-	-	-
Biomass and Waste	121	207	224	240	247	72	54	49	42	36	1.8	1.0	0.8	0.6
Other Renewables	0	0	0	0	0	0.0	0.0	0.0	0.0	0.1	-	-	-	-

Industry	33	103	130	173	215	100	100	100	100	100	100	**3.8**	**3.0**	**3.0**	**3.0**	**2.7**
Coal	11	27	29	31	33	34	26	22	18	15	2.9	0.9	0.8	0.7
Oil	5	30	44	69	93	14	29	34	40	43	6.3	4.8	4.7	4.1
Gas	0	9	13	18	22	1	8	10	10	10	11.9	5.4	4.2	3.5
Electricity	3	14	19	26	36	10	14	15	15	17	5.0	3.8	3.5	3.3
Heat	0	0	0	0	0	0	0	0	0	0	-	-	-	-
Renewables	14	23	25	28	31	42	22	19	16	14	1.6	1.4	1.2	1.1
Transport	15	34	44	58	76	100	100	100	100	100	**2.8**	**3.1**	**3.0**	**2.9**
Oil	7	33	42	56	72	46	98	97	97	94	5.3	3.0	2.9	2.8
Other Fuels	8	1	1	2	4	54	2	3	3	6	-7.4	6.9	5.8	6.7
Other Sectors	119	233	271	318	364	100	100	100	100	100	**2.2**	**1.9**	**1.7**	**1.6**
Coal	6	9	8	7	6	5	4	3	2	2	1.3	-1.7	-1.6	-1.5
Oil	5	23	35	47	58	4	10	13	15	16	5.1	5.5	4.1	3.4
Gas	0	1	1	2	3	0	0	0	1	1	12.7	9.4	7.6	5.6
Electricity	1	16	28	51	83	1	7	10	16	23	8.7	6.9	6.5	5.9
Heat	0	0	0	0	0	0	0	0	0	0	-	-	-	-
Renewables	107	184	199	211	215	90	79	73	66	59	1.8	1.0	0.8	0.6
Non-Energy Use	2	12	17	19	22						**5.5**	**4.7**	**2.8**	**2.3**
Electricity Generation (TWh)	61	598	878	1 362	2 004	100	100	100	100	100	**7.6**	**4.9**	**4.7**	**4.4**
Coal	26	424	540	828	1 285	43	71	62	61	64	9.4	3.1	3.8	4.0
Oil	5	27	38	48	44	9	4	4	4	2	5.3	4.5	3.2	1.7
Gas	0	60	103	210	312	1	10	12	15	16	17.9	7.1	7.2	6.1
Nuclear	1	19	49	71	110	2	3	6	5	5	9.4	12.4	7.5	6.4
Hydro	28	64	130	182	211	46	11	15	13	11	2.7	9.3	6.0	4.3
Biomass and Waste	0	2	9	9	20	0	0	1	1	1	-	22.5	9.4	8.9
Other Renewables	0	2	8	13	23	0	0	1	1	1	-	16.1	10.1	8.3
Traditional Biomass (included above)	107	184	199	211	215	59	34	30	25	21	1.8	1.0	0.8	0.5
Total Primary Energy Supply (excluding traditional biomass)	75	354	457	617	812						**5.1**	**3.2**	**3.1**	**3.0**

Reference Scenario: India

	2002	Capacity (GW) 2010	2020	2030	2002	Shares (%) 2010	2020	2030	Growth Rates (% per annum) 2002-2010	2002-2020	2002-2030
Total Capacity	116	162	252	365	100	100	100	100	4.3	4.4	4.2
Coal	69	84	127	197	59	52	51	54	2.6	3.5	3.8
Oil	5	9	12	11	4	5	5	3	7.3	4.7	2.6
Gas	13	22	45	72	11	13	18	20	6.3	7.1	6.2
Nuclear	3	6	9	14	2	4	4	4	12.4	7.5	6.4
Hydro	25	37	52	60	21	23	21	16	5.2	4.2	3.2
of which pumped storage	0	0	0	0	0	0	0	0	-	-	-
Renewables (excluding hydro)	2	5	6	11	2	3	3	3	10.2	6.4	6.2

Renewables Breakdown

	2002	Capacity (GW) 2010	2020	2030	2002	Shares (%) 2010	2020	2030	Growth Rates (% per annum) 2002-2010	2002-2020	2002-2030
Renewables (excluding hydro)	2	5	6	11	100	100	100	100	10.2	6.4	6.2
Biomass and Waste	0	1	1	3	18	33	23	28	19.0	8.0	7.9
Wind	2	3	5	7	81	66	75	63	7.5	6.0	5.3
Geothermal	0	0	0	0	0	1	1	1	31.6	17.8	12.8
Solar	0	0	0	1	1	1	0	8	0.0	0.0	13.5
Tide/Wave	0	0	0	0	0	0	0	0	-	-	-

	2002	Electricity Generation (TWh) 2010	2020	2030	2002	Shares (%) 2010	2020	2030	Growth Rates (% per annum) 2002-2010	2002-2020	2002-2030
Renewables (excluding hydro)	4	17	23	42	100	100	100	100	15.1	8.8	8.0
Biomass and Waste	2	9	9	20	43	54	41	47	17.6	8.4	8.3
Wind	2	8	13	20	56	44	56	47	12.2	8.7	7.3
Geothermal	0	0	1	1	0	2	3	2	-	-	-
Solar	0	0	0	2	0	0	0	4	26.4	12.4	22.1
Tide/Wave	0	0	0	0	0	0	0	0	-	-	-

Reference Scenario: India

	CO_2 Emissions (Mt)					Shares (%)					Growth Rates (% per annum)			
	1971	2002	2010	2020	2030	1971	2002	2010	2020	2030	1971-2002	2002-2010	2002-2020	2002-2030
Total CO_2 Emissions	**199**	**1 016**	**1 263**	**1 714**	**2 254**	**100**	**100**	**100**	**100**	**100**	**5.4**	**2.8**	**2.9**	**2.9**
change since 1990 (%)		*70.9*	*112.4*	*188.3*	*279.0*									
Coal	142	671	779	1 027	1 377	71	66	62	60	61	5.1	1.9	2.4	2.6
Oil	56	293	399	541	673	28	29	32	32	30	5.5	4.0	3.5	3.0
Gas	1	52	85	146	204	1	5	7	8	9	12.6	6.3	5.9	5.0
Power Generation and Heat Plants	**37**	**553**	**684**	**980**	**1 368**	**100**	**100**	**100**	**100**	**100**	**9.1**	**2.7**	**3.2**	**3.3**
Coal	31	505	609	853	1 198	85	91	89	87	88	9.4	2.4	3.0	3.1
Oil	5	23	29	34	32	13	4	4	4	2	5.0	3.0	2.4	1.2
Gas	1	26	46	92	139	2	5	7	9	10	12.9	7.5	7.4	6.2
Transformation, Own Use and Losses	**7**	**31**	**33**	**37**	**40**	**100**	**100**	**100**	**100**	**100**	**5.0**	**0.8**	**0.9**	**0.9**
Total Final Consumption	**156**	**432**	**546**	**698**	**845**	**100**	**100**	**100**	**100**	**100**	**3.4**	**3.0**	**2.7**	**2.4**
Coal	108	165	167	171	175	69	38	31	25	21	1.4	0.2	0.2	0.2
Oil	47	246	346	480	612	30	57	63	69	72	5.5	4.3	3.8	3.3
Gas	1	21	33	46	58	0	5	6	7	7	11.8	5.7	4.5	3.7
Industry	**67**	**218**	**269**	**349**	**422**	**100**	**100**	**100**	**100**	**100**	**3.9**	**2.6**	**2.6**	**2.4**
Coal	55	130	139	150	159	81	60	52	43	38	2.8	0.9	0.8	0.7
Oil	12	69	100	158	212	18	31	37	45	50	5.8	4.8	4.7	4.1
Gas	1	20	30	41	51	1	9	11	12	12	11.7	5.4	4.2	3.5
Transport	**48**	**94**	**119**	**157**	**201**	**100**	**100**	**100**	**100**	**100**	**2.2**	**3.0**	**2.9**	**2.8**
Oil	18	94	119	157	201	38	100	100	100	100	5.5	3.0	2.9	2.8
Other Fuels	30	0	0	0	0	62	0	0	0	0	-100.0	-	-	-
Other Sectors	**38**	**103**	**133**	**164**	**191**	**100**	**100**	**100**	**100**	**100**	**3.3**	**3.3**	**2.6**	**2.2**
Coal	24	35	28	21	16	62	34	21	13	8	1.3	-2.7	-2.7	-2.7
Oil	14	66	102	137	168	38	65	77	84	88	5.1	5.5	4.1	3.4
Gas	0	1	3	6	7	0	1	2	3	4	12.7	9.4	7.6	5.6
Non-Energy Use	**2**	**17**	**26**	**28**	**30**	**100**	**100**	**100**	**100**	**100**	**6.6**	**4.8**	**2.6**	**2.0**

Reference Scenario: Latin America

	Energy Demand (Mtoe)					Shares (%)					Growth Rates (% per annum)			
	1971	2002	2010	2020	2030	1971	2002	2010	2020	2030	1971-2002	2002-2010	2002-2020	2002-2030
Total Primary Energy Supply	**203**	**465**	**575**	**746**	**958**	**100**	**100**	**100**	**100**	**100**	**2.7**	**2.7**	**2.7**	**2.6**
Coal	7	21	25	32	43	3	5	4	4	4	3.8	2.0	2.3	2.5
Oil	108	216	260	325	403	53	46	45	44	42	2.2	2.3	2.3	2.3
Gas	18	90	131	202	296	9	19	23	27	31	5.2	4.9	4.6	4.4
Nuclear	0	5	6	7	7	0	1	1	1	1	-	0.5	1.7	1.1
Hydro	6	47	58	72	84	3	10	10	10	9	6.7	2.9	2.5	2.1
Biomass and Waste	63	83	90	100	109	31	18	16	13	11	0.9	1.1	1.0	1.0
Other Renewables	0	3	5	7	16	0	1	1	1	2	-	6.7	5.4	6.5
Power Generation and Heat Plants	**27**	**111**	**154**	**222**	**310**	**100**	**100**	**100**	**100**	**100**	**4.6**	**4.2**	**3.9**	**3.7**
Coal	2	7	9	14	21	6	6	6	6	7	4.7	3.6	4.0	4.2
Oil	14	21	21	19	13	50	19	13	9	4	1.3	0.1	-0.4	-1.5
Gas	5	24	48	92	153	17	21	31	41	49	5.4	9.2	7.8	6.9
Nuclear	0	5	6	7	7	0	5	4	3	2	-	0.5	1.7	1.1
Hydro	6	47	58	72	84	23	42	38	32	27	6.7	2.9	2.5	2.1
Biomass and Waste	1	5	8	12	16	5	5	5	6	5	4.5	5.9	4.8	4.0
Other Renewables	0	3	4	6	14	0	2	3	3	5	-	6.3	5.0	6.2
Other Transformation, Own Use and Losses	**32**	**62**	**70**	**83**	**101**						**2.2**	**1.5**	**1.6**	**1.8**
of which electricity	*2*	*14*	*18*	*24*	*33*						*6.4*	*3.6*	*3.3*	*3.2*
Total Final Consumption	**155**	**362**	**446**	**575**	**732**	**100**	**100**	**100**	**100**	**100**	**2.8**	**2.7**	**2.6**	**2.6**
Coal	3	10	11	13	15	2	3	2	2	2	3.7	1.7	1.7	1.7
Oil	78	181	226	292	373	50	50	51	51	51	2.7	2.8	2.7	2.6
Gas	6	43	56	78	104	4	12	13	14	14	6.6	3.5	3.4	3.2
Electricity	10	56	77	110	152	6	15	17	19	21	5.9	4.1	3.8	3.6
Heat	0	0	0	0	0	0	0	0	0	0	-	-	-	-
Biomass and Waste	58	73	77	81	87	38	20	17	14	12	0.7	0.6	0.6	0.6
Other Renewables	0	0	0	1	2	0.0	0.0	0.0	0.1	0.2	-	18.4	14.3	13.1

Industry	49	140	167	209	257	100	100	100	100	100	3.5	2.2	2.2	2.2
Coal	3	9	11	13	15	5	7	6	6	6	4.2	1.7	1.7	1.7
Oil	21	40	46	57	68	42	28	28	27	27	2.1	1.9	2.0	2.0
Gas	4	30	37	49	63	8	21	22	23	25	6.6	2.9	2.8	2.7
Electricity	5	25	33	46	62	10	18	20	22	24	5.5	3.4	3.4	3.3
Heat	0	0	0	0	0	0	0	0	0	0	-	-	-	-
Renewables	16	36	40	44	48	34	26	24	21	19	2.6	1.2	1.1	1.0
Transport	41	109	146	199	264	100	100	100	100	100	3.3	3.7	3.4	3.2
Oil	40	101	133	179	237	99	92	91	90	90	3.0	3.5	3.2	3.1
Other Fuels	1	9	14	20	26	1	8	9	10	10	9.4	5.9	4.7	4.1
Other Sectors	62	102	122	153	195	100	100	100	100	100	1.6	2.3	2.3	2.4
Coal	0	0	0	0	0	0	0	0	0	0	-3.8	-1.7	-1.7	-1.6
Oil	14	30	35	42	50	22	30	29	28	26	2.5	2.0	1.9	1.8
Gas	2	10	14	22	31	3	10	12	14	16	5.8	4.1	4.1	4.0
Electricity	5	31	44	64	89	7	30	36	42	46	6.3	4.7	4.2	3.9
Heat	0	0	0	0	0	0	0	0	0	0	-	-	-	-
Renewables	42	31	28	26	25	67	30	23	17	13	-1.0	-1.2	-1.0	-0.8
Non-Energy Use	4	10	12	14	17						3.4	1.8	1.8	1.8
Electricity Generation (TWh)	135	809	1 108	1 567	2 149	100	100	100	100	100	6.0	4.0	3.7	3.5
Coal	4	29	39	60	100	3	4	4	4	5	6.3	3.8	4.2	4.5
Oil	41	82	84	78	56	30	10	8	5	3	2.3	0.3	-0.2	-1.3
Gas	14	113	244	497	887	11	14	22	32	41	6.9	10.1	8.6	7.6
Nuclear	0	21	22	29	29	0	3	2	2	1	-	0.5	1.7	1.1
Hydro	73	541	678	838	978	54	67	61	53	46	6.7	2.9	2.5	2.1
Biomass and Waste	2	20	31	45	58	2	2	3	3	3	7.3	5.7	4.6	3.8
Other Renewables	0	4	11	20	41	0	0	1	1	2	-	14.9	9.9	9.0
Traditional Biomass (included above)	42	31	28	25	23	21	7	5	3	2	-1.0	-1.2	-1.1	-1.0
Total Primary Energy Supply (excluding traditional biomass)	161	434	548	721	935						3.3	2.9	2.9	2.8

Annex A - Tables for Reference Scenario Projections

Reference Scenario: Latin America

	2002	Capacity (GW) 2010	2020	2030	Shares (%) 2002	2010	2020	2030	Growth Rates (% per annum) 2002-2010	2002-2020	2002-2030
Total Capacity	195	270	382	528	100	100	100	100	4.1	3.8	3.6
Coal	6	8	12	21	3	3	3	4	4.0	4.0	4.4
Oil	31	32	30	23	16	12	8	4	0.6	-0.1	-1.1
Gas	35	76	147	255	18	28	39	48	10.4	8.4	7.4
Nuclear	3	3	4	4	1	1	1	1	0.0	1.5	1.0
Hydro	116	142	175	205	59	53	46	39	2.5	2.3	2.1
of which pumped storage	0	0	0	0	0	0	0	0	-	-	-
Renewables (excluding hydro)	5	8	13	21	2	3	3	4	7.4	5.9	5.6

Renewables Breakdown

	2002	Capacity (GW) 2010	2020	2030	Shares (%) 2002	2010	2020	2030	Growth Rates (% per annum) 2002-2010	2002-2020	2002-2030
Renewables (excluding hydro)	5	8	13	21	100	100	100	100	7.4	5.9	5.6
Biomass and Waste	4	5	7	9	82	64	57	45	4.0	3.8	3.4
Wind	0	2	5	8	8	28	36	40	25.1	14.9	11.7
Geothermal	1	1	1	2	10	8	7	9	4.6	3.8	5.5
Solar	0	0	0	1	0	0	0	5	0.0	0.0	33.5
Tide/Wave	0	0	0	0	0	0	0	0	-	-	-

	2002	Electricity Generation (TWh) 2010	2020	2030	Shares (%) 2002	2010	2020	2030	Growth Rates (% per annum) 2002-2010	2002-2020	2002-2030
Renewables (excluding hydro)	24	42	65	99	100	100	100	100	6.0	5.2	4.9
Biomass and Waste	20	31	45	58	84	74	69	58	4.5	4.1	3.6
Wind	1	7	14	25	3	16	22	26	26.4	16.6	13.0
Geothermal	3	4	6	14	13	10	9	14	3.7	3.5	5.2
Solar	0	0	0	2	0	0	0	2	1.9	0.9	32.6
Tide/Wave	0	0	0	0	0	0	0	0	-	-	-

Reference Scenario: Latin America

	CO$_2$ Emissions (Mt)					Shares (%)					Growth Rates (% per annum)			
	1971	2002	2010	2020	2030	1971	2002	2010	2020	2030	1971-2002	2002-2010	2002-2020	2002-2030
Total CO$_2$ Emissions	**365**	**854**	**1 084**	**1 457**	**1 929**	**100**	**100**	**100**	**100**	**100**	**2.8**	**3.0**	**3.0**	**3.0**
change since 1990 (%)		*42.7*	*81.3*	*143.5*	*222.4*									
Coal	23	76	88	117	159	6	9	8	8	8	3.9	1.9	2.4	2.7
Oil	299	592	716	900	1 118	82	69	66	62	58	2.2	2.4	2.4	2.3
Gas	42	185	280	440	652	11	22	26	30	34	4.9	5.3	4.9	4.6
Power Generation and Heat Plants	**61**	**147**	**210**	**324**	**481**	**100**	**100**	**100**	**100**	**100**	**2.9**	**4.6**	**4.5**	**4.3**
Coal	7	30	36	55	87	12	20	17	17	18	4.7	2.5	3.5	3.9
Oil	43	64	63	58	41	71	44	30	18	8	1.3	-0.2	-0.6	-1.6
Gas	11	53	111	211	353	17	36	53	65	73	5.3	9.7	8.0	7.0
Transformation, Own Use and Losses	**51**	**76**	**81**	**94**	**116**	**100**	**100**	**100**	**100**	**100**	**1.3**	**0.8**	**1.2**	**1.5**
Total Final Consumption	**253**	**631**	**793**	**1 038**	**1 332**	**100**	**100**	**100**	**100**	**100**	**3.0**	**2.9**	**2.8**	**2.7**
Coal	15	43	49	58	68	6	7	6	6	5	3.4	1.6	1.7	1.7
Oil	225	493	619	803	1 025	89	78	78	77	77	2.6	2.9	2.7	2.6
Gas	13	95	126	178	239	5	15	16	17	18	6.6	3.6	3.5	3.4
Industry	**87**	**208**	**247**	**309**	**380**	**100**	**100**	**100**	**100**	**100**	**2.9**	**2.2**	**2.2**	**2.2**
Coal	13	42	47	57	67	15	20	19	18	18	3.9	1.7	1.7	1.7
Oil	65	102	119	146	175	75	49	48	47	46	1.5	1.9	2.0	2.0
Gas	9	64	81	107	138	11	31	33	35	36	6.5	2.9	2.9	2.7
Transport	**114**	**296**	**393**	**536**	**711**	**100**	**100**	**100**	**100**	**100**	**3.1**	**3.6**	**3.3**	**3.2**
Oil	113	290	381	514	681	99	98	97	96	96	3.1	3.5	3.2	3.1
Other Fuels	1	7	12	21	30	1	2	3	4	4	7.0	7.8	6.8	5.5
Other Sectors	**46**	**110**	**134**	**170**	**214**	**100**	**100**	**100**	**100**	**100**	**2.9**	**2.5**	**2.5**	**2.4**
Coal	1	1	1	0	0	3	1	0	0	0	-2.5	-1.7	-1.7	-1.6
Oil	41	85	100	120	142	89	78	75	71	66	2.4	2.0	1.9	1.8
Gas	4	24	33	49	72	8	22	25	29	33	6.1	4.1	4.1	4.0
Non-Energy Use	**7**	**17**	**19**	**23**	**27**	**100**	**100**	**100**	**100**	**100**	**2.8**	**1.8**	**1.8**	**1.7**

Reference Scenario: Brazil

	Energy Demand (Mtoe)					Shares (%)					Growth Rates (% per annum)			
	1971	2002	2010	2020	2030	1971	2002	2010	2020	2030	1971-2002	2002-2010	2002-2020	2002-2030
Total Primary Energy Supply	70	188	228	295	372	100	100	100	100	100	**3.2**	**2.5**	**2.6**	**2.5**
Coal	2	13	14	18	22	4	7	6	6	6	5.5	1.1	1.7	1.9
Oil	28	88	109	139	172	40	47	48	47	46	3.8	2.7	2.6	2.4
Gas	0	12	18	35	59	0	6	8	12	16	16.4	5.2	6.1	5.8
Nuclear	0	4	4	6	6	0	2	2	2	2	-	0.6	3.1	2.0
Hydro	4	25	31	39	45	5	13	14	13	12	6.3	3.1	2.6	2.2
Biomass and Waste	35	46	51	58	65	51	25	23	19	18	0.9	1.3	1.2	1.2
Other Renewables	0	0	0	1	2	0	0	0	0	1	-17.9	181.6	65.9	42.9
Power Generation and Heat Plants	6	39	49	73	99	100	100	100	100	100	**6.2**	**3.0**	**3.6**	**3.4**
Coal	1	2	2	4	5	10	6	5	5	5	4.5	0.0	2.4	3.0
Oil	2	3	3	4	3	25	8	7	5	3	2.5	0.4	0.9	-0.6
Gas	0	3	5	15	32	0	7	10	21	32	-	7.9	10.4	9.5
Nuclear	0	4	4	6	6	0	9	8	9	6	-	0.6	3.1	2.0
Hydro	4	25	31	39	45	62	63	63	53	46	6.3	3.1	2.6	2.2
Biomass and Waste	0	2	4	5	6	2	6	7	7	6	9.4	5.1	3.9	3.2
Other Renewables	0	0	0	1	1	0	0	1	1	1	-	170.8	62.2	40.5
Other Transformation, Own Use and Losses	5	22	24	29	35						**4.5**	**1.4**	**1.7**	**1.8**
of which electricity	*1*	*6*	*8*	*10*	*13*						*6.8*	*2.7*	*2.8*	*2.7*
Total Final Consumption	63	160	195	249	310	100	100	100	100	100	**3.1**	**2.5**	**2.5**	**2.4**
Coal	1	6	7	9	11	1	4	4	4	4	7.1	2.2	2.2	2.2
Oil	25	80	100	129	161	39	50	51	52	52	3.9	2.9	2.7	2.5
Gas	0	7	11	16	23	0	5	5	7	7	13.3	5.1	4.7	4.1
Electricity	4	27	33	46	60	6	17	17	18	19	6.6	2.9	3.0	2.9
Heat	0	0	0	0	0	0	0	0	0	0	-	-	-	-
Biomass and Waste	34	40	44	48	55	53	25	23	19	18	0.6	1.1	1.0	1.1
Other Renewables	0	0	0	0	1	0.0	0.0	0.0	0.1	0.2	-	-	-	-

Industry	**18**	**69**	**82**	**103**	**127**	**100**	**100**	**100**	**100**	**100**	**4.4**	**2.2**	**2.3**	**2.2**
Coal	1	6	7	9	11	4	9	9	9	9	7.3	2.2	2.3	2.2
Oil	7	19	23	29	37	37	28	28	28	29	3.5	2.0	2.3	2.3
Gas	0	6	9	14	20	0	9	11	14	15	18.0	5.1	4.8	4.3
Electricity	2	13	15	20	24	11	19	19	19	19	6.3	2.1	2.4	2.3
Heat	0	0	0	0	0	0	0	0	0	0	–	–	–	–
Renewables	9	25	28	31	35	49	36	34	30	28	3.4	1.5	1.3	1.3
Transport	**14**	**50**	**68**	**90**	**116**	**100**	**100**	**100**	**100**	**100**	**4.1**	**4.0**	**3.4**	**3.1**
Oil	14	43	58	77	98	99	87	85	85	84	3.6	3.7	3.2	3.0
Other Fuels	0	7	10	14	18	1	13	15	15	16	12.2	5.5	4.1	3.7
Other Sectors	**29**	**37**	**41**	**50**	**61**	**100**	**100**	**100**	**100**	**100**	**0.8**	**1.3**	**1.6**	**1.8**
Coal	0	0	0	0	0	0	0	0	0	0	–	–	–	–
Oil	3	13	15	17	20	9	35	36	35	33	5.2	1.6	1.7	1.6
Gas	0	0	1	1	1	0	1	1	2	2	3.9	4.7	4.6	3.9
Electricity	2	14	18	26	35	6	37	44	52	58	7.0	3.6	3.6	3.4
Heat	0	0	0	0	0	0	0	0	0	0	–	–	–	–
Renewables	25	10	7	5	4	85	27	18	11	7	-2.9	-3.6	-3.4	-2.9
Non-Energy Use	**1**	**4**	**5**	**6**	**7**						**4.3**	**1.5**	**1.5**	**1.5**
Electricity Generation (TWh)	**52**	**345**	**441**	**611**	**808**	**100**	**100**	**100**	**100**	**100**	**6.3**	**3.1**	**3.2**	**3.1**
Coal	2	8	8	14	23	3	2	2	2	3	5.3	0.0	2.8	3.7
Oil	6	13	14	17	12	12	4	3	3	2	2.5	1.0	1.3	-0.2
Gas	0	13	21	77	179	0	4	5	13	22	–	6.2	10.4	9.8
Nuclear	0	14	15	24	24	0	4	3	4	3	–	0.6	3.1	2.0
Hydro	43	285	362	450	529	84	83	82	74	65	6.3	3.1	2.6	2.2
Biomass and Waste	1	11	17	23	28	1	3	4	4	3	9.7	5.1	3.9	3.2
Other Renewables	0	0	3	6	14	0	0	1	1	2	–	–	–	–
Traditional Biomass (included above)	25	10	7	5	4	35	5	3	2	1	-2.9	-3.8	-3.7	-3.6
Total Primary Energy Supply (excluding traditional biomass)	**45**	**178**	**220**	**290**	**368**						**4.5**	**2.7**	**2.8**	**2.6**

Annex A - Tables for Reference Scenario Projections

Reference Scenario: Brazil

	Capacity (GW)				Shares (%)				Growth Rates (% per annum)		
	2002	2010	2020	2030	2002	2010	2020	2030	2002-2010	2002-2020	2002-2030
Total Capacity	79	100	135	179	100	100	100	100	3.0	3.0	3.0
Coal	2	2	2	4	2	2	2	2	-0.1	1.3	2.7
Oil	4	5	5	4	5	5	4	2	2.1	1.0	-0.2
Gas	4	9	22	45	5	9	16	25	10.2	9.5	8.8
Nuclear	2	2	3	3	2	2	2	2	0.0	2.8	1.8
Hydro	64	78	97	114	82	78	72	64	2.5	2.3	2.1
of which pumped storage	0	0	0	0	0	0	0	0	-	-	-
Renewables (excluding hydro)	2	4	6	9	3	4	4	5	5.5	4.8	4.9

Renewables Breakdown

	Capacity (GW)				Shares (%)				Growth Rates (% per annum)		
	2002	2010	2020	2030	2002	2010	2020	2030	2002-2010	2002-2020	2002-2030
Renewables (excluding hydro)	2	4	6	9	100	100	100	100	5.5	4.8	4.9
Biomass and Waste	2	3	4	4	95	75	65	48	2.5	2.7	2.4
Wind	0	1	2	4	5	25	35	46	28.5	16.4	13.4
Geothermal	0	0	0	0	0	0	0	0	31.6	17.8	12.8
Solar	0	0	0	0	0	0	0	5	0.0	0.0	38.7
Tide/Wave	0	0	0	0	0	0	0	0	-	-	-

	Electricity Generation (TWh)				Shares (%)				Growth Rates (% per annum)		
	2002	2010	2020	2030	2002	2010	2020	2030	2002-2010	2002-2020	2002-2030
Renewables (excluding hydro)	11	20	29	42	100	100	100	100	5.7	4.7	4.4
Biomass and Waste	11	17	23	28	100	86	79	67	4.1	3.5	3.0
Wind	0	3	6	13	0	14	21	31	-	54.5	37.0
Geothermal	0	0	0	0	0	0	0	0	-	-	-
Solar	0	0	0	1	0	0	0	2	-	-	-
Tide/Wave	0	0	0	0	0	0	0	0	-	-	-

Reference Scenario: Brazil

	CO$_2$ Emissions (Mt)					Shares (%)					Growth Rates (% per annum)			
	1971	2002	2010	2020	2030	1971	2002	2010	2020	2030	1971-2002	2002-2010	2002-2020	2002-2030
Total CO$_2$ Emissions	**91**	**302**	**375**	**508**	**665**	**100**	**100**	**100**	**100**	**100**	**3.9**	**2.8**	**2.9**	**2.9**
change since 1990 (%)		*56.7*	*95.0*	*164.2*	*245.9*									
Coal	7	41	44	56	73	8	14	12	11	11	5.7	0.7	1.8	2.1
Oil	83	237	294	379	469	92	79	79	75	70	3.4	2.7	2.6	2.5
Gas	0	23	37	73	124	0	8	10	14	19	16.0	6.2	6.6	6.2
Power Generation and Heat Plants	**7**	**24**	**27**	**56**	**96**	**100**	**100**	**100**	**100**	**100**	**3.9**	**1.4**	**4.7**	**5.0**
Coal	3	11	9	13	20	35	46	32	24	21	4.9	-3.2	0.9	2.1
Oil	5	9	8	10	7	65	38	31	17	7	2.2	-1.3	0.2	-1.1
Gas	0	4	10	33	69	0	16	37	59	72	-	13.1	12.7	10.9
Transformation, Own Use and Losses	**5**	**20**	**22**	**26**	**32**	**100**	**100**	**100**	**100**	**100**	**4.3**	**1.3**	**1.6**	**1.7**
Total Final Consumption	**78**	**257**	**326**	**426**	**537**	**100**	**100**	**100**	**100**	**100**	**3.9**	**3.0**	**2.8**	**2.7**
Coal	4	26	31	39	48	5	10	10	9	9	6.2	2.2	2.2	2.2
Oil	74	217	273	354	443	95	84	84	83	82	3.5	2.9	2.8	2.6
Gas	0	14	21	33	46	0	6	6	8	9	20.0	5.0	4.8	4.3
Industry	**25**	**85**	**105**	**140**	**179**	**100**	**100**	**100**	**100**	**100**	**4.1**	**2.6**	**2.8**	**2.7**
Coal	4	26	31	39	48	14	31	30	28	27	6.6	2.2	2.3	2.2
Oil	21	46	54	69	87	85	54	51	50	49	2.5	2.0	2.3	2.3
Gas	0	14	20	31	44	0	16	19	22	24	19.7	5.1	4.8	4.3
Transport	**42**	**127**	**170**	**225**	**288**	**100**	**100**	**100**	**100**	**100**	**3.6**	**3.7**	**3.2**	**3.0**
Oil	42	127	170	225	288	100	100	100	100	100	3.6	3.7	3.2	3.0
Other Fuels	0	0	0	0	0	0	0	0	0	0	-	-	-	-
Other Sectors	**8**	**38**	**43**	**51**	**60**	**100**	**100**	**100**	**100**	**100**	**5.2**	**1.7**	**1.7**	**1.7**
Coal	0	0	0	0	0	0	0	0	0	0	-	-	-	-
Oil	8	37	42	49	57	97	98	97	97	96	5.3	1.6	1.7	1.6
Gas	0	1	1	2	2	0	2	3	3	4	-	4.7	4.6	3.9
Non-Energy Use	**4**	**8**	**9**	**10**	**11**	**100**	**100**	**100**	**100**	**100**	**2.4**	**1.5**	**1.5**	**1.4**

Annex A - Tables for Reference Scenario Projections

Reference Scenario: Middle East

	Energy Demand (Mtoe)					Shares (%)					Growth Rates (% per annum)			
	1971	2002	2010	2020	2030	1971	2002	2010	2020	2030	1971-2002	2002-2010	2002-2020	2002-2030
Total Primary Energy Supply	**51**	**407**	**524**	**695**	**809**	**100**	**100**	**100**	**100**	**100**	**6.9**	**3.2**	**3.0**	**2.5**
Coal	0	8	9	12	14	0	2	2	2	2	11.9	1.6	2.5	2.3
Oil	38	206	257	325	374	75	51	49	47	46	5.6	2.8	2.6	2.1
Gas	11	189	250	349	405	22	46	48	50	50	9.5	3.6	3.5	2.8
Nuclear	0	0	2	2	2	0	0	0	0	0	-	-	-	-
Hydro	0	1	2	3	3	1	0	0	0	0	5.0	6.5	4.1	3.1
Biomass and Waste	1	2	2	3	7	1	0	0	0	1	4.1	2.2	2.1	5.0
Other Renewables	0	1	1	2	3	0	0	0	0	0	-	6.8	5.8	5.5
Power Generation and Heat Plants	**8**	**118**	**153**	**198**	**244**	**100**	**100**	**100**	**100**	**100**	**9.3**	**3.3**	**2.9**	**2.6**
Coal	0	6	7	10	13	0	5	5	5	5	-	1.4	2.7	2.4
Oil	6	46	48	59	70	78	39	31	30	29	6.8	0.6	1.4	1.5
Gas	1	64	94	124	151	17	55	61	63	62	13.4	4.8	3.7	3.1
Nuclear	0	0	2	2	2	0	0	1	1	1	-	-	-	-
Hydro	0	1	2	3	3	4	1	2	2	1	5.0	6.5	4.1	3.1
Biomass and Waste	0	0	0	0	4	0	0	0	0	2	-	0.0	0.0	19.6
Other Renewables	0	0	0	1	1	0	0	0	0	0	-	52.9	26.2	19.1
Other Transformation, Own Use and Losses	**11**	**44**	**66**	**117**	**138**						**4.6**	**5.3**	**5.6**	**4.2**
of which electricity	*0*	*9*	*11*	*16*	*20*						*11.1*	*3.4*	*3.3*	*3.0*
Total Final Consumption	**35**	**289**	**363**	**461**	**532**	**100**	**100**	**100**	**100**	**100**	**7.1**	**2.9**	**2.6**	**2.2**
Coal	0	1	1	1	1	0	0	0	0	0	5.5	2.9	2.1	1.6
Oil	25	167	208	255	284	71	58	57	55	53	6.4	2.8	2.4	1.9
Gas	7	84	104	136	157	21	29	29	29	29	8.2	2.8	2.7	2.3
Electricity	2	35	47	65	85	6	12	13	14	16	9.7	3.6	3.5	3.2
Heat	0	0	0	0	0	0	0	0	0	0	-	-	-	-
Biomass and Waste	0	2	2	2	3	1	1	1	1	1	4.3	2.3	2.2	2.1
Other Renewables	0	1	1	1	2	0.0	0.3	0.3	0.3	0.4	-	4.0	4.0	4.0

Industry	12	92	113	142	159	100	100	100	100	100	100	6.8	2.6	2.4	2.0
Coal	0	1	1	1	1	1	1	1	1	1	1	5.5	2.9	2.1	1.6
Oil	5	39	46	51	53	41	43	41	36	33	-	7.0	1.9	1.4	1.1
Gas	6	45	57	77	88	50	49	50	54	55	-	6.7	2.9	3.0	2.4
Electricity	1	7	9	13	16	7	8	8	9	10	-	7.0	3.9	3.5	3.1
Heat	0	0	0	0	0	0	0	0	0	0	-	-	-	-	-
Renewables	0	0	0	0	0	0	0	0	0	0	-	4.5	1.5	1.5	1.5
Transport	12	72	94	120	139	100	100	100	100	100	100	6.0	3.4	2.9	2.4
Oil	12	72	94	120	139	100	100	100	100	100	100	6.0	3.4	2.9	2.4
Other Fuels	0	0	0	0	0	0	0	0	0	0	-	-	-	-	-
Other Sectors	10	118	146	185	218	100	100	100	100	100	100	8.3	2.7	2.5	2.2
Coal	0	0	0	0	0	0	0	0	0	0	-	-	-	-	-
Oil	7	49	58	71	77	70	41	40	38	35	-	6.5	2.2	2.1	1.6
Gas	1	39	48	59	69	14	33	33	32	32	-	11.3	2.7	2.4	2.1
Electricity	1	28	38	52	68	12	24	26	28	31	-	10.8	3.6	3.5	3.2
Heat	0	0	0	0	0	0	0	0	0	0	-	-	-	-	-
Renewables	0	2	3	3	4	4	2	2	2	2	-	5.6	2.5	2.5	2.6
Non-Energy Use	1	8	10	13	16							6.1	3.8	3.1	2.6
Electricity Generation (TWh)	27	512	679	940	1 214	100	100	100	100	100	100	9.9	3.6	3.4	3.1
Coal	0	29	33	48	60	0	6	5	5	5	-	-	1.6	2.9	2.7
Oil	20	207	216	264	314	72	40	32	28	26	-	7.9	0.6	1.4	1.5
Gas	4	259	392	579	775	15	51	58	62	64	-	14.4	5.3	4.6	4.0
Nuclear	0	0	0	6	6	0	0	0	1	1	-	-	-	-	-
Hydro	4	17	28	35	40	14	3	4	4	3	-	5.0	6.5	4.1	3.1
Biomass and Waste	0	0	0	0	5	0	0	0	0	0	-	-	0.0	0.0	18.5
Other Renewables	0	0	3	6	13	0	0	0	1	1	-	-	52.9	26.2	19.1
Traditional Biomass (included above)	0	1	2	2	2	1	0	0	0	0	0	4.2	1.7	1.6	1.6
Total Primary Energy Supply (excluding traditional biomass)	50	405	522	693	807							7.0	3.2	3.0	2.5

Annex A - Tables for Reference Scenario Projections

Reference Scenario: Middle East

	2002	Capacity (GW) 2010	2020	2030	Shares (%) 2002	2010	2020	2030	Growth Rates (% per annum) 2002-2010	2002-2020	2002-2030
Total Capacity	140	170	236	305	100	100	100	100	**2.5**	**3.0**	**2.8**
Coal	5	6	9	11	4	3	4	4	2.1	3.3	3.0
Oil	55	63	79	96	40	37	34	32	1.6	2.0	2.0
Gas	73	89	132	176	52	53	56	58	2.5	3.3	3.2
Nuclear	0	1	1	1	0	1	0	0	-	-	-
Hydro	6	10	12	14	4	6	5	5	6.3	4.0	3.0
of which pumped storage	0	0	0	0	0	0	0	0	-	-	-
Renewables (excluding hydro)	0	1	2	6	0	1	1	2	46.9	23.9	18.6

Renewables Breakdown

	2002	Capacity (GW) 2010	2020	2030	Shares (%) 2002	2010	2020	2030	Growth Rates (% per annum) 2002-2010	2002-2020	2002-2030
Renewables (excluding hydro)	0	1	2	6	100	100	100	100	**46.9**	**23.9**	**18.6**
Biomass and Waste	0	0	0	1	16	1	0	15	0.0	0.0	18.2
Wind	0	1	2	4	60	98	99	63	56.4	27.5	18.9
Geothermal	0	0	0	0	0	0	0	0	-	-	-
Solar	0	0	0	1	24	1	1	22	0.0	0.0	18.2
Tide/Wave	0	0	0	0	0	0	0	0	-	-	-

	2002	Electricity Generation (TWh) 2010	2020	2030	Shares (%) 2002	2010	2020	2030	Growth Rates (% per annum) 2002-2010	2002-2020	2002-2030
Renewables (excluding hydro)	0	3	7	18	100	100	100	100	**35.3**	**21.0**	**17.5**
Biomass and Waste	0	0	0	5	32	2	1	30	0.0	0.0	17.2
Wind	0	3	6	10	43	97	99	56	46.6	26.0	18.5
Geothermal	0	0	0	0	0	0	0	0	-	-	-
Solar	0	0	0	3	24	1	1	15	0.0	0.0	15.6
Tide/Wave	0	0	0	0	0	0	0	0	-	-	-

Reference Scenario: Middle East

	CO$_2$ Emissions (Mt)					Shares (%)					Growth Rates (% per annum)			
	1971	2002	2010	2020	2030	1971	2002	2010	2020	2030	1971-2002	2002-2010	2002-2020	2002-2030
Total CO$_2$ Emissions	127	1 080	1 345	1 741	1 999	100	100	100	100	100	7.2	2.8	2.7	2.2
change since 1990 (%)		*84.6*	*129.9*	*197.6*	*241.6*									
Coal	1	29	33	46	56	1	3	2	3	3	12.2	1.7	2.6	2.3
Oil	100	628	750	919	1 039	79	58	56	53	52	6.1	2.2	2.1	1.8
Gas	26	423	562	776	904	21	39	42	45	45	9.4	3.6	3.4	2.8
Power Generation and Heat Plants	22	317	396	513	619	100	100	100	100	100	9.0	2.8	2.7	2.4
Coal	0	25	28	40	49	0	8	7	8	8	-	1.4	2.7	2.4
Oil	19	142	149	183	217	86	45	38	36	35	6.8	0.6	1.4	1.5
Gas	3	151	220	290	353	14	47	55	57	57	13.4	4.8	3.7	3.1
Transformation, Own Use and Losses	26	129	160	244	269	100	100	100	100	100	5.3	2.8	3.6	2.7
Total Final Consumption	79	634	788	985	1 111	100	100	100	100	100	7.0	2.8	2.5	2.0
Coal	1	4	6	6	7	1	1	1	1	1	5.6	2.9	2.1	1.6
Oil	61	437	543	666	743	77	69	69	68	67	6.6	2.7	2.4	1.9
Gas	17	192	240	312	360	21	30	30	32	32	8.2	2.8	2.7	2.3
Industry	30	197	240	298	329	100	100	100	100	100	6.3	2.5	2.3	1.8
Coal	1	4	6	6	7	3	2	2	2	2	5.6	2.9	2.1	1.6
Oil	15	91	106	117	123	51	46	44	39	37	5.9	1.9	1.4	1.1
Gas	14	102	128	175	199	46	52	54	59	61	6.7	2.9	3.0	2.4
Transport	23	190	249	319	369	100	100	100	100	100	7.1	3.4	2.9	2.4
Oil	23	190	249	319	369	100	100	100	100	100	7.1	3.4	2.9	2.4
Other Fuels	0	0	0	0	0	0	0	0	0	0	-	-	-	-
Other Sectors	24	235	284	348	390	100	100	100	100	100	7.6	2.4	2.2	1.8
Coal	0	0	0	0	0	0	0	0	0	0	-	-	-	-
Oil	21	145	173	211	229	87	62	61	61	59	6.5	2.2	2.1	1.6
Gas	3	90	111	137	161	13	38	39	39	41	11.3	2.7	2.4	2.1
Non-Energy Use	2	11	15	20	23	100	100	100	100	100	5.8	3.8	3.1	2.6

Reference Scenario: Africa

	Energy Demand (Mtoe)					Shares (%)					Growth Rates (% per annum)			
	1971	2002	2010	2020	2030	1971	2002	2010	2020	2030	1971-2002	2002-2010	2002-2020	2002-2030
Total Primary Energy Supply	200	534	660	852	1 096	100	100	100	100	100	3.2	2.7	2.6	2.6
Coal	36	92	104	121	139	18	17	16	14	13	3.1	1.5	1.6	1.5
Oil	35	113	150	209	292	18	21	23	25	27	3.9	3.6	3.5	3.4
Gas	2	57	84	142	228	1	11	13	17	21	10.7	5.0	5.2	5.1
Nuclear	0	3	3	3	3	0	1	0	0	0	-	1.5	0.7	0.4
Hydro	2	6	7	7	10	1	1	1	1	1	3.9	1.8	0.8	1.7
Biomass and Waste	124	262	310	367	419	62	49	47	43	38	2.4	2.1	1.9	1.7
Other Renewables	0	1	1	2	4	0	0	0	0	0	-	7.8	6.1	7.0
Power Generation and Heat Plants	26	107	143	208	301	100	100	100	100	100	4.6	3.7	3.7	3.7
Coal	21	58	67	84	101	81	54	47	40	34	3.3	1.9	2.1	2.0
Oil	3	14	18	23	31	10	13	12	11	10	5.5	2.8	2.9	2.9
Gas	0	24	46	87	147	1	22	32	42	49	14.4	8.3	7.4	6.7
Nuclear	0	3	3	3	3	0	3	2	2	1	-	1.5	0.7	0.4
Hydro	2	6	7	7	10	7	6	5	4	3	3.9	1.8	0.8	1.7
Biomass and Waste	0	2	2	2	5	0	1	1	1	2	-	0.0	0.0	4.4
Other Renewables	0	1	1	1	2	0	1	1	1	1	-	3.2	2.8	4.8
Other Transformation, Own Use and Losses	14	66	77	100	136						5.1	2.0	2.4	2.6
of which electricity	*1*	*8*	*11*	*17*	*25*						*6.8*	*4.1*	*4.0*	*4.1*
Total Final Consumption	167	402	497	632	795	100	100	100	100	100	2.9	2.7	2.5	2.5
Coal	18	17	17	17	16	11	4	3	3	2	-0.2	-0.1	-0.1	-0.2
Oil	31	100	130	178	246	18	25	26	28	31	3.9	3.3	3.3	3.3
Gas	1	17	22	28	35	0	4	4	4	4	12.0	2.9	2.7	2.6
Electricity	7	33	47	72	112	4	8	9	11	14	5.1	4.4	4.4	4.4
Heat	0	0	0	0	0	0	0	0	0	0	-	-	-	-
Biomass and Waste	110	234	281	336	383	66	58	57	53	48	2.5	2.3	2.0	1.8
Other Renewables	0	0	0	1	2	0.0	0.0	0.1	0.1	0.2	-	-	-	-

Industry	33	84	97	112	130	100	100	100	100	100	3.0	1.9	1.6	1.6
Coal	11	15	15	15	14	34	18	16	13	11	1.0	-0.1	-0.2	-0.2
Oil	7	15	16	17	18	20	18	17	15	14	2.7	0.8	0.5	0.7
Gas	0	13	16	20	24	1	16	17	18	18	11.7	2.6	2.3	2.1
Electricity	4	16	21	29	41	13	19	22	26	32	4.2	3.7	3.5	3.5
Heat	0	0	0	0	0	0	0	0	0	0	-	-	-	-
Renewables	11	24	29	31	32	32	29	29	28	25	2.7	2.0	1.4	1.0
Transport	20	57	74	105	147	100	100	100	100	100	3.4	3.4	3.5	3.5
Oil	16	55	73	102	144	80	98	98	98	98	4.1	3.4	3.5	3.5
Other Fuels	4	1	2	2	3	20	2	2	2	2	-3.8	4.1	3.6	3.4
Other Sectors	113	257	318	407	507	100	100	100	100	100	2.7	2.7	2.6	2.5
Coal	4	2	2	2	2	3	1	1	1	0	-1.9	0.3	0.3	0.3
Oil	7	24	34	50	73	6	9	11	12	14	4.0	4.1	4.0	4.0
Gas	0	4	5	7	10	0	1	2	2	2	12.3	3.9	3.8	3.8
Electricity	2	17	25	42	69	2	7	8	10	14	6.5	5.0	5.2	5.2
Heat	0	0	0	0	0	0	0	0	0	0	-	-	-	-
Renewables	100	210	253	306	352	88	82	79	75	70	2.4	2.3	2.1	1.9
Non-Energy Use	1	5	8	9	11						4.6	4.7	3.3	2.8
Electricity Generation (TWh)	90	480	671	1 028	1 583	100	100	100	100	100	5.5	4.3	4.3	4.4
Coal	56	225	271	361	462	62	47	40	35	29	4.6	2.4	2.7	2.6
Oil	11	60	76	102	140	12	13	11	10	9	5.8	2.8	2.9	3.0
Gas	1	105	220	458	825	1	22	33	45	52	16.3	9.6	8.5	7.6
Nuclear	0	11	13	13	13	0	2	2	1	1	-	1.5	0.7	0.4
Hydro	22	74	86	86	118	25	15	13	8	7	3.9	1.8	0.8	1.7
Biomass and Waste	0	3	3	3	9	0	1	0	0	1	-	0.0	0.0	4.4
Other Renewables	0	1	3	6	16	0	0	1	1	1	-	13.1	9.1	9.5
Traditional Biomass (included above)	100	210	252	305	350	50	39	38	36	32	2.4	2.3	2.1	1.8
Total Primary Energy Supply (excluding traditional biomass)	100	324	407	547	745						3.9	2.9	2.9	3.0

Annex A - Tables for Reference Scenario Projections

Reference Scenario: Africa

	2002	Capacity (GW) 2010	2020	2030	2002	Shares (%) 2010	2020	2030	Growth Rates (% per annum) 2002-2010	2002-2020	2002-2030
Total Capacity	111	153	230	352	100	100	100	100	4.1	4.2	4.2
Coal	39	45	60	76	35	30	26	22	2.0	2.4	2.4
Oil	23	32	47	67	21	21	20	19	3.9	4.0	3.9
Gas	24	47	94	165	21	31	41	47	8.9	7.9	7.2
Nuclear	2	2	2	2	2	1	1	1	0.0	0.0	0.0
Hydro	22	26	26	35	20	17	11	10	1.8	0.8	1.6
of which pumped storage	0	0	0	0	0	0	0	0	-	-	-
Renewables (excluding hydro)	1	2	3	8	1	1	1	2	8.6	6.5	8.3

Renewables Breakdown

	2002	Capacity (GW) 2010	2020	2030	2002	Shares (%) 2010	2020	2030	Growth Rates (% per annum) 2002-2010	2002-2020	2002-2030
Renewables (excluding hydro)	1	2	3	8	100	100	100	100	8.6	6.5	8.3
Biomass and Waste	0	0	0	1	56	29	18	19	0.0	0.0	4.2
Wind	0	1	2	5	29	63	77	60	19.9	12.5	11.2
Geothermal	0	0	0	0	13	7	4	3	0.0	0.0	2.4
Solar	0	0	0	1	2	1	1	18	0.0	0.0	16.7
Tide/Wave	0	0	0	0	0	0	0	0	-	-	-

	2002	Electricity Generation (TWh) 2010	2020	2030	2002	Shares (%) 2010	2020	2030	Growth Rates (% per annum) 2002-2010	2002-2020	2002-2030
Renewables (excluding hydro)	4	6	9	26	100	100	100	100	4.4	4.1	6.3
Biomass and Waste	3	3	3	9	68	44	31	36	0.0	0.0	4.1
Wind	1	3	5	13	14	44	61	49	17.4	12.1	11.0
Geothermal	1	1	1	1	18	11	8	5	0.0	0.0	2.2
Solar	0	0	0	3	1	1	0	10	0.0	0.0	15.7
Tide/Wave	0	0	0	0	0	0	0	0	-	-	-

Reference Scenario: Africa

	CO₂ Emissions (Mt)					Shares (%)					Growth Rates (% per annum)			
	1971	2002	2010	2020	2030	1971	2002	2010	2020	2030	1971-2002	2002-2010	2002-2020	2002-2030
Total CO₂ Emissions	**266**	**764**	**969**	**1 343**	**1 861**	**100**	**100**	**100**	**100**	**100**	**3.5**	**3.0**	**3.2**	**3.2**
change since 1990 (%)		*41.4*	*79.2*	*148.4*	*244.3*									
Coal	161	315	354	423	493	60	41	37	32	26	2.2	1.5	1.7	1.6
Oil	100	325	423	589	832	38	42	44	44	45	3.9	3.4	3.4	3.4
Gas	5	125	192	330	536	2	16	20	25	29	10.6	5.5	5.6	5.3
Power Generation and Heat Plants	**93**	**348**	**454**	**650**	**903**	**100**	**100**	**100**	**100**	**100**	**4.4**	**3.4**	**3.5**	**3.5**
Coal	83	241	280	351	423	90	69	62	54	47	3.5	1.9	2.1	2.0
Oil	9	44	56	74	100	9	13	12	11	11	5.5	2.8	2.9	2.9
Gas	1	62	118	225	381	1	18	26	35	42	14.8	8.3	7.4	6.7
Transformation, Own Use and Losses	**8**	**41**	**53**	**89**	**157**	**100**	**100**	**100**	**100**	**100**	**5.4**	**3.2**	**4.4**	**4.9**
Total Final Consumption	**165**	**375**	**463**	**604**	**802**	**100**	**100**	**100**	**100**	**100**	**2.7**	**2.6**	**2.7**	**2.7**
Coal	77	74	73	72	70	47	20	16	12	9	-0.2	-0.1	-0.1	-0.2
Oil	87	268	346	476	661	53	71	75	79	82	3.7	3.3	3.2	3.3
Gas	1	34	43	56	71	1	9	9	9	9	12.7	3.0	2.8	2.7
Industry	**70**	**133**	**141**	**148**	**158**	**100**	**100**	**100**	**100**	**100**	**2.1**	**0.7**	**0.6**	**0.6**
Coal	49	66	65	64	62	70	49	46	43	39	0.9	-0.1	-0.2	-0.2
Oil	21	43	46	47	52	29	32	33	32	33	2.4	0.8	0.5	0.7
Gas	1	24	30	37	44	1	18	21	25	28	12.2	2.6	2.3	2.1
Transport	**57**	**148**	**194**	**274**	**386**	**100**	**100**	**100**	**100**	**100**	**3.2**	**3.4**	**3.5**	**3.5**
Oil	42	147	192	271	382	75	99	99	99	99	4.1	3.4	3.5	3.5
Other Fuels	14	2	2	3	4	25	1	1	1	1	-7.0	4.1	3.7	3.5
Other Sectors	**36**	**87**	**117**	**169**	**244**	**100**	**100**	**100**	**100**	**100**	**2.9**	**3.8**	**3.8**	**3.8**
Coal	14	8	8	8	8	39	9	7	5	3	-1.9	0.3	0.3	0.3
Oil	22	71	98	145	212	61	82	84	86	87	3.9	4.1	4.0	4.0
Gas	0	8	11	16	23	0	9	9	9	9	13.6	3.9	3.8	3.8
Non-Energy Use	**3**	**7**	**10**	**12**	**15**	**100**	**100**	**100**	**100**	**100**	**3.2**	**4.7**	**3.3**	**2.8**

ANNEX B

ENERGY PROJECTIONS: ASSESSMENT AND COMPARISON

SUMMARY

- This annex reviews the accuracy of past *World Energy Outlook* projections. It then provides a comparison of the *WEO-2004* projections with those of other organisations.

- Since 1993, IEA projections for global energy demand have been within 2.2% of the most recently reported data. Projections for electricity demand have fallen even closer to recorded values, while those for global oil demand have been within 2.3%. The accuracy of these projections grows out of the strong and predictable relationship between economic growth and energy usage.

- Economic growth remains the major determinant of energy demand. Energy price assumptions have had less influence on demand levels. That is fortunate from an analyst's point of view, because long-term oil-price forecasts have proved very tricky, reflecting the volatile nature of the oil market.

- Previous IEA projections have understated non-OPEC oil supply. This is in part due to the high oil prices in recent years coupled with technological developments that have increased the viability of some marginal oil reserves. The high prices also pushed non-OPEC producers to maximise their output.

- The comparison of the IEA's latest projections with those of other organisations reveals a broad consensus on the rate of long-term energy demand growth but shows there is less agreement on the evolution of the fuel mix.

- Most new energy demand has come, and will continue to come, from the developing world, where uncertainty about economic growth is at its highest.

Assessment of Past Projections

Overview of WEO Series

The *World Energy Outlook* (*WEO*) was published annually from 1993 to 1996 and has appeared biennially (in even-numbered years) since then. Its primary objective is to identify and quantify global trends in energy demand and supply. Since its first edition the *WEO* has undergone many changes. Data have been increasingly disaggregated by region, use and fuel. The projection period has lengthened. More analysis is now given to the energy needs of developing countries.

The IEA's World Energy Model is the principal tool used to generate the detailed *WEO* projections. The model is refined with each new *Outlook* to include the most recent data, to adjust assumptions and to update the treatment of other factors, such as technological developments, renewable energy sources and supply-side issues. Because of lags in data collection and analysis, the final energy data included are for two years earlier than the publication date of each *Outlook*.

An important characteristic of the Reference Scenario in each *WEO* is the standard assumption that government policies will *not* change over the *Outlook* period. Government policies invariably do, of course, change. For this reason *WEO* projections should not be taken as forecasts, but rather as a baseline vision of how energy markets might evolve if governments did nothing more or less than they have already committed themselves to do. Recent *WEOs* have included an Alternative Policy Scenario to assess the impact of a range of possible new energy and environmental policies, as well as possible technology advances and accelerated technology deployment.

Exogenous Variables

GDP Assumptions

Economic growth is by far the most important driver of energy demand. The link between energy demand and economic output remains roughly linear, despite some signs of its loosening. Only the oil price shocks of 1973 and 1979-1980, the very warm weather of 1990 and the high energy prices of 2000 have altered this relationship to any significant degree.

The *WEOs'* economic growth assumptions for the short to medium term are based largely on those prepared by the Organisation for Economic Co-operation and Development, the International Monetary Fund and the World Bank. Over the long term, growth in each region is assumed to converge to a particular long-term rate. This rate is dependent on demographic and productivity trends, macroeconomic conditions and the pace of technological change.

Assumptions for short–term world economic growth (through to 2000) were revised upwards in each edition of the *WEO* from 1994 until 1998. This was due initially to expectations of rapid recovery in the former Soviet Union and in Central and Eastern Europe. Later in the decade, projected growth was raised to account for an economic resurgence in Asia. Similarly, assumptions for economic growth through to 2010 have been revised upward in each edition of the *WEO* since 2000. These revisions followed the quicker-than-expected recovery of Asia from the economic crisis of 1997-1998 and progressively more optimistic assessments of growth prospects in China, the United Sates and the transition economies.

Table B.1 compares the GDP assumptions made in previous *WEOs* with the latest available data or most recent estimates. For the period leading up to 2000, economic growth was systematically underestimated in the *WEO* by an average one-half of a percentage point. The *WEOs*' projections for global annual long-term economic growth, beyond 2000, have remained fairly stable at around 3%. Compared to the latest long-term GDP estimates it appears that the historical *WEO* assumptions have underestimated future growth, but this is yet to be seen.

Table B.1: **WEO GDP Growth-Rate Assumptions**

World Energy Outlook	Assumption period	OECD*		World*	
		Initial assumption	Latest estimate	Initial assumption	Latest estimate
WEO-1993	1991-2010	2.4	2.6	3.0	3.4
WEO-1994	1991-2000	2.3	2.8	2.6	3.3
WEO-1995	1992-2000	2.6	2.9	3.0	3.5
WEO-1996	1993-2000	2.7	3.2	3.2	3.7
WEO-1998	1995-2000	2.6	3.3	3.6	3.8
WEO-2000	1997-2020	2.0	2.5	3.1	3.4
WEO-2002	2000-2030	2.0	2.2	3.0	3.1

*Average annual growth rate (%).

Because of the uncertainty surrounding projections of economic growth, several past *WEOs* have included separate models of the impact of alternative economic growth rates. The *WEO-1995's* high-growth scenario concluded that four-tenths of one percentage point in higher annual global GDP growth would translate into an 8% increase in energy demand by 2010. A low-growth scenario was also modelled. The two supplementary scenarios suggested that high and low economic growth have nearly symmetrical impacts on energy usage. Analogous scenarios included in the *WEO-1996* produced broadly similar results.

Oil Price Assumptions

The price of oil is an important determinant of demand and supply. The World Energy Model does not project or forecast the evolution of prices. The assumed price path reflects our best judgement about the prices needed to ensure sufficient supply to meet projected global demand. Past *WEOs* have assumed that the oil price would follow a smooth curve. This should not be interpreted as a prediction of stable prices, but rather as an indication of the trend line around which prices could fluctuate.

Up until this year, long-term oil price assumptions in real terms were revised downward in each successive edition of the *WEO* (Figure B.1). This has followed technological improvements which have reduced the cost of petroleum production and upward revisions to estimates of world resources. There has also been a change in the assumed shape of the price trend. Until 1996, it was thought that oil prices would edge continuously upward over time before levelling off. The view that an upward price trend was inevitable had been popular among analysts ever since the first oil shock. Since the *WEO-1996*, however, the assumed short-term trend has been for flat prices while prices would steadily increase in the medium to long term, in line with importers' increasing dependence on OPEC supply.

Before the *WEO-1996*, our assumptions tended to overestimate the actual price of crude oil. High oil prices in the aftermath of the first Gulf War were, in fact, followed by an era of relatively low prices. This resulted from

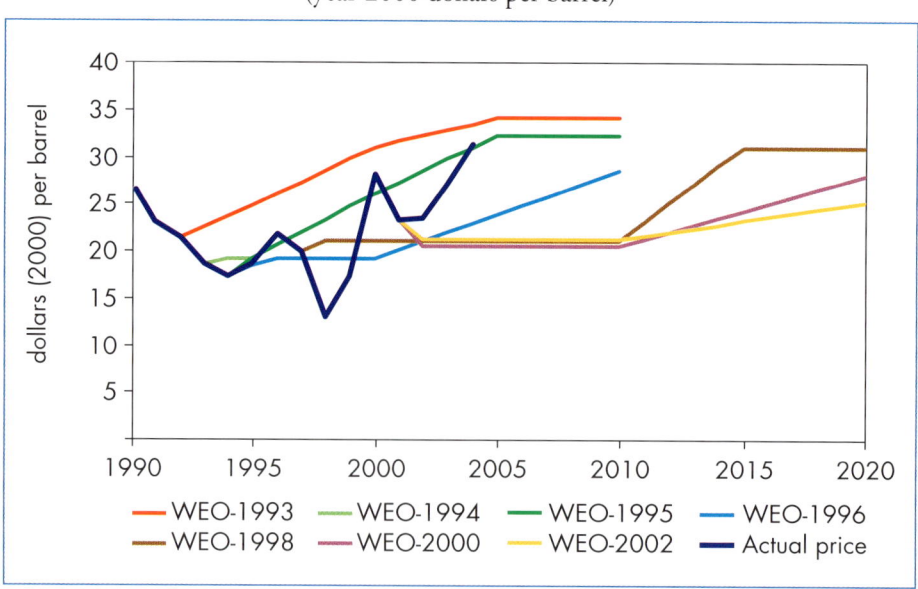

Figure B.1: **Past IEA Oil Import Price Assumptions**
(year-2000 dollars per barrel)

unwillingness on the part of producers to limit supply and deterred them from large investments in new capacity. Technical advances in oil exploration and increased competition from other fuels also helped dampen prices. Since the *WEO-1996*, the situation has turned around and our oil price assumptions have tended to be on the low side. The higher price trend can be attributed to the appearance of sustained OPEC cohesion since 1999 and ongoing global tensions.

Figure B.2 shows the average annual absolute deviation of previous *WEO* assumptions for the IEA crude oil import price from actual prices. On average, *WEO* oil price assumptions have been within 19% of the recorded level. Based on the IEA import price for 2003 of $27.10 per barrel in year-2000 dollars, this is equivalent to a deviation of $5.15. To date, the most accurate price projections were those in the *WEO-2000*, which were still 12% lower than reality.

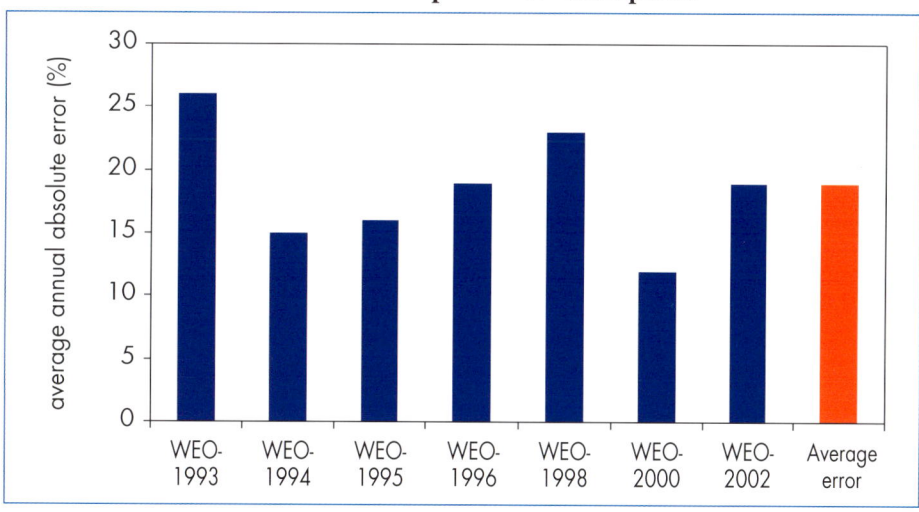

Figure B.2: **Average Annual Absolute Deviation of Past IEA Oil Import Price Assumptions***

* Measured from year of first assumption until 2003.
Note: Errors are defined as: [(assumed value − actual value)/assumed value] × 100.

Projections

Total Primary Energy Demand

Table B.2 shows the percentage deviations of our global energy demand projections from 1993 to 2002. On average, the projections have fallen within 2.2% of most recently reported data. Based on 2002 demand of 9.3 trillion tonnes of oil equivalent,[1] this represents an average deviation of 204 Mtoe.

1. This excludes non-commercial biomass. Non-commercial biomass is being included in the world energy balance for the first time in this edition of the *WEO*.

The accuracy of *WEO* demand projections has benefited from the reasonably steady course of world economic growth since 1993. The most accurate projections were those in the *WEO-1995*.

Table B.2: **Global Energy Demand: Per Cent Errors, 1993 to 2002**

Projection year	*World Energy Outlook* edition							Average absolute per cent error
	1993	1994	1995	1996	1998	2000	2002	
1993	1.5							1.5
1994	2.3	2.0						2.2
1995	1.5	1.2	0.7					1.1
1996	0.1	-0.1	-0.6	0.6				0.3
1997	0.8	0.7	0.2	1.8				0.9
1998	2.0	2.0	1.5	3.4	2.6			2.3
1999	1.7	1.8	1.3	3.6	2.9			2.3
2000	0.9	1.0	0.6	3.2	2.5	0.9		1.5
2001	2.7	3.0	2.5	5.1	4.2	2.6		3.4
2002	2.8	3.0	2.6	5.1	4.1	2.6	2.6	3.3
Average absolute per cent error	1.6	1.7	1.2	3.2	3.3	2.0	2.6	2.2

Note: Errors are defined as: [(projected value − actual value)/projected value] × 100.

Oil Demand

The IEA's past projections for global oil demand have fallen on average within 2.3% of the most recently reported data (Table B.3). Based on 2003 demand of 79.6 mb/d, this is equivalent to a deviation of 1.8 mb/d. Given the length of time that has passed, the *WEO-1994* has proved to be remarkably accurate. The very modest growth in world oil demand in 2002, caused by high prices and slow economic growth, led to the worst of our errors.

In recent years the surging pace of Chinese oil demand, driven by breakneck economic growth, has left many analysts scrambling to revise their earlier projections. As China has become heavily reliant on foreign oil, its rapid demand growth has had a significant effect on world oil markets and has been a key driver of escalating oil prices. As it happens, this trend was identified in the *WEO-1996*, which concluded that galloping economic growth in Asia, led by China, would be accompanied by rapid energy demand growth which in turn would have significant implications on world oil markets. It projected Chinese oil demand of 5.1 mb/d by 2003. With demand standing at the time at less than 3.5 mb/d, this appeared rather bullish. But this projection has proved to be remarkably close to actual demand of 5.5 mb/d.

Errors in past oil price assumptions have not translated into major errors in oil-demand projections partly owing to the high tax components in gasoline and diesel prices. In many countries, particularly in OECD Europe, end-users, especially car owners, do not "feel" the full impact of higher crude oil prices, because they result in smaller percentage increases in the price of fuel at the pump.

Table B.3: **Global Oil Demand: Per Cent Errors, 1993 to 2003**

Projection year	World Energy Outlook edition							Average absolute per cent error
	1993	1994	1995	1996	1998	2000	2002	
1993	3.5							3.5
1994	3.1	2.0						2.5
1995	2.3	1.4	-1.0					1.6
1996	1.1	0.3	-1.6	-0.5				0.9
1997	-0.5	-1.2	-2.7	-1.4				1.5
1998	-0.4	-0.9	-2.0	-0.5	-1.2			1.0
1999	3.0	2.7	2.0	3.7	2.9			2.8
2000	1.2	1.0	0.7	2.6	1.6	0.2		1.2
2001	2.6	2.6	2.6	4.5	3.4	2.3		3.0
2002	4.5	4.7	5.0	6.8	5.7	4.7	0.4	4.5
2003	2.3	2.8	3.4	5.3	4.0	3.3	-1.4	3.2
Average absolute per cent error	2.2	2.0	2.3	3.2	3.1	2.6	0.9	2.3

Note: Errors are defined as: [(projected value – actual value)/projected value] × 100.

Non-OPEC Oil Supply

The period since 1993 has seen dramatic changes in the source of oil supply, especially in non-OPEC production. Between the *WEO-1995* and the *WEO-2002*, projections for the market share of non-OPEC oil production in 2010 were revised upward from 51% to 60%. Despite these revisions, projections we made before the *WEO-2002* still tended to understate non-OPEC oil supply (Figure B.3). This underestimation of non-OPEC oil supply did not translate into significant errors in our overall estimates of oil supply because the World Energy Model calculates OPEC production as the difference between projected world oil demand and non-OPEC oil supply.

Early editions of the *WEO* expected OPEC's market share to increase rapidly based on the economics of oil production and exploration. Instead, non-OPEC supplies, especially from the North Sea, came into the market at prices much lower than those assumed by the *WEO*, and by the majority of analysts. In more recent years, sustained high prices have encouraged some non-OPEC producers to maximise their output. A good example of this has been the remarkable rebound in production in Russia since 2000. Technological advances in oilfield exploration, development and production have also buoyed prospects for some high-cost supplies, such as ultra-heavy crudes and deep offshore oil.

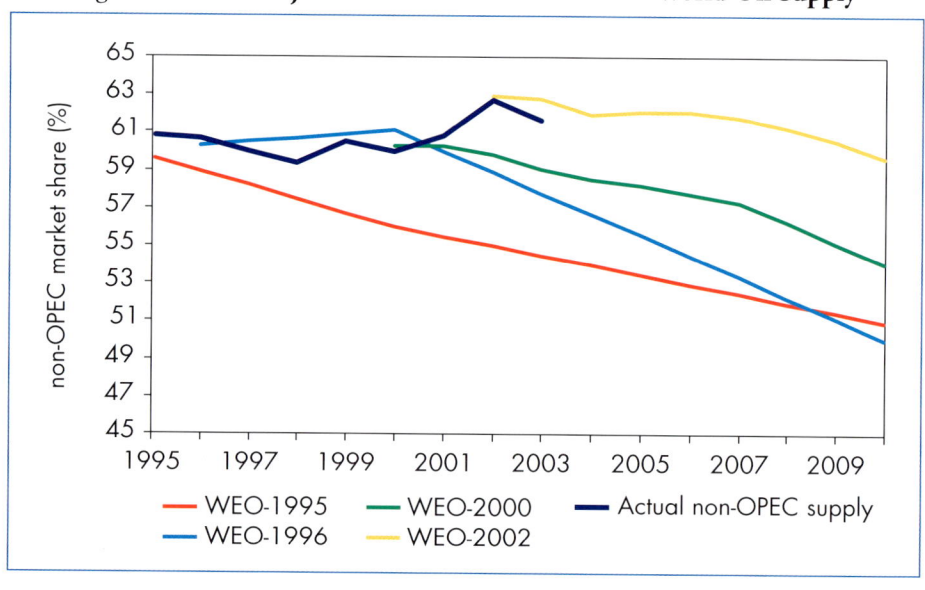

Figure B.3: **Past Projections of Non-OPEC Share of World Oil Supply**

Electricity Demand

Past *WEO* projections for electricity demand have done better than those for oil, falling within 1.5% of the outcome (Table B.4). Based on global demand in 2002 of 1 139 Mtoe, this is equivalent to an average deviation of just 17 Mtoe. This high level of accuracy is due in part to the particularly robust relationship that exists between electricity use and economic growth.

Table B.4: **Global Electricity Demand: Per Cent Errors, 1993 to 2002**

Projection year	\multicolumn{7}{c}{*World Energy Outlook* edition}	Average absolute per cent error
	1993	1994	1995	1996	1998	2000	2002	
1993	0.4							0.4
1994	0.3	0.9						0.6
1995	-0.6	0.0	-0.3					0.3
1996	-1.4	-0.8	-1.0	0.2				0.9
1997	-1.4	-0.8	-0.9	0.9				1.0
1998	-1.4	-0.8	-0.9	1.5	2.2			1.4
1999	-1.9	-1.3	-1.2	1.7	2.8			1.8
2000	-3.8	-3.2	-3.1	0.5	1.9	-1.9		2.4
2001	-1.7	-1.0	-0.9	2.6	4.1	-0.2		1.7
2002	-1.0	-0.2	-0.1	3.2	4.7	0.1	1.4	1.5
Average absolute per cent error	1.4	1.0	1.0	1.5	3.2	0.7	1.4	1.5

Note: Errors are defined as: [(projected value – actual value)/projected value] × 100.

Projection Comparison

We will now compare the latest *WEO* projections with those of other organisations in order to identify areas of general agreement as well as areas of high uncertainty. The subjects compared are global energy consumption, world oil demand, OPEC oil supply and the key underlying assumptions for economic growth and oil prices.

A number of organisations publish energy outlooks that are comparable to the *WEO*. This comparison has drawn on the "reference scenario" or "base case" from the following reports:

- *International Energy Outlook 2004*; US Department of Energy, Energy Information Administration (EIA).

- *Oil Outlook to 2025*; Organisation of the Petroleum Exporting Countries (OPEC).

- *Asia/World Energy Outlook (2004)*; Institute of Energy Economics, Japan (IEEJ).

- *World energy, technology and climate policy outlook (2003)*; European Commission (EC).

A number of other organisations make projections. Some of these are referred to below to broaden the comparison, including the Centre for Global Energy Studies (CGES) and Shell.

GDP Assumption

Although the time periods covered by the various GDP assumptions are slightly different, the generally held view is that long-term economic growth will average around 3% (Figure B.4). This implies a doubling of world economic output by around 2025. OPEC's GDP assumption, of 3.6% annual growth through 2025, is slightly higher than ours, and the IEEJ's assumption of 2.7% growth through 2020 is slightly lower. All the analyses see developing countries, especially China and India, as the main drivers of the world economy over the long term.

The assumed relationship between economic growth and energy use differs among the studies. In the current *WEO*, 1% of GDP growth entails a 0.53% increase in energy demand. In the IEEJ's *Asia/World Energy Outlook*, energy demand grows by 0.78% for each 1% hike in economic growth. This is illustrated in Figure B.4.

Figure B.4: **Global GDP Assumptions and Projected Energy Demand Growth Rates**

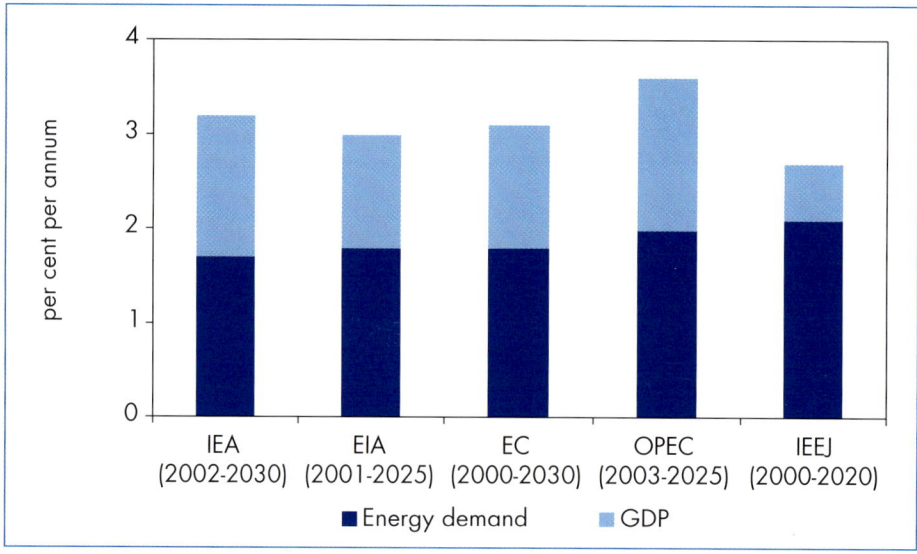

Oil Price Assumptions

The range of oil price estimates among different forecasting agencies is wide (Table B.5). A part of the difference can be attributed to technical factors, as the assumptions cover differing types or baskets of crude oil. The EIA, EC and

IEEJ all share the IEA's view that oil prices will drift higher in the long term, reflecting a gradual change in marginal production costs. OPEC's view differs from this, because the producers' organisation expects that its price-band concept will continue to operate successfully and that prices will stabilise and then remain constant in real terms just below $20 per barrel in year-2000 dollars. CGES see oil prices weakening in real terms over the projection period.

Table B.5: **Comparison of Long-Term Oil Price Assumptions**
(in year-2000 dollars per barrel)

Source	2010	2020	2030
IEA	22	26	29
EIA	23.3	25.1	
EC*	27.7	33.4	40.3
OPEC	19.3	19.3	
IEEJ	24.0	27.0	
CGES	20.5	15.1	

* Based on the average euro-dollar exchange rate for 2003 of 0.88.

Energy Demand Projections

Table B.6 compares world energy consumption projections up to 2020. A reasonable consensus exists on an annual rate of long-term energy demand growth of around 1.9%. The IEEJ study, with projected annual growth of 2.1%, sits above the norm, despite its moderate assumption for long-term economic growth.

Table B.6: **Growth in World Energy Consumption***

Fuel	IEA	EIA	EC	IEEJ
Oil	1.8	1.8	1.9	1.9
Gas	2.6	2.1	2.8	2.6
Coal	1.6	1.5	2.2	2.0
Nuclear	0.6	1.1	0.9	–
Renewable / Other	1.8	1.8	0.2	0.3
Total	1.9	1.8	1.9	2.1

* Average annual growth (%) for 2000-2020 except IEA which is 2002-2020.

The similarity in the overall energy consumption projections masks some important differences in the evolution of the fuel mix. In comparison to the IEA, the EIA expects much higher growth in nuclear energy and lower growth in natural gas. The EC and IEEJ see a very high rate of growth for coal consumption. In each study, natural gas is expected to be the fastest-growing fuel. The divergences, on the prospects for renewables and nuclear power, are due to the fact that demand for both is strongly influenced by government actions which are inherently difficult to foresee and open to much debate.

Oil Demand Projections

Table B.7 compares various long-term oil demand projections. Unlike the case of energy demand, there is considerable variance among the studies. The highest projection for demand in 2020 is the EIA's, at 110 mb/d, the lowest is the CGES "high price" scenario which sees demand at 90 mb/d. The similarity of the IEA's and OPEC's oil demand projections, despite very different GDP and price assumptions, may grow out of OPEC's optimistic view of the potential for technology to improve fuel efficiency. Shell's forecast is much lower than most others, largely because its "Dynamics as Usual" scenario projects a high share of new technologies in place in 2020. All studies foresee the growth in oil demand to be led by the transport sector in developing countries.

Table B.7: **Comparison of World Oil Demand Projections**

	Oil demand (mb/d)		
	2010	2020	2030
IEA	90	106	121
EIA	91	110	
OPEC	89	106	
IEEJ		102	
EC	87	104	120
CGES "high price"	82	90	
Shell*	85	95	

* Shell's Dynamics As Usual Scenario, 2000-2025.

ANNEX C

WORLD ENERGY MODEL 2004

Background

Since 1993, the IEA has provided long-term energy projections using a World Energy Model (WEM). The WEM underwent a significant transformation for the *WEO-2002*, extending the time horizon to 2030. The model has been further extended for the *WEO-2004*, including new and more detailed features and topics. These include:

- Twenty separately modelled countries and regions, including new, separate, models for OECD Asia, OECD Oceania and the European Union of twenty-five members in addition to regional aggregates developed in the *WEO-2002* (Figure C.2).

- More detailed sectoral representation of the industry, transport, residential, and services sectors for all major non-OECD countries, including detailed stock-transport models.

- A new model for renewable energies, both for power generation and for final uses.

- Improvements in the power generation model, with separate models for combined heat and power (CHP) plants, electricity-only plants and heat-only plants.

- Improvements in the oil and gas supply models, with more geographical detail for production and trade.

- A new coal supply model to cover world production and trade by region.

- Projections of electrification rates and traditional biomass use in the developing countries.

- A new model to evaluate the investment needed on the demand side to meet the energy reduction targets in the Alternative Policy Scenario.

A key reason for implementing these improvements was to develop the analysis for the Alternative Policy Scenario. In addition, a complex model was developed in the *World Energy Investment Outlook 2003* to evaluate investment needed in the fuel supply chain to satisfy projected energy demand over the next 30 years. This model has also been applied to the results of *WEO-2004*.

The WEM used to produce this *Outlook* is the eighth version of the model. It is designed to analyse:

- *Global energy prospects:* These include trends in demand, supply availability and constraints, international trade and energy balances by sector and fuel to 2030.
- *Environmental impact of energy use:* CO_2 emissions from fuel combustion are derived from the detailed projections of energy consumption.
- *Effects of policy actions or technological changes:* Alternative policy scenarios can be devised and run to analyse the impact of policy actions and technological developments on energy demand, supply, trade, investments and emissions. For the Alternative Policy Scenario, the WEM has been used to assess how the global energy market could evolve if countries around the world were to adopt a set of policies and measures that they are either currently considering or might reasonably be expected to implement over the projection period.

Model Structure

The WEM is a mathematical model made up of five main modules: *final energy demand; power generation; refinery and other transformation; fossil fuel supply and CO_2 emissions.* Figure C.1 provides a simplified overview of the structure of the model.

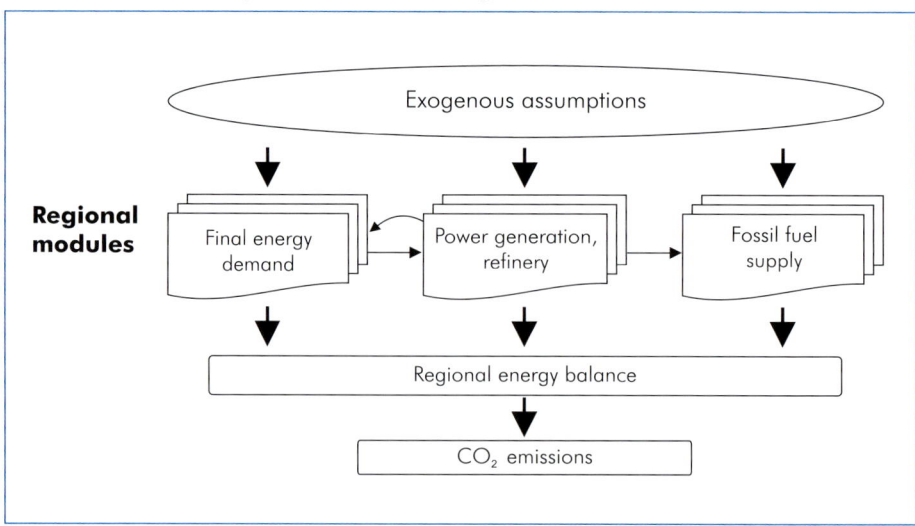

Figure C.1: **World Energy Model Overview**

Figure C.2: **World Energy Model Regions**

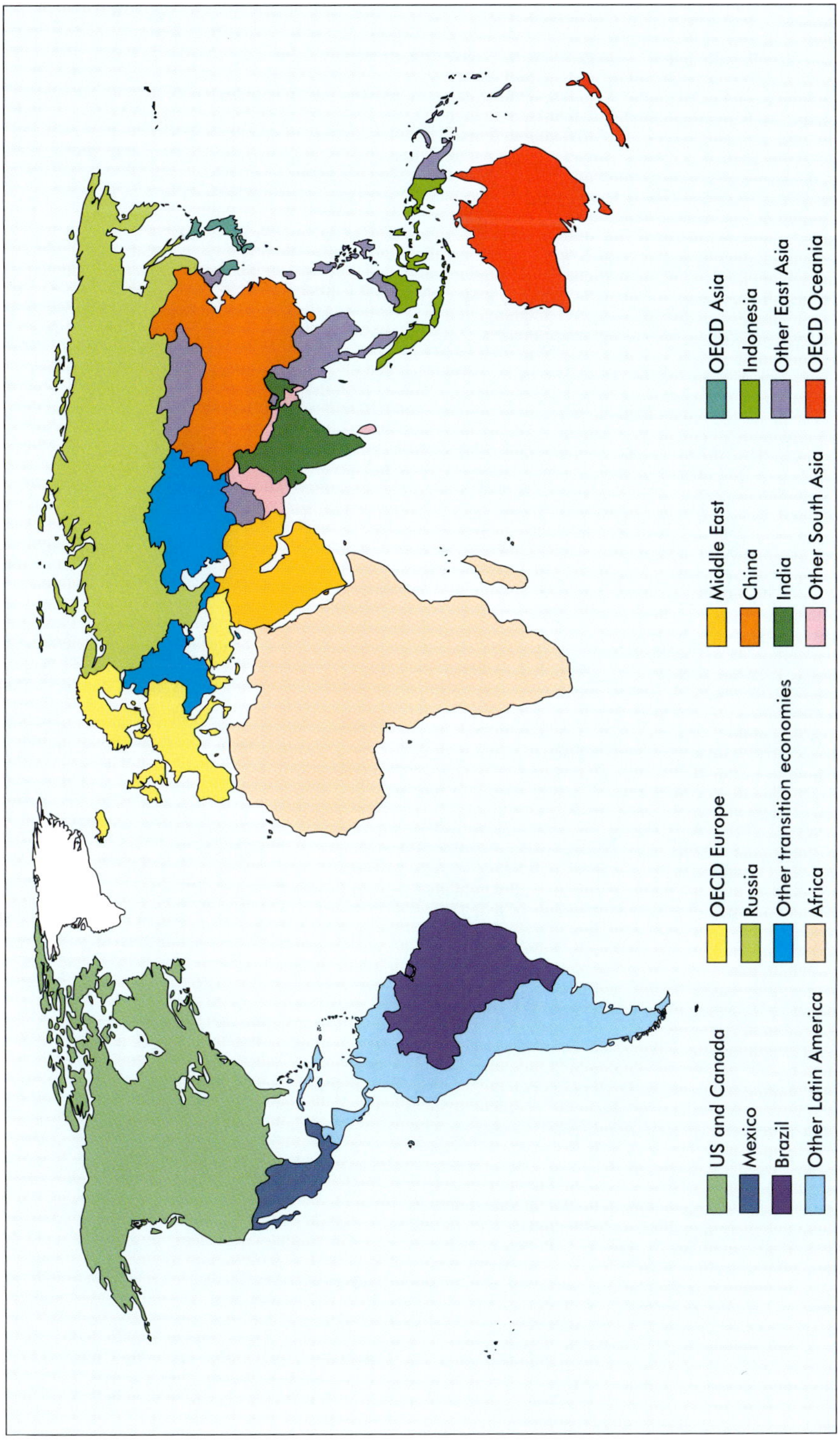

Note: A separate model for EU 25 has been added in the *WEO-2004*.

The main exogenous assumptions concern economic growth, demographics, international fossil fuel prices and technological developments. Electricity consumption and electricity prices dynamically link the final energy demand and power generation modules. Primary demand for fossil fuels serves as input for the supply modules. Complete energy balances are compiled at a regional level, and the CO_2 emissions of each region are then calculated using derived carbon factors.

Technical Aspects

The development and running of the WEM requires access to huge quantities of historical data on economic and energy variables. Most of the data are obtained from the IEA's own databases of energy and economic statistics. A significant amount of additional data from a wide range of external sources is also used.

The parameters of the demand-side modules' equations are estimated econometrically, usually using data for the period 1971-2002. Shorter periods are sometimes used where data are unavailable or significant structural breaks are identified. To take into account expected changes in structure, policy or technology, adjustments to these parameters are sometimes made over the *Outlook* period, using econometric and other modelling techniques. In regions such as the transition economies, where most data are available only from 1992, it has not been possible to use econometric estimation. In such cases, our results have been prepared by using assumptions based on cross-country analyses or expert judgement.

Simulations are carried out on an annual basis. Demand modules can be isolated and simulations run separately. This is particularly useful in the adjustment process and in sensitivity analyses of specific factors.

The WEM makes use of a wide range of software, including specific database management tools, econometric software and simulation programmes.

Description of the Modules

Final Energy Demand

The OECD regions and the major non-OECD regions have been modelled in greater sectoral and end-use detail than in previous editions. Specifically:

- Industry is separated into six sub-sectors allowing a more detailed analysis of trends and drivers in the industrial sector.
- Residential energy demand is separated into five end-uses by fuel.
- Services demand is modelled as three end-uses by fuel.
- Transport demand is modelled in detail by mode and fuel.

This level of detail in the data is not always available for all non-OECD regions. Nonetheless, disaggregation in non-OECD countries/regions has been increased for *WEO-2004*.

Total final energy demand is the sum of energy consumption in each final demand sector. In each sub-sector or end-use, at least six types of energy are shown: coal, oil, gas, electricity, heat and renewables. However, this aggregation conceals more detail. For example, the different oil products are modelled separately for the transport sector, and renewables are split into "biomass and waste" and "other renewables".

Within each sub-sector or end-use, energy demand is estimated as the product of an energy intensity and an activity variable. For example, the projection of the consumption of gas by a single household for water heating is multiplied by the projection of the number of households with water heating by gas to obtain the total residential consumption of gas for water heating.

In most of the equations, energy demand is a function of the following explanatory variables:

- *Activity variables:* This is often a GDP or GDP-per-capita variable. In many cases, however, a specific activity variable, which is usually driven by GDP, is used. For example, in the OECD regions demand in each industrial sub-sector is a function of the economic output of that sub-sector. In the transport sector, the vehicle stock, passenger-kilometres and tonne-kilometres of freight transported are used. In non-OECD regions, the situation differs from region to region, with more extensive detail for the main countries, such as Russia, China, India and Brazil. Demand-specific activity variables such as agricultural and iron and steel output are used less often in the non-OECD regions.

- *Prices:* End-user prices are calculated from assumed international energy prices. They take into account both variable and fixed taxes as well as transformation and distribution costs. For each sector, a representative price (usually a weighted average) is derived. This takes account of the product mix in final consumption and differences between countries. This representative price is then used as an explanatory variable – directly, with a lag, or as a moving average.

- *Other variables:* Other variables are used to take into account structural and technological changes, saturation effects or other important drivers (such as the gap between fuel efficiency in car manufacturers' tests and on the road).

Detailed capital stock models are integrated into the WEM model for the OECD and the main non-OECD regions in order to model the impact that capital stock turnover has on the penetration of new technology and equipment.

Industry Sector

The industrial sector in the OECD regions is split into six sub-sectors: *iron and steel, chemicals, paper and pulp, food and beverages, non-metallic minerals* and *other industry*. For the non-OECD regions, the breakdown is typically based on four instead of six sub-sectors.

The intensity of fuel consumption per unit of each sub-sector's output is projected on an econometric basis. The output level of each sub-sector is modelled separately and is combined with projections of its fuel intensity to derive the consumption of each fuel by sub-sector. This allows more detailed analysis of the drivers of demand and of the impact of structural change on fuel consumption trends.

The increased disaggregation also facilitates the modelling of alternative scenarios, where end-use shares and technology descriptions are applied in conjunction with capital stock turnover models to analyse in detail the impact of alternative policies or different choices of technology.

Transport Sector

For *WEO-2002*, the WEM fully incorporated a detailed bottom-up approach for the transport sector in all OECD regions. In *WEO-2004* the WEM model for transport has been extended to all major non-OECD regions (see Figure C.3).

Transport energy demand is split between *passenger* and *freight* and is broken down among *light duty vehicles, buses, trucks, rail, aviation* and *navigation*. Passenger cars and light trucks are subdivided by fuel used – gasoline, diesel, alternative fuels or hybrids of these. Freight trucks are divided between gasoline- and diesel-driven. The gap between test and on-road fuel efficiency is also projected.

For each region, activity levels for each mode of transport are estimated econometrically as a function of population, GDP and price. Additional assumptions to reflect passenger vehicle ownership saturation are also made. Transport activity is linked to price through elasticity of fuel cost per km, which is estimated for all modes except passenger buses and trains and inland navigation. This elasticity variable accounts for the "rebound" effect of increased car use that follows improved fuel intensity.

Energy intensity is projected by transport mode, taking into account changes in energy efficiency and fuel prices. Stock turnover is explicitly modelled in order to allow for the effects of fuel efficiency regulations for new cars on the energy intensity of the whole fleet.

Residential and Services Sectors

In *WEO-2002*, detail in the energy demand model for the residential and services sectors was significantly increased for the OECD regions. In *WEO-2004*, a similar increase in detail has been accomplished for major non-OECD regions (Figure C.4). For the other non-OECD regions, energy consumption in these sectors has been calculated econometrically for each fuel as a function of GDP, the related fuel price and the lag of energy consumption.

Figure C.3: **Structure of the Transport Demand Module**

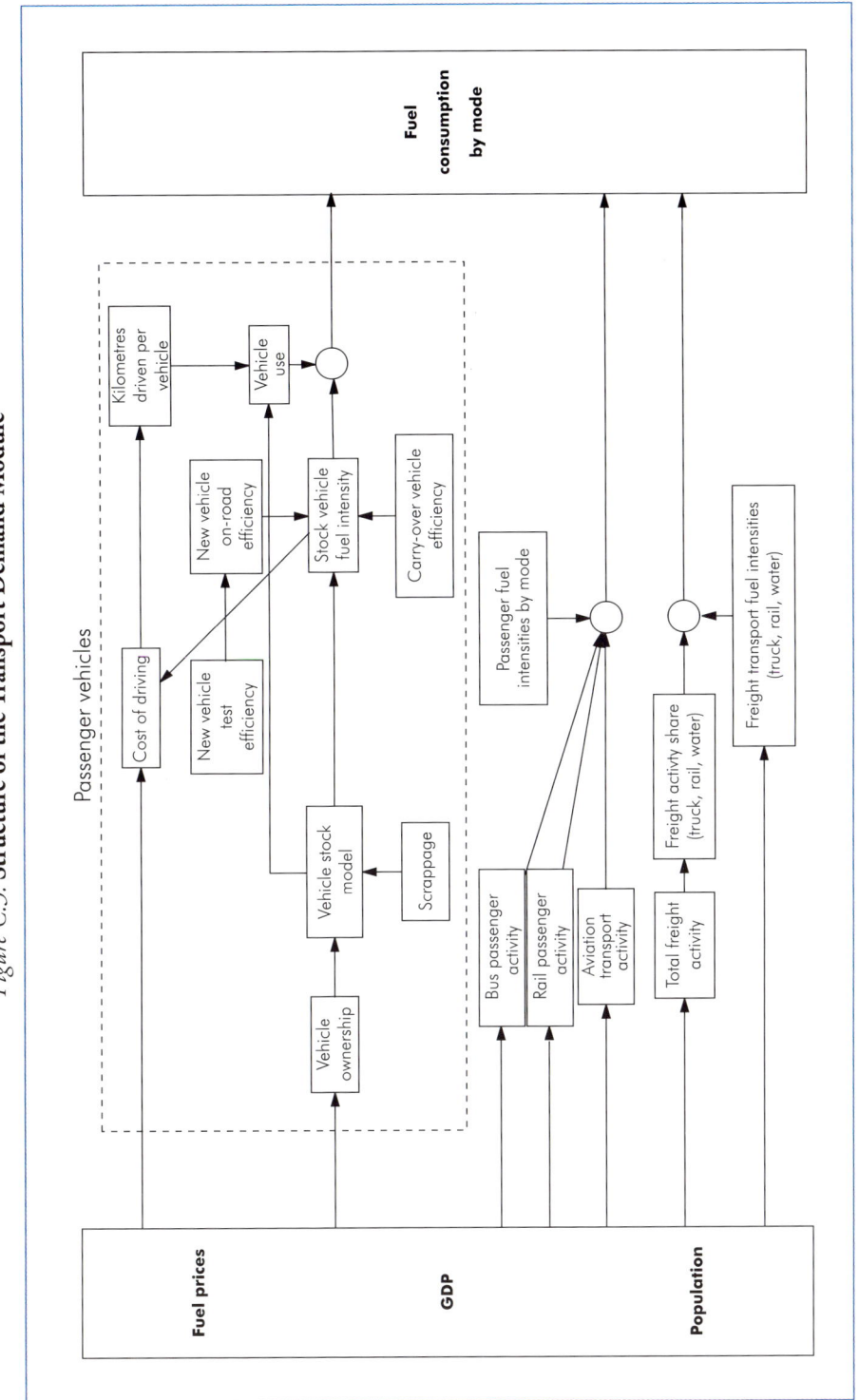

Annex C - World Energy Model 2004

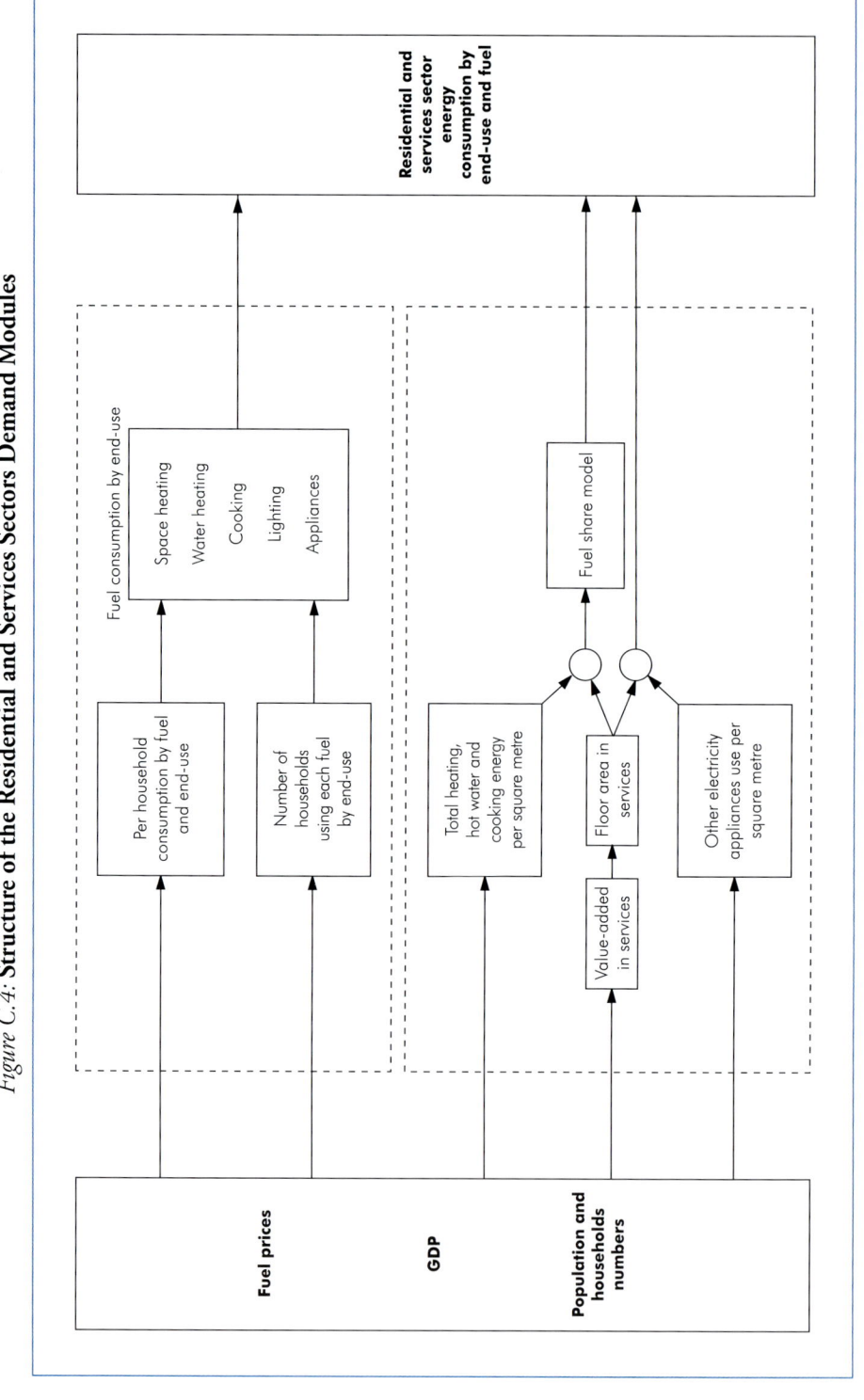

Figure C.4: **Structure of the Residential and Services Sectors Demand Modules**

For the OECD regions, the number of households using each fuel for water heating and space heating is projected econometrically, with some saturation limits on shares. The fuel intensity of space and water heating for each household is then estimated econometrically.

Lighting intensity and appliance intensity per household are then projected separately and combined with total household numbers to yield electricity demand for these end-uses. Detailed capital stock models are used to analyse the impact of new equipment standards and energy efficiency measures on individual appliances and on heating and cooling plant.

The services sector model splits consumption by fuel into three end-uses: heating, hot water and cooking use (HHC); personal computer use (including related equipment); and other electricity end-uses, including ventilation, space cooling and lighting. The total fuel demand for HHC is projected per square metre of floor area. Floor area in services is estimated as a function of value-added in the sector, which in turn is projected from GDP assumptions. The total demand for HHC is then allocated to two components: an "existing stock" model determines energy consumption based on historical shares of each fuel, while a portion of demand is allocated to "new stock" where fuel shares are a function of both relative prices and existing shares of each fuel.

Projections of PC-related electricity use and per-square-metre electricity use for the other electricity end-uses are combined to calculate their total electricity demand. The estimation of PC numbers is based on the growth in services sector employment and the size of the working population.

Electricity consumption in other end-uses is in most cases calculated econometrically on the basis of GDP and the electricity price.

Power Generation and Heat Plants

The power generation module calculates the following:

- Amount of electricity generated by each type of plant to meet electricity demand.
- Amount of new generating capacity needed.
- Type of new plants to be built.
- Fuel consumption of the power generation sector.
- Electricity prices.

The structure of the power generation module is described in Figure C.5. Electricity generation is calculated using the demand for electricity and taking into account electricity used by power plants themselves and system losses. The need for baseload, medium and peaking capacity is based on an assumed load curve. New generating capacity is the difference between total capacity requirements and plant retirements using assumed plant lives. When new

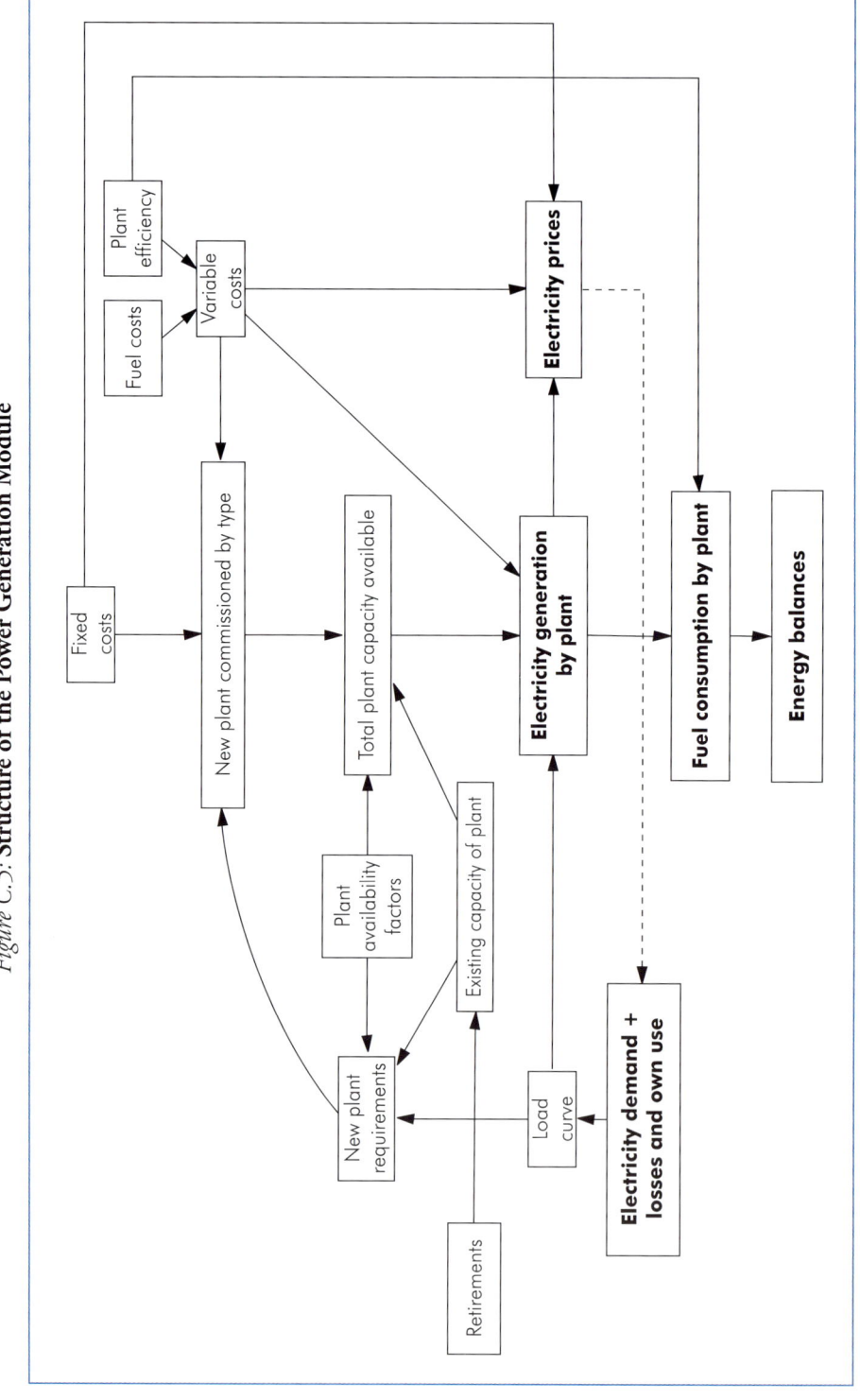

Figure C.5: **Structure of the Power Generation Module**

plant is needed, the model makes its choice on the basis of electricity generating costs, which combine capital, operating and fuel costs over the whole operating life of a plant, using a given discount rate, plant efficiency and plant utilisation rate. The model considers the following types of plant:

- Coal, oil and gas steam boilers.
- Combined-cycle gas turbine (CCGT).
- Open-cycle gas turbine (GT).
- Integrated gasification combined cycle (IGCC).
- Oil and gas internal combustion.
- Fuel cell.
- Nuclear.
- Biomass.
- Geothermal.
- Wind (onshore).
- Wind (offshore).
- Hydro (conventional).
- Hydro (pumped storage).
- Solar (photovoltaics).
- Solar (thermal).
- Tidal/wave.

Capacities for nuclear power plants are based on assumptions, which are in turn based on government plans. Where market conditions prevail, the assumptions are influenced by international fossil fuel prices.

Fossil fuel prices and efficiencies are used to rank plants in ascending order of their short-run marginal operating costs, allowing for assumed plant availability. Once the mix of generation plants has been determined, the fuel requirements are then deduced by plant type, using an assumed efficiency.

The marginal generation cost of the system is calculated, and this cost is then fed back to the demand model to determine the final electricity price.

The combined heat and power (CHP) option is considered for fossil-fuel and biomass plants. CHP, renewables and distributed generation are sub-modules of the power generation module. The CHP sub-module uses the potential for heat production in industry and buildings together with heat demand projections, which are estimated econometrically in the demand modules. The distributed generation (DG) sub-module is based on assumptions about market penetration of DG technologies.

Renewables module

The projections of renewable electricity generation were derived in a separate model. We have assessed the future deployment of renewable energies for electricity generation and the investment needed for such deployment.[1] The methodology is illustrated in Figure C.6.

The model uses a database of dynamic cost-resource curves. The development of renewables is based on an assessment of potentials and costs for each source (biomass, hydro, photovoltaics, solar thermal electricity, geothermal electricity, on- and offshore wind, tidal and wave) in each of twenty world regions. By defining financial incentives for the use of renewables and non-financial barriers in each market, as well as technical and societal constraints, the model calculates deployment as well as the resulting investment needs on a yearly base for each renewable source in each region. The model includes the concept of

---

Box C.1 **Long-Term Potential of Renewables**

The starting point for deriving future deployment of renewables is the assessment of long-term realisable potentials for each type of renewable and for each world region. The assessment is based on a review of the existing literature and on the refinement of available data. It includes the following steps:

1. The *theoretical* potentials for each region are derived. General physical parameters are taken into account to determine the theoretical upper limit of what can be produced from a certain energy, based on current scientific knowledge.

2. The *technical* potential can be derived from an observation of such boundary conditions as the efficiency of conversion technologies and the available land area to install wind turbines. For most resources, technical potential is a changing factor. With increased research and development, conversion technologies might be improved and the technical potential increased.

3. Long-term *realisable* potential is the fraction of the overall technical potential that can be actually realised. To estimate it, overall constraints like technical and economical feasibility, social acceptance, planning requirements and industrial growth are taken into consideration.

---

1. For a detailed description of this model – developed by Energy Economics Group (EEG) at Vienna University of Technology in co-operation with Wiener Zentrum für Energie, Umwelt und Klima – see Resch *et al.* (2004).

Figure C.6: **Mehtod of Approach for the Renewables Module**

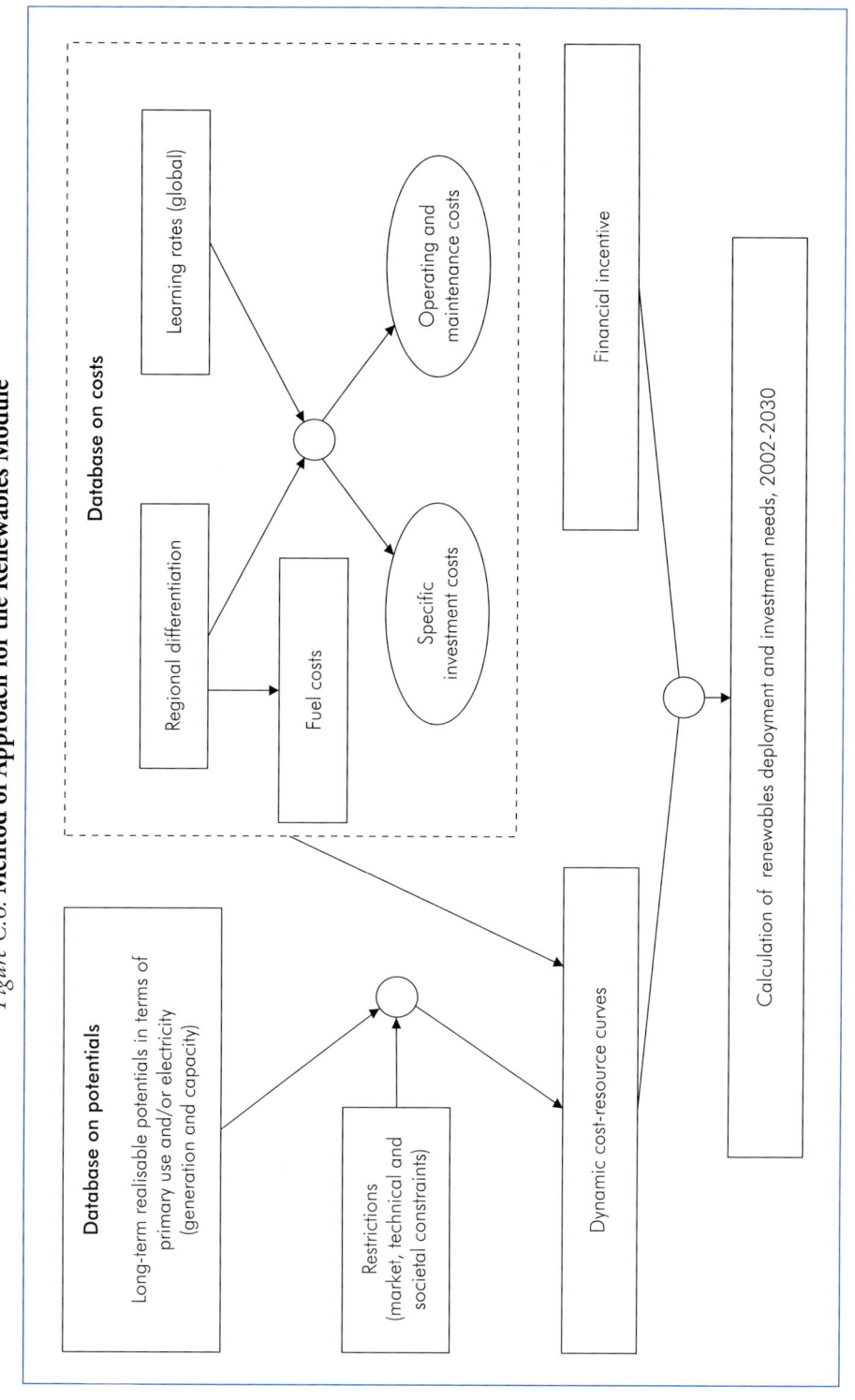

"technological learning". This concept holds that a certain increase in the production and sale of new technology will lead to a given decrease of the price.

The model uses *dynamic cost-resource curves*.[2] The approach consists of two parts: First, for each renewable source within each region, *static cost-resource curves*[3] are developed. For new plant, we determine long-term marginal generation costs. Realisable long-term potentials have been assessed for each type of renewable in each region.

Next, the model develops for each year a dynamic assessment of the previously described static cost-resource curves, consisting of:

Dynamic cost assessment: The dynamic adaptation of costs (in particular the investment and the operation and maintenance components) is based on the approach known as "technological learning". Learning rates are assumed by decade for specific technologies.

Dynamic restrictions: To derive realisable potentials for each year of the simulation, dynamic restrictions are applied to the predefined overall long-term potentials. Default figures are derived from an assessment of the historical development of renewables and the barriers they must overcome, which include:

- Market constraints: The penetration of renewables follows an S-curve pattern, which is typical of any new commodity.[4] Within the model, a polynomial function has been chosen to describe this impact – representing the market and administrative constraints by region.
- Technical barriers: Grid constraints are implemented as annual restrictions which limit the penetration to a certain percentage of the overall realisable potential.

CO$_2$ Emissions

For each region, sector and fuel, CO_2 emissions are calculated by multiplying energy demand by an implied carbon emission factor. Implied emission factors for coal, oil and gas differ between sectors and regions, reflecting the product mix. They have been calculated from year-2002 IEA emission data for all regions.

---

2. The concept of dynamic cost-resource curves in the field of energy policy modelling was originally devised for the research project Green-X, a joint European research project funded by the European Union's fifth Research and Technological Development Framework Programme – for details see www.green-x.at.
3. Renewable energy sources are characterised by limited resources. Costs rise with increased utilisation, as in the case of wind power. One tool to describe both costs and potentials is the (static) cost-resource curve. It describes the relationship between (categories of) available potentials (wind energy, biomass, hydropower) and the corresponding (full) costs of utilisation of this potential at a given point of time.
4. An S-curve shows relatively modest growth in the early stage of deployment, as the costs of technologies are gradually reduced. As this is achieved, there will be accelerating deployment. This will finally be followed by a slowing-down, corresponding to near-saturation of the market.

Fossil Fuel Supply

Oil module

The purpose of this module is to determine the level of oil production and trade in each region. Production is split into three categories:

- Non-OPEC.
- OPEC.
- Non-conventional oil production.

Total oil demand is the sum of regional oil demand, world bunkers and stock variation. OPEC conventional oil production is assumed to fill the gap between non-OPEC production and non-conventional and total world oil demand (Figure C.7).

Figure C.7: **Structure of Oil Supply Module**

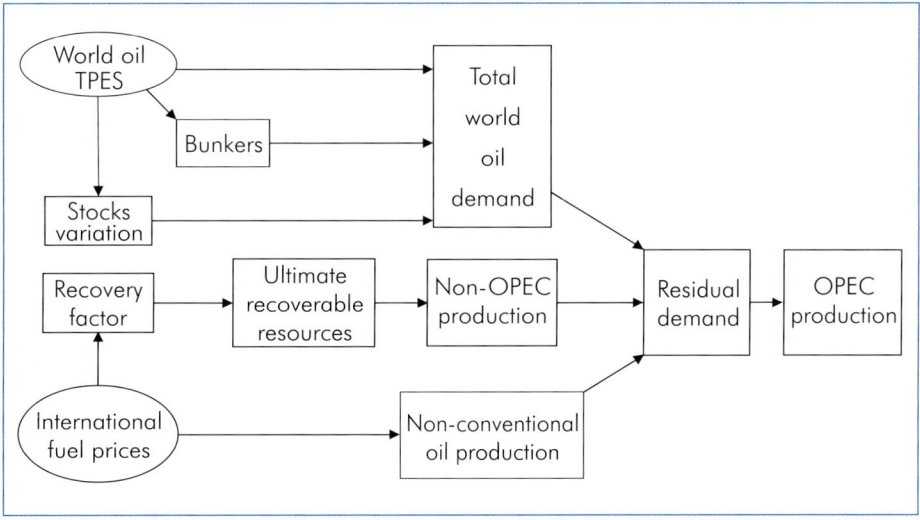

The derivation of non-OPEC production of conventional oil (crude and natural gas liquids) uses a combination of two different approaches. A short-term approach estimates production profiles based on a field-by-field analysis. A long-term approach involves the determination of production according to the level of ultimately recoverable resources and a depletion rate estimated by using historical data. Ultimately recoverable resources depend on a recovery factor. This recovery factor reflects reserve growth, which results from improvements in drilling, exploration and production technologies. The trend in the recovery rate is, in turn, a function of the oil price and of a technological improvement factor. Non-conventional oil supply is directly linked to the oil price. Higher oil prices bring forth greater non-conventional oil supply over time. The trade between regions is based on the extrapolation of market shares of major exporters.

Gas module

The gas module is similarly based on a resources approach. However, there are some important differences with the oil module. In particular, three regional gas markets are considered — North America, Europe and Asia — whereas oil is modelled as a single international market. Two country types are modelled: net importers and net exporters. Once gas production from each net-importing region is estimated, taking into account ultimately recoverable resources and depletion rates, the remaining regional demand is derived and then allocated to the net-exporting regions, again according to recoverable resources and depletion rates. Production in the net-exporting regions is consequently calculated from their own demand projections and exports needs. Trade is split between LNG and pipelines mainly according to the terms of existing long-term contracts, LNG and pipeline projects, transportation distances, and market shares of the different exporting regions.

Coal module

The coal module is a combination of a resources approach and an assessment of the development of domestic and international markets, based on the international coal price. Production, imports and exports are based on coal demand projections, and historical data on a country basis. Three markets are considered: coking coal, steam coal and brown coal. World coal trade, principally constituted of coking coal and steam coal, is separately modelled for the two markets and balanced on an annual basis.

Investment on the demand side

WEM energy consumption, end-use prices and income are used as an input to the calculations for the investments on the demand side.[5] The model uses a stock flow approach in modelling end-use energy demand. Flows are additions of equipment to meet growing service demands. Additions must also account for replacements due to retirements. Energy-related services are provided from the existing stocks of equipment, vehicles, and structures. The resulting energy services are a summation over all new and existing equipment vintages. Half of the new equipment put in place in the current year is assumed to be available for use in that year.

The energy intensity, capital investment intensity, and level of service demand are all functions of the energy price, cost of capital, and other prices and

---

5. These estimates are mostly based on a co-operative effort between the Argonne Laboratory in the United States and the IEA. For certain fuels and end-uses, not covered by the Argonne model, the costs are based on IEA estimates.

incomes. These, of course, vary by sector and by country or region. The formulas for energy and capital factor demands (associated with new additions to meet a given service demand) are derived from fitting the many technology options data to constant elasticity of substitution (CES) functional forms. The functions look like "isoquants" showing the incremental capital needed to reduce energy consumption of an appliance, piece of equipment, process, facility, or vehicle.[6]

Investment in the fuel supply chains

Projections of investment needs in the fuel supply chain are based on the methodology reported in the *World Energy Investment Outlook 2003*.[7]

---

6. Hanson and Laitner (2004).
7. IEA (2003).

ANNEX D

THE PRECARIOUS STATE OF ENERGY STATISTICS

Major challenges have arisen in recent years to the compiling of reliable energy statistics. Serious problems can be avoided, but only if nations and international organisations act quickly to reverse a series of negative trends.

Nowhere are the difficulties felt more sharply than in the International Energy Agency's Energy Statistics Division, where 25 full-time statisticians work year-round to produce authoritative statistics on every aspect of coal, electricity, gas, oil and renewable energy sources. These statistics, which provide a major input to the *World Energy Outlook*, cover 130 countries worldwide. Most time-series begin in 1960 for OECD countries and in 1971 for non-OECD countries.

Recently, however, maintaining the very high calibre of IEA statistics has become increasingly difficult, in many cases because national administrations have faced growing problems in maintaining the quality of their own statistics. Breaks in time-series and missing data have become frequent in some countries. These lapses compromise the completeness of our statistics. They could seriously affect any type of analysis, including modelling and forecasting.

Some reasons for the decline in the quality of energy statistics

1. **Energy market liberalisation:** Where, in the past, statisticians could obtain detailed information on gas or electricity from a single national utility company, they now have to survey tens, even hundreds, of companies. Moreover, the multiplication of companies has spawned confidentiality issues that add to the difficulty of collecting statistics.

2. **Additional data** have been required from statisticians in recent years, on:
   - *Renewable energy*, much of which comes from small and remote sources, such as wind turbines and solar collectors and about which information is difficult to assemble.
   - The results of *energy-saving* policies in many countries, where policy-makers need detailed information to monitor the evolution of their efforts – information that must often be disaggregated down to very detailed levels.
   - The socio-economic data required to deliver meaningful *energy-efficiency indicators*, which present similar challenges.
   - Estimations of *greenhouse gas emissions* needed in preparing national emissions inventories.

3. **Incomplete revisions:** For lack of funding, some countries cannot revise their entire time-series whenever there is a change in definitions or methodology. This can lead to breaks in the series that will have a negative effect on their use in econometric modelling and forecasting.

4. **Budget cuts:** In some countries, the number of statistical staff has been cut by half.

What can be done to reverse the trends?

Governments need to reassess the resources they devote to statistics and to readjust them, when necessary. Statistics and statisticians should be fully integrated into the energy-policy decision-making process of each country. Policy-makers need to become more aware of the limits and problems encountered by statisticians, and statisticians to understand better how the data they collect are used.

Procedures need to be adapted to the new energy environment featuring liberalisation, mergers and the rapid development of trade. The legal framework for statistics gathering needs to be reconsidered. There should be closer co-operation with the energy industry. Surveys should be made more consistent with the needs of data users.

To this end, together with five other international organisations dealing with energy statistics, the IEA launched in 2001 the Joint Oil Data Initiative (JODI), aimed at improving the quality of short-term oil data. Our partners are the Organisation of the Petroleum Exporting Countries (OPEC), the Statistical Office of the European Commission (Eurostat), the Asia Pacific Energy Research Centre (APERC), the Latin American Energy Organization (OLADE), and the Energy and Industry Statistics Section of the United Nations Statistics Division (UNSD). The initiative is a first step towards increasing co-operation between countries and international organisations, and improving communication among the international organisations themselves. We look forward to expanding these efforts to improve the reliability of wider energy statistics. Later this year, the IEA will organise an Energy Statistics Working Group to assess the relevance of the current energy questionnaires. We plan a similar meeting with international organisations.

Strengthening the expertise and experience of energy statisticians and rebuilding corporate memory are also key priorities. In line with this objective, the IEA, in co-operation with Eurostat, has prepared an *Energy Statistics Manual* which should help newcomers in the energy statistics field to have a better grasp on definitions, units, and methodology. Other initiatives should be considered for raising the level of expertise and the interest in the job, and therefore for raising the profile of statistics, a necessary condition for attracting and retaining highly motivated professionals.

The IEA has taken the unusual step of raising this issue in *World Energy Outlook 2004* because we believe there is an urgent need to preserve the reliability of our statistical base. We feel that more members of the energy community – especially policy-makers and legislators – should be aware of a looming crisis. We are convinced that the crisis can be met and overcome.

ANNEX E

DEFINITIONS, ABBREVIATIONS AND ACRONYMS

This annex provides general information on definitions, abbreviations and acronyms used throughout *WEO-2004*. Readers interested in obtaining more detailed information should consult the annual IEA publications *Energy Balances of OECD Countries; Energy Balances of Non-OECD Countries; Coal Information; Oil Information; Gas Information* and *Renewables Information*.

Fuel and Process Definitions

Coal
Coal includes all coal: both primary coal (including hard coal and lignite) and derived fuels (including patent fuel, brown-coal briquettes, coke-oven coke, gas coke, coke-oven gas and blast-furnace gas). Peat is also included in this category.

Oil
Oil includes crude oil, natural gas liquids, refinery feedstocks and additives, other hydrocarbons and petroleum products (refinery gas, ethane, liquefied petroleum gas, aviation gasoline, motor gasoline, jet fuels, kerosene, gas/diesel oil, heavy fuel oil, naphtha, white spirit, lubricants, bitumen, paraffin waxes, petroleum coke and other petroleum products).

Gas
Gas includes natural gas (both associated and non-associated with petroleum deposits but excluding natural gas liquids) and gas-works gas.

Biomass and Waste
Biomass includes solid biomass and animal products, gas and liquids derived from biomass, industrial waste and municipal waste.

Traditional Biomass
Traditional biomass refers mainly to non-commercial biomass use.

Other Renewables
Other renewables include geothermal, solar, wind, tide and wave energy for electricity generation. Direct use of geothermal and solar heat is also included in this category.

Heat

Heat is heat produced for sale. The large majority of the heat included in this category comes from the combustion of fuels, although some small amounts are produced from electrically-powered heat pumps and boilers. Heat not sold and consumed by autoproducers is not reported as heat consumption but as final consumption of the fuel used to produce it.

Nuclear

Nuclear refers to the primary heat equivalent of the electricity produced by a nuclear plant with an average thermal efficiency of 33%.

Hydro

Hydro refers to the energy content of the electricity produced in hydropower plants, assuming 100% efficiency.

Hydrogen Fuel Cell

A hydrogen fuel cell is a high-efficiency electrochemical energy conversion device that generates electricity and produces heat, with the help of catalysts. The chemical reaction involved is hydrogen + oxygen $\rightarrow$ water.

Energy Intensity

Energy intensity is a measure of total primary energy use per unit of gross domestic product.

Light Petroleum Products

Light petroleum products include liquefied petroleum gas, naphtha and gasoline.

Middle Distillates

Middle distillates include jet fuel, diesel and heating oil.

Heavy Petroleum Products

Heavy petroleum products include heavy fuel oil.

Other Petroleum Products

Other petroleum products include refinery gas, ethane, lubricants, bitumen, petroleum coke and waxes.

Hard Coal

Coal of gross calorific value greater than 5 700 kcal/kg on an ash-free but moist basis and with a mean random reflectance of vitrinite of at least 0.6. Hard coal is further disaggregated into coking coal and steam coal.

Coking Coal

Hard coal with a quality that allows the production of coke suitable to support a blast furnace charge.

Steam Coal

All other hard coal not classified as coking coal. Also included are recovered slurries, middlings and other low-grade coal products not further classified by type. Coal of this quality is also commonly known as thermal coal.

Brown Coal

Sub-bituminous coal and lignite. Sub-bituminous coal is defined as non-agglomerating coal with a gross calorific value between 4 165 kcal/kg and 5 700 kcal/kg. Lignite is defined as non-agglomerating coal with a gross calorific value less than 4 165 kcal/kg.

Coke-oven Coke

The solid product obtained from carbonisation of coal, principally coking coal, at high temperature. Semi-coke, the solid product obtained from the carbonisation of coal at low temperature is also included along with coke and semi-coke.

Peat

A combustible soft, porous or compressed fossil sedimentary deposit of plant origin with high water content (up to 90% in the raw state), easily cut, of light to dark brown colour.

Clean Coal Technologies (CCT)

Technologies designed to enhance the efficiency and the environmental acceptability of coal extraction, preparation and use.

Total Primary Energy Supply

Total primary energy supply is equivalent to primary energy demand. This represents inland demand only and, except for world energy demand, excludes international marine bunkers.

International Marine Bunkers

International marine bunkers cover those quantities delivered to sea-going ships of all flags, including warships. Consumption by ships plying in inland and coastal waters is not included.

Power Generation

Power generation refers to fuel use in electricity plants, heat plants and combined heat and power (CHP) plants. Both public plants and small plants that produce fuel for their own use (autoproducers) are included.

Total Final Consumption

Total final consumption is the sum of consumption by the different end-use sectors. TFC is broken down into energy demand in the following sectors: industry, transport, other (includes agriculture, residential, commercial and public services) and non-energy use. Industry includes manufacturing, construction and mining industries. In final consumption, petrochemical feedstocks appear under industry use. Other non-energy uses are shown under non-energy use.

Other Transformation, Own Use and Losses

Other transformation, own use and losses covers the use of energy by transformation industries and the energy losses in converting primary energy into a form that can be used in the final consuming sectors. It includes energy use and loss by gas works, petroleum refineries, coal and gas transformation and liquefaction. It also includes energy used in coal mines, in oil and gas extraction and in electricity and heat production. Transfers and statistical differences are also included in this category

Electricity Generation

Electricity generation shows the total amount of electricity generated by power plants. It includes own use and transmission and distribution losses.

Other Sectors

Other sectors include the residential, commercial and public services and agriculture sectors.

Regional Definitions

OECD Europe

OECD Europe consists of Austria, Belgium, the Czech Republic, Denmark, Finland, France, Germany, Greece, Hungary, Iceland, Ireland, Italy, Luxembourg, the Netherlands, Norway, Poland, Portugal, the Slovak Republic, Spain, Sweden, Switzerland, Turkey and the United Kingdom.

OECD North America

OECD North America consists of the United States of America, Canada and Mexico.

OECD Pacific

OECD Pacific consists of Japan, Korea, Australia and New Zealand.

OECD Asia

OECD Asia consists of Japan and Korea.

OECD Oceania

OECD Oceania consists of Australia and New Zealand.

Transition Economies

The transition economies include: Albania, Armenia, Azerbaijan, Belarus, Bosnia-Herzegovina, Bulgaria, Croatia, Estonia, the Federal Republic of Yugoslavia, the former Yugoslav Republic of Macedonia, Georgia, Kazakhstan, Kyrgyzstan, Latvia, Lithuania, Moldova, Romania, Russia, Slovenia, Tajikistan, Turkmenistan, Ukraine and Uzbekistan. For statistical reasons, this region also includes Cyprus, Gibraltar and Malta.

China

China refers to the People's Republic of China, including Hong Kong.

East Asia

East Asia includes: Afghanistan, Bhutan, Brunei, Chinese Taipei, Fiji, French Polynesia, Indonesia, Kiribati, Democratic People's Republic of Korea, Malaysia, Maldives, Myanmar, New Caledonia, Papua New Guinea, the Philippines, Samoa, Singapore, Solomon Islands, Thailand, Vietnam and Vanuatu.

South Asia

South Asia consists of Bangladesh, India, Nepal, Pakistan and Sri Lanka.

Latin America

Latin America includes: Antigua and Barbuda, Argentina, Bahamas, Barbados, Belize, Bermuda, Bolivia, Brazil, Chile, Colombia, Costa Rica, Cuba, Dominica, the Dominican Republic, Ecuador, El Salvador, French Guiana, Grenada, Guadeloupe, Guatemala, Guyana, Haiti, Honduras, Jamaica, Martinique, Netherlands Antilles, Nicaragua, Panama, Paraguay, Peru, St. Kitts-Nevis-Anguilla, Saint Lucia, St. Vincent-Grenadines and Suriname, Trinidad and Tobago, Uruguay, and Venezuela.

Africa

Africa comprises Algeria, Angola, Benin, Botswana, Burkina Faso, Burundi, Cameroon, Cape Verde, the Central African Republic, Chad, Congo, the Democratic Republic of Congo, Côte d'Ivoire, Djibouti, Egypt, Equatorial Guinea, Eritrea, Ethiopia, Gabon, Gambia, Ghana, Guinea, Guinea-Bissau, Kenya, Lesotho, Liberia, Libya, Madagascar, Malawi, Mali, Mauritania, Mauritius, Morocco, Mozambique, Niger, Nigeria, Rwanda, Sao Tome and Principe, Senegal, Seychelles, Sierra Leone, Somalia, South Africa, Sudan, Swaziland, the United Republic of Tanzania, Togo, Tunisia, Uganda, Zambia and Zimbabwe.

Middle East

The Middle East is defined as Bahrain, Iran, Iraq, Israel, Jordan, Kuwait, Lebanon, Oman, Qatar, Saudi Arabia, Syria, the United Arab Emirates and Yemen. It includes the neutral zone between Saudi Arabia and Iraq.

Annex I Parties to the Kyoto Protocol

Australia, Austria, Belarus, Belgium, Bulgaria, Canada, Croatia, the Czech Republic, Denmark, Estonia, the European Community, Finland, France, Germany, Greece, Hungary, Iceland, Ireland, Italy, Japan, Latvia, Liechtenstein, Lithuania, Luxembourg, Monaco, the Netherlands, New Zealand, Norway, Poland, Portugal, Romania, Russia, the Slovak Republic, Slovenia, Spain, Sweden, Switzerland, Turkey, Ukraine, the United Kingdom and the United States.

Asia

OECD Pacific, China, East Asia and South Asia.

Developing Asia

China, East Asia and South Asia.

European Union

The European Union consists of Austria, Belgium, Cyprus, the Czech Republic, Denmark, Estonia, Finland, France, Germany, Greece, Hungary, Ireland, Italy, Latvia, Lithuania, Luxembourg, Malta, the Netherlands, Poland, Portugal, the Slovak Republic, Slovenia, Spain, Sweden and the United Kingdom.

Organization of the Petroleum Exporting Countries (OPEC)

Algeria, Indonesia, Iran, Iraq, Kuwait, Libya, Nigeria, Qatar, Saudi Arabia, the United Arab Emirates and Venezuela.

Sub-Saharan Africa

Includes all African countries except North Africa (Algeria, Egypt, Libya, Morocco and Tunisia).

Abbreviations and Acronyms

bcm	billion cubic metres
b/d	barrels per day
boe	barrels of oil equivalent
CAFE	Corporate Average Fuel Economy
CBM	coal-bed methane
CCGT	combined-cycle gas turbine
CCS	carbon capture and storage
CDU	crude distillation unit
CHP	combined production of heat and power; sometimes, when referring to industrial CHP, the term co-generation is used
CNG	compressed natural gas
CO_2	carbon dioxide
DG	distributed generation
DoE	Department of Energy
EC	European Commission
EDI	Energy Development Index
EOR	enhanced oil recovery
EU	European Union

FDI	foreign direct investment
FSU	former Soviet Union
GDP	gross domestic product
GHG	greenhouse gas
Gt	gigatonnes (1 tonne × 10^9)
GTL	gas-to-liquids
GW	gigawatt (1 watt × 10^9)
GWh	gigawatt-hour
HDI	Human Development Index
IEA	International Energy Agency
IGCC	integrated gasification combined cycle
IMF	International Monetary Fund
IPP	independent power producer
kb/d	thousand barrels per day
kgoe	kilogrammes of oil equivalent
kW	kilowatt (1 watt × 1 000)
kWh	kilowatt-hour
LNG	liquefied natural gas
LPG	liquefied petroleum gas
mb/d	million barrels per day
MBtu	million British thermal units
mcm/d	million cubic metres per day
MDG	Millennium Development Goals
MSC	multiple service contract
mpg	miles per gallon
Mt	million tonnes
Mtoe	million tonnes of oil equivalent
MW	megawatt (1 watt × 10^6)
MWh	megawatt-hour
NGL	natural gas liquid
NO_x	nitrogen oxides

OECD	Organisation for Economic Co-operation and Development
OPEC	Organization of the Petroleum Exporting Countries
PPP	purchasing power parity
PPA	power purchasing agreement
R&D	research and development
RPS	renewables portfolio standards
RTO	Regional Transmission Organization (USA)
SO_2	sulphur dioxide
tcf	thousand cubic feet
tcm	trillion cubic metres
TFC	total final consumption
toe	tonne of oil equivalent
tonne	metric ton
TPES	total primary energy supply
TW	terawatt (1 watt × 10^{12})
TWh	terawatt-hour
UES	United Energy Systems
UNDP	United Nations Development Programme
WEM	World Energy Model
WEO	World Energy Outlook
WHO	World Health Organization
WTO	World Trade Organization

ANNEX F

REFERENCES

CHAPTER 1

American Chemistry Council (ACC) (2003), *Natural Gas Price Shocks and the Economy*, ACC, Arlington.

Asian Development Bank (ADB) (2004), *Key Indicators 2004*, ADB, Manilla.

BP (2004), *BP Statistical Review of World Energy 2004*, BP, London.

Development Bank of Japan (DBJ) (2004) *Statistical Yearbook 2004*, DBJ, Tokyo.

International Energy Agency (IEA) (2002), *World Energy Outlook 2002* OECD/IEA, Paris.

IEA (2003), *World Energy Investment Outlook: 2003 Insights*, OECD/IEA, Paris.

IEA (2004), "The Impact of High Oil Prices on the Global Economy", Economic Analysis Division Working Paper, May, OECD/IEA, Paris.

International Iron and Steel Institute (IISI) (2004), *World Steel in Figures*, IISI, Brussels.

International Monetary Fund (IMF) (2004), *World Economic Outlook: Advancing Structural Reforms*, April, IMF, Washington.

Organisation for Economic Co-operation and Development (OECD) (2004), *OECD Economic Outlook 75*, June, OECD, Paris.

United Nations Population Division (UNPD) (2003), *World Population Prospects: The 2002 Revision*, United Nations, New York.

US Geological Survey (USGS) (2004), *Mineral Commodity Summaries 2004*, USGS, Washington.

CHAPTER 2

International Energy Agency (2003), *World Energy Investment Outlook: 2003 Insights*, OECD/IEA, Paris.

CHAPTER 3

Appert, Olivier (2004), "Investment and Profitability of the Oil and Gas Sector", 9th International Energy Forum, 22-24 May, Amsterdam.

Baqi, Mahmoud M. Abdul and Nansen G. Saleri (2004), "Fifty Year Crude Oil Supply Scenarios: Saudi Aramco's Perspective", Presentation to Center for Strategic and International Studies (CSIS), February 24, Washington.

BP (2004), *BP Statistical Review of World Energy 2004*, BP, London.

Comité Professionnel du Pétrole (CPDP) (2003), *Pétrole 2003*, CPDP, Rueil-Malmaison.

Cupcic, F. (2003), *Journées du Pétrole 2003*, Paris.

Dyni, J.R. (2003), *Geology and Resources of Some World Oil-shale Deposits*, U.S. Geological Survey, Denver Federal Center, Colorado.

Energy Information Administration (EIA) (2004), *World Oil Transit Chokepoints Country Analysis Brief*, www.eia.doe.gov/cabs/choke.html.

Energy Intelligence Group, Inc. (2004), *The Oil Supply Dilemma: The Financial Position and Investment Needs of Major Oil Exporters*, Energy Intelligence Group, Inc., New York.

Hiller, K. (1999) "Verfügbarkeit von Erdöl", in *Erdöl, Erdgas, Kohle*, Vol. 115, February, OilGasPublisher, Hamburg.

International Energy Agency (IEA) (2002), *World Energy Outlook 2002*, OECD/IEA, Paris.

IEA (2003), *World Energy Investment Outlook: 2003 Insights*, OECD/IEA, Paris.

National Energy Board (2004), *Canada's Oil Sands: Opportunities and Challenges to 2015*, National Energy Board, Calgary.

Odell, Peter R. (2004), *Why Carbon Fuels Will Dominate the 21st Century's Global Energy Economy*, Multi-Science Publishing, Brentwood.

Oil and Gas Journal, various issues, Penwell Corporation, Oklahoma.

Organization of the Petroleum Exporting Countries (OPEC) (2003), *OPEC Annual Statistical Bulletin 2003*, OPEC, Vienna.

Ruud, Morten (2004), "Performance and Technology", Hydro Oil & Energy/Oil & Energy Business Area Seminar 2 – 3 June, Bergen.

Securities and Exchange Commission (SEC) (2003), Regulation S-X article 4-10, SEC.

Shell (2001), *Energy Needs, Choices and Possibilities*, Shell, London.

Society of Petroleum Engineers, World Petroleum Congress and American Association of Petroleum Geologists, SPE/WPC/AAPG (2000), *Petroleum Resources Classification and Definitions*, SPE, Richardson.

United States Geological Survey (USGS) (2000), *World Petroleum Assessment 2000*, USGS, Washington.

UTILIS Energy (2004), *Oil Sands – Alberta 2004*, UTILIS Energy, New York.

World Oil, various issues, Gulf Publishing Company, Houston.

CHAPTER 4

Cedigaz (2004), *Natural Gas in the World*, Institut Français du Pétrole, Rueil-Malmaison.

European Commission (EC) (2004), "Third benchmarking report on the implementation of the internal electricity and gas market", Draft Working Paper, DG TREN, Brussels.

International Energy Agency (IEA) (2003), *World Energy Investment Outlook: 2003 Insights*, OECD/IEA, Paris.

IEA (2004), *Security of Gas Supply in Open Markets: LNG and Power at a Turning Point*, OECD/IEA, Paris.

National Petroleum Council (NPC) (2003), *Balancing Natural Gas Policy: Fuelling the Demands of a Growing Economy*, NPC, Washington.

United States Geological Survey (USGS) (2000), *World Petroleum Assessment 2000*, USGS, Washington.

CHAPTER 5

BP (2004), *BP Statistical Review of World Energy 2004,* BP, London.

International Energy Agency (IEA) (2001), *World Energy Outlook: 2001 Insights: Assessing Today's Supplies to Fuel Tomorrow's Growth*, OECD/IEA, Paris.

IEA (2002), *World Energy Outlook 2002*, OECD/IEA, Paris.

IEA (2003a), *World Energy Investment Outlook: 2003 Insights*, OECD/IEA, Paris.

IEA (2003b), *Coal Information 2003*, OECD/IEA, Paris.

IEA (2004), *International Coal Trade – Contributing to Sustainable Energy Supply*, OECD/IEA, Paris, forthcoming.

McCloskey (2003), *Coal Report 65*, The McCloskey Group Ltd, Hampshire.

World Coal Institute (WCI) (2003), *The Role of Coal as an Energy Source*, WCI, London.

World Energy Council (WEC) (2003), *Survey of Energy Resources*, WEC, London.

CHAPTER 6

Autorità per l'Energia Elettrica e il Gas (AEEG) and Commission for the Regulation of Energy (CRE) (2004), *Report on the Events of September 28th, 2003, Culminating in the Separation of the Italian Power System from the Other UCTE Networks*, www.autorita.energia.it/

Commissariat à l'Energie Atomique (CEA) (2003), *Les centrales nucléaires dans le monde*, CEA, France.

Elkraft System (2003), *Power Failure in Eastern Denmark and Southern Sweden on 23 September 2003, Final Report on the Course of Events*, Elkraft System, Denmark (also available at http://eng.elkraft-system.dk/).

International Energy Agency (IEA) (2003a), *Power Generation Investment in Electricity Markets*, OECD/IEA, Paris.

IEA (2003b), *World Energy Investment Outlook: 2003 Insights*, OECD/IEA, Paris.

IEA (2004), *Security of Gas Supply in Open Markets: LNG and Power at a Turning Point*, OECD/IEA, Paris.

IEA Clean Coal Centre (IEA CCC) (2002), *Greenhouse Gas Emission Reductions by Technology Transfer to Developing Countries*, IEA CCC, United Kingdom.

IEA CCC (2003), *Improving Efficiencies of Coal-Fired Power Plants in Developing Countries*, IEA CCC, United Kingdom.

Lawrence Berkeley National Laboratory (Berkeley Lab) (2004), *China Energy Databook v. 6.0*, Berkeley Lab, United States.

Platts (2003), *World Electric Power Plants Database*, Platts.

Swiss Federal Office of Energy (2003), *Report on the Blackout in Italy on 28 September 2003*, Swiss Federal Office of Energy, Switzerland.

Union for the Co-ordination of Transmission of Electricity (UCTE) (2004), *Final Report of the Investigation Committee on the 28 September 2003 Blackout in Italy*, www.ucte.org

US-Canada Power System Outage Task Force (2004), *Final Report on the August 14, 2003 Blackout in the United States and Canada: Causes and Recommendations*, April, https://reports.energy.gov/BlackoutFinal-web.pdf.

World Nuclear Association (2004), *Nuclear Power in China*, www.world-nuclear.org.

CHAPTER 7

International Energy Agency (IEA) (2001), *World Energy Outlook: 2001 Insights: Assessing Today's Supplies to Fuel Tomorrow's Growth*, OECD/IEA, Paris.

IEA (2003), *World Energy Investment Outlook: 2003 Insights*, OECD/IEA, Paris.

IEA (2004a), *Projected Costs of Generating Electricity, Update 2004*, OECD/IEA, Paris.

IEA (2004b), *Biofuels for Transport: An International Perspective*, OECD/IEA, Paris.

The Energy Research Institute (TERI) (2003), *Information Digest on Energy and Environment*, TERI, New Delhi.

CHAPTER 8

Cambridge Energy Research Associates (CERA) (2004), *Riding the Tiger: the Global Impact of China's Energy Quandary*, CERA, Cambridge.

International Energy Agency (IEA) (2002a), *World Energy Outlook 2002*, OECD/IEA, Paris.

IEA (2002b), *Developing China's Natural Gas Market: the Energy Policy Challenges*, Paris: OECD/IEA, Paris.

IEA (2003), *World Energy Investment Outlook: 2003 Insights*, OECD/IEA, Paris.

Tata Energy Research Institute (TERI) (2003), *TEDDY 2002/2003: TERI Energy Data Directory and Yearbook*, TERI, New Delhi.

United Nations Population Division (UNPD) (2003), *World Population Prospects: The 2002 Revision*, United Nations, New York.

CHAPTER 9

BP (2004), *BP Statistical Review of World Energy 2004*, BP, London.

Brunswick UBS (2004), *The Russian Hydrocarbon Resource Base*, Global Equity Research Report, Moscow.

Cedigaz (2004), *Natural Gas in the World,* Institut Français du Pétrole, Rueil Malmaison.

Government of the Russian Federation (2003), *Energy Strategy of the Russian Federation for the Period up to 2020*, Ministry of Energy, Moscow.

International Energy Agency (IEA) (2001), *World Energy Outlook: 2001 Insights: Assessing Today's Supplies to Fuel Tomorrow's Growth*, OECD/IEA, Paris.

IEA (2002a), *Russia Energy Survey 2002*, OECD/IEA, Paris.

IEA (2002b), *World Energy Outlook 2002*, OECD/IEA, Paris.

IEA (2003a), *World Energy Investment Outlook: 2003 Insights*, OECD/IEA, Paris.

IEA (2003b), *Renewables in Russia*, OECD/IEA, Paris.

IEA (2004a), *Security of Gas Supply in Open Markets: LNG and Power at a Turning Point*, OECD/IEA, Paris.

IEA (2004b), *Electricity Sector Reform in Russia*, OECD/IEA, Paris, forthcoming.

IEA (2004c), *Coming in from the Cold: District Heating Policy in the Transition Economies*, OECD/IEA, Paris

Oppenheimer Technical Assistance Consultants (OTAC) (2004), "Policies for Reforming the Heating and Hot Water Sub-sector in Russia", Interim Report for GOF Project, 14 March.

Organisation for Economic Co-operation and Development (OECD) (2004a), *Economic Survey of Russia*, OECD, Paris.

OECD (2004b), *Annual National Accounts*, OECD, Paris.

Renaissance Capital (2004), *Gazprom: Don't Miss it!*, Renaissance Capital, Moscow.

World Bank (2003), *Global Development Finance*, World Bank, Washington.

World Bank (2004), *Russian Economic Report*, February, World Bank, Washington.

United Financial Group (UFG) (2004), *Russian Oils 2004: The Empire Strikes Back*, UFG, Moscow.

United States Geological Survey (USGS) (2000), *World Petroleum Assessment*, June, USGS, Washington.

CHAPTER 10

Ayres, Robert and Benjamin Warr (2003), *Accounting for Growth: The Role of Physical Work*, Centre for the Management of Environmental Resources, INSEAD, Fontainebleau.

Collins, Susan and Barry Bosworth (1996), "Economic Growth in East Asia: Accumulation versus Assimilation", *Brookings Papers on Economic Activity*, fall 1996, The Brookings Institution, Washington.

International Energy Agency (IEA) (2002), *World Energy Outlook 2002*, OECD/IEA, Paris.

IEA (2003), *World Energy Investment Outlook: 2003 Insights*, OECD/IEA, Paris.

IEA/United Nations Environment Programme (UNEP) (2004), *Energy Subsidies: Lessons Learned in Assessing their Impact and Designing their Reform*, Greenleaf Publishing, Sheffield.

United Nations Development Programme (UNDP) (2004), *Human Development Report 2004*, UNDP, Washington.

United Nations Development Programme (UNDP), United Nations Department of Economic and Social Affairs (UNDESA) and World Energy Council (WEC) (2004), *World Energy Assessment – Overview: 2004 Update*, UNDP, New York.

United Nations Population Division (2003), *World Population Prospects: The 2002 Revision*, United Nations, New York.

World Bank (2004), *World Development Indicators 2004*, World Bank, Washington.

World Bank/World LP Gas Association (WLPGA) (2002), *The Role of LP Gas in Meeting the Goals of Sustainable Development*, WLPGA, Paris.

CHAPTER 11

Hanson, Donald A. and John A. "Skip" Laitner (2004), *Estimating Energy Efficiency Investments in the AMIGA Modeling System*, Argonne National Laboratory, Argonne, Illinois.

International Energy Agency (IEA) (2002), *World Energy Outlook 2002*, OECD/IEA, Paris.

IEA (2003), *Energy Policies of IEA Countries 2003*, OECD/IEA, Paris.

IEA (2004), *Biofuels for Transport: An International Perspective*, OECD/IEA, Paris.

ANNEX B

Centre for Global Energy Studies (CGES) (2003), *Annual Oil Market Forecast and Review 2003*, CGES, London.

Energy Information Administration (EIA) (2004), *International Energy Outlook 2004*, U.S. Department of Energy, Washington.

European Commission (EC) (2003), *World energy, technology and climate policy outlook 2030*, EC, Brussels.

International Energy Agency (IEA) (1993), *World Energy Outlook 1993*, OECD/IEA, Paris.

IEA (1994), *World Energy Outlook 1994*, OECD/IEA, Paris

IEA (1995), *World Energy Outlook 1995*, OECD/IEA, Paris.

IEA (1996), *World Energy Outlook 1996*, OECD/IEA, Paris.

IEA (1998), *World Energy Outlook 1998*, OECD/IEA, Paris.

IEA (2000), *World Energy Outlook 2000*, OECD/IEA, Paris

IEA (2002), *World Energy Outlook 2002*, OECD/IEA, Paris.

Institute of Energy Economics, Japan (2004), *Asia/World Energy Outlook*, presented at 385th Forum on Research Works, 10 March, Tokyo, Japan.

Organization of the Petroleum Exporting Countries (OPEC) (2004), *Oil Outlook to 2025*, presented at 9th International Energy Forum, May, Amsterdam.

Shell International (2001), *Energy Needs, Choices and Possibilities: Scenarios to 2050*, Shell International Limited, London.

ANNEX C

Resch, Gustav *et al.* (2004), "Forecasts of the Future Deployment of Renewable Energy Sources for Electricity Generation – a World-Wide Approach", Working Paper published by Energy Economics Group, Vienna University of Technology (available at http://eeg.tuwien.ac.at).

Hanson, Donald A. and John A. "Skip" Laitner (2004), *Estimating Energy Efficiency Investments in the AMIGA Modeling System*, Argonne National Laboratory, Argonne, Illinois.

IEA (2003), *World Energy Investment Outlook: 2003 Insights*, Paris: OECD.

The Online Bookshop

International Energy Agency

All IEA publications can be bought online on the IEA Web site:

www.iea.org/books

You can also obtain PDFs of all IEA books at 20% discount.

Books published in 2002 and before
- with the exception of the statistics publications -
can be downloaded in PDF, free of charge,
on the IEA website.

IEA BOOKS

Tel: +33 (0)1 40 57 66 90
Fax: +33 (0)1 40 57 67 75
E-mail: books@iea.org

International Energy Agency
9, rue de la Fédération
75739 Paris Cedex 15, France

CUSTOMERS IN NORTH AMERICA

Extenza-Turpin Distribution
56 Industrial Park Drive
Pembroke,
MA 02359, USA
Toll free: +1 (800) 456 6323
Fax: +1 (781) 829 9052
oecdna@extenza-turpin.com

www.extenza-turpin.com

You can also send your order to your nearest OECD sales point or through the OECD online services:
www.oecd.org/bookshop

CUSTOMERS IN THE REST OF THE WORLD

Extenza-Turpin
Stratton Business Park,
Pegasus Drive, Biggleswade,
Bedfordshire SG18 8QB, UK
Tel.: +44 (0) 1767 604960
Fax: +44 (0) 1767 601640
oecdrow@extenza-turpin.com

www.extenza-turpin.com

OIL MARKET REPORT

Each month, the first access to data on supply, demand, stocks, prices and refinery activity

Since its appearance in 1983, the International Energy Agency's Oil Market Report has become the definitive source for information on the world oil market fundamentals, covering supply, demand, OECD stocks, prices and OECD and selected non-OECD trade.

The OMR provides the most extensive, up-to-date statistical data available on current world oil market trends. It is the first and exclusive source to present official government statistics from all OECD countries and data from non-OECD countries.

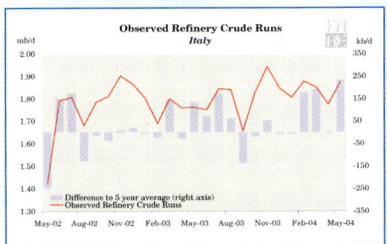

The main market movements of the month are highlighted in a convenient summary, while detailed analysis explains recent market developments and provides an insight into the months ahead. It is the only regular short-term analysis of the oil industry based on information obtained from the IEA's extensive network of contacts with governments and industry.

The OMR provides both historical data from previous months and forecasts for the year ahead. Featuring tables, graphs and statistics, it provides all the data and analysis necessary to track the oil market and identify trends in production, consumption, inventories in OECD countries and prices for both crude and products.

To subscribe electronically, please see our Web site at
www.oilmarketreport.org
Annual subscription rate for single electronic copy: €1750

IEA PUBLICATIONS, 9, rue de la Fédération, 75739 PARIS CEDEX 15
PRINTED IN FRANCE BY STEDI
(61 2004 25 1P1) ISBN 92-64-1081-73 - 2004